Leishmania

After the Genome

Peter J. Myler and Nicolas Fasel

Caister Academic Press

Caister Academic Press
Norfolk, UK

www.caister.com

British Library Cataloguing-in-Publication Data
A catalogue record for this book is available from the British Library

ISBN: 978-1-904455-28-8

Printed and bound in Great Britain

Contents

Contributors

Nathalie Acestor
Seattle Biomedical Research Institute
Seattle, WA
USA

nathalie.acestor@sbri.org

Jenefer Blackwell
Cambridge Institute for Medical Research
Wellcome Trust/MRC Building
Addenbrooke's Hospital Site
Cambridge
UK

jmb37@cam.ac.uk

Mitali Chatterjee
Department of Pharmacology
Institute of Postgraduate Medical Education & Research
Kolkata
India

mitali4315@dataone.in

Conan Chow
Centre de Recherche en Infectiologie et Division de Microbiologie
Faculté de Médecine
Université Laval
Ste-Foy, QC
Canada

conan.chow@crchul.ulaval.ca

David P. De Souza
Department of Biochemistry and Molecular Biology
The Bio21 Institute of Molecular Science and Biotechnology
University of Melbourne
Parkville, Victoria
Australia

d.desouza@pgrad.unimelb.edu.au

Jolyne Drummelsmith
Centre de Recherche en Infectiologie et Division de Microbiologie
Faculté de Médecine
Université Laval
Ste-Foy, QC
Canada

jolyne.drummelsmith@crchul.ulaval.ca

Carole Dumas
Centre de Recherche en Infectiologie et Division de Microbiologie
Faculté de Médecine
Université Laval
Ste-Foy, QC
Canada

Carole.Dumas@crchul.ulaval.ca

Miriam A. Ellis
Department of Biochemistry and Molecular Biology
The Bio21 Institute of Molecular Science and Biotechnology
University of Melbourne
Parkville, Victoria
Australia

ellism@unimelb.edu.au

Karima El Fadili
Centre de Recherche en Infectiologie et Division de Microbiologie
Faculté de Médecine
Université Laval
Canada

karima.el-fadihi.l@ulaval.ca

Amal El Fadili-Kündig
Department of Biochemistry
Faculty of Biology and Medicine
University of Lausanne
Epalinges
Switzerland

amal.kundig@unil.ch

Nicolas Fasel
Department of Biochemistry
Faculty of Biology and Medicine
University of Lausanne
Epalinges
Switzerland

Nicolas.Fasel@unil.ch

Aude Foucher
Centre de Recherche en Infectiologie et Division de Microbiologie
Faculté de Médecine
Université Laval
Ste-Foy, QC
Canada

aunde.foucher@hotmail.com

James W. Goding
The Walter and Eliza Hall Institute of Medical Research and Monash University Medical School
Melbourne
Australia

Jim.Goding@med.monash.edu.au

Iveth Gonzalez
Department of Biochemistry
Faculty of Biology and Medicine
University of Lausanne
Epalinges
Switzerland

Iveth.Gonzalez@unil.ch

David J. Gregory
Centre for the Study of Host Resistance at the Research Institute of the McGill University Health Centre
and Departments of Microbiology and Immunology
McGill University
Montreal, QC
Canada

david.gregory@mail.mcgill.ca

Emanuela Handman
The Walter and Eliza Hall Institute of Medical Research and Monash University Medical School
Melbourne
Australia

handman@wehi.edu.au

Anupam Jhingran
School of Life Sciences
Jawaharlal Nehru University
New Delhi
India

anupam1981@gmail.com

Shaden Kamhawi
Laboratory of Malaria and Vector Research
National Institute of Allergy and Infectious Diseases
National Institutes of Health
Bethesda, MD
USA

skamhawi@nih.gov

Pascal Launois
WHO-IRTC
Department of Biochemistry
Quartier UNIL-Epalinges
Epalinges
Switzerland

Pascal.launois@unil.ch

Phillip Lawyer
Intracellular Parasite Biology Section
Laboratory of Parasitic Diseases
National Institute of Allergy and Infectious Diseases
National Institutes of Health
Bethesda, MD
USA

plawyer@nih.gov

Danielle Légaré
Centre de Recherche en Infectiologie et Division de Microbiologie
Faculté de Médecine
Université Laval
Ste-Foy, QC
Canada

Danielle.Legare@crchul.ulval.ca

Philippe Leprohon
Centre de Recherche en Infectiologie et Division de Microbiologie
Faculté de Médecine
Université Laval
Ste-Foy, QC
Canada

phil_leprohon@hotmail.com

Vladimir A. Likic
Department of Biochemistry and Molecular Biology
The Bio21 Institute of Molecular Science and Biotechnology
University of Melbourne
Parkville, Victoria
Australia

vlikic@unimelb.edu.au

Malcolm J. McConville
Department of Biochemistry and Molecular Biology
The Bio21 Institute of Molecular Science and Biotechnology
University of Melbourne
Parkville, Victoria
Australia

malcolmm@unimelb.edu.au

Diane McMahon-Pratt
Yale School of Public Health
Yale University
New Haven, CT
USA

Diane.McMahon-Pratt@yale.edu

François McNicoll
Centre de Recherche en Infectiologie et Division de Microbiologie
Faculté de Médecine
Université Laval
Ste-Foy, QC
Canada

francois.mcnicoll@crchul.ulaval.ca

Rentala Madhubala
School of Life Sciences
Jawaharlal Nehru University
New Delhi
India

madhubala@mail.jnu.ac.in

Slavica Masina
Department of Biochemistry
Faculty of Biology and Medicine
University of Lausanne
Epalinges
Switzerland

Slavica.Masina@unil.ch

Paul A.M. Michels
Research Unit for Tropical Diseases
de Duve Institute and Laboratory of Biochemistry
Université Catholique de Louvain
Brussels
Belgium

paul.michels@uclovain.be

Michaela Müller
Centre de Recherche en Infectiologie et Division de Microbiologie
Faculté de Médecine
Université Laval
Ste-Foy, QC
Canada
G1V 4G

michaela.muller@crchul.ulaval.ca

Peter J. Myler
Seattle Biomedical Research Institute
307 Westlake Ave N.
Suite 500
Seattle
WA 98109-5219
USA

peter.myler@sbri.org

Thomas Naderer
Department of Biochemistry and Molecular Biology
The Bio21 Institute of Molecular Science and Biotechnology
University of Melbourne
Parkville, Victoria
Australia

naderert@unimelb.edu.au

Martin Olivier
Department of Microbiology and Immunology
Lyman Duff Medical Building
McGill University
Montreal, QC
Canada

martin.olivier@mcgill.ca

Fred R. Opperdoes
Research Unit for Tropical Diseases
de Duve Institute and Laboratory of Biochemistry
Université Catholique de Louvain
Brussels
Belgium

fred.opperdoes@uclovain.be

Marc Ouellette
Infectious Diseases Research Centre
CHUQ
Pavillon CHUL
Ste-Foy, QC
Canada

marc.ouellette@crchul.ulaval.ca

Barbara Papadopoulou
Faculty of Medicine
Department of Medical Biology
Laval University
Infectious Diseases Research Center
CHUL Research Center (CHUQ)
Ste-Foy, QC
Canada

barbara.papadopoulou@crchul.ulaval.ca

Anthony T. Papenfuss
The Walter and Eliza Hall Institute of Medical
Research and Monash University Medical School
Melbourne
Australia

papenfuss@wehi.edu.au

James Pyke
Department of Biochemistry and Molecular Biology
The Bio21 Institute of Molecular Science and
Biotechnology
University of Melbourne
Parkville, Victoria
Australia

jpyke@unimelb.edu.au

Julie E. Ralton
Department of Biochemistry and Molecular Biology
The Bio21 Institute of Molecular Science and
Biotechnology
University of Melbourne
Parkville, Victoria
Australia

julieer@unimelb.edu.au

Annie Rochette
Centre de Recherche en Infectiologie et Division de
Microbiologie
Faculté de Médecine
Université Laval
Ste-Foy, QC
Canada

annie.rochette@crchul.ulaval.ca

David Sacks
Intracellular Parasite Biology Section
Laboratory of Parasitic Diseases
National Institute of Allergy and Infectious Diseases
National Institutes of Health
Bethesda, MD
USA

dsacks@nih.gov

Eleanor C. Saunders
Department of Biochemistry and Molecular Biology
The Bio21 Institute of Molecular Science and
Biotechnology
University of Melbourne
Parkville, Victoria
Australia

e.saunders@pgrad.unimelb.edu.au

Fleur M. Sernee
Department of Biochemistry and Molecular Biology
The Bio21 Institute of Molecular Science and
Biotechnology
University of Melbourne
Parkville, Victoria
Australia

fms@unimelb.edu.au

Terence P. Speed
The Walter and Eliza Hall Institute of Medical
Research and Monash University Medical School
Melbourne
Australia

speed@wehi.edu.au

Fabienne Tacchini-Cottier
WHO-IRTC
Department of Biochemistry
Quartier UNIL-Epalinges
Epalinges
Switzerland

fabienne.tacchini-cottier@unil.ch

Baptiste Vergnes
Centre de Recherche en Infectiologie et Division de Microbiologie
Faculté de Médecine
Université Laval
Ste-Foy, QC
Canada

baptiste.vergnes@mpl.ird.fr

Mary E. Wilson
Department of Internal Medicine
The University of Iowa
Iowa City, IA
USA

mary-wilson@uiowa.edu

Dan Zilberstein
Faculty of Biology
Israel Institute of Technology
Haifa
Israel

danz@techunix.technion.ac.il

Other Books of Interest

Vibrio cholerae: Genomics and Molecular Biology	2008
Pathogenic Fungi: Insights in Molecular Biology	2008
Helicobacter pylori: Molecular Genetics and Cellular Biology	2008
Corynebacteria: Genomics and Molecular Biology	2008
Staphylococcus: Molecular Genetics	2008
Archaea: New Models for Prokaryotic Biology	2008
RNA and the Regulation of Gene Expression	2008
Legionella Molecular Microbiology	2008
Molecular Oral Microbiology	2008
Epigenetics	2008
Animal Viruses: Molecular Biology	2008
Segmented Double-Stranded RNA Viruses	2008
Acinetobacter Molecular Microbiology	2008
Pseudomonas: Genomics and Molecular Biology	2008
Microbial Biodegradation: Genomics and Molecular Biology	2008
The Cyanobacteria: Molecular Biology, Genomics and Evolution	2008
Coronaviruses: Molecular and Cellular Biology	2007
Real-Time PCR in Microbiology: From Diagnosis to Characterisation	2007
Bacteriophage: Genetics and Molecular Biology	2007
Candida: Comparative and Functional Genomics	2007
Bacillus: Cellular and Molecular Biology	2007
AIDS Vaccine Development: Challenges and Opportunities	2007
Alpha Herpesviruses: Molecular and Cellular Biology	2007
Pathogenic *Treponema*: Molecular and Cellular Biology	2007
PCR Troubleshooting: The Essential Guide	2006
Influenza Virology: Current Topics	2006
Microbial Subversion of Immunity: Current Topics	2006
Cytomegaloviruses: Molecular Biology and Immunology	2006
Papillomavirus Research: From Natural History To Vaccines and Beyond	2006
Epstein Barr Virus	2005
HIV Chemotherapy: A Critical Review	2005
Probiotics and Prebiotics: Scientific Aspects	2005
Foodborne Pathogens: Microbiology and Molecular Biology	2005

Preface

The order Kinetoplastida includes a large number of pathogenic parasite species, which infect a wide range of hosts, including humans. Over the past 10 years, an international consortium of more than 200 people in 47 laboratories from 25 countries on six continents has worked together to determine the genome sequence of several of these organisms. In July 2005, analysis of the genomes from three species (*Leishmania major*, *Trypanosoma brucei* and *Trypanosoma cruzi*) was published in *Science*, and the post-genomic era for these organisms began. Several other genome projects are now in various stages of completion and analysis, and a number of genome-wide studies are under way in order to assemble the information for a 'systems biology' approach. The aim of this book is to present, in some detail, how the availability of these genome sequences has informed and changed our understanding of the biology of the various species of *Leishmania* that cause leishmaniasis, one of the most important human parasitic diseases. In addition, we have sought to elucidate what are the implications of this new knowledge in terms of discovering novel diagnostics, chemotherapeutic agents and vaccines to control these organisms. In this book, we have limited most of the chapters to discussion of *Leishmania*-specific aspects of trypanosomatid biology and pathology, since a recent publication (*Trypanosomes: after the genome*, edited by Barry, McCulloch, Mottram and Acosta-Serrano) describing insights from the *T. brucei* genome has already illuminated many facets of molecular and cellular biology shared by the 'Trityps'.

In the first chapter, Madhubala *et al.* discuss the epidemiology and diagnosis of leishmaniasis. Three major forms of leishmaniasis exist, depending on the parasite species but also on the immune response of the patient: cutaneous (CL), mucocutaneous (MCL) and visceral (VL) leishmaniasis. Variation can be also observed such as diffuse cutaneous lesihmaniais (DCL) or in the case of visceral leishmaniasis, post kala-azar dermal leishmaniasis (PKDL). An extremely serious new manifestation of disease is *Leishmania*-HIV co-infection, since cases are reported with increasing frequency. Although *Leishmania*-HIV co-infection has been mainly described in Southern Europe until now, its further development in Asia, Africa and South America is even more frightening considering the importance of both diseases on these continents. Because of the diversity of leishmanial disease types, no single diagnosis and treatment will be suitable for all epidemiological situations, and field diagnosis is difficult in the absence of simpler and less invasive tests with good sensitivity and specificity. Recently developed nucleic acid-based methods offer an advantage but require sophisticated equipment, while serodiagnosis has been associated with a cross-reactivity problem. Nonetheless, many recombinant antigens have helped make these tests more specific. These newer approaches offer great promise in making the diagnosis specific and sensitive.

Chapter 2 provides a brief overview of what we have recently learned about *Leishmania* genome structure and content. This includes chromosome number and size, identification and annotation of protein-coding and RNA genes, as well as their organization into polycistronic gene clusters and unusual transcription. There is also a discussion of telomere structure and retroelements. Finally, some details are presented of *Leishmania*-restricted and species-specific genes identified by comparison of the *L. major*, *L. infantum* and *L. braziliensis* genome sequences.

As a result of the unusual polycistronic transcription of protein-coding genes and subsequent *trans*-splicing, developmental regulation of mRNA levels is determined not at the level of transcription as in other organisms, but post-transcriptionally. In Chapter 3, Papadopoulou *et al.* describe what is known about sequences located in the 3′-untranslated regions that control mRNA stability and translation. This regulation is complex and involves the coordination of different mechanisms that can be independently triggered by environmental signals inducing differentiation of promastigotes to amastigotes within mammalian macrophages. Regulation of protein abundance at translational and post-translational levels is also associated with stage-specific gene expression.

Data mining of genomes has provided a valuable tool for the identification, comparative analysis and functional definition of the *Leishmania* proteome. Chapter 4 describes how large-scale proteomic studies using mass spectrometry are being used to dissect not only protein expression, regulation and function, but also biological processes involving complex protein–protein interactions, post-translational regulation events and post-translational modifications. Access to databases representing the kinetoplastid genomes was, is and will be more and more an essential tool to support these technological developments. These approaches should be valuable for the design of original alternatives to prevent or treat leishmaniasis.

Chapter 5, authored by McConville *et al.*, continues this technological emphasis by providing an overview of *Leishmania* metabolism and describing new analytical approaches that are being used to directly characterize the *Leishmania* metabolome. Metabolite profiling is an emerging field in functional genomics that is both complimentary to transcript and protein profiling approaches, and a powerful tool in its own right for defining metabolic processes and annotating gene function. Major metabolic pathways in *Leishmania* that appear to be essential for parasite survival and/or differentiation in the mammalian host are highlighted.

Leishmania has a digenetic life cycle, successively differentiating into the promastigote or insect vector-infecting form, and amastigote form that infects mammals (including dogs, rodents and humans). In Chapter 6, Zilberstein *et al.* summarize the current knowledge about how promastigotes differentiate into amastigotes when surrounded by a phagolysosome milieu. A host-free differentiation system can be achieved *in vitro* by shifting cultured promastigotes to a lysosome-like environment where promastigotes start to differentiate into amastigotes only when exposed to high temperature and low pH. The relevance of this axenic system to *in vivo* development within the host macrophage is discussed.

The genome sequencing projects provided substantial insight in the metabolic capacities of *Leishmania* and revealed important differences from the other trypanosomes, as well as the mammalian host. These differences could be exploited as potential targets for developing new high-selectivity drugs, and this is the focus of Chapter 7, by Opperdoes and Michels. Several examples of validated drug targets, such as enzymes and transporters involved in carbohydrate and energy metabolism, are discussed in detail. Finally genomic information has provided a possible explanation for the mode of action of the newly introduced anti-leishmanial drug Miltefosine.

Since the 1940s, the pentavalent antimony compounds pentostam and glucantime have been the mainstays of anti-leishmanial therapy. However, their disadvantages include parenteral mode of administration, long duration of treatment (several weeks), and severe (but generally reversible) toxic effects. Resistance to commonly used anti-*Leishmania* drugs such as pentavalent antimonials is increasing throughout the endemic regions. It is probably only a matter of time until resistance develops against the second-line parenteral drugs, including pentamidine and amphotericin B. In Chapter 8, Ouellette *et al.* describe our understanding of resistance mechanisms induced under laboratory conditions. The current challenge is to understand resistance mechanisms occurring in the field, and whole-genome approaches will be useful in this endeavour. In any case, the arising problem of drug resistance is emerging in most part of the world.

Leishmania are polarized flagellated cells with surface membrane domains of distinct composition and function. Specific surface proteins are distributed on the flagellum, the flagellar pocket and around the cell body, and they contribute to the functional specialization of these regions. For years, these cell surface molecules have been studied in great detail and major contributions in glycobiology have been made thanks to studies in *Leishmania* and *Trypanosoma*. For example, GPI anchoring of cell surface antigens was first described in *Leishmania*, well before its importance was recognized in higher eukaryotes. Handman *et al.*, describe, in Chapter 9, how the genome sequences, combined with the rapid development of algorithms for use in bioinformatics, are providing tools of extraordinary power for the identification of the genes encoding all membrane proteins. However, they caution that large-scale 'systems' approaches to studying membrane proteins is limited by our lack of knowledge concerning the remarkably diverse post-translational modifications seen in *Leishmania*.

Leishmania promastigotes live extracellularly in the alimentary tract of the insect vector, the female phlebotomine sandfly (genus *Phlebotomus* in the Old World and *Lutzomyia* in the New World). Of the more than 400 phlebotomine species described, fewer than 50 are known to be involved in the transmission cycle of these parasites, and some vectors species are highly restricted to the species of *Leishmania* that they transmit in nature. In Chapter 10, Sacks *et al.*, review the natural habitat of *Leishmania* amastigotes and promastigotes during their transformation, growth, differentiation and migration in the alimentary tract of their competent vectors, and the barriers to survival that are encountered in refractory flies. *Leishmania* species-, strain- and stage-specific molecules that are known or postulated to permit the development of transmissible infections to proceed are described in some detail, with a view toward the development of a successful transmission blocking vaccine(s).

An enormous amount of work has been performed during the last 20 years to dissect the immune responses of mice to *Leishmania* infection, both at the innate and acquired immune response level. From the initial interactions between receptor proteins on the macrophage membrane and *Leishmania* surface molecules, the parasite modulates the activities of the macrophage's intracellular signalling pathways for its own ends. As a result, infected macrophages become non-responsive to activating stimuli such as IFNγ, and a number of mechanisms interfere with antigen presentation. In Chapter 11, Olivier and Gregory discuss the molecular details of these processes and their important contribution to the survival of *Leishmania* in the host.

A large number of studies in murine models of *L. major* infection indicate that induction of a T helper-1 (Th1) immune response is protective, whereas a Th2 immune response is not. A clear dichotomy is observed between cytokine production by draining lymph node cells of susceptible versus resistant mouse strains. In Chapter 12, Tacchini-Cottier and Launois explore the role of cytokines such as γ-interferon (γ-IFN) and interleukin-4 (IL-4) in experimental infection with different strains and species of *Leishmania*. Th1 cells secrete activators of cell-mediated immunity such as γ-IFN, inducing activation of the infected phagocytes (principally macrophages) and consequent production of nitric oxide (NO), which rapidly kills intracellular parasites. Th2 cells secrete cytokines, such as IL-4, that promote humoral responses which stimulate the production of parasite-specific antibodies. But IL-4 down-regulates the expression of the IL-12 receptor on potentially protective Th1 cells, rendering them unresponsive to IL-12, and production of γ-IFN and NO is inhibited, preventing the killing of intracellular parasites. However, it is becoming increasingly clear, in both human and mice, that the role of cytokines differs according to the *Leishmania* species. Understanding these differences should contribute to the design of better and more efficient therapies.

Despite the large amount of information we have on the immune response in animal models and in humans, the numerous attempts to develop a vaccine against leishmaniasis, and the powerful techniques that are available to researchers, no safe and efficient vaccine has been obtained to date (as is the case for most of the human parasitic diseases). In Chapter 13, Blackwell *et al.* review the literature on *Leishmania* vaccine development, including specific defined vaccine candidates, adjuvants and delivery systems. They also discuss two recent studies using genome-based approaches to screen for novel vaccine candidates that identified 20 currently being further evaluated for vaccine development.

After a few years working with this fascinating parasite, every 'leishmaniac' scientist is aware of the parasites' diversity, complex biology and influence on the immune system of the host. Access to genome sequence from three *Leishmania* species is a major step in eliminating one barrier in the understanding of the biology of the parasite and offers new avenues to fight this parasite. However, it does not immediately solve urgent problems such as outbreak of the disease, public health problems and does not directly provide a defined treatment or vaccine. Thus, much work remains to be done in the field, clinic and laboratory, using traditional biochemical, molecular and immunological methods, as well as modern state-of-the-art high-throughput systems biology approaches, if the promise wrought by the genome sequence projects is to come to fruition.

Peter Myler and Nicolas Fasel
October 2007

Leishmaniasis: Epidemiological Trends and Diagnosis

1

Anupam Jhingran, Mitali Chatterjee and Rentala Madhubala

Abstract

Human leishmaniasis consists of a spectrum of diseases ranging from simple self-limiting or asymptomatic cutaneous forms to a horribly disfiguring, debilitating mucocutaneous form and a fatal, if untreated, visceral form. Infection may be classified into three clinical syndromes, namely cutaneous leishmaniasis (Oriental sore), mucocutaneous leishmaniasis (espundia) and visceral leishmaniasis (kala-azar). Clinical manifestation of the disease depends on the species involved. The symptoms range from cutaneous ulcers to fatal visceral lesions. Because of the diversity of its epidemiological situations, no single diagnosis, treatment and control will be suitable for all. Field diagnosis is difficult in the absence of simpler and less invasive tests with good sensitivity and specificity. The existing parasitological tests from marrow aspirates and splenic aspirates require are invasive and unsuitable for all patients. Recently developed nucleic acid based methods offer an advantage, but require sophisticated equipment. Serodiagnosis has long been in use, but is associated with a cross-reactivity problem. Nonetheless, many recombinant antigens have helped make these tests more specific. Thus, the field of diagnosis is improving and newer approaches offer great promise in making the diagnosis specific and sensitive.

Introduction

Leishmaniasis, a vector-borne disease caused by intramacrophage protozoa, is caused by approximately 21 species of the genus *Leishmania* (order *Kinetoplastida*). It is generally transmitted by 30 species of sandflies, either *Phlebotomus* (Old World) or *Lutzomyia* (New World), but other rare modes of transmission include congenital and parenteral via blood transfusion, needle sharing or rarely through laboratory accident (Magill, 1995; Herwaldt, 2001).

The disease endemicity extends to over 88 countries, 72 being in the developing world while 13 belong to the category of least developed countries, but its public health impact remains grossly neglected. Another stumbling block is that declaration of disease being mandatory in only 32 of the 88 countries, a substantial number of cases remain unrecorded, resulting in its true burden remaining largely invisible (Guerin *et al.*, 2002; Murray *et al.*, 2005). Globally, the population at risk is 350 million, the overall prevalence being 12 million, with 2 million new cases occur annually, while the disability-adjusted life years (DALY) burden is considered to be 860,000 men and 1.2 million women (TDR, 2003–04). The estimated annual incidence of the visceral form is around 500,000 in 61 countries with 90% of these cases being confined to five countries, namely India (especially the state of Bihar and its adjoining states), Bangladesh, Nepal (Terai region), Sudan and North Eastern Brazil (WHO, 2000). With regard to the cutaneous form, 1.0–1.5 million cases are reported annually and 90% of the cases are limited to eight countries, six being in the Old World (Afghanistan, Algeria, Iran, Iraq, Saudi Arabia, Syria) and two in the New World (Brazil, Peru). In contrast 90% of the mucocutaneous form is focused in Bolivia, Brazil and Peru (WHO, 2000).

Since 1993, *Leishmania* endemic regions have expanded considerably, primarily due to unplanned economic development and the increasing incidence of HIV–VL co-infections. For example, the incidence of CL in Brazil has progressively increased from 21,000 cases in 1998 to 30,550 cases in 1999 and over 35,000 cases in 2000. A similar trend has been reported from Kabul, Afghanistan, where the number of cases was 14,200 in 1994, but by 1999 had risen dramatically to 200,000 (TDR, 2003–04).

Causative species and clinical manifestations

The disease is known to have a wide range of symptoms ranging from a simple, apparently innocuous, cutaneous lesion; through the more severe mucocutaneous form; to the fatal, if left untreated, visceral form. Cutaneous leishmaniasis (CL) occurs in both the Old World (where it is classically caused by *L. tropica*, *L. major* and *L. aethiopica*, although *L. infantum* and *L. donovani* have also been implicated) and New World (where it is usually caused by the *L. mexicana* species complex and *Vianna* subgenus, but also *L. major* like organisms and *L. chagasi*). Generally lesions develop from papules to nodules to ulcerative lesions having a central depression and an indurated border that subsequently over time result in atrophic scars. On the other hand, another subset of lesions does not ulcerate but instead persist as nodules or plaques.

Variation in the disease pattern includes mucosal leishmaniasis caused by the *Vianna* subgenus, generally *L. (V) braziliensis*, but also may be caused by *L. (V) panamensis*, *L. (V) guyanensis* and *L. amazonensis*. This occurs following hematogenous or lymphatic dissemination of amastigotes from the skin to the nasopharyngeal mucosa, culminating in nasopharyngeal destruction involving the nose, mouth and throat.

Another variation can be leishmaniasis recidivans typically caused by *L. tropica* and sometimes *L. major* and is found in Iran, Iraq and their neighbouring countries. It is generally a chronic, oligoparasitic variant of Old World cutaneous leishmaniasis and presents as a solitary facial lesion that heals centrally but gradually enlarges over time. Another variant is the chronic, progressive polyparasitic variant that manifests as diffuse cutaneous leishmaniasis or DCL. It is caused by *L. aethiopica* in the Old World and *L. mexicana* species complex in the New World, and comprises disseminated, but non-ulcerative, skin lesions (Grevelink and Lerner, 1996).

The visceral form or visceral leishmaniasis (VL) is caused by the *L. donovani* complex that includes *L. donovani* and *L. infantum* in the Old World and *L. chagasi* in the New World; rarely *L. tropica* (Old World) and *L. amazonensis* (New World) have been implicated. VL encompasses a broad range of clinical signs characterized by irregular bouts of fever, substantial weight loss, hepatosplenomegaly with splenomegaly predominating and pancytopenia (that includes anaemia, thrombocytopenia, leucopenia, and neutropenia, along with a relative lymphocytosis and monocytosis) associated hypergammaglobulinaemia concomitant with hypoalbuminaemia. Importantly, this systemic infection is generally fatal if left untreated (Herwaldt, 1999).

Variations in the disease profile include post-kala-azar dermal leishmaniasis (PKDL), a dermal manifestation of the visceral form that randomly develops after resolution of VL and has been termed as the 'unhappy sequel' of life-saving drug treatment (Dye and Wolport, 1988). The dermal location of PKDL makes it more accessible to the sandfly and it has been suggested that PKDL patients possibly serve as reservoir hosts of infection, especially in the intervening periods between outbreaks (Zijlstra *et al.*, 2003). In Kenya, three independent studies showed that PKDL occurred in 1%, 6% and 30% of VL patients, while in Sudan 56% of the kala-azar patients developed PKDL (Zijlstra *et al.*, 1994, 1995), whereas in India, roughly 10% of cases move to PKDL (Ramesh and Mukherjee, 1995).

Another variant of VL is viscerotropic leishmaniasis, an oligoparasitic syndrome with non-specific manifestations caused by visceral infection with *L. tropica* that was reported in US servicemen serving in the Persian Gulf conflict in the 1990s (Magill *et al.*, 1993).

Localized cutaneous leishmaniasis (LCL) in India is mainly due to *Leishmania tropica*, endemic in the deserts of Rajasthan. Recently, Himachal Pradesh has been identified as a new endemic focus for the disease where 161 cases of CL were seen between May 2001 and December 2003 (Sharma *et al.*, 2005). Importantly, *L. donovani* was identified as the predominant causative species along with *L. tropica,* indicating a novel shift in the causative species. A similar scenario has been reported from Sri Lanka, where five isolates from CL patients typed as *L. donovani* zymodeme MON-37. Associated risk factors included close proximity to jungles, potential reservoir hosts in the environment, lack of awareness, as well as inadequate knowledge and application of preventive measures (Siriwardena *et al.*, 2003; Karunaweera *et al.*, 2003).

In HIV-positive individuals, VL is commonly caused by *L. infantum* or *L. donovani.* Other *Leishmania* species like *L. braziliensis, L. aethiopica, L. tropica* and *L. major* have also been reported as the causative species. Another feature of VL in HIV co-infected individuals is the variability of *L. infantum* zymodemes. Amongst 150 isolates studied, a total of 17 zymodemes were identified and importantly, from a single HIV-VL co-infected person, several *Leishmania* zymodemes have been identified (Paredes *et al.*, 2003). Dermatropic variants of *L. (vianna) braziliensis, L. mexicana* and *L. amazonensis* have been ascribed to visceral disease in HIV-VL co-infected individuals; conversely, viscerotropic *Leishmania* spp variants like *L. chagasi* and *L. infantum* have been reported in cutaneous lesions from co-infected patients (Paredes *et al.*, 2003).

Transmission characteristics and expanding geographical distribution

As VL occurring in the Indian subcontinent and East Africa (Sudan, Kenya and Ethiopia) has no known animal reservoir, it is termed an anthroponotic VL. In India, 'kala-azar' as it is locally referred to, raged across the Eastern part of the country in three major epidemics between 1880 and 1950 that were further exacerbated by contributory factors including famine as well as epidemics of malaria and influenza (Dye and Wolpert, 1988). The availability of antimonial drugs in 1920 and later an effective insecticide spraying under the aegis of a 'National Malaria Eradication Programme' halted the transmission dynamics. However, in 1963, cessation of insecticidal spraying caused the disease to re-emerge in the 1970s when an estimated 600,000 cases were reported and was followed by another outbreak in 1993 when an estimated 250,000 cases were reported (Peters, 1981; Sundar *et al.*, 2000).

In Sudan, the main endemic zones include the southern and eastern regions of the White Nile and Upper Nile states where several epidemics have occurred since 1904 when the first case of VL was reported in Africa. Exclusive anthroponotic transmission dynamics have not been clearly demonstrated as heavy transmission has been reported from relatively scarcely populated areas and animal reservoirs have been implicated (Zijsltra *et al.*, 1994, 1995). In more recent times, a major decade-long epidemic of VL occurred from 1984–1994 in the Western Upper Nile province in Southern Sudan (Seaman *et al.*, 1996; Zijlstra and el Hassan, 2001). This area was considered as a non-endemic zone for VL and owing to the ongoing civil war, the disease remained neglected until 1988 when estimated counts were that 100,000 deaths had occurred in a population of 300,000. In 1997, the number of cases again exploded showing a 400% increase, further accentuated by large population movement triggered by civil unrest, which resulted in the epidemic spreading to neighbouring countries of Eritrea and Ethiopia.

Zoonotic VL (ZVL) caused by *L. donovani infantum* occurs in the Mediterranean VL foci, Western Africa, tropical and subtropical America, Northern China, Western Asia (Iran) and Pakistan, with domestic dogs being the main animal reservoir for transmission (Boeleart *et al.*, 2000). More recently, ZVL has gained considerable attention in view of the increasing incidence of VL appearing as an opportunistic infection in AIDS, primarily in the Mediterranean region (Desjeux, 1996). The HIV-VL infection has been termed a 'deadly gridlock' where HIV and VL are locked in a vicious circle of mutual reinforcement as both infections synergistically reinforce their impact on the immune system, thus resulting in an exponential increase in disease morbidity and mortality (WHO, 1996).

This has greatly expanded the human reservoir resulting in a major shift in the epidemiology of VL in the Mediterranean region from a traditionally zoonotic disease to a person-to-person transmission that is either vector-borne or an artificially created transmission via exchange of infected needles; the latter mode predominating in over 73% of HIV-VL co-infected individuals. The pocket of HIV-VL co-infected individuals are mainly from Spain, Italy, France and Portugal, since up to 1998, 1440 of the first 1700 cases reported to the WHO, were from these countries (Alvar *et al.*, 1997). According to data from the WHO, 33 countries have reported HIV-VL co-infections. In Southern Europe, 25–70% of adult VL cases are related to HIV and 1.5–9.0% of HIV cases suffer from newly acquired or reactivated VL (WHO, 1997). With the advent of the highly active anti-retroviral therapy (HAART), the incidence of VL in HIV infected patients has decreased from 11.6 ± 1.2 per 10,000 person years in 1996 to 6.3 ± 0.7 per 10,000 person. Unfortunately, the benefits of HAART are available only to 5% or less of HIV-infected patients worldwide (UNAIDS, 2002). In the Americas, cases have been reported from Brazil where the incidence has increased from 0.8 cases per 100,000 inhabitants in 1986 to 10.5 cases per 100,000 inhabitants in 1997. In East Africa, cases of HIV-VL have been reported, but available information is possibly a gross underestimate that could be improved by increasing the surveillance network. In 1998, a WHO/Joint United Nations programme on HIV/AIDS (UNAIDS) was launched. It included 28 institutions in their purview and is expected to help in mapping the geographic distribution and disease profile of HIV-VL co-infections (WHO, 1997).

In recent years, several parts of the world have seen an alarming increase in the number of cases of all forms of leishmaniasis. This spread is known to be associated with the economic development and iatrogenic environmental changes that have resulted in increased human exposure to the sandfly vector. These changes include establishing new settlements, unplanned urbanization, deforestation, extracting timber, mining, building of dams, widening of areas under cultivation, creating new irrigation schemes, and expanding reconstruction in primary forests such as the Amazon and excessive migration from rural to urban areas. In more recent times, the war in Iraq has seen 522 cases of cutaneous leishmaniasis and five cases of visceral leishmaniasis reported among US military personnel (Elston and Miller, 2004; Zapor and Moran, 2005).

Apart from this, emergence of *Leishmania*/HIV coinfections have been a major cause for concern amongst intravenous drug users. Also at risk are the large population of seasonal workers, refugees, sexworkers and truck drivers. The disease duo produces a cumulative effect, as both of them target the same immune cells thus exponentially increasing disease severity and complications. This increasing interface between leishmaniasis in more developed countries and the threat of co-infections has prompted growing interests in the disease. Rapid methods for diagnosis and species identification are needed, as are therapies, prophylactic and control measures that are effective, safe, affordable and can be easily administered.

Diagnosis

Diagnosis has been relatively static until recent years due to the lack of funds and primarily because the disease is generally associated with poor countries. Consequently, in many areas, the diagnosis of leishmaniasis is based on inefficient tests along with clinical evaluation. In fact, in areas where prevalence of cutaneous leishmaniasis (CL) is high, patients showing characteristic lesions of CL are often empirically provided anti-leishmanial treatment, in the absence of any diagnostic tests. Such situations can prove fatal, as treatment involves antimonials, whose toxicity is well established. An example of such a situation has been reported where a child from an endemic region for leishmaniasis was misdiagnosed based on microscopic detection of an intramacrophage organism, received antimonials and later succumbed to fungal infection (Reed, 1996).

Therefore correct clinical diagnosis is vital before starting treatment. A good diagnosis not only helps in prompt initiation of therapy but can also shorten the treatment schedule. Development of vaccines and vector control strategies are also being considered as potential areas for leishmaniasis control, but like diagnostics also need development and to date, no effective vaccine is available. The first generation (whole killed parasites with or without BCG as adjuvant) has been tested in humans. Newer approaches include surface antigens (gp63, lipophosphoglycan), a promastigote antigen derived from *L. amazonensis* (46 kDa) enzyme receptor (LACK); they have shown protection in mice but human studies are yet to be done

Vector/reservoir control strategies have been implemented in many areas and in many instances found to be effective. Although DDT spraying has helped reduce the number of Phlebotomine sandflies, it has also resulted in production of resistant mosquitoes (Bora, 1999). In zoonotic foci where dogs are the main reservoir of parasites for vectors, control strategies based on culling of infected dogs have been criticized, since susceptible young dogs quickly replace the eliminated dogs. Also anti-leishmanial therapy doesn't work well with dogs (Gramiccia *et al.*, 1992). We therefore could say that an efficient case management based on early diagnosis and treatment is the key to limit morbidity and prevent mortality.

One of the earliest ways of diagnosing leishmaniasis is the direct detection method. These methods include identification of tissue amastigotes, in cutaneous, mucosal or lymphoid tissue, culture of *Leishmania* promastigotes from tissues, or concentrated white cells, or the use of molecular tools for parasite detection. With the development of increasingly improved technology, immunodiagnostic methods are becoming simpler and less expensive. These methods include serology and intradermal skin test. In serodiagnostics, the presence of anti-leishmanial antibody is the primary criterion for defining infection; however intradermal skin tests rely on strong T-cell response of the bodies' immune system.

The response of the immune system has been utilized for other cellular diagnostic assays as well, which involve measurement of cytokines and *in vitro* proliferation of T-cells. In spite of the availability of the above-mentioned diagnostic methods, the search for more rapid, better and user-friendly tools are still on, some of which will be discussed here.

Diagnostic methods

Parasite detection methods

Visual demonstration of the parasite has been considered as the 'gold standard' for diagnosis, wherein the tissue sample is stained and examined microscopically for intramacrophage amastigotes. To ensure that the visualized structures are indeed amastigotes, an experienced observer should look for the characteristic size (2–4 μm in diameter), shape (round to oval) and internal organelles (the nucleus and kinetoplastid). In particular the kinetoplast should be visualized, a rod-shaped, specialized mitochondrial structure that contains extranuclear DNA in catenated maxicircles and minicircles. With Giemsa staining, the cytoplasm is pale-blue and nucleus or kinetoplast pinkish red or violet-blue. The samples could be taken from liver, lymph nodes, bone marrow or spleen, splenic aspiration being the most sensitive though risky. The parasitological detection of CL is made either by demonstrating parasites in the skin lesions or by culture of these specimens and direct demonstration of the promastigote.

Lymph node aspiration

This is easiest to perform and comfortable for patients and is therefore recommended for use both in the field and hospitals. The aspirates from the lymph nodes, most commonly inguinal, is sucked through a needle and ejected onto the microscopic glass slide for examination after fixing and staining. Parasites can also be observed in aspirates from axillary, cervical and epitrochlear nodes. However, the sensitivity is reported to be a mere 52–58% (Siddig *et al.*, 1988; Zijlstra *et al.*, 1992).

Bone marrow aspiration

This is the most painful procedure and unsuitable for field diagnosis and large-scale screening. The sample is taken from the posterior superior iliac crest or from sternum and processed as above. The sensitivity broadly ranges from 52% to 85% (Siddig *et al.*, 1988; Zijlstra *et al.*, 1992; Chowdhury *et al.*, 1993; Bryceson *et al.*, 1996).

Splenic aspiration

This is considered to be the most sensitive and the sample is the splenic pulp, which is processed and stained for examination. However, a major limitation is the mandatory technical expertise and the risk of fatal haemorrhage, especially in patients with low platelet counts or in whom the prothrombin time is significantly prolonged (WHO, 1990). The reported sensitivity is highest, being 90–97% (Siddig *et al.*, 1988; Zijlstra *et al.*, 1992).

Other tissues

Parasites can also be found in liver, skin (Rab *et al.*, 1992), cerebrospinal fluid (Prasad *et al.*, 1996) and blood (Martinez *et al.*, 1993). Of these, blood is routinely used in diagnosing HIV co-infected patients (Medrano *et al.*, 1993). Skin biopsies should be done for diagnosing PKDL and lymph-node biopsies can be an alternative (Ghosh *et al.*, 1995). Molina *et al.*, (1992) demonstrated the parasite in sandflies that were allowed to feed on blood drawn from an AIDS patient suspected with leishmaniasis through a membrane of chicken skin. This method, referred to as xenodiagnostic, later confirmed that sandflies easily become infected after indirectly feeding on the blood on HIV and *Leishmania* co-infected patients as early as within 48 hours (Molina *et al.*, 1992).

As a novel method demonstrating the parasite, antibody probes against a small subunit of ribonucleotide reductase of the parasite were used. These probes could detect the parasite inside the macrophage and potentially offer a better way of staining and counting based on fluorescence detection (Chang *et al.*, 2004).

Parasites could be easily cultured under laboratory conditions. The solid medium used for growing parasites is the classical Novy–MacNeal–Nicolle (NNN) medium supplemented with 20–30% rabbit blood or Schneider's insect medium supplemented with 10% (v/v) fetal calf serum (FCS).

Culturing has an advantage over the above methods in that viability can also be assessed and this helps to follow up the chemotherapeutic response. Sensitivity is also higher, since lower numbers of parasites can be detected (Lightner *et al.*,1983; Sinha *et al.*, 1993). Moreover, the cultured parasite could be subjected to isoenzyme analysis and other methods for species identification, which makes the choice of therapy easy. Cultures could also be used for *in vitro* drug resistance study in patients not responding to therapy. The contamination problem caused by tissue samples can be overcome by animal inoculation, though it is time-consuming and not practical for rapid diagnosis.

Recently a micro-culture method (MCM) has been introduced for the diagnosis of cutaneous leishmaniasis, which utilizes microcapillaries for culturing of parasites and is reported to take lesser time than traditional culture methods (TCM) (Allahverdiyev *et al.*, 2004). Culture-based diagnosis is not routinely done for mucocutaneous leishmaniasis, due to the very low number of parasites, which further reduces sensitivity of the test (Davidson, 1998).

Serological methods

The commonly used serological methods are targeted towards circulating antibodies against leishmania proteins present in the patients' blood, the underlying principle being the antigen – antibody interaction. Earlier serological methods like direct agglutination tests (DAT) and immunofluorescence tests (IFAT) were based on whole cell promastigotes serving as the antigen. However, the use of whole cell extracts decreases specificity of the tests and batch-to-batch variation causes lack of standardization for antigen preparation. More recently, enzyme-linked immunosorbent assays (ELISA) have become popular as they are more sensitive, specific and require less antigen. Improvements and research in the field of serodiagostics have paved the way to improve assay specificity and reproducibility. The use of gp63 antigen can help distinguish VL and CL, as antibody against gp63 is not found in CL patients (Shreffler *et al.*, 1993). Another study has reported that antibodies against *L. infantum* peroxiredoxin can serve as a possible marker for monitoring efficacy of the treatment (Santarem *et al.*, 2005).

Serological tests that can differentiate different forms of the disease have also been developed. In one recent report, based on the antibody titre and the antigen recognition pattern of the patient sera, the authors were able to differentiate patients of visceral leishmaniasis from post kala-azar dermal leishmaniasis (Saha *et al.*, 2005).

Direct agglutination tests (DAT)

These tests are based on the fact that whenever the antigen and a polyvalent antibody (against that particular antigen) are incubated, there is formation of visible precipitate-like structures. The formalin-fixed antigen is incubated with the patient sera which has polyvalent IgM antibodies. The reaction goes for up to 48 hours, and the results are analysed visually.

A large number of tests can be done simultaneously as these are done on microtitre plates, thus DAT is very useful under field conditions where large-scale screening is required. The production of freeze-dried antigen has obviated the need for a refrigerator, making transportation easy (Meredith *et al.*, 1995). Furthermore, the improvement in protein technology has resulted in identification of some immunodominant *Leishmania*-specific antigens, which has helped overcome the associated cross-reactivity (Harith *et al.*, 1995).

The need for larger amount of antigen still remains a problem and the inability of this test to discriminate between past and current infection, as antibody titres remain high, years after the infection to give a false positivity (Hailu, 1990).

Immunofluorescence tests (IFAT)

The basic principle here is the same as for other serological tests. The incubation of antigen and antibody is done on microscopic glass slides and the interaction is visualized with the help of fluorescently labelled secondary antibody. Since the process involves the use of fluorescence microscopy, this method remains impractical for application in the field setting.

Enzyme-linked immunosorbent assays (ELISA)

These tests are done on microtitre ELISA plates in a manner similar to the above tests and the results are read spectrophotometrically by measuring the optical density. This requires a laboratory with proper equipment, but samples can be collected on filter paper, making it suitable for large epidemiological surveys. Dot-ELISA is an advancement over the conventional methods as it can be visually ascertained and has been used for canine leishmaniasis (Dietze *et al.*, 1995). Fucose-mannose ligand (FML) ELISA is another advancement where antibodies are detected against FML, a glycoprotein fraction present on promastigotes and amastigotes. FML-ELISA showed 100% sensitivity and 96% specificity and could be used to detect subclinical infection (Palatnik-de-sousa *et al.*, 1995). ELISA can also be used for detecting the circulating antigen in the patients' body by using laboratory raised anti-leishmanial antibody and has been applied in the polyethylene glycol (PEG)-ELISA. Here the circulating immune complexes (CIC) are first precipitated by PEG, dissociated with glycine-HCl buffer and the bound antigen is detected by labelled antibody. In one study, it was able to detect leishmaniasis in 88–100% of patients (Azazy *et al.*, 1994).

More recently, several recombinant antigens like rGBP from *L. donovani*, rORFF from *L. infantum*, rgp63, rK9, rK26, rK39 from *L. chagasi* have been developed and tested (Burns *et al.*, 1993; Martin *et al.*, 1998; Jensen *et al.*, 1999; Raja *et al.*, 1999; Bhatia *et al.*, 1999; Rajasekariah *et al.*, 2001).

Similarly, other studies have found antigens like LmSTI1 (Webb *et al.*, 1996), TSA (Webb *et al.*, 1998), and A2 protein specific for CL and VL patients (Carvalho *et al.*, 2002). Recombinant cysteine proteinases are proteins for which sera from patients with active and recovered CL are positive (Rafati *et al.*, 2001). MCL has been found to be associated with LeIF antigen (Skeiky *et al.*, 1995) and, finally LmSTI1, GBP, and HASPB1 generate antibodies detectable in patients with post-kala-azar dermal leishmaniasis (PKDL) (Alce *et al.*, 1999).

Other serological tests

Many other tests have also been used, conventional tests being countercurrent immunoelectrophoresis (CIEP), which is based on the electrophoretic mobility of oppositely charged antigen and antibodies on agar and assessing their interaction. Another approach is complement fixation which is based on fixing the complement with the antigen-antibody complexes and measuring the lysis of antibody-coated RBCs by the free complement factors. Immunoblot tests, based on antigen-antibody interaction on a nitrocellulose filter, gel diffusion and indirect haemagglutination have also been attempted but since they are cumbersome and less sensitive, they have been superseded by newer tests with greater sensitivity and specificity (Weigle *et al.*, 1987; Willams, 1995; Singh, 1999).

Immunochromatographic strip tests

Here the antigen is immobilized on the strip (as a spot or band) and the serum is allowed to move through it by capillary action. The appearance of a visible band at the region of immobilized antigen indicates a positive reaction. A study for the diagnosis of canine visceral leishmaniasis has evaluated the sensitivity of recombinant antigens K39 and K26, and has found them to be 96% sensitive and 100% specific (da Costa *et al.*, 2003). The recombinant K39 is a 39 amino acid repeat within the 230-kDa LcKin protein. It is found in several species of *Leishmania* and is used to diagnose VL caused by *L. donovani*, *L. chagasi* and *L. infantum* (Badaro *et al.*, 1996). Since antibodies against rK39 are absent in patients of CL and MCL, it is specific for VL (Burns *et al.*, 1993). Besides rK39, the hydrophilic antigen rK26 have also been reported to have specificity towards a single leishamanial species (Bhatia *et al.*, 1999). The rK39 dipstick, a serological test based on recombinant leishmanial antigen is being used in active case detection, and is under consideration for use in the elimination programme on the Indian subcontinent (Sundar *et al.*, 1998, 2002).

Latex agglutination test (KAtex)

This test is based on detecting the presence of antigen in the urine of kala-azar patients. Latex beads coated with polyclonal antibody against *Leishmania* antigen are incubated with the patients' urine. The antigen–antibody complex formed (if any) can be detected visually. The test is easy to perform and doesn't need any equipment and is thus suitable for field conditions. A recent report has shown the specificity of the method to be 99%, although sensitivity needs to be increased (Sunder *et al.*, 2005).

Nucleic acid based detection

Advances in molecular biology have resulted in the development of more specific and sophisticated ways of detecting parasites. More recently with the advent of PCR, the theoretical sensitivity of detection has gone down to the presence of a single parasite. PCR could be done on any sample to amplify the DNA. Many PCR-based methods are targeted towards kinetoplast minicircles, which are present at thousands of copies per cell (Aviles *et al.*, 1999; Bhattacharyya *et al.*, 1996; Nuzum *et al.*, 1995). An added advantage of PCR is its use in species-specific discrimination of infection, based on identification of conserved sequence elements and subsequent designing of primers for that region (Salotra *et al.*, 2001).

Although PCR has been the basis of many methods involved in species-specific diagnosis like PCR-RFLP (Marfurt *et al.*, 2003), PCR-SSCP (Lewin *et al.*, 2002), and sequence analysis (El Tai *et al.*, 2001), they all require many steps of processing of the DNA, which invariably increases the need for equipment and reagents and the risk of contamination. Recently, a study combining PCR with fluorescence resonance energy transfer/melting curve analysis (FRET/MCA) claims to overcome such limitation as it requires only a real-time thermal cycler with the ability of detecting fluorogenic probes (Quispe-Tintaya *et al.*, 2005). The use of fluorogenic real time PCR using SSU-rRNA gene added with complete automation has made quantification of parasite burden possible (Singh and Sivakumar, 2003). A PCR–ELISA technique developed recently is reported to be sufficiently sensitive and specific for use as a diagnostic test in case of mucocutaneous leishmaniasis and showed 83.3% detection in blood samples of clinically diagnosed cases (Pinero *et al.*, 1999).

Different DNA sequences in the genome of *Leishmania*, such as the ITS region, gp63 locus, telomeric sequences, 18s and SSU-rRNA, and both conserved and variable region of kinetoplast DNA (kDNA) minicircles are being used by various workers (El Tai *et al.*, 2001; Pizzuto *et al.*, 2001; Wortman *et al.*, 2001; Chiurillo *et al.*, 2001; Monroy Ostria *et al.*, 2002).

The samples can be from splenic aspirates, bone marrow aspirates or blood. In bone marrow, the reported sensitivity of detection is nearly 100%. Blood samples are easy to handle and since here the sample recovery is non-invasive this approach can be used for immunocompromised patients as well. PCR also finds its application in epidemiological studies. However, the principal drawbacks associated with PCR are the expense, availability of reagents and mandatory accessibility to sophisticated equipment that makes it impractical for field conditions.

Tests for cellular immunity

Lymphocyte proliferation test

Patients with mucosal or cutaneous leishmaniasis, and those with sub-clinical or cured VL, have circulating leukocytes capable of responding to leishmanial antigens *in vitro*. In this assay, sensitized peripheral blood lymphocytes of the patients are cultured in presence of *Leishmania* antigen for 5 days. Any proliferation of the lymphocytes is recorded by radioactive thymidine incorporation into the DNA of dividing cells. Another way of diagnosing through this cellular assay is to measure the cytokine levels produced by the proliferating lymphocytes and compare those from cultures where no antigen is added (Reed *et al.*, 1990).

In active VL, no proliferation is found after stimulation with *Leishmania* antigen. The mechanism of this immunosuppression is not clear; the presence of suppressor cells, serum factors and absence of antigen reactive cells in peripheral blood have all been suggested. These assays are relatively complex, time-consuming and expensive, and their application in routine diagnosis is definitely limited, although they can be used as an aid to diagnosis and as a test for exposure.

Intradermal skin tests

This is also known as leishmanin skin test (LST), which measures the delayed type hypersensitivity of the individual in terms of induration of the skin in reaction to an intradermal injection of *Leishmania* antigen. This test is useful in diagnosing cutaneous and mucosal leishmaniasis. The induration is measured by standard approaches such as the ball-point pen method. Here, from both sides, a line is drawn on the skin toward the injection site until resistance caused by induration is met. This is repeated at a perpendicular angle. Both diameters of indurations are measured in millimetres with a ruler and a mean is recorded. A value of more than 5 mm is considered positive.

The intradermal skin test relies on the strong T-cell responses that usually accompany cutaneous or mucosal leishmaniasis, as well as cured VL. The most common form of leishmanial skin test preparation used consists of whole cell promastigotes suspended in phenol-saline, as described by Buss (1929), although such preparation have been shown to be less effective than more standardized soluble preparations (Reed *et al.*, 1985). The obvious advantages of soluble preparations are increased safety, stability and the ability to standardize based on protein concentration. The simple and low technology requirement of applying and interpreting the intradermal skin test makes it versatile and useful.

Conclusions and perspectives

Leishmaniasis is considered as one of the most neglected diseases in terms of its research and development, an important reason being its prevalence in poor countries, like India, Bangladesh and Sudan. Recent economic developments and urbanization, has resulted in connecting leishmaniasis with developing countries, and this has led to people considering the problem more seriously. Despite a great deal of effort by different groups to limit the disease, we are yet to get a suitable vaccine; moreover the impact of vector control programmes has been limited. Therefore case detection and management have become the most critical factors in control programs.

Diagnosis has always been a problem for health care workers, especially because patients in remote areas have poor access to health services, and treating patients on the basis of clinical symptoms is risky due to potential side effects of the drugs.

Since the time of the discovery of *Leishmania* parasites, microscopic demonstration in body tissue has been considered as the main diagnostic measure. Various tissue aspirates have been used, with splenic aspirates are considered as the most sensitive, but a mandatory requirement is that patients to be checked for prothrombin time and platelet count and moreover is contraindicated in highly anaemic patients. Alternatively, bone marrow aspirates and lymph node aspirates have also been used, but have lower sensitivity. Bone marrow aspiration, being invasive, finds limited use. Thus, though being highly specific, direct detection of parasites suffers the drawbacks of limited use under field conditions as they are invasive and require technical expertise. A novel ultrasound-guided lymph node aspiration has been demonstrated as a less invasive and less risky procedure, but requires more sophisticated instruments (Reus *et al.*, 2005).

As non-invasive and easy-to-use methods, serological tests have been assessed, but all of them suffer the common disadvantage of cross-reactivity with other pathogens, inability to discriminate between current and past infection, thus interfering with follow-up of therapy. Moreover, they have been found to be unreliable in immunocompromised patients (Gradoni *et al.*, 1993). Tests such as ELISA and IFAT are very sensitive, but require sophisticated instruments. For any diagnostic test to be successful, it is important that the scales tilt in favour of the 'technology user' rather than the 'technology developer'. In this regard, DAT has a distinct advantage as it doesn't require a high-facility laboratory and is therefore most suitable for mass screening. A modification of DAT, referred to as FAST (fast agglutination screening test) considered to be more rapid and much easier has also been tested for large-scale screening (Silva *et al.*, 2005). The advancement towards making of freeze-dried antigen (Abdallah *et al.*, 2004) and rK39 antigen has accelerated the application of serodiagnostic methods in far-off remote areas with relatively basic facilities. Multiple epitope-based antigens have also been designed and tested for their potential application in diagnostics. (Boarino *et al.*, 2005).

More recently, molecular tools have also come up and the application of PCR-based methods for parasite detection has further enhanced the sensitivity to the level of one parasite. A recent study has reported that PCR can detect the presence of parasites even before the appearance of clinical symptoms (Lachaud *et al.*, 2000). The requirement of materials and reagents is the only limiting factor for their field use. Once this is overcome, molecular methods would also be able to identify the species variation amongst the population thus making therapy species specific.

The assays based on *Leishmania*-specific cell-mediated immunity include intradermal skin testing and lymphocyte proliferation assays, but these methods like other serological methods cannot distinguish between past and present infections. They are well suited for cutaneous and mucosal leishmaniasis. To date only few antigens have been identified that can stimulate the T-cells. The application of recombinant/synthetic antigen would perhaps have a great impact in the area as diagnosis could be tailored for different species.

In summary, molecular methods have greatly advanced the diagnosis of leishmaniasis in the past five years. Specific sensitive DNA sequences are available for use in PCR, and others will become available in future. Recombinant antigens have been produced for use in both serological and cellular assays, demonstrating the potential for developing standardized assays that are rapid, easy to perform, and with increased sensitivity and specificity. The next few years will see the development of improved PCR and immunology-based assays for various forms of leishmaniasis and will have an important impact on management of the disease.

References

Abdallah, K.A., Nour, B.Y., Schallig, H.D., Mergani, A., Hamid, Z., Elkarim, A.A., Saeed, O.K., and Mohamadani, A.A. (2004). Evaluation of the direct agglutination test based on freeze-dried *Leishmania donovani* promastigotes for the serodiagnosis of visceral leishmaniasis in Sudanese patients. Trop. Med. Int. Health. 9, 1127–1131.

Alce, T.M., Gokool, S., McGhie, D., Stager, S., and Smith, D.F. (1999). Expression of hydrophilic surface proteins of infective stages of *Leishmania donovani*. Mol. Biochem. Parasitol. *102*, 191–196.

Allahverdiyev, A.M., Uzun, S., Bagirova, M., Durdu. M., and Memisoglu, H.R. (2004). A sensitive new microculture method for diagnosis of cutaneous leishmaniasis. Am. J. Trop. Med. Hyg. *70*, 294–297.

Alvar, J., Canavate, C., Gutierrez-Solar, B., Jimenez, M., Laguna, F., Lopez-Velez, R., Molina, R., and Moreno, J. (1997). *Leishmania* and human immunodeficiency virus coinfection, the first 10 years. Clin. Microbiol. Rev. *10*, 298–319.

Aviles, H., Belli, A., Armijos, R., Monroy, F.P., and Harris, E. (1999). PCR detection and identification of *Leishmania* parasites in clinical specimens in Ecuador, a comparison with classical diagnostic methods. J. Parasitol. *85*, 181–187.

Azazy, A.A., Devaney, E., and Chance, M.L. (1994). A PEG-ELISA for the detection of *Leishmania donovani* antigen in circulating immune complexes. Trans. R. Soc. Trop. Med. Hyg. *88*, 62–66.

Badaro, R., Benson, D., Eulalio, M.C., Freire, M., Cunha, S., Netto, E.M., Pedral, S.D., Madureira, C., Burns,

J.M., Houghton, R.L., David, J.R., and Reed S.G. (1996). rK39, a cloned antigen of *Leishmania chagasi* that predicts active visceral leishmaniasis. J. Infect. Dis. *173*, 758–761.

Bhattacharyya, R., Das, K., Sen, S., Roy, S., and Majumder, H.K. (1996). Development of a genus specific primer set for detection of *Leishmania* parasites by polymerase chain reaction. FEMS Microbiol. Lett. *135*, 195–200.

Bhatia, A., Daifalla, N.S., Jen, S., Badaro, R., Reed, S.G., and Skeiky, Y.A. (1999). Cloning, characterization and serological evaluation of K9 and K26, two related hydrophilic antigen of *Leishmania chagasi*. Mol. Biochem. Parasitol. *102*, 249–261.

Boarino, A., Scalone, A., Gradoni, L., Ferroglio, E., Vitale, F., Zanatta, R., Giuffrida, M.G., and Rosati, S. (2005). Development of recombinant chimeric antigen expressing immunodominant B epitopes of *Leishmania infantum* for serodiagnosis of visceral leishmaniasis. Clin. Diagn. Lab. Immunol. *12*, 647–653.

Boelaert, M., Criel, B., Leeuwenburg, J., Van Damme, W., Le Ray, D., and Van der Stuyft, P. (2000). Visceral leishmaniasis control, a public health perspective. Trans. R. Soc. Trop. Med. Hyg. *94*, 465–471. Review.

Bora, D. (1999). Epidemiology of visceral leishmaniasis in India. Natl. Med. J. India *12*, 62–68.

Bryceson, A. (1996). Leishmaniasis. In Manson's tropical diseases, 20th edition, Cook, G. C. (editor) London, W.B. Saunders, pp 1213–1243.

Burns, J.M. Jr., Shrefffler, W.G., Benson, D.R., Ghalib, H.W., Badaro, R., and Reed, S.G. (1993). Molecular characterization of kinesin related antigen of *Leishmania chagasi* that detects specific antibody in African and American visceral leishmaniasis. Proc. Natl. Acad. Sci. USA *90*, 775–779.

Buss, G. (1929). Untersuchungen mit *Leishmania* vakzine. Arch. F. Schiffs. N. Trop. Hyg. *33*, 65–72.

Carvalho, F.A., Charest, H., Tavares, C.A., Matlashewski, G., Valente, E.P., Rabello, A., Gazzinelli, R.T., and Fernandes, A.P. (2002). Diagnosis of American visceral leishmaniasis in humans and dogs using the recombinant *Leishmania donovani* A2 antigen. Diagn. Microbiol. Infect. Dis. *43*, 289–295

Chang, S.C., Kuzmenok, O., Chiang, S.C., and Lee, S.T. (2004). An improved method for detection of *Leishmania* amastigotes by an antibody probe against the small subunit of leishmanial ribonucleotide reductase. Parasitol. Res. *94*, 243–245.

Chiurillo, M.A., Sachdeva, M., Dole, V.S., Yepes, Y., Miliani, E., Vazquez, L., Rojas, A., Crisante, G., Guevara, P., Anez, N., Madhubala, R., and Ramirez, J.L. (2001). Detection of *Leishmania* causing visceral leishmaniasis in the old and new worlds by a polymerase chain reaction assay based on telomeric sequence. Am. J. Trop. Med. Hyg. *65*, 573–582.

Chowdhury, M.S., Al Masum, A., Al Karim, E., SemiaoSantos, S., Rahman, K.M., Ar Rashid, H., and El Harith, A. (1993). Applicability of direct agglutination test at a rural health setting in Bangladesh and feasibility of antigen production. Arch. Inst. Pasteur Tunis. *70*, 333–344.

da Costa, R.T., Franca, J.C., Mayrink, W., Nascimento, E., Genaro, O., and Campos-Neto, A. (2003). Standardization of a rapid immunochromatographic test with the recombinant antigens K39 and K26 for the diagnosis of canine visceral leishmaniasis. Trans. R. Soc. Trop. Med. Hyg. *97*, 678–682.

Davidson, R. N. (1998). Practical guide for treatment of leishmaniasis. Drugs *56*, 1009–1018.

Desjeux, P. (1996). leishmanias is Public health aspects and control. Clin. Dermatol. *14*, 417–423.

Dietze, R., Falqueto, A., Valli, L.C., Rodriques, T.P., Boulos, M., and Corey, R. (1995). Diagnosis of canine visceral leishmaniasis with a dot-enzyme linked immunosorbent assay. Am. J. Trop. Med. Hyg. *53*, 40–42.

Dye, C. and Wolpert, D.M. (1988). Earthquakes, influenza and cycles of Indian kala-azar. Trans. R. Soc. Trop. Med. Hyg. *82*, 843–850.

El, Tai. N.O., El, Fari. M., Mauricio, I., Miles, M.A., Oskam, L., El, Safi, S.H., Presber, W.H., and Schonian, G. (2001). *Leishmania donovani*; intraspecific polymorphisms of Sudanese isolates revealed by PCR-based analyses and DNA sequencing. Exp. Parasitol. *97*, 35–44.

Elston, D.M. and Miller, S.D. (2004). Leishmaniasis acquired in the Iraqi theater of operations, lessons learned. Cutis *74*, 253–5.

Ghosh, M.K., Nandy, A., Addy, M., Maitra, T.K., and Ghose, A.C. (1995). Subpopulations of T-lymphocytes in the peripheral blood, dermal lesions and lymphnodes of post Kala-azar dermal leishmaniasis patients. Scand. J. Immunol. *41*, 11–17.

Gradoni, L., Sclone, A. and Gramiccia, M. (1993). HIV-leishmania coinfections in Italy; serological data as an indication of the sequence of acquisition of the two infections. Trans. R. Soc. Trop. Med. Hyg. *87*, 94–96.

Gramiccia, M., Gradoni, L., and Orsini, S. (1992). Decreased sensitivity to meglumine antimoniate (glucantime) of *Leishmania infantum* isolated from dogs after several courses of drug treatment. Ann. Trop. Med. Parasitol. *86*, 613–620.

Grevelink, S.A. and Lerner, E.A. (1996). Leishmaniasis, J. Am. Acad. Dermatol. *34*, 257–272.

Guerin, P.J., Olliaro, P., Sundar, S., Boelaert, M., Croft, S.L, Desjeux, P., Wasunna, M.K., and Bryceson, A.D. (2002). Visceral leishmaniasis, current status of control, diagnosis, and treatment, and a proposed research and development agenda. Lancet Infect. Dis. *2*, 494–501.

Hailu, A. (1990). A pre and post treatment antibody levels in visceral leishmaniasis. Trans. R. Soc. Trop. Med. Hyg. *84*, 673–675.

Harith, A.E., Chowdury, S., and Al-Masum, A. (1995). Evaluation of cleaving agents other than trypsin in direct agglutination test for further improving diagnosis of visceral leishmaniasis. J. Clin. Microbiol. *33*, 1984–1988.

Herwaldt, B.L. (1999). Leishmaniasis. Lancet. *354*, 1191–1199.

Herwaldt, B.L. (2001). Laboratory-acquired parasitic infections from accidental exposures. Clin. Microbiol. Rev.*14*, 659–688.

Jensen, A.T., Gasim, S., Moller, T., Ismail, A., Gaafar, A., Kemp, M., El Hassan, A.M., Kharazmi, A., Alce, T.M., Smith, D.F., and Theander, T.G (1999). Serodiagnosis of *Leishmania donovani* infections, assessment of enzyme-linked immunosorbent assays using recombinant *L. donovani* gene B protein (GBP) and a peptide sequence of *L. donovani* GBP. Trans. R. Soc. Trop. Med. Hyg. *93*, 157–160.

Karunaweera, N.D., Pratlong, F., Siriwardane, H.V., Ihalamulla, R.L., and Dedet, J.P. (2003). Sri Lankan cutaneous leishmaniasis is caused by *Leishmania donovani* zymodeme MON-37.Trans. R. Soc. Trop. Med. Hyg. *97*, 380–381.

Lachaud, L., Dereure, J., Chabbert, E., Reynes, J., Mauboussin, J.M., Oziol, E., Dedet, J.P., and Bastien, P. (2000). Optimized PCR using patient blood samples for diagnosis and follow-up of visceral leishmaniasis, with special reference to AIDS patients. J. Clin. Microbiol. 38, 236–240.

Lewin, S., Schonian, G., El, Tai, N. Oskam, L., Bastien, P., and Presber, W. (2002). Strain typing in *Leishmania donovani* by using sequence-confirmed amplified region analysis. Int. J. Parasitol. *32*, 1267–1276.

Lightner, L.K., Chulay, J.D., and Bryceson, A.D.M. (1983). Comparison of microscopy and culture in the detection of *Leishmania donovani* from splenic aspirates. Am. J. Trop. Med. Hyg. 32, 296–299.

Magill, A.J. (1995). Epidemiology of the leishmaniases. Dermatol. Clin. *13*, 505–523.

Magill, A.J., Grogl, M., Gasser, R.A.Jr. Sun, W., and Oster, C.N. (1993). Visceral Infection caused by *Leishmania tropica* in veterans of Operation Desert Storm. New Engl. J. Med.*328*, 1383–1387.

Marfurt, J., Niederwieser, I., Makia, N.D., Beck, H.P., and Felger, I. (2003). Diagnostic genotyping of old and new world *Leishmania* species by PCR-RFLP. Diagn. Microbiol. Infect. Dis. *46*, 115–124.

Martin, S.K., Thuita-Harun, L., Adoyo-Adoyo, M., and Wasunna, K.M. (1998). A diagnostic ELISA for visceral leishmaniasis, based on antigen from media conditioned by *Leishmania donovani* promastigotes. Ann. Trop. Med. Parasitol. 92, 571–577.

Martinez, P., de la Vega. E., Laguna, F., Laguna, F., Soriano, V., Puente, S., Moreno, V., Sentchordi, M.J., Garcia-Aquado, C., Gonzalez-Lahoz, J. (1993). Diagnosis of visceral leishmaniasis in HIV infected individuals using peripheral blood smears. AIDS *7*, 227–230.

Medrano, F.J., Jimenez-Mejias, E., Calderon, E., Calderon, E., Regordan, C., and Leal, M. (1993). An easy and quick method for diagnosis of visceral leishmaniasis in HIV-1 infected individuals. AIDS 7, 1399.

Meredith, S.E., Kroon, N.C., Sandorp, E., Seaman, J., Goris, M.G., van Ingen, C.W., Oosting, H., Schoone, G.J., Terpstra, W.J., and Oskam, L. (1995). Leish-kit, a stable direct Agglutination test based on freeze dried antigen for serodiagnosis of visceral leishmaniasis. J. Clin. Microbiol. 33, 1742–1745.

Molina, R., Lopez-Velez, R., Gutierrez-Solar, B., Jimenez, M.I., and Alvar, J. (1992). Isolation of *Leishmania infantum* from the blood of a patient with AIDS using sandflies. Trans. R. Soc. Trop. Med. Hyg. *86*, 516.

Monroy Ostria, J.,, and Sanchez-Tejeda G. (2002). Molecular probes and the polymerase chain reaction for detection and typing of *Leishmania* species in Mexico. Trans. R. Soc. Trop. Med. Hyg. 96(Suppl. 1), S101–104.

Murray, H.W., Berman, J.D., Davies, C.R., and Saravia, N.G. (2005). Advances in leishmaniasis. Lancet *366*, 1561–1577.

Nuzum, E., White, F., Thakur, C., Dietze, R., Wages, J., Grogl, M., and Berman, J. (1995). Diagnosis of symptomatic visceral leishmaniasis by use of the polymerase chain reaction on patient blood. J. Infect. Dis *171*, 751–754.

Palatnik-de-sousa, C.B., Gomes, E.M., Paraguai-de-sousa, E., Palatnik, M., Leez, K., and Borojevic, R. (1995). *Leishmania donovani*, titration of antibodies to fucose mannose ligand as an aid in diagnosis and prognosis of visceral leishmaniasis. Trans. R. Soc. Trop. Med. Hyg. *89*, 390–393.

Paredes, R., Munoz, J., Diaz, I., Domingo, P., Gurgui, M., and Clotet, B. (2003). Leishmaniasis in HIV infection. J. Postgrad. Med. *49*, 39–49.

Peters, W. (1981). The treatment of kala-azar – new approaches to an old problem. Indian. J. Med. Res. 73 Suppl. 1–18.

Pinero, J., Martnez, E., Pacheco, R., Aragon, Z., De Armas, F., Del Castillo, A., and Valladares, B. (1999). PCR-ELISA for diagnosis of mucocutaneous leishmaniasis. Acta. Trop. *73*, 21–29.

Pizzuto, M., Piazza, M., Senese, D., Scalamogna, C., Calattini, S., Corsico, L., Persico, T., Adriani, B., Magni, C., Guaraldi, G., Gaiera, G., Ludovisi, A., Gramiccia, M., Galli, M., Moroni, M., Corbellino, M., and Antinori, S. (2001). Role of PCR in diagnosis and prognosis of visceral leishmaniasis in patients co-infected with human immunodeficiency virus type 1. J. Clin. Microbiol. *39*, 357–361.

Prasad, L.S.N., and Sen, S. (1996). Migration of *Leishmania donovani* amastigotes in cerebrospinal fluid. Am. J. Trop. Med. Hyg. *55*, 652–654.

Quispe-Tintaya, K.W., Laurent, T., Decuypere, S., Hide, M., Banuls, A.L., De Doncker, S., Rijal, S., Canavate, C., Campino, L., and Dujardin, J.C. (2005). Fluorogenic assay for molecular typing of the *Leishmania donovani* complex, taxonomic and clinical applications. J. Infect. Dis. *192*, 685–692.

Rab, M.A., Hassan, M., Bux, D., Mahmood, M.T., and Evans, D.A. (1992). The isolation and cultivation of *Leishmania infantum* from apparently normal skin of visceral leishmaniasis of northern Pakistan. Trans. R Soc. Trop. Med. Hyg. *86*, 620–621.

Rafati, S., Salmanian, A.H., Hashemi, K., Schaff, C., Belli, S., and Fasel, N. (2001). Identification of *Leishmania major* cysteine proteinases as targets of immune response in humans. Mol. Biochem. Parasitol. *113*, 35–43.

Raj, V.S., Ghosh, A., Dole, V.S., Madhubala, R., Myler, P.J., and Stuart, K.D. (1999). Serodiagnosis of leishmaniasis with recombinant ORFF antigen. Am. J. Trop. Med. Hyg. *61*, 482–487.

Rajasekariah, G.H., Ryan, J.R., Hillier, S.R., Yi, L.P., Stiteler, J.M., Cui, L., Smithyman, A.M., and Martin, S.K. (2001). Optimisation of an ELISA for the serodiagnosis of visceral leishmaniasis using *in vitro* derived promastigotes antigens. J. Immunol. Methods. *252*, 105–119.

Ramesh, V., and Mukherjee, A. (1995). Post-kala-azar dermal leishmaniasis. Int J. Dermatol. *34*, 85–91. Review.

Reed, S.G., Badaro, R., Masur, H., Carvalho, E.M., Lorenco, R., Lisboa, A., Teixeira, R., Johnson, W.D.Jr., and Jones, T.C. (1986). Selection of a skin test antigen for American visceral leishmaniasis. Am. J. Trop. Med. Hyg. *35*, 79–85.

Reed, S.G., Carvalho, E.M., Sherbert, C.H., Sampaio, D.P., Russo, D.M., Bacelar, O., Pihl, D.L., Scott, J.M., Barral, A., Grabstein, K.H., and Johnson, W.D.Jr. (1990). In vitro responses to *Leishmania* antigens by lymphocytes from patients with leishmaniasis and chagas disease. J. Clin. Invest. *85*, 690–696.

Reed, S.G. (1996). Diagnosis of leishmaniasis. Clin. Dermatol. *14*, 471–478.

Reus, M., Garcia, B., Vazquez, V., Morales, D., Fuster, M., and Sola, J. (2005). Visceral leishmaniasis, diagnosis by ultrasound-guided fine needle aspiration of an axillary node. Br. J. Radiol. *78*, 158–160.

Saha, S., Mazumdar, T., Anam, K., Ravindran, R., Bairagi, B., Saha, B., Goswami, R., Pramanik, N., Guha, S. K., Kar, S., Banerjee, D., and Ali, N. (2005). *Leishmania* promastigote membrane antigen-based enzyme-linked immunosorbent assay and immunoblotting for differential diagnosis of Indian post-kala-azar dermal leishmaniasis. J. Clin. Microbiol. *43*, 1269–1277.

Salotra, P., Sreenivas, G., Pogue, G.P., Lee, N., Nakhasi, H.L., Ramesh, V., and Negi, N.S. (2001). Development of a species-specific PCR assay for detectin of *Leishmania donovani* in clinical samples from patients with kala-azar and post kala-azar dermal leishmaniasis. J. Clin. Microbiol. 39, 849–854

Santarem, N., Tomas, A., Ouaissi, A., Tavares, J., Ferreira, N., Manso, A., Campino, L., Correia, J.M., and Cordeiro-da-Silva, A. (2005). Antibodies against a *Leishmania infantum* peroxiredoxin as a possible marker for diagnosis of visceral leishmaniasis and for monitoring the efficacy of treatment. Immunol. Lett. *101*, 18–23.

Seaman, J., Mercer, A.J., and Sondorp, E. (1996). The epidemic of visceral leishmaniasis in western Upper Nile, southern Sudan, course and impact from 1984 to 1994. Int. J. Epidemiol. 25, 862–871.

Sharma, N.L, Mahajan, V.K., Kanga, A., Sood, A., Katoch, V.M., Mauricio, I., Singh, C.D., Parwan, U.C., Sharma, V.K., and Sharma, R.C. (2005). Localized cutaneous leishmaniasis due to *Leishmania donovani* and *Leishmania tropica*, preliminary findings of the study of 161 new cases from a new endemic focus in Himachal Pradesh, India. Am. J. Trop. Med. Hyg. *72*, 819–824.

Shreffler, W.G., Burns, J.M., Badaro, R., Ghalib, H.W., Butten, L.L., Mcmaster, W.R., and Reed, S.G. (1993). Antibody responses of visceral leishmaniasis patients to rgp63, a recombinant major surface protein of *Leishmania* species. J. Infect. Dis. *167*, 426–430.

Siddig, M., Ghalib, H.W., Shillington, D.C., and Peterson, E.A. (1988). Visceral leishmaniasis in the Sudan, comparative parasitological methods of diagnosis. Trans. R. Soc. Trop. Med. Hyg. *82*, 66–68.

Silva, E.S., Schoone, G.J., Gontijo, C.M., Brazil, R.P., Pacheco, R.S., and Schallig, H.D. (2005). Application of Direct Agglutination Test (DAT) and Fast Agglutination Screening Test (FAST) for sero-diagnosis of visceral leishmaniasis in endemic area of Minas Gerais, Brazil. Kinetoplastid. Biol. Dis. *4*, 4

Singh, S. (1999). Diagnostic and Prognostic markers of anti-kala-azar therapy and vaccination. Proc. V. R. Table Conf. Series. No. 5. Gupta S and Sood OP (Eds). New Delhi, Ranbaxy Science Foundation, pp. 95–114.

Singh, S. and Sivakumar, R. (2003). Recent advances in diagnosis of leishmaniasis. J. Postgrad Med. *49*, 55–60. Review.

Sinha, R., Datta, U. and Sehgal, S. (1993). Importance of bone marrow culture for diagnosis of Kala-azar. Scand. J. Infect. Dis. *25*, 787–789.

Siriwardena, H.V., Udagedara, C.U., and Karunaweera, N.D. (2003). Clinical features, risk factors and efficacy of cryotherapy in cutaneous leishmaniasis in Sri Lanka. Ceylon Med. J. *48*, 10–12.

Skeiky, Y.A., Guderian, J.A., Benson, D.R., Bacelor, O., Carvalho, E.M., Kubin, M., Badaro, R., Trinchieri, G., and Reed, S.G. (1995). A recombinant *Leishmania* antigen that stimulates human peripheral blood mononuclear cells to express a Th1 type cytokine profile and to produce interleukin 12. J. Exp. Med. *181*, 1527–1537.

Sundar, S., Reed, S.G., Singh, V.P., Kumar, P.C., and Murray, H.W.(1998). Rapid accurate field diagnosis of Indian visceral leishmaniasis. Lancet *351*, 563–565.

Sundar, S., More, D.K., Singh, M.K., Singh, V.P., Sharma, S., Makharia, A., Kumar, P.C., and Murray, H.W.(2000). Failure of pentavalent antimony in visceral leishmaniasis in India, report from the center of the Indian epidemic. Clin. Infect. Dis. *31*, 1104–1107.

Sundar S, Sahu M, Mehta H, Gupta A, Kohli U, Rai M, Berman JD, Murray HW. (2002). Noninvasive management of Indian visceral leishmaniasis, clinical application of diagnosis by K39 antigen strip testing at a kala-azar referral unit. Clin. Infect. Dis. 35, 581–586.

Sundar, S., Sahu, M., Mehta, H., Gupta, A., Kohli, U., Rai, M., Berman, J.D., and Murray, H.W. (2005). Detection of leishmanial antigen in the urine of patients with visceral leishmaniasis by a latex agglutination test. Am. J. Trop. Med. Hyg. 73, 269–271.

Tropical disease research, progress 2003-(2004). Seventeenth Programme Report of the UNICEF/UNDP/World Bank/WHO Special Programme for Research and Training in Tropical Diseases.

UNAIDS, HIV-related opportunistic diseases, UNAIDS technical update. (1998). www.unaids.org.

Webb, J.R., Campos-Neto, A., Ovendale, P.J., Martin, T.I., Stromberg, E.T., Badaro, R., and Reed, S.G. (1998). Human and murine immune responses to novel *Leishmania major* recombinant protein encoded by the members of a multicopy gene family. Infect. Immun. 66, 3279–3289.

Webb, J.R., Kaufmann, D., Campos-Neto, A., and Reed, S.G. (1996). Molecular cloning of a novel protein antigen of *Leishmania major* that elicits a potent immune response in experimental murine leishmaniasis. J. Immunol. *157*, 5034–5041.

Weigle, K.A., de Davalos, M., Heredia, P., Molineros, R., Saravia, N.G., and D'Alessandro, A. (1987). Diagnosis of cutaneous and mucocutaneous leishmaniasis in Columbia, a comparison of seven methods. Am. J. Trop. Med. Hyg. *36*, 489–496.

WHO (1990). Control of the Leishmaniases. Report of a WHO Expert Committee. Geneva, World Health Organization, Technical Report Series, no.793.

WHO (1996). Leishmania/HIV co-infection. Epidemiologic analysis of 692 retrospective cases. Wkly. Epidemiol. Rec.72, 49–54.

WHO (1997). Annex 3. Burden of disease in DALYS by cause, sex and mortality stratum in WHO regions, estimates for (2001). In The World Health Report, Geneva, WHO *2002*, 192–197. (www.who.int/whr/2002/whr2002_annex3.pdf).

WHO (2000). Fact Sheet No. 116, Rev. May (2000).

Williams, J.E. (1995). *Leishmania* trypanosoma,. In Medical Parasitology. A Practical Approach. In Gillespie, SH Hawkey PM (Eds). London, Oxford University Press.

Wortmann, G., Sweeney, C., Houng, H.S., Aronson, N., Stiteler, J., Jackson, J., and Ockenhouse, C. (2001). Rapid diagnosis of leishmaniasis by fluorogenic polymerase chain reaction. Am. J. Trop. Med. Hyg. *65*, 583–587.

Zapor, M.J., and Moran, K.A. (2005). Infectious diseases during wartime. Curr. Opin. Infect. Dis. *18*, 395–399.

Zijlstra, E.E., Ali, M.S., El Hassan, A.M., El Toum, I.A., Satti, M., Ghalib, H.W., and Kager, P.A. (1992). Kala-Azar, a comparative study of parasitological methods and the direct agglutination test in diagnosis, Trans. R. Soc. Trop. Med. Hyg. *86*, 505–507.

Zijlstra, E.E., el-Hassan, A.M., Ismael, A., and Ghalib, H.W. (1994). Endemic kala-azar in eastern Sudan, a longitudinal study on the incidence of clinical and subclinical infection and post-kala-azar dermal leishmaniasis. Am. J. Trop. Med. Hyg. *51*, 826–836.

Zijlstra, E.E, el-Hassan, A.M., and Ismael, A. (1995). Endemic kala-azar in eastern Sudan, post-kala-azar dermal leishmaniasis. Am J Trop Med Hyg. 52, 299–305.

Zijlstra, E.E., and el-Hassan, A.M. (2001, Leishmaniasis in Sudan. Visceral leishmaniasis. Trans. R. Soc. Trop. Med. Hyg. 95 Suppl. 1, S27–58.

Zijlstra, E.E, Musa, A.M., Khalil E.A., el-Hassan, I.M., and el-Hassan, A.M. (2003). Post-kala-azar dermal leishmaniasis. Lancet Infect. Dis. 3, 87–98.

Genome Structure and Content

2

Peter J. Myler

Abstract

In the last 2 years, the genomes of three *Leishmania* species (*L. major*, *L. infantum* and *L. braziliensis*) have been sequenced, revealing more than 8300 protein-coding and 900 RNA genes. Almost 40% of protein-coding genes fall into 662 families containing between two and 500 members. Most of the smaller gene families are tandem arrays of one to three genes, while the larger gene families are often dispersed in tandem arrays at different loci throughout the genome. Each of the 35 or 36 chromosomes are organized into a small number of gene clusters of tens-to-hundreds of genes on the same DNA strand. These clusters can be organized in head-to-head (divergent) or tail-to-tail (convergent) fashion, with the latter often separated by tRNA, rRNA and/or snRNA genes. Transcription of protein-coding genes initiates bi-directionally in the divergent strand-switch regions between gene clusters and extends polycistronically through each gene cluster before terminating in the strand-switch region separating convergent clusters. *Leishmania* telomeres are usually relatively small, consisting of a few different types of repeat sequence. Evidence can be found for recombination between several different groups of telomeres. The *L. major* and *L. infantum* genomes contain only ~50 copies of inactive degenerated *Ingi*/L1Tc-related elements (DIREs), while *L. braziliensis* also contains several telomere-associated transposable elements (TATEs) and spliced leader-associated (SLACs) retroelements. The *Leishmania* genomes share a conserved core proteome of ~6200 genes with the related trypanosomatids *Trypanosoma brucei* and *Trypanosoma cruzi*, but there are ~1000 *Leishmania*-specific genes (LSGs), which are mostly randomly distributed throughout the genome. There are relatively few (~200) species-specific differences in gene content between the three sequenced *Leishmania* genomes, but ~8% of the genes appear to be evolving at different rates between the three species, indicative of different selective pressures that could be related to disease pathology. While these genome sequences have vastly increased our knowledge of *Leishmania* genome content and organization, much work remains to be done, since ~65% of protein-coding genes currently lack functional assignment.

Genome size and chromosome number

In 1996, the *Leishmania* Genome Network (LGN) chose *L. major* MHOM/IL/81/Friedlin (LmjF) as the representative strain for complete genome sequencing. The LmjF genome sequencing effort used a chromosome-by-chromosome approach, with the sequencing divided between Seattle Biomedical Research Institute (SBRI), the Welcome Trust Sanger Institute (WTSI), and consortium of European laboratories (EuLeish). The sequencing strategy was initially clone-based, with a 9216-clone cosmid library (Ivens *et al.*, 1998), and a 10,394-clone BAC library used for this process. Clone contig maps for the entire genome were constructed using FPC software at Imperial College (cosmid clones) and SBRI in collaboration with the Vancouver BC Genome Centre (BAC clones). The maps were refined by end-sequencing of BAC and cosmid clones, and comparison of this sequence to the genomic sequence to obtain a final map. After initial sequencing of chr1 (Myler *et al.*, 1999), and chr3 (Worthey *et al.*, 2003) at SBRI, and chr4 at WTSI; the sequencing approach was supplemented by shotgun sequencing of chromosomal DNA isolated by pulse-field gel electrophoresis (PFGE). The sequencing was completed in 2005 with all 36 individual chromosomal sequences represented as single contiguous sequences, although small amount of sub-telomeric sequence and telomeric hexamer repeats are missing from some chromosomes (Ivens *et al.*, 2005). The accuracy of sequence assemblies was initially verified by restriction enzyme analysis (Myler *et al.*, 1999), but subsequently an optical map based on *Xba*I digestion (Zhou *et al.*, 2004) was used to identify gaps and correct misassemblies. The optical and sequence maps showed high concordance; although most chromosomes contained one or two large regions of potentially misassembled tandem repeats, particularly at chromosome ends. The current assembly of the LmjF genome (version 5.2) can be found at GeneDB (http://www.genedb.org/genedb/leish/), with the sequence accounting for 32,816,678 base pairs (bp) (Table 2.1).

The various *Leishmania* species have very similar genomic arrangements, generally consisting of 34 to 36 chromosome pairs (numbered 1–36 from smallest to largest) ranging in size from ~250 kilobases (kb) to ~4 megabases (Mb). The Old World species (*L. major* and *L. donovani* groups) have 36 chromosomes (Wincker *et al.*, 1996), while the New World species have only 34 (*L. mexicana*/*L. amazonensis* group) or 35 (*L. braziliensis* group), resulting from fusion of chromosomes 8 + 29 and 20 + 36 in the former case, and 20+34 in the latter (Britto *et al.*, 1998). While the *Leishmania* genomes are generally diploid, varying degrees of aneuploidy can occur (see below), and small amplified DNA molecules are frequently observed in response to drug selection or nutritional stress (Coderre *et al.*, 1983; Kink and Chang, 1987; Beverley *et al.*, 1988; Detke *et al.*, 1989; Petrillo-Peixoto and Beverley, 1989; Beverley, 1991; Tripp *et al.*, 1991; Hanson *et al.*, 1992; Tripp *et al.*, 1992; Wilson *et al.*, 1992; Segovia, 1994; Olmo *et al.*, 1995; Segovia and Ortiz, 1997).

Chromosome 1 is triploid in LmjF and one homologue apparently underwent a size change during laboratory cultivation, presumably due to unequal recombination in the repetitive sub-telomeric region at one end (Sunkin *et al.*, 2000). Indeed, this chromosome appears to be particularly prone to copy number variation, since tetraploidy has also been observed (Martinez-Calvillo *et al.*, 2005). In addition, at least two other chromosomes (20 and 31) also appear to be tetraploid, based on fluorescence staining intensity of chromosomal DNA on pulse-field gel electrophoresis (PFGE) and over-representation of BAC clones in whole genome shotgun libraries (see Table 2.1). In addition to the partial aneuploidy, three chromosomes (1, 2 and 27) show large-scale allelic differences, due to changes in the number of sub-telomeric, spliced leader (SL) RNA, and ribosomal (r) RNA repeats, respectively. However, apart from these differences, there are remarkably few (<0.1%) sequence differences between homologous chromosomes.

Gene content

Protein-coding genes

A combination of gene-finding algorithms such as TESTCODE (Fickett, 1982), GLIMMER (Salzberg *et al.*, 1998; Delcher *et al.*, 1999), CODONUSAGE (Staden and McLachlan, 1982), GENESCAN (Tiwari *et al.*, 1997; Aggarwal and Ramaswamy, 2002), and HEXAMER (Claverie and Bougueleret, 1986) was used to identify potential protein-coding genes within the finished LmjF chromosomal sequence. These gene predictions were refined by manual examination, including comparison with gene predictions from the *T. brucei* 927 (Berriman *et al.*, 2005) and *T. cruzi* CL-Brener (El-Sayed *et al.*, 2005a) genome projects, and the number of likely protein-coding genes stood at 8272 when the LmjF sequence was published (Ivens *et al.*, 2005). Subsequent comparison with the recently sequenced *L. infantum* and *L. braziliensis* genomes (Peacock *et al.*, 2007) has increased this number to the current 8378, including 74 pseudogenes. Since these numbers are undergoing continual modification, the most recent information can be found at the GeneDB website. The protein-coding genes show mean and median CDS lengths of 1901 and 1407 bp,

Table 2.1 *Leishmania major* Friedlin gene content

Chromosome	Size (bp)	PGCs	Protein-coding genes		RNA genes	Gene density (genes/Mb)
			Intact	Pseudogene		
1	268 984	2	85	0	0	316
2	355 714	2	74	2	63	214
3	384 518	3	95	1	1	250
4	472 856	2	126	3	0	273
5	465 823	5	124	1	34	268
6	516 874	4	134	1	0	261
7	596 348	5	127	2	1	216
8	574 972	5	132	3	0	235
9	573 441	4	169	1	12	296
10	570 864	4	143	4	3	258
11	582 575	3	135	3	10	237
12	675 347	3	142	1	0	212
13	654 604	4	170	0	0	260
14	622 648	3	152	5	1	252
15	629 514	4	164	3	25	265
16	714 659	4	171	3	1	243
17	684 831	3	157	1	3	231
18	739 751	2	171	0	48	231
19	702 212	2	177	0	2	252
20	742 551	3	176	2	78	240
21	772 974	6	225	1	6	292
22	716 608	5	169	3	14	240
23	772 567	7	196	2	27	256
24	840 950	5	237	1	7	283
25	912 849	4	252	2	66	278
26	1 091 579	2	274	2	199	253
27	1 130 447	6	263	6	113	238
28	1 160 128	6	314	3	0	273
29	1 212 674	5	296	1	2	245
30	1 403 454	4	395	4	9	284
31	1 484 336	5	332	2	6	225
32	1 565 925	4	411	4	0	265
33	1 583 673	6	367	3	47	234
34	1 866 754	7	483	0	14	259
35	2 090 491	6	535	3	76	257
36	2 682 183	10	731	1	43	273
Total	32 816 678	155	8304	74	911	255

respectively, with an average density of 252 genes per Mb. Slightly less than half (47.9%) of the genome is protein-coding, with a mean inter-CDS length of 2045 bp. The G+C content of coding and non-coding regions are 62.5% and 57.3%, respectively. The protein-coding genes are evenly spread between chromosomes, with a density range of only 212 (chr12) to 316 (chr1) genes per Mb on individual chromosomes (see Table 2.1). Only two genes (*LmjF29.2600*, encoding a polyA polymerase and *LmjF07.0340*, encoding an RNA helicase) have been shown to contain introns (Mair *et al.*, 2000; Ivens *et al.*, 2005), although another two genes (*LmjF29.2000* and *LmjF32.0850*) have some characteristics that suggest they also may undergo *cis*-splicing.

A combination of database searches and protein sequence analyses was used to annotate the predicted protein-coding genes. However, only 35% could be assigned putative biological function(s) based on experimental characterization (3.7%) or sequence similarity to proteins with known function in other organisms (31.5%). Most (2802) of these genes have been annotated with Gene Ontology (GO) Function, while fewer have been assigned GO Process (2153) or GO Component (1123) annotation. The remaining predicted genes encode hypothetical proteins, with some (5.7% of the total) having similarity to proteins of unknown function in non-kinetoplastid organisms, most (50.7%) showing sequence similarity to proteins found only in other trypanosomatids, and 8.3% being found only in *Leishmania*. The last two groups may represent genes that have novel kinetoplastid- or *Leishmania*-specific functions or are sufficiently diverged as to have no significant sequence similarity to their functional homologues in other organisms. The gene distribution on the chromosomes indicates that *Leishmania* genes with related functions generally do not cluster into prokaryote-like operons. Exceptions to this rule include clusters of tandem repeats (see below) and a locus on chr16 containing six genes involved in pyrimidine biosynthesis (see Chapter 7).

Gene families

Tribe-MCL clustering (Enright *et al.*, 2002) of the LmjF predicted protein sequences using the most-relaxed criteria revealed 662 families containing 3083 sequences (Ivens *et al.*, 2005). Most of the smaller gene families are tandem arrays of one to three genes, presumably arising from repeat expansion. These include a large number of gene duplications, particularly for ribosomal proteins and translation initiation factors, perhaps reflecting the need for higher expression levels of these proteins. In contrast, individual members of the larger gene families are often dispersed throughout the genome, although many also occur in tandem arrays at different loci. The largest protein family contains 492 loosely related sequences mostly consisting of hypothetical proteins. Other large families include 189 kinesin-related proteins, 60 protein kinases, 46 amastins, 32 CMGC protein kinases, 32 promastigote surface antigens (PSA-2/GP46), 29 RNA helicases, 27 ATPase/serine peptidases, 25 protein phosphatases, 25 tuzins, 24 STE protein kinases, and 23 amino acid permeases.

RNA genes

Almost 1000 genes encoding structural RNAs were identified in the LmjF genome (Table 2.1). There is a tandem array of 18S, 5.8S, and 28S ribosomal RNA genes on chr27 (Yan *et al.*, 1999), while the 5S *rRNA* genes are found as dispersed single copies on chr5, chr9 (three loci), chr11 (three loci), chr15, chr21 (two loci), and chr23. The 5S *rRNA* locus on chr5 also contains a gene encoding a 7SL RNA involved in protein translocation across membranes as part of the signal recognition particle (SRP). This RNA has a truncated Alu domain, suggesting that its function may be taken over by a tRNA-like molecule that is a *bona fide* RNA of the trypanosome SRP complex (Liu *et al.*, 2003).

The LmjF genome contains 83 tRNA genes, representing 45 of the 61 possible anticodons. This includes one apparent pseudogene (*LmjF31.TRNApseudo.01* with anticodon CTT) and three Tyr_{GTA} tRNAs which contain introns. Of the 16 anticodons not represented by tRNA genes, eight can be covered by wobble in the third position, while the other eight require modification of cognate anticodons.

Six snRNAs (U1–U6) involved in RNA splicing have been found in all three trypanosomatid genomes. The U1 and U3 genes are co-localized in a tRNA cluster on chr23, while the other snRNAs are distributed throughout the genomes, again associated with tRNA genes. A tandem array of *SLRNA* genes that encode the 39-nt 5′ exon trans-spliced onto all trypanosomatid mRNAs (Perry and Agabian, 1991) is located on chr2 and the size of this array differs between haplotypes (Antoniazi *et al.*, 2000).

SnoRNAs are involved in 2′-*O*-methylation and pseudouridylation of rRNA, SL RNA, and snRNA by formation of canonical guide RNA-target duplexes. C/D snoRNAs guide methylation and can be identified by their complementarity to the targets RNA to be modified. Analysis of the LmjF genome revealed 455 genes encoding 62 different C/D snoRNAs at 23 loci on chromosomes 5, 14, 15, 18, 19, 20, 22, 23, 25, 26, 27, 30, 33, 34, 35 and 36 (Liang *et al.*, 2007). Within in each locus, the snoRNA genes were clustered in up to 18 tandem repeats of one to five genes. Fifteen of these loci also contained 241 genes encoding 37 different H/ACA snoRNAs, which direct the pseudouridylation of target RNAs.

Gene organization/transcription

One of the most remarkable findings from the trypanosomatid genome projects was the discovery that the protein-coding genes are organized in a manner peculiar to trypanosomatid genomes: they are arranged into long polycistronic gene clusters (PGCs) of tens-to-hundreds of genes on the same DNA strand (Myler *et al.*, 1999; Worthey *et al.*, 2003). The PGCs can be organized in head-to-head (divergent) or tail-to-tail (convergent) fashion, with the latter often separated by tRNA, rRNA and/or snRNA genes. In addition, a number of long clusters of protein-coding genes are interrupted by RNA genes, and are thus split into two or more PGCs organized in a tail-to-head (sequential) fashion. In LmjF, there are 155 PCGs, containing 2–171 genes and spanning up to 817 kb. Individual chromosomes contain between two and ten PGCs. There are 11 instances in which single genes occur on the opposite strand to the adjacent genes. Eight of these are telomere-proximal, and are usually (7/8) are oriented towards the telomere. In another four cases, the telomere-proximal cluster contained only two or three genes (all oriented towards the telomere). Indeed, for the vast majority of telomeres (55/72), the last gene (and PGC) is oriented toward the chromosome end.

Interestingly, these gene clusters are associated with consistent biases in nucleotide base composition. Plots of cumulative purine excess, GC skew, and AT skew exhibit characteristic 'inverted V-shape' curves that are negatively correlated with the PGC coding strand (McDonagh *et al.*, 2000). This phenomenon is often observed in Eubacteria and Archaea, where it is associated with replication (Lobry, 1996), but is generally not seen at a genome scale in eukaryotes, although local skews linked to transcription have been reported (Green *et al.*, 2003; Aerts *et al.*, 2004). As the bias is greater in weakly selected positions (i.e. third base codon positions and inter-CDS regions), and there is no heterogeneity in synonymous codon usage, the most likely explanation is transcription-associated mutation biases (Lobry and Sueoka, 2002), associated with the unique large PGCs in trypanosomes. Interestingly, however, while the bias in AT skew in Tb927 and TcBr is also negatively correlated with the gene coding strand, GC skew is positively correlated with gene direction. In addition, the magnitude of the GC skew is greater in Tb927 and TcBr than in LmjF. While an explanation for this inter-genus difference is not yet apparent, it suggests that there may be fundamental differences in DNA repair and/or transcriptional processes between the different organisms.

Very little is known about the promoters that drive RNA polymerase (RNAP) II-mediated expression of protein-coding genes in trypanosomatids. Putative promoters for the *actin* (Ben Amar *et al.*, 1991) and *hsp70* (Lee, 1996) genes have been described in *T. brucei*, but their authenticity remains controversial (McAndrew *et al.*, 1998). Consequently, the observation that episomal molecules are transcribed promiscuously on both strands in *Leishmania* (Curotto de Lafaille *et al.*, 1992), and the presence of antisense transcripts in some regions of the *Leishmania* genome (Kapler *et al.*, 1990; Belli *et al.*, 2003), has led to the hypothesis that in trypanosomatids RNAP II has very low specificity, and that transcription can initiate indiscriminately within the genome (Swindle and Tait, 1996; McAndrew *et al.*, 1998). However, nuclear run-on analyses on LmjF chr1 indicated that there is substantially greater transcription of the protein-coding strand than of the non-coding strand and that RNAP II transcription initiates bi-directionally in the strand-switch region (Martinez-Calvillo *et al.*, 2003). Similar analyses on chr3 indicate that transcription initiates bidirectionally within the strand-switch region between the first two genes at the left end and initiates unidirectionally away from the telomere upstream of last gene at the right end (Worthey *et al.*, 2003; Martinez-Calvillo *et al.*, 2004). There is also an RNAP III promoter immediately upstream of the tRNA gene on the 'top' strand within the convergent strand-switch region. Transcription of the tRNA terminates (as measured by 3′ RACE) just downstream of the tRNA in a T-tract, as is typical of tRNA terminators. RNAP II-mediated transcription of the protein-coding genes also terminates within the tRNA region. Transfection studies with constructs containing the chr1 divergent strand-switch region (Martinez-Calvillo *et al.*, 2003) and the chr3 convergent strand-switch region (Martinez-Calvillo *et al.*, 2004) confirm their importance in transcription initiation and termination, respectively, although other factors are also likely to be involved.

Thus, it appears that transcription of protein-coding genes initiates bidirectionally in the divergent strand-switch regions between gene clusters and extends polycistronically through each gene cluster before terminating in the strand-switch region separating convergent clusters. Consequently there are only a few initiation sites (promoters) on each chromosome, and perhaps fewer than 100 in the entire genome. This organization is possible because the polycistronic RNA is subsequently processed by concomitant *trans*-splicing and polyadenylation, which results in a 39-nt capped spliced leader (SL) sequence being added to the 5′ end of each mRNA and a 3′ poly(A) tail to the upstream mRNA (Perry and Agabian, 1991). Unlike polycistronic genes in prokaryotes, trypanosomatid PGCs generally do not contain only genes with related functions, and there is currently no evidence for regulation of their expression at the level of transcription initiation. Consequently, trypanosomatid gene expression appears to be regulated post-transcriptionally, at the level of RNA turnover, translation, and/or protein stability (Clayton, 2002).

Repetitive sequences

In contrast to most other eukaryotes (including other trypanosomatids), the *Leishmania* genome contains relatively little (<10%) repetitive sequence, which is composed primarily of tandem arrays of protein-coding and RNA genes (see above), telomeric and subtelomeric repeats, as well as degenerate retrotransposons (see below).

Telomeres

Leishmania telomeres are usually relatively uncomplicated, with PGCs of protein-coding genes extending to within 1–6 kb of the chromosome termini. In LmjF, the chromosome ends consist of a 'repeated-repeat' structure made up of tandem arrays of multiple telomeric hexamer (ACCCTA) repeats, interspersed with five copies of non-perfect tandem repeats of an octameric telomere-associated sequence (TAS)-A, and a single-copy 62-bp TAS-B sequence (Chiurillo *et al.*, 2000; Sunkin *et al.*, 2000). The TAS-A and TAS-B sequences differ between species (Fu and Barker, 1998), and *L. braziliensis* telomeres also contain telomere-associated transposable elements (Peacock *et al.*, 2007). Six LmjF telomeres also contain 0.7–25 kb of distinctive combinations of *Leishmania* sub-telomeric repeats (LST-R) (Sunkin *et al.*, 2000), present in different copy numbers on different telomeres, as well as sub-telomeric non-repetitive sequences (STNR).

There appear to have been several recombination events between telomeres, which resulted in sharing of telomere-proximal protein-coding genes as well as sub-telomeric sequences (Fig. 2.1). For example, an apparent recombination within the seventh protein-coding gene from the telomere of chr27R and chr2R resulted in their sharing 20 kb of their sequence (Fig. 2.1A). Two recombination events have resulted in their sharing the most telomere-proximal gene (which encodes a putative peptidyl dipeptidase) and ~0.7 kb upstream of the LST-RJ region with chr1aR. Interestingly, chr19R and chr5R both contain related genes (encoding putative β-galactofuranosyltransferases) between the LST-R and TAS regions.

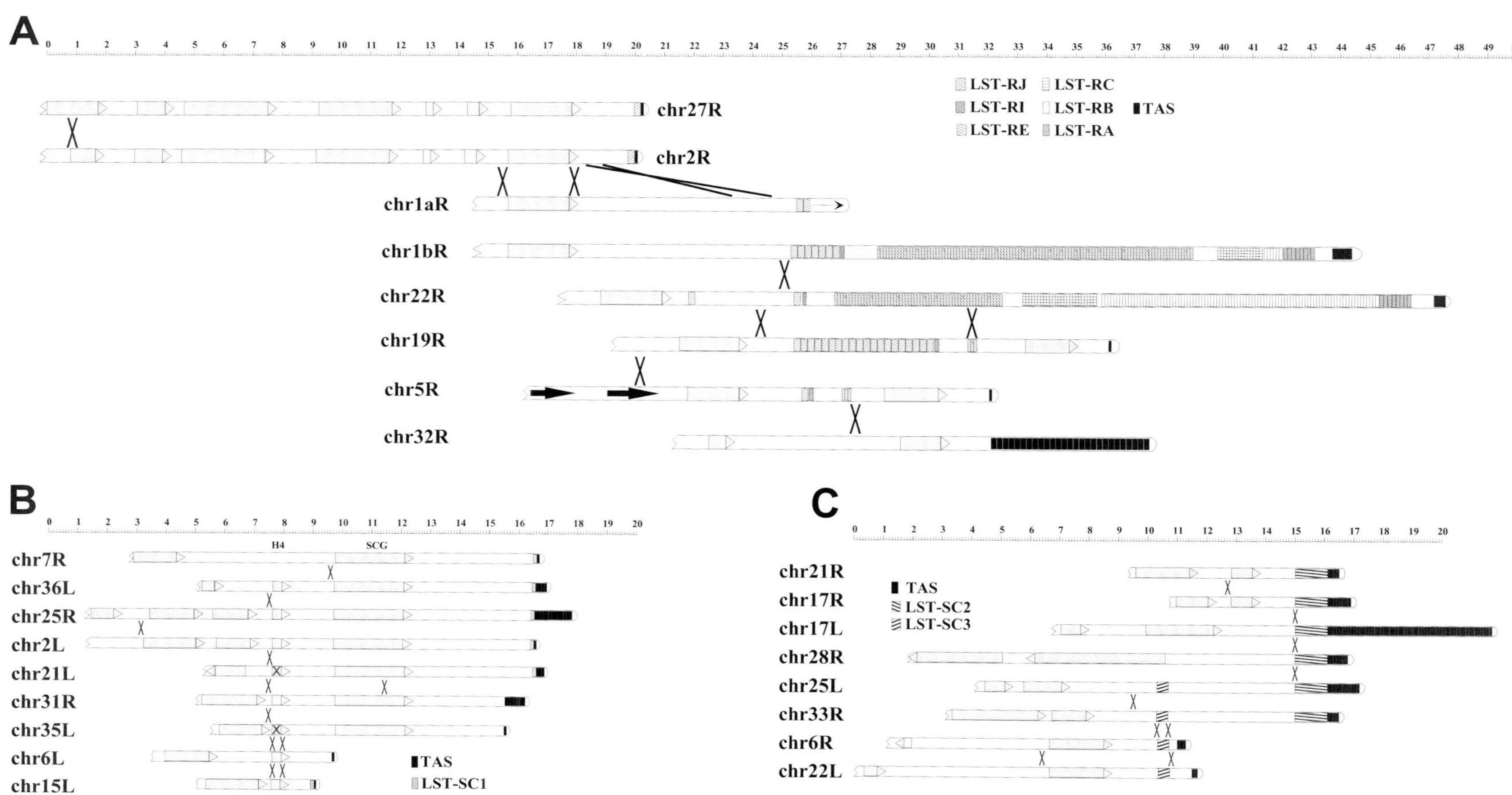

Figure 2.1 Recombination between LmjF telomeres. Telomeres in three different recombination groups are shown in panels A, B and C. Telomere-proximal protein-coding genes are indicated by light grey boxes, with arrowheads indicating the direction of transcription. Telomere associated (TAS) and *Leishmania* sub-telomeric repeats (LST-R) sequences are shown by patterned boxes, as indicated in each panel. Potential recombination sites are indicated by Xs. The arrows on chr5R indicate a tandem repeat, while the arrowhead on chr1a indicates that this telomere has not been completely sequenced and probably also contains TAS repeats. The size scales at the top of each panel are in kilobases.

The most telomere-proximal gene (*SCG*) on chr7R, chr36L, chr25R, chr2L, chr21L, chr31R, and chr35L encodes phosphoglycan-β-1,3-galactosyltransferase 3 (Dobson *et al.*, 2003), usually associated with an adjacent histone *H4* gene (or pseudogene). These telomeres also appear to have undergone recombination events, since the first four of these telomeres share the STNR sequences between the SCG gene and the TAS, while the last two share a different STNR sequence (Fig. 2.1B). The 5′ recombination boundary differs between telomeres, with that in chr7R being immediately upstream of the *SCG3* gene, while those in chr36L, chr21L, chr31R and chr35L occurring just upstream of the *H4* gene, and the recombination between chr25R and chr2L occurring upstream of the fourth protein-coding gene. Two additional telomeres, chr6L and chr15L, have the *H4* gene as the last gene before the STNR sequence, but otherwise show no similarity to each other of the other telomeres of this group, except for a ~200-bp conserved sequence (LST-CS1) adjacent to the TAS region on several telomeres.

A third group of telomeres (chr21R, chr17R, chr17L, chr25L, chr28R and chr33R) contain a 1.1 kb conserved sub-telomeric sequence (LST-CS2), which is adjacent to the TAS region (see Fig. 2.1C). In addition, chr17R and chr21R share an STNR sequence and a telomere-proximal protein-coding gene encoding a conserved hypothetical protein. Two other telomeres, chr25L and chr33R, share substantial STNR sequence, including a ~250 bp sequence (LST-CS3) also found on two other telomeres (chr6R and chr22L). It is also interesting to note that the two telomeres of chr10 share all of their sub-telomeric sequence, as well as the three telomere-proximal protein-coding genes.

Retroelements

In contrast to the *T. brucei* and *T. cruzi* genomes that contain hundreds of retroelements of several different types (El-Sayed *et al.*, 2005b; El-Sayed *et al.*, 2005a), the LmjF genome contains only 52 inactive degenerated *Ingi*/L1Tc-related elements (DIREs) (Ghedin *et al.*, 2004). DIREs are also found in *T. brucei* (73 elements) and *T. cruzi* (257 elements). Sequence similarity in the reverse transcriptase domains of DIREs, *Ingi* (*T. brucei*), and L1Tc (*T. cruzi*), implies that they have a common ancestor that presumably was active in retrotransposition. Retroelements are often present at synteny break points in all three genomes, suggesting that they may be involved in chromosome evolution. Interestingly, *L. braziliensis* contains 20–30 telomere-associated transposable elements (TATEs) and several spliced leader-associated (SLACs) retroelements not found in the *L. major* and *L. infantum* genomes (Peacock *et al.*, 2007).

Leishmania-specific genes

The trypanosomatid genomes contain a conserved core proteome of ~6200 genes and show a remarkable degree of synteny considering their divergence 200–500 million years ago (El-Sayed *et al.*, 2005b). Nevertheless, ~1000 *Leishmania* genes have no orthologue in *T. brucei* or *T. cruzi*. While these *Leishmania*-specific genes (LSGs) are by and large randomly distributed throughout the genome, most synteny breaks have an adjacent LSG. Some of the LSGs are responsible for the key *Leishmania*-specific metabolic processes (e.g. peptidases, transporters, glycoconjugate biosynthesis components), but the majority have unknown functions. Of particular interest are two LSGs (*LmjF33.1740* and *LmjF33.1750*) in a convergent strand-switch region of chr33 that both encode proteins containing macrophage migration inhibition factor (MIF) domains. Phylogenetic analysis suggests that these genes entered the *Leishmania* genome as a result of horizontal gene transfer from bacteria, and it is tempting to speculate that one, or both, may play a role in modulating the host macrophage response to promote parasite survival (Ivens *et al.*, 2005).

Species-specific differences

Specific species of *Leishmania* are associated with the different types of disease (Herwaldt, 1999; Cunningham, 2002; Dedet *et al.*, 1999), but this relationship is complex, with different types of diseases sometimes being seen within the same species group (Dedet *et al.*, 1999). The New World and Old World species probably diverged 20–100 million years ago following the break-up of Gondwanaland. Thus, it seems reasonable to speculate that species-specific differences in gene content might be responsible for at least some of the differences in disease pathology observed between species group. Indeed, monoclonal antibodies (Grimaldi and McMahon-Pratt, 1996; Grisard *et al.*, 2000; Romero *et al.*, 2002), PCR analyses (Grisard *et al.*, 2000; Uezat *et al.*, 2001; Schonian *et al.*, 2001a; Schonian *et al.*, 2001b), kDNA heterogeneity (Lu *et al.*, 1994) and RFLP analysis (Akman *et al.*, 2000) have been used to discriminate species and strains of *Leishmania*, indicating that detectible genetic differences do exist. Furthermore, there is already direct evidence that such genetic differences can have functional significance. In *L. donovani*, the *A2* gene occurs in four loci per diploid genome, which contains different numbers (1–7) of tandemly arranged copies of the gene, interspersed with another gene (*A2rel*) (Zhang and Matlashewski, 2001). A2 proteins contain variable numbers of a 10-amino-acid repeat, which have been implicated in aggregation with HSP100 (Clos *et al.*, 2001). The *L. donovani A2* genes are differentially expressed in amastigotes (Charest *et al.*, 1996), and are essential for virulence *in vivo* (Zhang and Matlashewski, 1997). However, the LmjF diploid genome contains only four copies of the *A2* gene (one per locus), the proteins lack all but one copy of the 10-amino acid repeat, and are apparently not expressed at any detectable level (Zhang *et al.*, 2003). Transfer of the *L. donovani* A2 gene into *L. major* parasites leads to an increased ability to survive in the spleen of BALB/c mice (Zhang and Matlashewski, 2001), and inhibition of its ability to establish a cutaneous infection in susceptible BALB/c or resistant C57BL6 mice (Zhang *et al.*, 2003), suggesting a role in the visceralization of infection associated with *L. donovani*.

The recent sequencing of the *L. infantum* and *L. braziliensis* genomes (Peacock *et al.*, 2007) presents an unprecedented opportunity for genome-wide comparison with the *L. major* genome in order to search for species-specific differences in gene content. In general, sequence conservation between species is high, with the average amino acid identity between the Old World species (*L. major* and *L. infantum*) being 92%, and the average identity between Old World and New World (*L. braziliensis*) species being 77%. Surprisingly, only a modest number of species-specific genes were found: 5 in *L. major*, 26 in *L. infantum* and 47 in *L. braziliensis*. Another 32, 6 and 98 genes, respectively, are absent from *L. major*, *L. infantum* or *L. braziliensis*, but present in the other two species. In the vast majority (~80%) of these cases, the species-specific differences are due to pseudogene formation, although *L. braziliensis* does contain 26 genes that are not present in the other two *Leishmania* species. In addition, another 29 genes are found only as pseudogenes in two or three species, of which 10 are present as intact genes in *T. brucei* or *T. cruzi*.

Of the 224 species-restricted genes, only 48 could be assigned putative function of the basis of sequence homology. Perhaps the most interesting result of these analyses is the finding of *L. braziliensis* contains intact orthologues of the *Argonaute* (*LbrM11_V2.03600*) and *Dicer* (*LbrM23_V2.0390*) genes implicated in RNA interference (RNAi), as well as an additional gene (*LbrM25_V2.1020*) encoding a protein with RNase III motifs. All three genes are absent or present as pseudogenes in *L. major* and *L. infantum*. This finding explains the failure of previous attempts to detect RNAi activity in *L. major* (Robinson and Beverley, 2003), and raises hopes that it will be found in *L. braziliensis*. The presence of RNA viruses (Stuart *et al.*, 1992) and transposable elements (see above) in the *Viannia* sub-genus (which includes *L. braziliensis*) may have provided the selective pressure for the retention of an active RNAi system.

Other *L. braziliensis*-specific genes include those encoding a methyltransferase (*LbrM04_V2.1180*), an adenine phosphoribosyltransferase (*LbrM26_V2.0120*), a β-galactosyltransferase (*LbrM20_V2.0480*), a galactokinase (*LbrM35_V2.3650*), and an amino acid transporter (*LbrM36_V2.1500*). *L. infantum*-specific genes include *LinJ14_V3.0020*, which encodes a phosphatidylinositol 3-kinase; *LinJ22_V3.0220* (putative methylenetetrahydrofolate dehydrogenase); *LinJ25_V3.2500* (putative glutathionylspermidine synthase); *LinJ24_V3.1510* (multidrug resistance protein), *LinJ31_V3.1140* (putative n-acyl-l-amino acid amidohydrolase); *LinJ31_V3.2420* (putative p-nitrophenylphosphatase), and *LinJ36_V3.0640*, which encodes a putative phosphatidylinositol-phosphatidylcholine transfer protein. The yeast orthologue of the last gene (*sec14*) encodes a cytosolic factor that is required for export of secretory proteins from the Golgi (Bankaitis *et al.*, 1989). Only two of the five *L. major*-specific genes have putative function: *LmjF13.1450*, which encodes an EF hand protein; and *LmjF35.3910*, which encodes a cysteine peptidase.

Both the *L. infantum* and *L. braziliensis* genomes contain an intact gene (*LinJ08_V3.0560* and *LbrM08_V2.0590*, respectively) encoding a putative cyclopropane-fatty-acyl-phospholipid synthase (CPFS), which is absent in the *L. major* genome. Phylogenetic analyses suggest that this gene has been acquired by horizontal gene transfer from bacteria (Peacock *et al.*, 2007). Cyclopropination has been implicated in the structural integrity of the plasma membrane in *E. coli* and the cell wall complex in *Mycobacterium tuberculosis*, as well as modulation of the innate immune response (Rao *et al.*, 2005). Interestingly, the *L. braziliensis* genome lacks the multigene *HASP/SHERP* locus that encodes stage-specific surface proteins in *L. major* and *L. infantum* (Knuepfer *et al.*, 2001).

While there are relatively few species-specific differences in gene content between the three sequenced *Leishmania* genomes, ~8% of the genes appear to be evolving at different rates between the three species (Peacock *et al.*, 2007), suggesting different selective pressures that could be related to disease pathology. In addition, microarray-based expression analysis suggests considerable differences in mRNA expression pattern between *L. major* and *L. donovani* (unpublished data), suggesting that species-specific genes may not be the only factors in determining disease outcome.

Acknowledgements

I would like to thank Matt Berriman and others at the Wellcome Trust Sanger Institute for providing their manuscript describing the *L. major–L. infantum–L. braziliensis* comparative sequence analysis prior to publication.

References

Aerts, S., Thijs, G., Dabrowski, M., Moreau, Y., and De Moor, B. (2004). Comprehensive analysis of the base composition around the transcription start site in Metazoa. BMC. Genomics 5, 1–10.

Aggarwal, G. and Ramaswamy, R. (2002). *Ab initio* gene identification: prokaryote genome annotation with Genescan and Glimmer. J. Biosci. *27*, 7–14.

Akman, L., Aksu, H.S., Wang, R.Q., Ozensoy, S., Ozbel, Y., Alkan, Z., Ozcel, M.A., Culha, G., Ozcan, K., Uzun, S., Memisoglu, H.R., and Chang, K.P. (2000). Multi-site DNA polymorphism analyses of *Leishmania* isolates define their genotypes predicting clinical epidemiology of leishmaniasis in a specific region. J. Eukaryot. Microbiol *47*, 545–554.

Antoniazi, S., Lima, H.C., and Cruz, A.K. (2000). Overexpression of miniexon gene decreases virulence of *Leishmania major* in BALB/c mice *in vivo*. Mol. Biochem. Parasitol. *107*, 57–69.

Bankaitis, V.A., Malehorn, D.E., Emr, S.D., and Greene, R. (1989). The *Saccharomyces cerevisiae* SEC14 gene encodes a cytosolic factor that is required for transport of secretory proteins from the yeast Golgi complex. J. Cell Biol. *108*, 1271–1281.

Belli, S., Monnerat, S., Schaff, C., Masina, S., Noll, T., Myler, P.J., Stuart, K.D., and Fasel, N. (2003). Sense and antisense transcripts in the histone H1 (HIS-1) locus of *Leishmania major*. Int. J. Parasitol. 33, 965–975.

Ben Amar, M.F., Jefferies, D., Pays, A., Bakalara, N., Kendall, G., and Pays, E. (1991). The actin gene promoter of *Trypanosoma brucei*. Nucleic Acids Res. *19*, 5857–5862.

Berriman, M., Ghedin, E., Hertz-Fowler, C., Blandin, G., Lennard, N.J., Bartholomeu, D., Renauld, H.J., Caler, E., Hamlin, N., Haas, B., Harris, B.R., Hannick, L., Barrell, B., Donelson, J., Hall, N., Fraser, C.M., Melville, S.E., El-Sayed, N., Böhme, U.C., Shallom, J., Aslett, M., Hou, L., Atkin, B., Barron, A.J., Bringaud, F., Brooks, K., Cherevach, I., Chillingworth, T., Churcher, C., Clark, L.N., Corton, C.H., Cronin, A., Davies, R., Doggett, J., Djikeng, A., Feldblyum, T., Fraser, A., Goodhead, I., Hance, Z., Harper, A.D., Hauser, H., Hostetler, J., Jagels, K., Johnson, D., Johnson, J., Jones, C., Kerhornou, A., Koo, H., Larke, N., Larkin, C., Leech, V., Line, A., MacLeod, A., Mooney, P., Moule, S., Mungall, K., Norbertczak, H., Ormond, D., Pai, G., Peterson, J., Quail, M.A., Rajandream, M.A., Reitter, C., Sanders, M., Schobel, S., Sharp, S., Simmonds, M., Simpson, A.J., Tallon, L., Turner, C.M., Tait, A., Tivey, A., Van Aken, S., Walker, D., Wanless, D., White, B., White, O., Whitehead, S., Wortman, J., Barry, J.D., Fairlamb, A.H., Field, M.C., Gull, K., Landfear, S., Marcello, L., Martin, D.M., Opperdoes, F., Ullu, E., Whickstead, B., Alsmark, C., Arrowsmith, C., Carrington, M., Embley, T.M., Ivens, A., Lord, A., Morgan, G.M., Peacock, C.S., Rabbinowitsch, E., Salzberg, S., Wang, S., Woodward, J., and Adams, M.D. (2005). The genome of the African trypanosome, *Trypanosoma brucei*. Science *309*, 416–422.

Beverley, S.M. (1991). Gene amplification in *Leishmania*. Annu. Rev. Microbiol. *45*, 417–444.

Beverley, S.M., Ellenberger, T.E., Iovannisci, D.M., Kapler, G.M., Petrillo-Peixoto, M., and Sina, B.J. (1988). Gene amplification in *Leishmania*. In The Biology of Parasitism, P.T. Englund and A. Sher, eds. (New York: Alan R. Liss, Inc.), pp. 431–448.

Britto, C., Ravel, C., Bastien, P., Blaineau, C., Pagès, M., Dedet, J.P., and Wincker, P. (1998). Conserved linkage groups associated with large-scale chromosomal rearrangements between Old World and New World *Leishmania* genomes. Gene *222*, 107–117.

Charest, H., Zhang, W.W., and Matlashewski, G. (1996). The developmental expression of *Leishmania donovani* A2 amastigote-specific genes is post-transcriptionally mediated and involves elements located in the 3′-untranslated region. J. Biol. Chem. *271*, 17081–17090.

Chiurillo, M.A., Beck, A., deVos, T., Myler, P.J., Stuart, K., and Ramirez, J.-L. (2000). Cloning and characterization of *Leishmania donovani* telomeres. Exp. Parasitol. *94*, 248–258.

Claverie, J.M. and Bougueleret, L. (1986). Heuristic informational analysis of sequences. Nucleic Acids Res. *14*, 179–196.

Clayton, C.E. (2002). Life without transcriptional control? From fly to man and back again. EMBO J. *21*, 1881–1888.

Clos, J., Klaholz, L., Kroemer, M., Krobitsch, S., and Lindquist, S. (2001). Heat shock protein 100 and the amastigote stage-specific A2 proteins of *Leishmania donovani*. Med. Microbiol. Immunol. (Berl.) *190*, 47–50.

Coderre, J.A., Beverley, S.M., Schmike, R.T., and Santi, D.V. (1983). Overproduction of a bifunctional thymidylate synthetase-dihydrofolate reductase and DNA amplification in methotrexate-resistant *Leishmania tropica*. Proc. Natl. Acad. Sci. USA *80*, 2132–2136.

Cunningham, A.C. (2002). Parasitic adaptive mechanisms in infection by *Leishmania*. Exp. Mol. Pathol. *72*, 132–141.

Curotto de Lafaille, M.A., Laban, A., and Wirth, D.F. (1992). Gene expression in *Leishmania*: Analysis of essential 5′ DNA sequences. Proc. Natl. Acad. Sci. USA *89*, 2703–2707.

Dedet, J.P., Pratlong, F., Lanotte, G., and Ravel, C. (1999). Cutaneous leishmaniasis. The parasite. Clin. Dermatol. *17*, 261–268.

Delcher, A.L., Harmon, D., Kasif, S., White, O., and Salzberg, S.L. (1999). Improved microbial gene identification with GLIMMER. Nucleic Acids Res. *27*, 4636–4641.

Detke, S., Katakura, K., and Chang, K.-P. (1989). DNA amplification in arsenite-resistant *Leishmania*. Exp. Cell Res. *180*, 161–170.

Dobson, D.E., Scholtes, L.D., Valdez, K.E., Sullivan, D.R., Mengeling, B.J., Cilmi, S., Turco, S.J., and Beverley, S.M. (2003). Functional identification of galactosyltransferases (SCGs) required for species-specific modifications of the lipophosphoglycan adhesin controlling *Leishmania major*–sandfly interactions. J. Biol. Chem. *278*, 15523–15531.

El-Sayed, N.M.A., Myler, P.J., Bartholomeu, D., Nilsson, D., Aggarwal, G., Tran, A.-N., Ghedin, E., Worthey, E.A., Delcher, A.L., Blandin, G., Westenberger, S.J., Caler, E., Cerqueira, G., Branche, C., Haas, B., Anapuma, A., Arner, E., Åslund, L., Attipoe, P., Bontempi, E., Bringaud, F., Burton, P., Cadag, E., Campbell, D.A., Carrington, M., Crabtree, J., Darban, H., Da Silveira, J.F., de Jong, P., Edwards, K., Englund, P.T., Fazelina, G., Feldblyum, T., Ferella, M., Frasch, A.C., Gull, K., Horn, D., Hou, L., Huang, Y., Kindlund, E., Klingbeil, M., Kluge, S., Koo, H., Lacerda, D., Levin, M.J., Lorenzi, H., Louie, T., Machado, C.R., McCulloch, R., McKenna, A., Mizuno, Y., Mottram, J.C., Nelson, S., Ochaya, S., Osoegawa, K., Pai, G., Parsons, M., Pentony, M., Pettersson, U., Pop, M., Ramirez, J.L., Rinta, J., Robertson, L., Salzberg, S.L., Sanchez, D.O., Seyler, A., Sharma, R., Shetty, J., Simpson, A.J., Sisk, E., Tammi, M.T., Tarleton, R., Teixeira, S., Van Aken, S., Vogt, C., Ward, P.N., Wickstead, B., Wortman, J., White, O., Fraser, C.M., Stuart, K.D., and Andersson, B. (2005a). The genome sequence of *Trypanosoma cruzi*, etiological agent of Chagas' disease. Science *309*, 409–415.

El-Sayed, N.M.A., Myler, P.J., Blandin, G., Berriman, M., Crabtree, J., Aggarwal, G., Caler, E., Renauld, H., Worthey, E.A., Hertz-Fowler, C., Ghedin, E., Peacock, C., Bartholomeu, D.C., Haas, B.J., Tran, A.-N., Wortman, J.R., Alsmark, U.C.M., Angiuoli, S., Anupama, A., Badger, J., Bringaud, F., Cadag, E., Carlton, J.M., Cerqueira, G.C., Creasy, T., Delcher, A.L., Djikeng, A., Embley, T.M., Hauser, C., Ivens, A.C., Kummerfeld, S.K., Pereira-Leal, J.B., Nilsson, D., Peterson, J., Salzberg, S.L., Shallom, J., Silva, J.C., Sundaram, J., Westenberger, S., White, O., Melville, S.E., Donelson, J.E., Andersson, B., Stuart, K.D., and Hall, N. (2005b). Comparative genomics of trypanosomatid parasitic protozoa. Science *309*, 404–409.

Enright, A.J., Van Dongen, S., and Ouzounis, C.A. (2002). An efficient algorithm for large-scale detection of protein families. Nucleic Acids Res. *30*, 1575–1584.

Fickett, J.W. (1982). Recognition of protein-coding regions in DNA sequences. Nucleic Acids Res. *10*, 5303–5318.

Fu, G. and Barker, D.C. (1998). Characterization of *Leishmania* telomeres reveal unusual telomeric repeats and conserved telomere-associated sequence. Nucleic Acids Res. *26*, 2161–2167.

Ghedin, E., Bringaud, F., Peterson, J., Myler, P., Berriman, M., Ivens, A., Andersson, B., Bontempi, E., Eisen, J., Angiuoli, S., Wanless, D., Von Arx, A., Murphy, L., Lennard, N., Salzberg, S., Adams, M.D., White, O., Hall, N., Stuart, K., Fraser, C.M., and El-Sayed, N.M.A. (2004). Gene synteny and evolution of genome architecture in trypanosomatids. Mol. Biochem. Parasitol. *134*, 183–191.

Green, P., Ewing, B., Miller, W., Thomas, P.J., and Green, E.D. (2003). Transcription-associated mutational asymmetry in mammalian evolution. Nat. Genet. *33*, 514–517.

Grimaldi, G. and McMahon-Pratt, D. (1996). Monoclonal antibodies for the identification of New World *Leishmania* species. Mem. Inst. Oswaldo Cruz *91*, 37–42.

Grisard, E.C., Steindel, M., Shaw, J.J., Ishikawa, E.A., Carvalho, P., Eger, M., Toma, H.K., Lima, J.H., Romanha, A.J., and Campbell, D.A. (2000). Characterization of *Leishmania* sp. strains isolated from autochthonous cases of human cutaneous leishmaniasis in Santa Catarina State, southern Brazil. Acta Trop. *74*, 89–93.

Hanson, S., Beverley, S.M., Wagner, W., and Ullman, B. (1992). Unstable amplification of two extrachromosomal elements in alpha-difluoromethylornithine-resistant *Leishmania donovani*. Mol. Cell. Biol. *12*, 5499–5507.

Herwaldt, B.L. (1999). Leishmaniasis. Lancet *354*, 1191–1199.

Ivens, A.C., Lewis, S.M., Bagherzadeh, A., Zhang, L., Chang, H.M., and Smith, D.F. (1998). A physical map of the *Leishmania major* Friedlin genome. Genome Res. *8*, 135–145.

Ivens, A.C., Peacock, C.S., Worthey, E.A., Murphy, L., Aggarwal, G., Berriman, M., Sisk, E., Rajandream, M.-A., Aert, R., Anupama, A., Apostolou, Z., Attipoe, P., Bason, N., Bauser, C., Beck, A., Beverley, S.M., Bianchettin, G., Borzym, K., Bothe, G., Bruschi, C.V., Collins, C., Cadag, E., Ciarloni, L., Clayton, C., Coulson, R.M.R., Cronin, A., Cruz, A.K., Davies, R.M., De Gaudenzi, J., Dobson, D.E., Duesterhoeft, A., Fazelinia, G., Fosker, N., Frasch, A.C., Fraser, A., Fuchs, M., Gabel, C., Goble, A., Goffeau, A., Harris, D., Hertz-Fowler, C., Hilbert, H., Horn, D., Huang, Y., Klages, S., Knights, A., Kube, M., Larke, N., Litvin, L., Lord, A., Louie, T., Marra, M., Masuy, D., Matthews, K., Michaeli, S., Mottram, J.C., Mueller-Auer, S., Munden, H., Nelson, S., Norbertczak, H., Oliver, K., O'Neil, S., Pentony, M., Pohl, T.M., Price, C., Purnelle, B., Quail, M.A., Rabbinowitsch, E., Reinhardt, R., Rieger, M., Rinta, J., Robben, J., Robertson, L., Ruiz, J.C., Rutter, S., Saunders, D., Schäfer, M., Schein, J., Schwartz, D.C., Seeger, K., Seyler, A., Sharp, S., Shin, H., Sivam, D., Squares, R., Squares, S., Tosato, V., Vogt, C., Volckaert, G., Wambutt, R., Warren, T., Wedler, H., Woodward, J., Zhou, S., Zimmermann, W., Smith, D.F., Blackwell, J., Stuart, K.D., Barrell, B., and Myler, P.J. (2005). The genome of the kinetoplastid parasite, *Leishmania major*. Science *309*, 436–442.

Kapler, G.M., Zhang, K., and Beverley, S.M. (1990). Nuclease mapping and DNA sequence analysis of transcripts from the dihydrofolate reductase-thymidylate synthase (R) region of *Leishmania major*. Nucleic Acids Res. *18*, 6399–6408.

Kink, J.A. and Chang, K.-P. (1987). Tunicamycin-resistant *Leishmania mexicana amazonensis*: Expression of virulence associated with an increased activity of N-acetylglucosaminlytransferase and amplification of its presumptive gene. Proc. Natl. Acad. Sci. USA *84*, 1253–1257.

Knuepfer, E., Stierhof, Y.D., McKean, P.G., and Smith, D.F. (2001). Characterization of a differentially expressed protein that shows an unusual localization to intracellular membranes in *Leishmania major*. Biochem. J. 356, 335–344.

Lee, M.G.S. (1996). An RNA polymerase II promoter in the *hsp70* locus of *Trypanosoma brucei*. Mol. Cell. Biol. *16*, 1220–1230.

Liang, X.H., Hury, A., Hoze, E., Uliel, S., Myslyuk, I., Apatoff, A., Unger, R., and Michaeli, S. (2007). Genome-wide analysis of C/D and H/ACA-like small nucleolar RNAs in *Leishmania major* indicates conservation among trypanosomatids in the repertoire and in their rRNA targets. Eukaryot. Cell 6, 361–377.

Liu, L., Ben Shlomo, H., Xu, Y.X., Stern, M.Z., Goncharov, I., Zhang, Y., and Michaeli, S. (2003). The trypanosomatid signal recognition particle consists of two RNA molecules, a 7SL RNA homologue and a novel tRNA-like molecule. J. Biol. Chem. *278*, 18271–18280.

Lobry, J.R. (1996). A simple vectorial representation of DNA sequences for the detection of replication origins in bacteria. Biochimie *78*, 323–326.

Lobry, J.R. and Sueoka, N. (2002). Asymmetric directional mutation pressures in bacteria. Genome Biol. *3*, research0058.1–research0058.14.

Lu, H.-G., Zhong, L., Guan, L.-R., Qu, J.-Q., Hu, X.-S., Chai, J.-J., Xu, Z.-B., Wang, C.-T., and Chang, K.-P. (1994). Separation of Chinese *Leishmania* isolates into five genotypes by kinetoplast and chromosomal DNA heterogeneity. Am. J. Trop. Med. Hyg. *50*, 763–770.

McAndrew, M., Graham, S., Hartmann, C., and Clayton, C. (1998). Testing promoter activity in the trypanosome genome: isolation of a metacyclic-type VSG promoter, and unexpected insights into RNA polymerase II transcription. Exp. Parasitol. *90*, 65–76.

McDonagh, P.D., Myler, P.J., and Stuart, K.D. (2000). The unusual gene organization of *Leishmania major* chromosome 1 may reflect novel transcription processes. Nucleic Acids Res. 28, 2800–2803

Mair, G., Shi, H., Li, H., Djikeng, A., Aviles, H.O., Bishop, J.R., Falcone, F.H., Gavrilescu, C., Montgomery, J.L., Santori, M.I., Stern, L.S., Wang, Z., Ullu, E., and Tschudi, C. (2000). A new twist in trypanosome RNA metabolism: *cis*-splicing of pre-mRNA. RNA. 6, 163–169.

Martinez-Calvillo, S., Nguyen, D., Stuart, K., and Myler, P.J. (2004). Transcription initiation and termination on *Leishmania major* chromosome 3. Eukaryot. Cell 3, 506–517.

Martinez-Calvillo, S., Stuart, K., and Myler, P.J. (2005). Ploidy changes associated with disruption of two adjacent genes on *Leishmania major* chromosome 1. Int. J. Parasitol. 35, 419–429.

Martinez-Calvillo, S., Yan, S., Nguyen, D., Fox, M., Stuart, K.D., and Myler, P.J. (2003). Transcription of *Leishmania major* Friedlin chromosome 1 initiates in both directions within a single region. Mol. Cell *11*, 1291–1299.

Myler, P.J., Audleman, L., deVos, T., Hixson, G., Kiser, P., Lemley, C., Magness, C., Rickell, E., Sisk, E., Sunkin, S., Swartzell, S., Westlake, T., Bastien, P., Fu, G., Ivens, A., and Stuart, K. (1999). *Leishmania major* Friedlin chromosome 1 has an unusual distribution of protein-coding genes. Proc. Natl. Acad. Sci. USA 96, 2902–2906.

Olmo, A., Arrebola, R., Bernier, V., González-Pacanowska, D., and Ruiz-Pérez, L.R. (1995). Coexistence of circular and multiple linear amplicons in methotrexate-resistant *Leishmania*. Nucleic Acids Res. 23, 2856–2864.

Peacock, C. S., Seeger, K., Harris, D., Murphy, L., Ruiz, J. C., Quail, M. A., Peters, N., Adlem, E., Tivey, A., Aslett, M., Kerhornou, A., Ivens, A., Fraser, A., Rajandream, M. A., Carver, T., Norbertczak, H., Chillingworth, T., Hance, Z., Jagels, K., Moule, S., Ormond, D., Rutter, S., Squares, R., Whitehead, S., Rabbinowitsch, E., Arrowsmith, C., White, B., Thurston, S., Bringaud, F.,

Baldauf, S. L., Faulconbridge, A., Jeffares, D., Depledge, D. P., Oyola, S. O., Hilley, J. D., Brito, L. O., Tosi, L. R., Barrell, B., Cruz, A. K., Mottram, J. C., Smith, D. F., and Berriman, M. (2007) Comparative genomic analysis of three *Leishmania* species that cause diverse human disease. Nature Genet. 39, 839–847.

Perry, K. and Agabian, N. (1991). mRNA processing in the Trypanosomatidae. Experientia *47*, 118–128.

Petrillo-Peixoto, M.L. and Beverley, S.M. (1989). Amplification of a new region of DNA in an unselected laboratory stock of *L. tarentolae*: The T region. J. Protozool. *36*, 257–261.

Rao, V., Fujiwara, N., Porcelli, S.A., and Glickman, M.S. (2005). *Mycobacterium tuberculosis* controls host innate immune activation through cyclopropane modification of a glycolipid effector molecule. J. Exp. Med *201*, 535–543.

Robinson, K.A. and Beverley, S.M. (2003). Improvements in transfection efficiency and tests of RNA interference (RNAi) approaches in the protozoan parasite *Leishmania*. Mol. Biochem. Parasitol. *128*, 217–228.

Romero, G.A., Ishikawa, E., Cupolillo, E., Toaldo, C.B., Guerra, M.V., Paes, M.G., Macedo, V., and Shaw, J.J. (2002). Identification of antigenically distinct populations of *Leishmania* (*Viannia*) *guyanensis* from Manaus, Brazil, using monoclonal antibodies. Acta Trop. *82*, 25–29.

Salzberg, S.L., Delcher, A.L., Kasif, S., and White, O. (1998). Microbial gene identification using interpolated Markov models. Nucleic Acids Res. *26*, 544–548.

Schonian, G., El, F., Lewin, S., Schweynoch, C., and Presber, W. (2001a). Molecular epidemiology and population genetics in *Leishmania*. Med. Microbiol. Immunol. (Berl.) *190*, 61–63.

Schonian, G., Schnur, L., El, F., Oskam, L., Kolesnikov, A.A., Sokolowska, K., and Presber, W. (2001b). Genetic heterogeneity in the species *Leishmania tropica* revealed by different PCR-based methods. Trans. R. Soc. Trop. Med. Hyg. 95, 217–224.

Segovia, M. (1994). *Leishmania* gene amplification: a mechanism of drug resistance. Ann. Trop. Med. Parasitol. *88*, 123–130.

Segovia, M. and Ortiz, G. (1997). LD1 amplification in *Leishmania*. Parasitol. Today *13*, 342–348.

Staden, R. and McLachlan, A.D. (1982). Codon preference and its use in identifying protein-coding regions in long DNA sequences. Nucleic Acids Res. *10*, 141–156.

Stuart, K.D., Weeks, R., Guilbride, L., and Myler, P.J. (1992). Molecular organization of the *Leishmania* RNA virus 1. Proc. Natl. Acad. Sci. USA *89*, 8596–8600.

Sunkin, S.M., Kiser, P., Myler, P.J., and Stuart, K.D. (2000). The size difference between *Leishmania major* Friedlin chromosome one homologues is localized to sub-telomeric repeats at one chromosomal end. Mol. Biochem. Parasitol. *109*, 1–15.

Swindle, J. and Tait, A. (1996). Trypanosomatid genetics. In Molecular Biology of Parasitic Protozoa, D.F.Smith and M.Parsons, eds. (Oxford, UK: Oxford University Press), pp. 6–34.

Tiwari, S., Ramachandran, S., Bhattacharya, A., Bhattacharya, S., and Ramaswamy, R. (1997). Prediction of probable genes by Fourier analysis of genomic sequences. Comput. Appl. Biosci. *13*, 263–270.

Tripp, C.A., Myler, P.J., and Stuart, K. (1991). A DNA sequence (LD1) which occurs in several genomic organizations in *Leishmania*. Mol. Biochem. Parasitol. *47*, 151–160.

Tripp, C.A., Wisdom, W.A., Myler, P.J., and Stuart, K. (1992). A multicopy, extrachromosomal DNA in *Leishmania infantum* contains two inverted repeats of the 27.5 kb LD1 sequence and encodes numerous transcripts. Mol. Biochem. Parasitol. *55*, 39–50.

Uezat, H., Takei, K., Marunp, M., Khaskhel, N.M., Nonak, S., Oshiro, M., Kariya, K.I., Katakura, K., Mimor, T., Gomez, E.A., Furuya, M., Shamsuzzaman, S.M., and Hashiguch, Y. (2001). Detection of species of the subgenus *Leishmania* parasites using polymerase chain reaction and southern blotting. J. Dermatol. *28*, 475–480.

Wilson, K., Beverley, S.M., and Ullman, B. (1992). Stable amplification of a linear extrachromosomal DNA in mycophenolic acid-resistant *Leishmania donovani*. Mol. Biochem. Parasitol. *55*, 197–206.

Wincker, P., Ravel, C., Blaineau, C., Pages, M., Jauffret, Y., Dedet, J., Bastien, P., and Dedet, J.P. (1996). The *Leishmania* genome comprises 36 chromosomes conserved across widely divergent human pathogenic species. Nucleic Acids Res. *24*, 1688–1694.

Worthey, E., Martinez-Calvillo, S., Schnaufer, A., Aggarwal, G., Cawthra, J., Fazelinia, G., Fu, G., Hassebrock, M., Hixson, G., Ivens, A.C., Kiser, P., Marsolini, F., Rickell, E., Salavati, R., Sisk, E., Sunkin, S.M., Stuart, K.D., and Myler, P.J. (2003). *Leishmania major* chromosome 3 contains two long 'convergent' polycistronic gene clusters separated by a tRNA gene. Nucleic Acids Res. *31*, 4201–4210.

Yan, S., Lodes, M.J., Fox, M., Myler, P.J., and Stuart, K. (1999). Characterization of the *Leishmania donovani* ribosomal RNA promoter. Mol. Biochem. Parasitol. *103*, 197–210.

Zhang, W.W. and Matlashewski, G. (2001). Characterization of the A2-A2rel gene cluster in *Leishmania donovani*: involvement of A2 in visceralization during infection. Mol. Microbiol. 39, 935–948.

Zhang, W.-W. and Matlashewski, G. (1997). Loss of virulence in *Leishmania donovani* deficient in an amastigote-specific protein, A2. Proc. Natl. Acad. Sci. USA *94*, 8807–8811.

Zhang, W.-W., Mendez, S., Ghosh, A., Myler, P., Ivens, A., Clos, J., Sacks, D.L., and Matlashewski, G. (2003). Comparison of the *A2* gene locus in *L. donovani* and *L. major* and its control over cutaneous infection. J. Biol. Chem. *278*, 35508–35515.

Zhou, S., Kile, A., Kvikstad, E., Bechner, M., Severin, J., Forrest, D., Runnheim, R., Churas, C., Anantharaman, T.S., Myler, P., Vogt, C., Ivens, A., Stuart, K., and Schwartz, D.C. (2004). Shotgun optical mapping of the entire *Leishmania major* Friedlin genome. Mol. Biochem. Parasitol. *138*, 97–106.

Regulation of Gene Expression in *Leishmania* Throughout a Complex Digenetic Life Cycle

3

Barbara Papadopoulou, Michaela Müller, Annie Rochette, François McNicoll, Carole Dumas and Conan Chow

Abstract

In comparison with all other organisms, from *Escherichia coli* to man, *Leishmania* and related trypanosomatid protozoan parasites show unique features with respect to their regulation of gene expression in response to changes in their environment. The recent completion of the *L. major*, *L. infantum* and *L. braziliensis* genome projects indicates that protein-coding genes are organized as large polycistronic units in a head-to-head or tail-to-tail manner. RNA polymerase II transcribes long polycistronic messages in the absence of defined RNA pol II promoters. As the result of polycistronic transcription, mRNA synthesis requires post-transcriptional control, which involves 5′ *trans*-splicing of a 39-nt capped leader RNA and 3′ cleavage polyadenylation. Several examples in *Leishmania* support the notion that developmental regulation of mRNA levels is determined post-transcriptionally by sequences located in the 3′-untranslated regions (3′UTR) that usually control mRNA stability and translation. Post-translational regulation is also associated with stage-specific gene expression. The lack of promoter elements for RNA pol II and the unusually long 3′UTR sequences provide the molecular basis for this type of control. Controls at translational and post-translational levels could play major roles in the differentiation of *Leishmania*. Stage-specific post-transcriptional regulation is complex and involves the coordination of different mechanisms that can be independently triggered by environmental signals inducing differentiation of promastigotes to amastigotes within mammalian macrophages.

Introduction

Leishmania is an early-branching unicellular eukaryote (Sogin *et al.*, 1993) that belongs to the order *Kinetoplastida* and family *Trypanosomatidae*. These organisms possess a plethora of biochemical, genetic and morphological features that are either unique to the group (e.g. mitochondrial minicircles with extensively bent DNAs, uridine insertion/deletion editing, glycosomes and acidocalcisomes) or are used to a greater extent than in most other organisms (e.g. polycistronic transcription and *trans*-splicing, glycosyl phosphatidylinositol anchoring of membrane proteins) (Campbell *et al.*, 2003). At least 20 *Leishmania* spp. cause leishmaniasis, a vector-borne parasitic disease with a large spectrum of clinical manifestations ranging from self-resolving skin lesions to life-threatening visceral diseases. Parasites of the genus *Leishmania* are, according to the World Health Organization (WHO), endemic in 88 countries and are found in both tropical and subtropical regions. More than 12 million people are infected worldwide with 2 million new cases every year, and there is a risk of at least 350 million people being infected (Murray *et al.*, 2005).

The 32.8-megabase haploid genome of *Leishmania major* Friedlin strain has been recently sequenced and predicts 8370 protein-coding genes, 39 pseudogenes, and 911 RNA genes (Ivens *et al.*, 2005). Of the ~8300 protein-coding genes, only 36% can be ascribed a putative function. Protein-coding regions in *Leishmania* are not interrupted by introns with the single exception so far of the gene encoding poly(A) polymerase (Mair *et al.*, 2000a). Typically, genes are grouped together into long, strand-specific, polycistronic clusters that are co-transcribed to yield polycistronic pre-mRNAs (Myler *et al.*, 1999; Worthey *et al.*, 2003; Ivens *et al.*, 2005). Processing of polycistronic transcripts involves two RNA-processing reactions, namely, *trans*-splicing, which adds a small capped RNA of 39 nucleotides, the spliced leader RNA (SL RNA), to the 5′ terminus of all known protein-encoding RNAs, and 3′ cleavage-polyadenylation (Ullu *et al.*, 1996). Several pieces of experimental evidence support the model that *trans*-splicing and polyadenylation are mechanistically coupled in these parasites and share common regulatory signals, which are mainly polypyrimidine-rich stretches within the intergenic regions (LeBowitz *et al.*, 1993; Matthews *et al.*, 1994).

Leishmania, as well as other trypanosomatids, seem to have lost the ability to regulate transcription initiation by RNA polymerase II (RNAP II) and despite considerable efforts no canonical promoters for RNAP II-transcribed protein-encoding genes have been identified so far (Martinez-Calvillo *et al.*, 2003; Worthey *et al.*, 2003; Martinez-Calvillo *et al.*, 2004). Moreover, the typical general transcription factors, such as the TATA binding protein and TFII B, are highly divergent in the Trypanosomatidae family. The lack of control at the level of transcription initiation might be explained both by the mode of polycistronic transcription in these organisms and the considerable regulatory needs of these parasites during their complex life cycle, which may be more efficiently accomplished through downstream mechanisms that involve RNA and protein regulatory processes. Thus, regulation of gene expression in these organisms occurs exclusively at the post-transcriptional level and could involve pre-mRNA processing, mRNA stability and translation, protein stability and post-translational control (Clayton, 2002). Numerous RNA-binding proteins are encoded in the *Leishmania* genome (Ivens *et al.*, 2005), consistent with active post-transcriptional regulation of gene expression.

Leishmania parasites exist in two major developmental stages: free-living flagellated promastigotes in the alimentary tract of the sandfly vector and intracellular amastigotes residing in the phagolysosomes of macrophages in the mammalian host. Amastigotes divide by binary fission within phagolysosomes and eventually burst free from infected macrophages, evading the defence mechanisms induced by the host, thereby spreading the disease. *Leishmania* encounters drastic environmental changes during developmental switches from promastigotes to amastigotes and vice versa. These developmental stages display distinct morphologic and metabolic characteristics consistent with a highly regulated level of differential protein expression. To date, several stage-specific genes have been identified in *Leishmania*; however, the molecular mechanisms governing developmental gene regulation in this organism are still not well understood. Generally, sequences within 3′-untranslated regions (3′UTRs) that act post-transcriptionally have been shown to play a key role in regulating differentially expressed transcripts in *Leishmania* (Argaman *et al.*, 1994; Charest *et al.*, 1996; Quijada *et al.*, 2000; Wu *et al.*, 2000; Kelly *et al.*, 2001; Zilka *et al.*, 2001; Boucher *et al.*, 2002; Myung *et al.*, 2002; Folgueira *et al.*, 2005; McNicoll *et al.*, 2005). Post-transcriptional regulation mechanisms associated with such 3′UTR sequences may involve alternating rates of mRNA decay and mRNA translation.

General description of the topic

Polycistronic transcription and pre-mRNA processing in *Leishmania*

The first observation for a polycistronic transcription unit in *Kinetoplastida* was reported for the variant-specific surface glycoprotein (VSG) mRNAs in *Trypanosoma brucei* (Kooter *et al.*, 1984; Johnson *et al.*, 1987). Fifteen years later, the complete nucleotide sequence of *L. major* Friedlin chromosome 1 showed that transcription of the entire chromosome initiates bidirectionally between two divergent gene clusters of 29 and 50 genes, respectively (Myler *et al.*, 1999). The completion of *Leishmania* genome sequencing has revealed that most genes in this organism are organized into 133 clusters of tens to hundreds of protein-coding genes with unrelated predicted functions on the same DNA strand (Berriman *et al.*, 2005; El-Sayed *et al.*, 2005; Ivens *et al.*,

2005). The clusters can span up to 1259 kb and are separated by 0.9- to 14-kb divergent or convergent strand-switch regions, which show an unusual base composition (Ivens *et al.*, 2005). Experimental evidence suggests that polycistronic transcription by RNAP II initiates bidirectionally within the divergent strand-switch regions and terminates within the convergent strand-switch regions which often contain tRNA, rRNA, and/or snRNA genes (Martinez-Calvillo *et al.*, 2003; Worthey *et al.*, 2003; Martinez-Calvillo *et al.*, 2004).

RNA polymerase II transcription initiation in the absence of defined RNAP II promoter elements

As with other eukaryotes, *Leishmania* genes are transcribed by at least three different types of RNA polymerases (pol). RNA pol I and III transcribe the rRNA and tRNA genes, respectively, whereas RNA pol II transcribes protein-coding genes. However, as opposed to other eukaryotes, transcription in the kinetoplastid protozoa shows substantial differences such as polycistronic transcription, no qualitative regulated transcription initiation for most protein-coding genes, a paucity of RNAP II promoters, transcription of protein-coding genes by RNAP I, a dependence of small nuclear RNA gene transcription on an upstream tRNA gene, and the synthesis of mitochondrial tRNAs in the nucleus (Campbell *et al.*, 2003).

The recently published genomes of *Leishmania major*, *T. brucei*, and *T. cruzi*, clearly outline that while trypanosomatid homologues are present for all subunits of eukaryotic RNAP II, many of the basal transcription factors (e.g. the largest subunit of TFIIA, TFIIB, TFIIE, TFIIF, several of the TFIIH subunits, and the TATA-box-binding protein-associated factors found within TFIID) are missing (Berriman *et al.*, 2005; El-Sayed *et al.*, 2005; Ivens *et al.*, 2005), which suggests that the mechanisms regulating RNAP II-directed transcription are distinct from those operating in other eukaryotes. Very little is known about the sequences involved in transcription initiation of protein-coding genes in *Leishmania*. Transcriptional analysis of chromosome 1 by graded doses of UV-irradiation followed by nuclear run-on assays, localized the initiation sites to a <100 bp region within the strand switch region (Martinez-Calvillo *et al.*, 2003). Transfection studies support the presence of a bidirectional promoter within the strand-switch region, but suggest that other factors are also involved in RNAP II transcription. In the case of chromosome 3, nuclear run-on analyses indicated that specific transcription initiates bidirectionally between the single subtelomeric gene and the adjacent 67-gene cluster, and between the two convergent clusters of 67 and 30 genes (Worthey *et al.*, 2003; Martinez-Calvillo *et al.*, 2004). Although RNAP II appears to initiate transcription at the boundaries between divergent polycistrons, or between polycistrons and regions transcribed by other RNA polymerases (Martinez-Calvillo *et al.*, 2003; Worthey *et al.*, 2003; Martinez-Calvillo *et al.*, 2004; Ivens *et al.*, 2005), attempts to identify consensus promoter sequences have failed. In the absence of easily identifiable promoters, it has been proposed that RNAP II may either initiate transcription randomly within a stretch of DNA or there may be a few distantly spaced promoters (Horn, 2001; Martinez-Calvillo *et al.*, 2003). A discrete transcription initiation site may not be necessary for mRNA synthesis since the sequences upstream of each transcribed coding region are trimmed from the nascent transcript when a short m(7)G-capped RNA is added during mRNA maturation. Moreover, experiments with transfected plasmids have demonstrated that RNAP II can initiate in the absence of a specific kinetoplastid promoter (Bellofatto *et al.*, 1991; Papadopoulou *et al.*, 1994; Marchetti *et al.*, 1998). These findings, along with the polycistronic gene organization and lack of RNAP II initiation sites, are consistent with post-transcriptional control mechanisms being the primary determinants of trypanosomatid gene expression. So, how does the *Leishmania* RNAP II transcribe genes in the absence of well-defined promoter elements? Part of the answer is that because of the organization of the protein-coding genes as part of polycistronic transcription units in *Leishmania*, transcription initiation is not a significant rate-limiting factor in mRNA production. While polycistronic transcription appears energetically wasteful, it may actually help to ensure parasite survival in the face of dynamic environmental changes.

Our best knowledge of the proteins essential for RNAP II-dependent transcription comes mainly from the genes encoding the SL precursor RNA, SL-RNA. The SL-RNA genes have clearly identifiable RNAP II promoters with a short consensus initiator element (Luo *et al.*, 1999; Mair *et al.*, 2000b; Gilinger *et al.*, 2001). The SL-RNA gene promoter is TATA-less, B-recognition element-less, and contains a bipartite upstream sequence element and an element close to the transcription initiation site (Schimanski *et al.*, 2004). The best-characterized transcription initiation complex in trypanosomatids assembles PBP-1 and PBP-2 and binds to the –60 and –30 elements of the *Leptomonas seymouri* SL-RNA gene promoter (Luo *et al.*, 1997; Das *et al.*, 2003). A partially characterized small nuclear RNA-activating protein complex (SNAP(c)) and the TATA-binding protein related factor 4 have also been identified to bind specifically to the SL-RNA gene promoter upstream sequence element and have been shown to be essential for SL-RNA gene transcription *in vitro* (Schimanski *et al.*, 2005).

Transcription termination of RNAP II at the end of polycistronic regions has not been characterized yet. Transient-transfection studies support the role of the tRNA-gene region as a transcription terminator for RNAP II but also for RNAP III and RNAP I (Martinez-Calvillo *et al.*, 2004).

Antisense transcription

Endogenously expressed antisense RNAs are known to exist in *Leishmania* (Kapler *et al.*, 1989; Belli *et al.*, 2003; Monnerat *et al.*, 2004). Strand-specific nuclear run-on assays showed that a low level of non-specific antisense transcription probably takes place over the entire chromosome 1 of *L. major* Friedlin (Martinez-Calvillo *et al.*, 2003). Sequence analysis of the *L. major* 34 kb locus located on chr27 revealed 115 ORFs on either DNA strand. Low levels of transcription were detected in the bottom strand and stable transcripts were identified for four ORFs on this strand that were not predicted to be protein-coding (Monnerat *et al.*, 2004). Northern blot hybridizations of *Leishmania* RNA probed with histone HIS-1 strand-specific riboprobes revealed sense and antisense transcripts that were polyadenylated and developmentally regulated. When the HIS-1.2 coding region was replaced with the neomycin phosphotransferase gene (NEO), antisense transcription of NEO was unaffected, indicating that the regulatory elements controlling antisense transcription were located outside of the HIS-1.2 coding region. These results demonstrate not only that gene expression in *Leishmania* is regulated post-transcriptionally from the sense strand of genomic DNA, but that the complementary strand of DNA also contains sequences that could drive expression of open reading frames from the antisense strand of DNA (Belli *et al.*, 2003). In *T. brucei*, bidirectional transcription does occur (Liniger *et al.*, 2001), but it is unlikely to be widespread as this organism has a constitutively active RNA interference system that destroys double-stranded RNAs (Djikeng *et al.*, 2001). In the absence of RNAi in *L. major* and *L. infantum*, it is very unlikely that the antisense RNAs could silence the sense transcripts through RNA degradation after base-pairing. In *Leishmania*, antisense approaches work poorly and at its best only small effects are observed (Chen *et al.*, 2000; Somanna *et al.*, 2002). An exception is the antisense inhibition of *L. donovani A2* mRNA, which resulted in reduced parasite virulence (Zhang *et al.*, 1997).

The reasons for the existence of antisense transcripts are puzzling in the absence of any RNAi machinery in *Leishmania* (Robinson *et al.*, 2003). The lack of RNAi in these parasites may be a reflection of a successful viral attack on these organisms at some point in their evolutionary past.

Chromatin remodelling

Leishmania contains multiple copies of the four core (H2A, H2B, H3 and H4) and linker H1 histone genes, which package chromosomal DNA into nucleosomes in eukaryotes and regulate access of the RNAP II transcription complexes. The *Leishmania* genome encodes also a number of enzymes involved in histone modification that may influence transcription, replication, recombination and repair. These include two families of acetyltransferases, at least three families of methyltransferases and all three known classes of histone deacetylases (Ivens *et al.*, 2005). Several genes possibly involved in these processes, like histone H1, have shown differences in their level of expression between *Leishmania* life stages (Noll *et al.*, 1997). In higher eukaryotes, there is increasing evidence that epigenetic factors, such as histones and modified bases, influence gene expression. Chemical modification of these factors can restructure chromatin and lead to gene activation or silencing. In trypanosomes, expression of some developmentally regulated genes, such as VSG, could be controlled by an epigenetic mechanism (Horn, 2001). For instance, chromatin remodelling, histone acetyltransferases, and base J (DiPaolo *et al.*, 2005) may play a role in transcription/silencing of the VSG expression site in *T. brucei* (van Leeuwen *et al.*, 1998; Navarro *et al.*, 1999). In *Leishmania*, the putative role of chromatin remodelling, if any, in the activation or silencing of developmentally regulated genes has not been investigated yet. However, given the lack of transcription initiation control in these organisms and the accumulating evidence that stage-specific transcripts seem to be synthesized at similar levels between the two life stages of the parasite, stage-specific regulation by chromatin remodelling may not occur.

Pre-messenger RNA processing

As mentioned above, protein-coding genes in *Leishmania* are transcribed polycistronically by RNAP II, and polycistrons can be many tens of kilobases long (Myler *et al.*, 1999; Worthey *et al.*, 2003; Ivens *et al.*, 2005). Individual mature mRNAs are generated from polycistronic precursors by 5′ *trans* splicing of a 39-nt capped leader RNA and 3′ polyadenylation.

Trans-splicing

The process of *trans*-splicing was first discovered in trypanosomes more than 25 years ago, when it was found that the VSG mRNAs in *T. brucei* carry a common 39-nucleotide (nt) sequence, namely, the spliced leader (SL) sequence (Boothroyd *et al.*, 1982). It was later established that all *Trypanosoma* mRNAs undergo *trans*-splicing (Agabian, 1990). *Trans*-splicing was later found in nematodes, euglenoids, trematodes, and more recently in chordates (Liang *et al.*, 2003). *Trans*-splicing proceeds through a two-step trans-esterification reaction, analogous to *cis*-splicing but forming a Y structure instead of a lariat intermediate (Murphy *et al.*, 1986; Sutton *et al.*, 1986). The SL sequence was found to be provided by a small capped RNA, the SL-RNA (Milhausen *et al.*, 1984). In trypanosomatid protozoan parasites, the SL sequence is 39 to 41 nt in size and followed by an intron of variable length that can be folded into two stem-loops separated by a single-stranded region (Campbell *et al.*, 2003). In addition to playing an essential role for mRNA maturation, *trans*-splicing also serves to cap mRNAs. The trypanosomatid SL-RNA carries a unique cap structure among eukaryotes, the cap-4 structure m(7)Gpppm(2)(6)AmpAmpCmpm(3)Um, which consists of 7-methylguanosine (m^7G) followed by 2′-O methylations on the first four ribose moieties and unique base methylations on the first adenine and the fourth uracil (Perry *et al.*, 1987; Freistadt *et al.*, 1988; Bangs *et al.*, 1992). Cap-4 addition is post-transcriptional, however, and there is still a debate on whether the complete cap-4 modification is cotranscriptional (Mair *et al.*, 2000b) or exclusively post-transcriptional (Zeiner *et al.*, 2003). Following capping, the SL-RNA is exported to the cytoplasm, processed, and re-imported into the nucleus before it is used as a splice donor to form mRNAs from pre-mRNA polycistronic transcripts. SL-RNA capping is essential for *in vivo trans*-splicing but it may be also required in mRNA export, stability and translation (Zeiner *et al.*, 2003).

Trans-splicing signals are often U-rich polypyrimidine (poly(Y)) tracts, which precede AG splice acceptor sites 50–100 nucleotides in average upstream of the translational start codon. The elimination of the native splice acceptor site leads to splicing at the next AG downstream and the accumulation of Y-branched intermediate, supporting a scanning model by splicing factors (Hummel *et al.*, 2000). We have shown that in *Leishmania*, the minimal requirement for efficient *trans*-splicing is an AG 3′ splice site preceded by a synthetic poly(Y) tract (Papadopoulou *et al.*, 1994). In *T. brucei* the poly(Y) tract was also identified as a major determinant for accurate *trans*-splicing (Huang *et al.*, 1991b; Matthews *et al.*, 1994; Schurch *et al.*, 1994; Siegel *et al.*, 2005). The position of the poly(Y) tract relative to the 3′ splice site is important, and an AC dinucleotide at positions −3 and −4 can lead to a 20-fold decrease in *trans*-splicing (Siegel *et al.*, 2005). Although the poly(Y) tract definitely contributes to the selection of the 3′ splice site, other determinants that affect splicing efficacy may be involved. These determinants may account for the different *trans*-splicing efficacies of various genes (Huang *et al.*, 1991a; Matthews *et al.*, 1994).

Polyadenylation

There is no consensus polyadenylation signal in trypanosomatid mRNAs such as the AAUAAA sequence in higher eukaryotes (Borst, 1986; Agabian, 1990). Evidence obtained from a small number of loci suggests that polyadenylation occurs within a short region 100–400 nt upstream of the next polypyrimidine *trans*-splice signal (LeBowitz *et al.*, 1993; Hug *et al.*, 1994; LeBowitz, 1994; Matthews *et al.*, 1994; Schurch *et al.*, 1994; Vassella *et al.*, 1994). It was recently reported that in 89% of all available cDNA sequences from *T. brucei*, polyadenylation usually occurs at an A residue located between 80 and 300 nt from a downstream poly(Y) tract (Benz *et al.*, 2005). For 87% of the poly(A) sites, the downstream poly(Y) tracts fell within the range of 8–17 nt seen for most 5′UTRs. Accurate choice of the poly(A) site requires a downstream 3′ splice site.

Mechanistic coupling of trans-splicing and polyadenylation?

Several lines of evidence raised the intriguing possibility that in trypanosomatids poly(A) addition is coupled to *trans*-splicing of the downstream gene. For example, in *T. brucei*, polycistronic transcription of the HSP70 pre-mRNA appears to be processed co-transcriptionally by cleavage for poly(A) addition (Huang *et al.*, 1991b). It has been reported that *trans*-splicing is co-transcriptional and precedes polyadenylation in the processing of *Trypanosoma* tubulin pre-mRNAs (Ullu *et al.*, 1993). A preferential splicing at the inverted polyadenylation site of the VSG mRNA provides further evidence for coupling between *trans*-splicing and polyadenylation (Berberof *et al.*, 1995). Polyadenylation of a procyclic acidic repetitive protein transcript usually occurred at A residues, which is approximately 100-nt upstream of a *trans*-splicing acceptor signal (Hug *et al.*, 1994). In addition, for the *Leishmania* dihydrofolate reductase-thymidine synthetase pre-mRNA, poly(A) site selection is specified by sequences spanning the 3′ splice site region (LeBowitz *et al.*, 1993). For the *T. brucei*-tubulin 3′ splice site region, a poly(Y) tract governs both *trans*-splicing and polyadenylation (Matthews *et al.*, 1994).

Polypyrimidine tracts within intergenic regions of *Leishmania* and *Trypanosoma* transcripts play a major role in the coupling of these processes, since mutating the poly(Y) tract led to an aberrant poly(A) site choice (Hug *et al.*, 1994). The data available so far suggest that the context of a poly(Y) tract may be critical for determining both splicing and polyadenylation. In particular, poly(Y) tracts immediately upstream of ORFs appeared to be preferred (Hartmann *et al.*, 1998; Lopez-Estrano *et al.*, 1998). There are several major gaps in our understanding of pre-mRNA processing in trypanosomatids. How are intergenic regions of pre-mRNAs accurately recognized by the pre-mRNA-processing machinery? What distinguishes these regions from other regions in the pre-mRNA? In particular, how does the selection of the correct 3′ splice site upstream of the translation initiator start codon takes place? It has been suggested that the factors required for 3′ end cleavage and polyadenylation are associated with the pre-mRNA after the 3′ splice site region has been marked by the spliceosome machinery (Matthews *et al.*, 1994). The finding that the location of poly(A) site moves in concert with the 3′ splice site in *Leishmania* supports a model in which the factor(s) responsible for cleaving the pre-mRNA at the poly(A) site can bind only a certain short distance away from the 3′ splice site. After cleavage at the 3′ splice site, the same factor could direct polyadenylation upstream. In *Leishmania*, whenever the position of the downstream splice site was altered, the poly(A) site was shifted 400–500 nucleotides upstream of the new splice site (LeBowitz *et al.*, 1993). In contrast, when the wild-type poly(A) site was eliminated, the downstream splice site was unaffected and polyadenylation was maintained 200–500 nucleotides upstream of the splice site. Thus, poly(A) site selection in *Leishmania* is specified by the position of the downstream splice acceptor and suggests a scanning model for poly(A) site selection after splice site recognition (LeBowitz *et al.*, 1993).

Global analysis of gene expression throughout the digenetic life cycle of *Leishmania*

During its complex life cycle, *Leishmania* experiences drastic environmental changes. In the lumen of the digestive tract of female sandflies, *Leishmania* exist as extracellular promastigotes that go through several differentiation stages. Inside the sandfly gut, promastigotes are exposed to neutral pH and widely fluctuating temperatures around 25°C. After metacyclogenesis, infective forms called metacyclic promastigotes, accumulate in the interior parts of the digestive tract, from where they can be transferred to the mammalian host during a blood meal (Sacks *et al.*, 2001). Metacyclic promastigotes attach to macrophages through receptor-mediated binding (Antoine *et al.*, 2004). Once the parasites are ingested, they localize within the phagolysosomes, where they begin their differentiation into a non-flagellated amastigote form. This process takes several days and occurs within parasitophorous vacuoles where amastigotes continue to multiply and spread throughout macrophages of the reticuloendothelial system while clinical symptoms develop (Antoine *et al.*, 2004). Amastigotes reside in acidic phagolysosomes, experience near-constant temperatures around 37–39°C and encounter host factors that subject them to oxidative, proteolytic, and metabolic stresses. *Leishmania* are able to withstand these conditions and undergo morphological and physiological changes (MacFarlane *et al.*, 1990; Glaser *et al.*, 1991; Turco *et al.*, 1992; Zilberstein *et al.*, 1994; Barak *et al.*, 2005) that are mainly orchestrated by the differential expression of a variety of genes. Identification of genes preferentially expressed during the intracellular growth of *Leishmania* would provide a better insight into mechanisms controlling stage-specific gene expression and intracellular survival of the parasite. Moreover, amastigote-specific genes represent promising molecular targets for the development of new therapeutic strategies. Therefore, considerable efforts have been directed towards the identification and characterization of developmentally regulated genes in *Leishmania*.

Using several molecular and biochemical approaches, including microarray and proteomic comparative analyses, several stage-specific genes and proteins have been identified (Tables 3.1 and 3.2; Fig. 3.1). DNA microarrays containing 11,484 PCR products from a randomly amplified genomic library of *L. major* Friedlin hybridized with fluorescent RNA probes made from procyclics or metacyclic promastigotes demonstrated that ~15% of the PCR fragments showed statistically significant ($P<0.01$) changes in expression (from 1.6- to 5-fold) during this developmental transition (Saxena *et al.*, 2003). Hybridization of the same microarrays with RNA isolated from *L. major* promastigotes and lesion-derived amastigotes showed that only about 1% of the mRNA expression varied between the two life cycle stages (Akopyants *et al.*, 2004). Recent analysis of the *L. donovani* transcriptome using the above DNA microarrays revealed an ordered progression of transient or permanent changes in gene expression during axenic differentiation (Saxena *et al.*, 2007). Analysis of mRNA expression profiles of *L. major* promastigotes and lesion derived amastigotes using a full-genome DNA oligonucleotide microarrays representing 8160 genes showed a low degree of differential mRNA expression with 1.4% of changes in amastigotes and 1.5% in promastigotes (Leifso *et al.*, 2007). Using the same whole-genome oligonucleotide microarrays, 3.2% of *L. mexicana* genes were differentially regulated in promastigotes and lesion-derived amastigotes (Holzer *et al.*, 2006). Another study using microarrays containing 2183 full-length randomly selected cDNAs from *L. major* that were screened with *L. major* RNA from procyclic, metacyclic and lesion-derived amastigote populations showed a higher percentage of stage-specific gene expression in amastigotes (~35%) compared to metacyclics (~12%) (Almeida *et al.*, 2004). This modest fraction of genes whose mRNAs are altered in abundance throughout the life cycle seems to be much lower than in other eukaryotes and may reflect the absence of regulation at the level of transcription initiation in this parasite. Although important

Table 3.1 Transcriptomic analyses of stage-specific gene expression in *Leishmania* spp.

	Approach				
	Random genomic microarrays		cDNA microarrays	Oligonucleotide microarrays	
	L. major[1]	*L. major*[2]	*L. major*[3]	*L. mexicana*[4]	*L. major*[5]
Genes tested	~11 000 PCR clones*	~11 000 PCR clones*	1843 cDNAs	~8160 genes	~8160 genes
Promastigotes	141 (1.5%)	32 (0.3%)	131 (7.1%)	250 (3.0%)	125 (1.5%)
Amastigotes	ND	64 (0.6%)	263 (14.2%)	38 (0.5%)	115 (1.4%)
Metacyclics	472 (5.1%)	51 (0.5%)	36 (2%)	ND	ND

[1]Saxena *et al.* (2003); [2]Akopyants *et al.*, 2004; [3]Almeida *et al.* (2004); [4]Holzer *et al.* (2006); [5]Leifso *et al.* (2007).

*Representing approximately two-thirds of the genome.

ND, no data.

Table 3.2 Proteomic analyses of stage-specific gene expression in *Leishmania* spp.

	L. donovani[1]	*L. mexicana*[2]	*L. panamensis*[3]	*L. infantum*[4]
Detected spots	~2000	~2000	~850	~2200
Promastigotes	35 (1.8%)	18 (0.9%)	45 (5.3%)	138 (6.1%)
Amastigotes	65 (3.3%)	65 (3.3%)	75 (8.8%)	283 (12.4%)
Metacyclics	ND	38 (1.9%)	ND	ND

[1]Bente *et al.* (2003); [2]Nugent *et al.* (2004); [3]Walker *et al.* (2006); [4]McNicoll *et al.* (2006).

ND, no data.

differences are anticipated in the mechanisms of stage-specific regulation between *Leishmania* and *Trypanosoma* species, DNA microarray analyses in the latter species showed in average 2% differences in transcript levels between the bloodstream and procyclic forms (Diehl *et al.*, 2002; Minning *et al.*, 2003; Brems *et al.*, 2005), which are similar to those observed in *Leishmania*.

Comparative proteome maps of developmental life stages of *Leishmania* indicated that ~10–12% of the proteins analysed using two-dimensional gel electrophoresis (2D) are differentially expressed in axenically grown amastigotes (El-Fakhry *et al.*, 2002; Bente *et al.*, 2003; Nugent *et al.*, 2004; McNicoll *et al.*, 2006; Walker *et al.*, 2006) (Table 3.2). We have recently applied a comprehensive approach consisting of protein pre-fractionation, global proteomics and targeted DNA microarrays to analyse stage differentiation in *L. infantum* (McNicoll *et al.*, 2006). We showed that protein pre-fractionation allows the detection and identification of a large number of novel differentially expressed protein isoforms in *L. infantum*. Using mid-range two-dimensional gels, we detected over 2200 protein isoforms in each developmental stage. Of these, 6.1% were strongly increased or appeared unique in the promastigote stage, while the relative amounts of 12.4% were increased in amastigotes. Amastigote-specific protein isoform and mRNA expression trends correlated modestly (53%), while no correlation was found for promastigote-specific spots. Even where direction of regulation was similar, changes were more modest at the RNA level than protein level. Many proteins were present in multiple spots, suggesting that

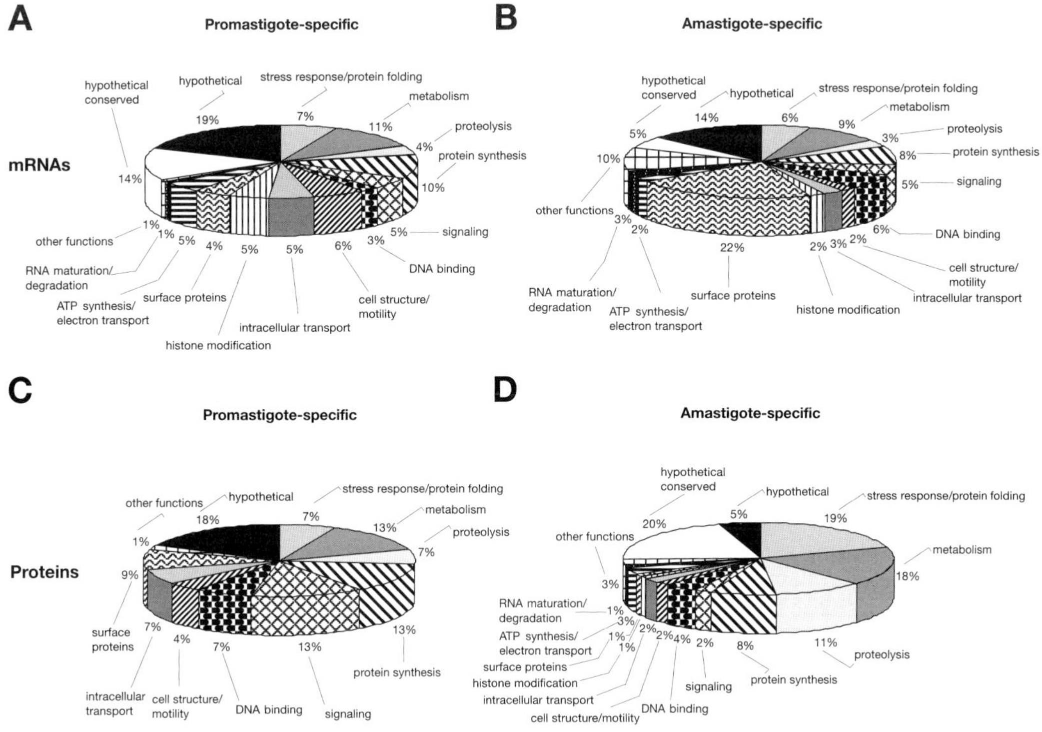

Figure 3.1 Global representation of stage-specific mRNA and protein expression in *Leishmania*. The top panel (A and B) illustrates functionally related groups to which promastigote- or amastigote-specific genes identified from different microarray studies (Saxena *et al.*, 2003; Akopyants *et al.*, 2004; Almeida *et al.*, 2004; Holzer *et al.*, 2006) belong. The lower panel (C and D) represents stage-specific protein expression per functional category as summarized from different proteomic studies (Bente *et al.*, 2003; Nugent *et al.*, 2004; Walter *et al.*, 2006; McNicoll *et al.*, 2006). The numbers represent the percentage of stage-specific genes or proteins for each category. The observed differences between transcriptomic and proteomic studies are possibly due to the poor degree of correlation between mRNA and stage-regulated protein species, the differences in expression between *Leishmania* species, and the differences in the type of experimental conditions used between studies.

post-translational modification is extensive in this organism. In several cases, different isoforms appeared to be specific to different life cycle stages. Our results suggest that post-transcriptional control at translational and post-translational levels could play a major role in differentiation of *Leishmania* parasites. The digitonin prefractionation protocol was also applied for a comparative proteomic analysis of the promastigote and amastigote life cycle stages of *L. infantum* and has allowed the identification of novel proteins expressed in a stage-specific manner (Foucher *et al.*, 2006). Quantitative proteomics using the isotope coded affinity tag (ICAT) technology and mass spectrometry were applied to *L. infantum* promastigote and axenic amastigote protein extracts and of the 91 distinct proteins identified, 8% were differentially expressed in amastigotes and 20% were differentially expressed in the promastigote stage (Leifso *et al.*, 2007). Most of the stage-specific identified proteins fall into five main functionally related groups (Bente *et al.*, 2003; Nugent *et al.*, 2004; McNicoll *et al.*, 2006); Fig. 3.1). These functional categories are: (i) stress response and protein folding (e.g. heat, oxidative stress); (ii) energy metabolism (e.g. ATP synthesis, electron transport); (iii) protein synthesis (e.g. amino acid metabolism, translation factors); (iv) cell cycle and signalling; and (v) proteolysis. In comparison, many of the identified proteins have unknown function and on this basis, they form the largest group.

Mechanisms of developmental gene regulation in *Leishmania*

To date, very little is known about the molecular mechanisms controlling developmental gene regulation in *Leishmania*. Developmentally regulated genes in *Leishmania* are co-transcribed with constitutive genes as part of the same polycistronic pre-mRNA unit. As mentioned above, there are no well-established examples of RNAP II promoters in these organisms, which could regulate transcription rates. Moreover, run-on experiments indicate no differences in the synthesis of several stage-specific transcripts between the two developmental life stages of the parasite (Wu *et al.*, 2000; Boucher *et al.*, 2002). Thus, in the absence of transcriptional control at the level of initiation, stage-specific regulation of gene expression in *Leishmania* is mediated almost exclusively by post-transcriptional mechanisms.

Due to the systematic selection by the splicing/polyadenylation machinery for the closest 3′ splice acceptor to the translational start site, *Leishmania* mRNAs have in general short 5′UTRs that most likely lack internal ribosome entry sites (IRES) and upstream ORFs. In other eukaryotes, IRES and upstream ORFs regulate translation of mRNAs in response to various environmental stimuli (Meijer *et al.*, 2002; Komar *et al.*, 2005). In contrast to other eukaryotes, most *Leishmania* mRNAs contain long and highly structured 3′UTRs (in average 1.5–2 kb; (Ivens *et al.*, 2005)). Sequences within 3′UTRs play a key role in regulating differentially expressed genes in *Leishmania* (Argaman *et al.*, 1994; Charest *et al.*, 1996; Quijada *et al.*, 2000; Wu *et al.*, 2000; Kelly *et al.*, 2001; Zilka *et al.*, 2001; Boucher *et al.*, 2002; Myung *et al.*, 2002; Folgueira *et al.*, 2005; McNicoll *et al.*, 2005). To date, no 5′UTR elements have been implicated in stage-specific regulation in *Leishmania* (McNicoll *et al.*, 2005). Several post-transcriptional regulation mechanisms could be applicable to these parasites, such as differential processing of polycistronic transcripts, alternating rates of mRNA decay, alternating levels of mRNA translation, post-translational modifications and protein degradation.

Stage-specific pre-mRNA processing by differential trans-splicing and/or polyadenylation

To date, only few cases have been reported where differential stage-specific mRNA abundance involves differences in pre-mRNA processing. It has been reported that a 120-nt regulatory element present in the intergenic region of the metacyclic-specific lysosomal cysteine protease genes *CPB1* and *CPB2* modulates maturation of polycistronic pre-mRNA processing probably via the binding of a stage-specific *trans*-acting factor to the 120-nt element. This modulation leads to increased expression of *CPB1* and *CPB2* in metacyclics but conversely to down-regulation of expression of these transcripts in amastigotes (Brooks *et al.*, 2001). More recently, alternative *trans*-splicing was demonstrated in *T. cruzi* for the LYT1 gene that functions in the lytic pathway (Manning-Cela *et al.*, 2002). Previous work showed that the phosphoglycerate kinase PGK A, B, and C mRNA levels in *T. brucei* differ, although they are transcribed as a polycistronic unit (Gibson *et al.*, 1988). It was shown that the position of the polypyrimidine tract relative to the branch point and 3′ splice site between PGK A, B, and C resulted in very different levels of *trans*-splicing efficiency (Siegel *et al.*, 2005). In addition, it was shown that a second AG site in the 5′ UTR enhances *trans*-splicing in constructs that contain a suboptimum sequence (AC) immediately upstream of the 3′ splice site (Siegel *et al.*, 2005). These results suggest that *trans*-splicing can play an important role in post-transcriptional gene control.

Stage-specific mRNA degradation

Very little is known about the mechanisms that control stability/degradation of developmentally regulated transcripts in *Leishmania*. Several line of evidence support the idea that 3′UTRs are possibly involved in mRNA stability/degradation (Aly *et al.*, 1994; Brittingham *et al.*, 2001; Mishra *et al.*, 2003; Larreta *et al.*, 2004; Purdy *et al.*, 2005), but *cis*-acting elements or factors participating in this process have still not been identified in *Leishmania*. AU-rich elements (AREs) in the 3′UTRs have been involved in ARE-mediated mRNA degradation in *Trypanosoma* species. The mechanism of mRNA degradation in *T. brucei* was studied for *VSG, EP1* and *PGKB* developmentally regulated genes and involves deadenylation, followed by degradation in both 5′ to 3′ and 3′ to 5′ directions (Berberof *et al.*, 1995; Irmer *et al.*, 2001; Haile *et al.*, 2003). Regulation of genes encoding mucin-like proteins that are expressed predominantly in the vertebrate forms of *T. cruzi* is also determined by an ARE (Di Noia *et al.*, 2000; D'Orso *et al.*, 2001). The resemblance between the AU-rich elements in kinetoplastids and those in mammalian mRNAs is indicative of conservation of this degradation mechanism. However, we cannot exclude the possibility that other, non-ARE elements, could mediate mRNA decay through the binding of stage-regulated factors. We have recently identified a new class of extinct retroposons, predominantly found within 3′UTRs of a large number of *Leishmania* transcripts, and showed that members of this family play a role in mRNA destabilization (Bringaud *et al.*, 2007). Mechanistically, RNA can be cleaved internally or degraded from either the 5′ end or the 3′ end (Parker *et al.*, 2004; Houseley *et al.*, 2006). More recently, it has been shown that microRNAs (miRNAs) can downregulate large numbers of target mRNAs (Lim *et al.*, 2005). MiRNAs are normally processed by two members of the RNase-III family of enzymes, Drosha and Dicer (Pillai, 2005). However, neither of these enzymes has an orthologue in *Leishmania*, and RNA interference (RNAi) does not seem to work in this parasite (Robinson *et al.*, 2003) with the exception of *L. braziliensis* (Peacock *et al.*, 2007). Rapid degradation of amastigote-specific transcripts in promastigotes or alternatively of promastigote-specific transcripts in the amastigote stage could occur through deadenylation followed either by decapping or by 3′ to 5′ exosome-mediated digestion. Ribonucleases and associated factors can be specifically regulated and can target key RNAs that are critical in the developmental switch of *Leishmania*. Most of the enzymes and proteins involved in mRNA degradation in other eukaryotes are well conserved in *Leishmania* (unpublished data). The control of mRNA stability is also intimately linked with translational processes (Coller *et al.*, 2005). It is therefore possible that translational repression of specific mRNAs facilitates their degradation and alternatively enhanced translation of some mRNAs could increase their stability (Gebauer *et al.*, 2003; Mazumder *et al.*, 2003; Deana *et al.*, 2005; Segal *et al.*, 2006). Recent preliminary data from our laboratory indicate that amastigote-specific transcripts can accumulate in the promastigote stage following treatment with cycloheximide or puromycin, inhibitors of translation elongation and initiation respectively (M. Müller, unpublished). The rapid degradation of these RNAs in promastigotes may be due to the binding of stage-specific *trans*-acting factor(s) that are absent or inactivated in amastigotes.

Stage-specific translational control

DNA microarray studies in *Leishmania* (Saxena *et al.*, 2003; Akopyants *et al.*, 2004; Holzer *et al.*, 2006; Leifso *et al.*, 2007) suggest that while differences in transcript abundance can be measured during parasite differentiation, changes are not as pronounced as in organisms that regulate gene expression by transcription initiation. The percentage of genes whose abundance of mRNAs varies throughout the life cycle (only ~1–3.5%) is much lower than that of proteins (~10–12%) (El-Fakhry *et al.*, 2002; Bente *et al.*, 2003; Nugent *et al.*, 2004; McNicoll *et al.*, 2006). Moreover, we recently showed that amastigote-specific protein isoform and mRNA expression trends correlated modestly (53%), while almost no correlation was found for promastigote-specific spots (McNicoll *et al.*, 2006). These important differences between a slight mRNA overexpression and the presence/absence of the gene product on 2D gels suggest that other mechanisms are also contributing to stage-regulated gene expression. Furthermore, my group and others have shown that several stage-specific transcripts are regulated at the level of translation (Zilka *et al.*, 2001; Boucher *et al.*, 2002; Larreta *et al.*, 2004; Folgueira *et al.*, 2005; McNicoll *et al.*, 2005). Altogether, these observations outline the importance of translational and post-translational controls in regulating stage-specific gene expression in these parasites.

Translational regulation provides several advantages, such as rapidity of response, reversibility, fine-tuning of protein abundance, and potential for spatial control that are critical in the adaptive response to various environmental stresses (de Moor *et al.*, 2001). This rapidity and versatility enable cells to modulate membrane composition, activate signal transduction pathways, regulate general and specific mRNA translational efficiency, and adjust needs for energy metabolism and nutrient requirements when changing conditions dictate changing needs. Several stage-specific proteins identified by comparative proteomic analyses (Bente *et al.*, 2003; Nugent *et al.*, 2004; McNicoll *et al.*, 2006) fall within these categories and could very well be regulated at the level of translation.

Translational regulation can be subdivided into global and transcript specific. Global control enables cells to utilize strategies that offer temporal, and possibly spatial, co-regulation of many, or most, expressed transcripts. Such control is used in cellular responses to a threatening stress, such as viral infection, ultraviolet irradiation, hypoxia and nutrient starvation (Clemens, 2005; Komar *et al.*, 2005; Schroder *et al.*, 2005; Wouters *et al.*, 2005). By analysing polysome profiles, we showed that global mRNA translation is decreased in *Leishmania* during differentiation from promastigote to amastigote forms or under a combined heat shock and acidic pH environment (M. Laverdière, unpublished). Intracellular amastigotes divide more slowly than promastigotes and their metabolic needs may change during the adaptation period within the hostile environment of the phagolysosomes, which could explain the overall downregulation of translation. This reduction in global translation, allows cells to conserve resources and to initiate a reconfiguration of gene expression to effectively manage stress conditions. The possible role of eIF2-alpha phosphorylation in this process, as it has been shown in other eukaryotes experiencing stressful conditions (Wek *et al.*, 2006), is now being investigated in our lab. Whereas diverse environmental stimuli can reduce global protein synthesis in *Leishmania*, translation of specific transcripts can be enhanced. In transcript-specific control, the synthesis of a small ensemble of functionally related proteins is regulated, usually, in response to metabolic perturbation or during development. In *Leishmania*, 3′UTRs are almost exclusively implicated in this process, and distinct classes of 3′UTR elements that possibly contribute to translational control in these parasites have been identified (Boucher *et al.*, 2002; McNicoll *et al.*, 2005; Rochette *et al.*, 2005) (Table 3.2).

Role of 3′UTRs in regulating stage-specific gene expression in Leishmania

Several amastigote-specific transcripts in *Leishmania* are regulated by sequences within 3′UTRs that affect either mRNA stability (Aly *et al.*, 1994; Brittingham *et al.*, 2001; Mishra *et al.*, 2003; Larreta *et al.*, 2004; Purdy *et al.*, 2005) or translation (Zilka *et al.*, 2001; Boucher *et al.*, 2002; Larreta *et al.*, 2004; Folgueira *et al.*, 2005; McNicoll *et al.*, 2005). The A2 developmental gene expression during promastigote-to-amastigote cytodifferentiation is mediated through differential RNA stability and involves the A2 mRNA 3′UTR (Charest *et al.*, 1996). Regulation of genes encoding the major surface protease (MSPL) of *L. chagasi* is mediated by a 202-nucleotide region within the MSPL 3′UTR which confers an increase in mRNA stability (Purdy *et al.*, 2005). Stage-specific accumulation of *L. major* leishmanolysin genes (GP63) is under the influence of sequences within the 3′UTR (Kelly *et al.*, 2001). The translational efficiencies of both types of HSP70 transcripts (HSP70-I and HSP70-II) at the different temperatures that the parasite encounters in the insect (26°C) or in the mammalian host (37°C) are controlled by sequences in the 3′UTR. 3′UTRI-bearing transcripts (HSP70-I) were found associated with ribosomes in promastigotes at normal and heat-shock temperatures, whereas the HSP70-II transcripts appear to be preferentially translated at heat-shock temperatures but not at 26°C (Folgueira *et al.*, 2005). Translational control of HSP83 in *Leishmania* originates also from its 3′UTR (Zilka *et al.*, 2001; Larreta *et al.*, 2004). An element located between positions 201 and 472 of the 3′UTR was identified which is essential for increasing translation of the CAT-HSP83 reporter RNA at 33–37°C. This region confers preferential translation during heat shock even in less stable transcripts (Zilka *et al.*, 2001). As discussed below, translational regulation of the amastin transcripts is mediated by conserved elements with their 3′UTR (Boucher *et al.*, 2002; McNicoll *et al.*, 2005; Rochette *et al.*, 2005).

Sequences within the 3′UTRs have also been involved in the regulation of promastigote-specific transcripts. A regulatory element contained within 10 nucleotides of the 3′UTR, termed the PFR regulatory element is necessary for the 10-fold regulation of PFR2 mRNA levels in promastigotes. This element destabilizes the PFR2 mRNA in amastigotes, hence acting as a negative regulatory element (Mishra *et al.*, 2003). MSP (GP63) and PSA (GP46) are abundant glycolipid-anchored proteins on the surface of the promastigote form of most *Leishmania* species. The steady state levels of mRNAs for both glycoproteins are regulated post-transcriptionally by 3′UTRs which results in a 30-fold increase when *L. chagasi* promastigotes grow *in vitro* from logarithmic to stationary phase (Myung *et al.*, 2002).

Widespread 3′UTR retroelements regulate gene expression in Leishmania*: a new paradigm of gene regulation in eukaryotes*

We have recently characterized a multigene family in *Leishmania* encoding the amastin surface proteins (Rochette *et al.*, 2005) that are related to *T. cruzi* amastins (Teixeira *et al.*, 1994) and are expressed predominantly in the amastigote stage. Amastin surface proteins represent the largest developmentally regulated gene family reported so far in *Leishmania* comprising up to 45 members. Most importantly, amastin transcripts share common 3′UTR elements that are implicated in stage-specific regulation of gene expression. The vast majority of the *Leishmania* amastin transcripts carry a conserved 450-nt element in their 3′UTR (Rochette *et al.*, 2005). We showed that this element regulates stage-specific gene expression at the level of translation (Boucher *et al.*, 2002; McNicoll *et al.*, 2005). More specifically, the 450-nt element enhances mRNA translation by increasing its association with heavy sedimenting polysomes, which is a hallmark of highly translated mRNAs (McNicoll *et al.*, 2005). We also showed that regulation through the 450-nt element is triggered in response to heat shock, which constitutes the first environmental change that the parasite encounters upon its entry in the mammalian host. An intriguing feature about this element is its widespread distribution within the *Leishmania* genome and its presence among several amastigote-specific transcripts (Boucher *et al.*, 2002). In collaboration with Dr Bringaud (University of Bordeaux, France) we found that this 3′UTR element belongs to a new family of extinct short retroposons, small interspersed degenerate retroposons (SIDER)1, comprising more than 785 sequences across the *L. major* genome (Bringaud *et al.*, 2007; A. Rochette, unpublished). The recent completion of the Tritryp genome projects confirmed the presence of LTR retrotransposons and retroposons (non-LTR retrotransposons), but no DNA transposons (Berriman *et al.*, 2005; El-Sayed *et al.*, 2005; Ivens *et al.*, 2005). Retroposons constitute the most abundant transposable elements (TE) described in the genome of *T. cruzi* and *T. brucei* (~3% of nuclear genome), while no potentially active TE have been characterized to date in *L. major* (El-Sayed *et al.*, 2005; Bringaud *et al.*, 2006).

Two of the amastin transcripts bear a distinct 3′UTR element of 500–550 nt (Rochette *et al.*, 2005), which is also widespread within the *Leishmania* genome as suggested by BLAST searches (Table 3.3). Interestingly, this 3′UTR element is part of another subfamily of SIDER retroposons, SIDER2, which are also widely distributed in the *Leishmania* genome (1073 copies) but are distinct from the SIDER1 retroposons.

Table 3.3 Widely distributed retroelements in the 3’UTR of *Leishmania* transcripts that are involved in post-transcriptional regulation of gene expression

Leishmania retroelements	BLAST hits in *L. major*	Average size (bp)	Chromosomal location	Putative regulatory function
SIDER1*	785	500	3’UTR	mRNA translation
SIDER2**	1073	500	3’UTR	mRNA stability

*Previously identified as a conserved 450-nt 3’UTR element (Boucher *et al.*, 2002).

**See reference by Bringaud *et al.*, 2007.

Our data support that members of this second retroposon family act as negative regulators of gene expression by promoting mRNA destabilization (Bringaud *et al.*, 2007). Both retroposon families are present almost exclusively within 3′-untranslated regions of *Leishmania* mRNAs in contrast to *T. brucei* where most retroposons have been mainly found in subtelomeric regions. The specific 3′UTR location of these retroelements suggest that *Leishmania*, but not *Trypanosoma* species, has domesticated, maintained and possibly expanded these short retroposons to fulfil critical regulatory functions required for their specialized lifestyle. The widespread genomic distribution of these retroposons raises the interesting possibility that numerous *Leishmania* transcripts encoding a wide repertoire of functionally diverse proteins may be regulated by a similar mechanism in response to specific environmental stimuli. A coordinated regulation of multiple genes might provide a great advantage to the parasite for a rapid adaptation throughout its digenetic complex life cycle.

3′UTR elements identified so far in *Leishmania* are distinct from other known regulatory elements reported previously in eukaryotes, including the related *Trypanosoma* species (Berberof *et al.*, 1995; Furger *et al.*, 1997; Hotz *et al.*, 1997; Coughlin *et al.*, 2000; Di Noia *et al.*, 2000). We also found that several developmentally regulated transcripts in *Leishmania* bear more than one regulatory elements in their 3′UTR ((McNicoll *et al.*, 2005; Rochette *et al.*, 2005); M. Müller, unpublished; Fig. 3.2), which suggests that regulation of gene expression in this parasite is complex and controlled by a variety of mechanisms.

Post-translational control

Stage-specific post-translational modifications

Post-translational modifications of proteins, like phosphorylation, methylation, acetylation or glycosylation, change the functionality of the protein, while others like ubiquitination target the protein to the proteasome-dependent degradation pathway, thereby acting on the half-life time of the protein (Coux, 2002). Although, post-translational modifications of proteins are also known in trypanosomatids, there are very few reports of post-translational regulation of protein abundance in these organisms. In *L. infantum*, the differences in gene expression of the GP63 protease family depending on the phase of parasite growth seem to be attributed to a differential pattern of glycosylation, which correlates with the different infective promastigote forms of the parasite (Gonzalez-Aseguinolaza *et al.*, 1997). Changes in protein phosphorylation or protein kinase activities have been detected in response to various environmental changes throughout the life cycle of *Leishmania*. For example, it has been shown that heat stress induces modulation of protein phosphorylation in virulent promastigotes of *L. donovani* (Salotra *et al.*, 2000) and that acidic pH stress induces

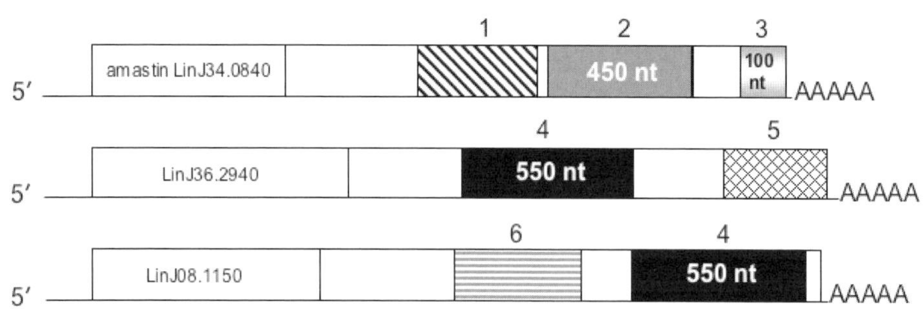

Figure 3.2 Stage-specific gene expression in *Leishmania* is often regulated by multiple distinct elements within 3′UTRs. These elements are involved either in mRNA stability/degradation or mRNA translation. Element 1 is possibly involved in mRNA degradation specifically in promastigotes; elements 2 (SIDER1) and 3 act in an additive manner to induce amastin mRNA translation in response to heat stress; element 4 (SIDER2) is involved in mRNA destabilization; elements 5 and 6 are distinct and seem to positively regulate either translation or mRNA processing in *L. infantum* amastigotes.

protein tyrosine phosphorylation in *L. pifanoi* (Rivero-Lezcano *et al.*, 1997). A marked increase in the phosphorylation pattern of the endogenous membrane proteins was observed during *L. major* differentiation from logarithmic promastigotes to virulent metacyclic forms (Hermoso *et al.*, 1993). An *in vitro* phosphorylation assay demonstrated that the pattern of serine and threonine phosphorylated proteins was regulated during the life cycle of *L. major* (Dell *et al.*, 1994). A promastigote-specific mitogen-activated protein kinase kinase homologue found to be critically involved in the regulation of flagellar assembly and cell size in *L. mexicana* is likely to be regulated by post-translational mechanisms such as phosphorylation (Wiese *et al.*, 2003).

Our recent comparative proteomic analysis between axenic *L. infantum* promastigote and amastigote life cycle stages showed that many proteins were present in multiple spots, suggesting that post-translational modification is extensive in this organism. In several cases, different isoforms appeared to be specific to different life stages. There were 168 protein matches to entries in *Leishmania* peptide databases but these matches were derived from 71 genes which corresponds to an average of 2.3 protein isoforms per gene (McNicoll *et al.*, 2006). Interestingly, a recent proteomic survey of the procyclic form of the related parasite *T. brucei* also identified a large number of protein isoforms at 2.6 per gene (Jones *et al.*, 2006). While these results reveal that stage-specific post-translational modifications may be more widespread in trypanosomatids than previously appreciated, the importance of these modifications to stage differentiation or other areas of parasite biology remains to be elucidated.

Protein degradation

Proteasomes are large multimeric self-compartmentalizing proteases, which play a crucial role in the clearance of misfolded proteins, breakdown of regulatory proteins, processing of proteins by specific partial proteolysis, cell cycle control as well as preparation of peptides for immune presentation (Gille *et al.*, 2003). In parasitic protozoa, including *L. mexicana* (Robertson, 1999), *T. brucei* (Hua *et al.*, 1996; Yao *et al.*, 1999; Wang *et al.*, 2003), and *T. cruzi* (Gonzalez *et al.*, 1996; de Diego *et al.*, 2001), a typical eukaryotic proteasome has been biochemically purified. Peptidases represent 2% of the protein-coding genes in trypanosomatids, similar to what is observed in other eukaryotes. Also, numerous enzymes that are part of ubiquitin conjugation are present in the *Leishmania* genome, which suggests an active system of non-lysosomal cytosolic protein degradation (Ivens *et al.*, 2005). Interestingly, in addition to the 20S proteasome, *Leishmania* and trypanosomes contain hslV, a proteasome-like heat shock protein encoded by the heat shock locus V, and HslU homologue, which are prevalent in bacteria (Couvreur *et al.*, 2002; Gille *et al.*, 2003). *Leishmania* and trypanosomes are the only eukaryotes to possess a proteasome and at the same time have the HslVU complex, which may have been acquired from an alpha-proteobacterial progenitor of mitochondria (Gille *et al.*, 2003). Contrary to other eukaryotes, *Leishmania* and trypanosomes have an ornithine decarboxylase (ODC) protein lacking the C-terminal domain, which contains the degradation signal for rapid turnover by the 26S proteasome, and ODC is therefore metabolically stable in these parasites (Svensson *et al.*, 1997). Thus, despite the high similarities of the *Leishmania* proteasome to proteasomes from other eukaryotes, several differences may exist which could be attributed to the specific needs of this parasite throughout its complex life cycle.

Approximately 10% of the differentially expressed proteins in *L. infantum* axenic amastigotes are involved in proteolysis (McNicoll *et al.*, 2006; Fig. 3.1). Interestingly, we found three proteins directly involved in ubiquitin-dependent protein catabolism. While two were overexpressed in the amastigote stage (proteasome non-ATPase regulatory subunit, and proteasome activator protein pa26), one was found as distinct isoforms in each stage of the parasite (proteasome $\alpha 1$ subunit). Three proteins involved in this process were also found preferentially expressed in *L. mexicana* amastigotes (Nugent *et al.*, 2004), and two were found in *L. donovani* amastigotes (Bente *et al.*, 2003). It has been suggested that multiple proteasome forms may also exist in different life stages of *T. brucei* (Wang *et al.*, 2003). Differential gene expression of several important components of the *Leishmania* proteasome in amastigotes could be explained by the high level of misfolded proteins generated either by heat shock or by other severe stresses within the phagolysosomes of mammalian macrophages, as well as by expected changes in cell cycle control during parasite differentiation. It is reasonable to hypothesize that proteins specifically expressed in one developmental life cycle stage of the parasite are either degraded in a stage-specific manner when the parasite switches to the other life-stage or are simply not synthesized, given that there is less or no mRNA available. However, very little is known about stage-specific protein degradation in *Leishmania*. To date, there are only a few reports on proteasome-dependent degradation of proteins in trypanosomatids. In *T. brucei*, cyclin degradation is strongly suspected to be dependent on proteasome activity (Van Hellemond *et al.*, 2000) but the molecular understanding of the cell cycle in trypanosomatids remains very incomplete. In *L. major*, the proteasome might be involved in the cell cycle regulation of the MCAK-like Kin-13 kinesin (Dubessay *et al.*, 2006).

Environmental signals and their effect on stage-regulated gene expression and on parasite differentiation

During the transition of *Leishmania* promastigotes into the obligatory intracellular amastigotes, the parasite is subjected to extreme environmental changes, including a sharp rise in temperature, a drop in extracellular pH, an increased exposure to oxygen and nitrogen-reactive species, an intense extracellular proteolytic activity, and a nutritional starvation. These external signals are critical in activating many regulatory mechanisms affecting global and specific gene expression in these parasites and in triggering *Leishmania* cell differentiation even when cells are raised in a host-free (axenic) culture system (Zilberstein *et al.*, 1994; Garlapati *et al.*, 1999; Sereno *et al.*, 2001). By sequential exposure to elevated temperature for 24 hours followed by acidic pH, it was found that heat was responsible for the growth arrest and acidic pH for the release of parasites from growth arrest which allowed the subsequent differentiation into amastigotes (Barak *et al.*, 2005).

Heat shock can cause misfolding of a fraction of cell proteins, thus altering gene regulation at the level of transcription and translation (Schroder *et al.*, 2005). Using polysome profiling analysis, we showed that only a few hours of exposure of *L. infantum* promastigotes to elevated temperature (from 25°C to 37°C) resulted in a significant decrease of global mRNA translation (M. Laverdière, unpublished) in agreement with what has been reported in other eukaryotes (Schroder *et al.*, 2005). Translation of an amastin mRNA carrying the SIDER1 element is stimulated by heat shock (McNicoll *et al.*, 2005). Translation of the HSP70 and HSP83 transcripts increases at least by 10-fold upon temperature elevation (Quijada *et al.*, 2000). The *L. infantum* HSP83 transcripts show a temperature-dependent accumulation that is controlled by a post-transcriptional mechanism involving sequences located in the 3′UTR (Zilka *et al.*, 2001; Larreta *et al.*, 2004). The *L. major* homologue of the yeast stress-inducible protein STI1 can be upregulated by a shift in culture temperature from 26 to 37°C (Webb *et al.*, 1997). Chaperones that are important in the folding of proteins during parasite differentiation like HSP100 (Krobitsch *et al.*, 1998) and CPN10 (Zamora-Veyl *et al.*, 2005) are either transiently upregulated or more expressed in amastigotes. It has been reported that a shift in temperature may play a role in signal transduction events in morphogenetic transformation of *L. donovani* promastigotes that involves mobilization of calcium in the cytoplasm. The intracellular free Ca^{2+} concentration during the conversion of promastigotes to axenic amastigotes increased by a 10-fold at the initial 6–12 hours followed by declining levels until 60 hours and thereafter a return to constant concentrations (Sarkar *et al.*, 1995).

Acidic pH has been documented to elicit signal transduction pathways in some lower organisms that can survive acidic conditions, including acid-tolerant bacteria and fungi. Bacteria and fungi possess a myriad of acid resistance systems that can help them to overcome the challenge posed by different acidic environments such as the use of proton pumps, the protection or repair of macromolecules, cell membrane changes, production of alkali, induction of pathways by transcriptional regulators, alteration of metabolism, and the role of cell density and cell signalling (Cotter *et al.*, 2003; Konno *et al.*, 2006). The survival of *Leishmania* parasites within acidic phagolysosomes proves that they adapt quickly to drastically changing conditions possibly by sensing acidic environment via still uncharacterized signalling pathways. Although *Leishmania* amastigotes multiply in an acidic intracellular compartment called the parasitophorous vacuole, the cytoplasm of amastigotes is regulated at near-neutral pH by a plasma membrane H^+-ATPase-mediated proton extrusion system (Glaser *et al.*, 1988; Zilberstein *et al.*, 1989). Differences in the mechanisms for regulation of intracellular pH were noted between amastigote and promastigote forms. Amastigotes maintained their intracellular pH neutral over a wide range of external pH, whereas promastigotes were unable to maintain a neutral pH under acidic conditions (Marchesini *et al.*, 2002). Exposure to acidic pH has been found to induce the accumulation of a number of developmentally regulated transcripts. For example, the stability of reporter gene transcripts bearing the A2 3′UTR was increased in cells incubated at low pH (Charest *et al.*, 1996). Similarly, accumulation of amastin transcripts was induced upon growth in acidic pH (McNicoll *et al.*, 2005).

The nutrient environment encountered by amastigotes is not well understood but the parasitophorous vacuole can fuse with endosomes, phagosomes and autophagosomes, suggesting that a diverse range of macromolecules will be present. The parasitophorous vacuole is a hydrolytic compartment in which such material will be rapidly degraded to low molecular weight components, which are typical substrates for membrane transporters. Amastigote surface transporters must compete for these substrates with equivalent host transporters in the membrane of the parasitophorous vacuole (Burchmore *et al.*, 2001). Amastigote uptake systems for glucose, amino acids, nucleosides and polyamines are optimally active at acidic pH (Zilberstein *et al.*, 1993; Burchmore *et al.*, 2001). Promastigote uptake systems are kinetically distinct and function optimally at a more neutral environmental pH, indicating that membrane transport activity is developmentally regulated. Several proteins that are involved in transport or glucose catabolism are expressed specifically in promastigotes, indicating the importance of this energy source for this developmental stage (Cairns *et al.*, 1989; Langford *et al.*, 1992; Burchmore *et al.*, 1998).

L. donovani promastigotes introduced into the bloodstream by sandfly vectors are exposed to reactive oxygen species like H_2O_2 during phagocytosis by the host macrophage. It has been shown that H_2O_2 can induce promastigote apoptosis-like death (Das *et al.*, 2001). The peroxidoxin LcPxn1 (Barr *et al.*, 2001) and iron superoxide dismutase LeFeSODB1 (Plewes *et al.*, 2003), which are important enzymes in the process of oxygen reactive component detoxification in the phagolysosome are more expressed in amastigotes.

Developmental regulation of non-coding RNAs in *Leishmania*

Non-coding RNAs (ncRNAs) are a major class of RNAs that vary in size from a few nucleotides to several thousand nucleotides and have been identified and characterized both in prokaryotes and eukaryotes (Storz *et al.*, 2005). The ncRNAs other than tRNAs, ribosomal RNAs and spliceosomal RNAs, participate in a variety of cellular functions including transcription, RNA processing or stability, RNA modification, RNA interference, mRNA translation, imprinting, DNA methylation and X chromosome dosage compensation (reviewed in (Eddy, 2001; Mattick *et al.*, 2005). Several of these ncRNAs in humans have links to a number of diseases (Mattick *et al.*, 2005).

At the completion of the genome sequencing project for *L. major* Friedlin, 911 RNA genes have been identified and these RNA genes fall into the categories of tRNAs, rRNAs, SL-RNAs, snRNAs, snoRNAs and srpRNAs (Ivens *et al.*, 2005). To the best of our knowledge, there are no reports on RNA genes that are regulated in a stage-specific manner with the only exception of the SL RNA (Lamontagne *et al.*, 1999). We have reported previously that the SL RNA is developmentally polyadenylated in *L. donovani* amastigotes and that this type of regulation is rapidly induced upon growth of the parasite in acidic pH conditions. Polyadenylated forms of the SL transcript have also been identified in poly(A)+ RNA fractions from short stumpy bloodstream forms of *T. brucei* (Pellé *et al.*, 1993). This process may be related to the adenylation and deadenylation control that take place in the maturation of several snRNAs and 7SL RNA (Perumal *et al.*, 2002).

Only one class of the genomic mini-exon copies which bears an additional 15-nt sequence ending by four T residues at its 3′ end is polyadenylated in a stage-specific manner. The additional 15-nt sequence at the 3′ end of the poly(A)+ SL RNA is rich in pyrimidines and it contains an AG splice acceptor site immediately preceded by a poly(Y) tract. This splice acceptor site may be recognized by splicing factors that could activate polyadenylation. As mentioned above, *trans*-splicing and polyadenylation are coupled in trypanosomatids (LeBowitz *et al.*, 1993; Matthews *et al.*, 1994; Vassella *et al.*, 1994). The poly(A)+ SL RNA represents 12–16% of the total SL transcript synthesized in amastigotes and is 2.5- to 3-fold more stable than the poly(A)- transcript. It is still not known whether the poly(A)+ SL RNA has any specific regulatory function. It is possible that the polyadenylated SL transcript cannot be used as a donor of the 39-nt spliced leader sequence to the 5′ end of all pre-mRNAs. This should have as a consequence to decrease *trans*-splicing in amastigotes. Interestingly, it has been shown that *l. tarentolae* SL transcripts that were malformed by more than one nucleotide at their 3′ ends were unable to participate in *trans*-splicing or in acquiring a proper cap 4 methylation (Sturm *et al.*, 1998). Because amastigotes replicate more slowly than promastigotes, less SL RNA might be required for pre-mRNA processing.

Using a differential hybridization strategy, we have identified a new class of non-coding RNAs (ncRNAs) ranging from ~300 to 600 nt in size which are expressed specifically in the amastigote stage and are unique to *Leishmania* (Dumas *et al.*, 2006). These ncRNA genes are part of subtelomeric tandem repeats that are transcribed both in sense and antisense orientation by RNA polymerase II. It is not clear why antisense ncRNAs are made, but it is possible that these may act as regulators of the sense transcripts by either competing for binding to the same factor(s) or base-pairing the sense RNAs hence changing their secondary structure. In contrast to other ncRNAs which usually utilize polyadenylation-independent pathways to produce entry sites for 3′–5′ trimming and to allow correct 3′-end maturation (Fatica *et al.*, 2000; Morlando *et al.*, 2002), these *Leishmania* ncRNAs can be polyadenylated. It is generally assumed that polyadenylation stabilizes the 3′-end of the mRNA (Beelman *et al.*, 1995), it facilitates transport of the mRNA to the cytoplasm (Huang *et al.*, 1996) and it increases the efficiency of translation initiation (Tarun *et al.*, 1996). Both the sense and antisense ncRNAs co-localize within the cytosol of amastigote cells and are part of a small RNP complex that has similar sedimentation coefficients to the SL RNA RNP complex (Dumas *et al.*, 2006). Using polysome profiling studies, we showed that these ncRNAs are mainly associated with the 40S ribosomal subunit, which suggests that they might play a role in translational regulation (C. Dumas, unpublished). Similar ribosomal distributions have been observed for the bacterial ncRNA DsrA (Worhunsky *et al.*, 2003) and the dendritic non-coding BC1 RNA (Zalfa *et al.*, 2003), which are known to inhibit translation of specific mRNAs. Further studies to characterize the molecular interactions of these stage-specific non-coding *Leishmania* RNAs with proteins or other RNA molecules and the ribosomes should provide new insights into the putative role that these ncRNAs may play in the control of stage-specific gene expression in *Leishmania*.

Concluding remarks

Leishmania have no defined RNA pol II promoters and have probably lost the ability to regulate gene expression at the level of transcription initiation. Protein-coding genes in this organism are transcribed polycistronically by RNAP II, and individual mature mRNAs are generated from polycistronic precursors by 5′ *trans* splicing of a 39-nt capped leader RNA and 3′ polyadenylation. Thus, regulation of gene expression in *Leishmania* occurs exclusively at the post-transcriptional level and involves pre-mRNA processing, mRNA stability and translation, protein stability and post-translational control. During the transition of *Leishmania* from promastigotes into the obligatory intracellular amastigotes, the parasite is subjected to extreme environmental changes, which are critical in activating many regulatory mechanisms affecting global and specific gene expression in these parasites and in triggering *Leishmania* differentiation. Generally, sequences within 3′UTRs that act post-transcriptionally have been shown to play a key role in regulating differentially expressed transcripts in *Leishmania*.

A variety of 3′UTR regulatory elements has been identified among stage-specific transcripts. We have identified two distinct classes of genome-wide conserved 3′UTR retroelements (SIDERs) and showed that they are involved in regulation of gene expression in *Leishmania*. Members of the SIDER1 subfamily are likely to enhance mRNA translation, whereas members of the SIDER2 subfamily act at the level of mRNA stability. The widespread genomic distribution of SIDERs suggests that a large number of *Leishmania* transcripts may be co-regulated in response to a specific environmental signal. This unusual type of regulation could provide a great advantage to the parasite for a rapid adaptation within its mammalian host. Stage-specific gene expression in *Leishmania* is complex and, in many cases, involves more than one regulatory element within a given 3′UTR. These distinct elements regulate either mRNA stability/degradation or mRNA translation and could act either in an additive or a competitive manner to modulate gene expression in response to different growth conditions or differentiation signals. Moreover, non-coding developmentally regulated RNAs, known to participate in a variety of cellular regulatory functions in other organisms, have been identified in *Leishmania* as part of small RNP complexes. Their putative role in regulating gene expression remains however, as yet unclear. Global proteomics and transcriptomic studies in different *Leishmania* species demonstrate a low degree of differential mRNA expression but a higher degree of regulation at the translational and post-translational levels. Using proteomics, several stage-specific protein isoforms have been isolated for a single gene suggesting post-translational modifications being important in stage-specific regulation of gene expression. This situation does not appear to be rare, and might explain a considerable degree of apparent discrepancies between transcriptomic and proteomic studies.

Clearly, mechanisms of stage-specific gene regulation in *Leishmania* are multi-layered and complex and a holistic view of each gene product, including the expression of mRNA and all protein isoforms, is required for a clearer understanding of stage-specific gene expression in this organism.

References

Agabian, N. (1990). *Trans* splicing of nuclear pre-mRNAs. Cell *61*, 1157–1160.

Akopyants, N.S., Matlib, R.S., Bukanova, E.N., Smeds, M.R., Brownstein, B.H., Stormo, G.D., and Beverley, S.M. (2004). Expression profiling using random genomic DNA microarrays identifies differentially expressed genes associated with three major developmental stages of the protozoan parasite *Leishmania major*. Mol. Biochem. Parasitol. *136*, 71–86.

Almeida, R., Gilmartin, B.J., McCann, S.H., Norrish, A., Ivens, A.C., Lawson, D., Levick, M.P., Smith, D.F., Dyall, S.D., Vetrie, D., *et al.* (2004). Expression profiling of the *Leishmania* life cycle: cDNA arrays identify developmentally regulated genes present but not annotated in the genome. Mol. Biochem. Parasitol. *136*, 87–100.

Aly, R., Argaman, M., Halman, S., and Shapira, M. (1994). A regulatory role for the 5′ and 3′ untranslated regions in differential expression of hsp83 in *Leishmania*. Nucleic Acids Res. 22, 2922–2929.

Antoine, J.C., Prina, E., Courret, N., and Lang, T. (2004). *Leishmania* spp.: on the interactions they establish with antigen-presenting cells of their mammalian hosts. Adv. Parasitol. *58*, 1–68.

Argaman, M., Aly, R., and Shapira, M. (1994). Expression of heat shock protein 83 in *Leishmania* is regulated post-transcriptionally. Mol. Biochem. Parasitol. *64*, 95–110.

Bangs, J.D., Crain, P.F., Hashizume, T., McCloskey, J.A., and Boothroyd, J.C. (1992). Mass spectrometry of mRNA cap 4 from trypanosomatids reveals two novel nucleosides. J. Biol. Chem. *267*, 9805–9815.

Barak, E., Amin-Spector, S., Gerliak, E., Goyard, S., Holland, N., and Zilberstein, D. (2005). Differentiation of *Leishmania* donovani in host-free system: analysis of signal perception and response. Mol. Biochem. Parasitol. *141*, 99–108.

Barr, S.D., and Gedamu, L. (2001). Cloning and characterization of three differentially expressed peroxidoxin genes from *Leishmania chagasi*. Evidence for an enzymatic detoxification of hydroxyl radicals. J. Biol. Chem. *276*, 34279–34287.

Beelman, C.A., and Parker, R. (1995). Degradation of mRNA in eukaryotes. Cell *81*, 179–183.

Belli, S., Formenton, A., Noll, T., Ivens, A., Jacquet, R., Desponds, C., Hofer, D., and Fasel, N. (1999). *Leishmania major*: histone H1 gene expression from the sw3 locus. Exp. Parasitol. *91*, 151–160.

Belli, S.I., Monnerat, S., Schaff, C., Masina, S., Noll, T., Myler, P.J., Stuart, K., and Fasel, N. (2003). Sense and antisense transcripts in the histone H1 (HIS-1) locus of *Leishmania major*. Int. J. Parasitol. *33*, 965–975.

Bellofatto, V., Torres-Munoz, J.E., and Cross, G.A. (1991). Stable transformation of Leptomonas seymouri by circular extrachromosomal elements. Proc. Natl. Acad. Sci. USA *88*, 6711–6715.

Bente, M., Harder, S., Wiesgigl, M., Heukeshoven, J., Gelhaus, C., Krause, E., Clos, J., and Bruchhaus, I. (2003). Developmentally induced changes of the proteome in the protozoan parasite *Leishmania donovani*. Proteomics 3, 1811–1829.

Benz, C., Nilsson, D., Andersson, B., Clayton, C., and Guilbride, D.L. (2005). Messenger RNA processing sites in *Trypanosoma brucei*. Mol. Biochem. Parasitol. *143*, 125–134.

Berberof, M., Vanhamme, L., Tebabi, P., Pays, A., Jefferies, D., Welburn, S., and Pays, E. (1995). The 3′-terminal region of the mRNAs for VSG and procyclin can confer stage specificity to gene expression in *Trypanosoma brucei*. EMBO J. *14*, 2925–2934.

Berriman, M., Ghedin, E., Hertz-Fowler, C., Blandin, G., Renauld, H., Bartholomeu, D.C., Lennard, N.J., Caler, E., *et al.* (2005). The genome of the African trypanosome *Trypanosoma brucei*. Science. *309*, 416–422.

Boothroyd, J.C., and Cross, G.A. (1982). Transcripts coding for variant surface glycoproteins of *Trypanosoma brucei* have a short, identical exon at their 5′ end. Gene. *20*, 281–289.

Borst, P. (1986). Discontinuous transcription and antigenic variation in trypanosomes. Annu. Rev. Biochem. *55*, 701–732.

Boucher, N., Wu, Y., Dumas, C., Dube, M., Sereno, D., Breton, M., and Papadopoulou, B. (2002). A common mechanism of stage-regulated gene expression in *Leishmania* mediated by a conserved 3′-untranslated region element. J. Biol. Chem. *277*, 19511–19520.

Brems, S., Guilbride, D.L., Gundlesdodjir-Planck, D., Busold, C., Luu, V.D., Schanne, M., Hoheisel, J., and Clayton, C. (2005). The transcriptomes of *Trypanosoma brucei* Lister 427 and TREU927 bloodstream and procyclic trypomastigotes. Mol. Biochem. Parasitol. *139*, 163–172.

Bringaud, F., Ghedin, E., Blandin, G., Bartholomeu, D.C., Caler, E., Levin, M.J., Baltz, T., and El-Sayed, N.M. (2006). Evolution of non-LTR retrotransposons in the trypanosomatid genomes: *Leishmania major* has lost the active elements. Mol. Biochem. Parasitol. *145*, 158–170.

Bringaud, F., Müller, M., Cerqueira, G.C., Smith, M., Rochette, A., El-Sayed, N.M., Papadopoulou, B., and Ghedin, E. (2007). Members of a large retroposon family are determinants of post-transcriptional gene expression in *Leishmania*. PLoS Pathog. *3*, 1291–1307.

Brittingham, A., Miller, M.A., Donelson, J.E., and Wilson, M.E. (2001). Regulation of GP63 mRNA stability in promastigotes of virulent and attenuated *Leishmania chagasi*. Mol. Biochem. Parasitol. *112*, 51–59.

Brooks, D.R., Denise, H., Westrop, G.D., Coombs, G.H., and Mottram, J.C. (2001). The stage-regulated expression of *Leishmania mexicana* CPB cysteine proteases is mediated by an intercistronic sequence element. J. Biol. Chem. *276*, 47061–47069.

Burchmore, R.J., and Barrett, M.P. (2001). Life in vacuoles – nutrient acquisition by *Leishmania* amastigotes. Int. J. Parasitol. *31*, 1311–1320.

Burchmore, R.J.S., and Landfear, S.M. (1998). Differential regulation of multiple glucose transporter genes in *Leishmania mexicana*. J. Biol. Chem. *273*, 29118–29126.

Cairns, B.R., Collard, M.W., and Landfear, S.M. (1989). Developmentally regulated gene from *Leishmania* encodes a putative membrane transport protein. Proc. Natl. Acad. Sci. USA *86*, 7682–7686.

Campbell, D.A., Thomas, S., and Sturm, N.R. (2003). Transcription in kinetoplastid protozoa: why be normal? Microbes Infect. *5*, 1231–1240.

Charest, H., Zhang, W.W., and Matlashewski, G. (1996). The developmental expression of *Leishmania donovani* A2 amastigote-specific genes is post-transcriptionally mediated and involves elements located in the 3′-untranslated region. J. Biol. Chem. *271*, 17081–17090.

Chen, D.Q., Kolli, B.K., Yadava, N., Lu, H.G., Gilman-Sachs, A., Peterson, D.A., and Chang, K.P. (2000). Episomal expression of specific sense and antisense mRNAs in *Leishmania amazonensis*: modulation of gp63 level in promastigotes and their infection of macrophages in vitro. Infect. Immun. *68*, 80–86.

Clayton, C.E. (2002). Life without transcriptional control? From fly to man and back again. EMBO J. *21*, 1881–1888.

Clemens, M.J. (2005). Translational control in virus-infected cells: models for cellular stress responses. Semin. Cell. Dev. Biol. *16*, 13–20.

Coller, J., and Parker, R. (2005). General translational repression by activators of mRNA decapping. Cell *122*, 875–886.

Cotter, P.D., and Hill, C. (2003). Surviving the acid test: responses of gram-positive bacteria to low pH. Microbiol. Mol. Biol. Rev. *67*, 429–453.

Coughlin, B.C., Teixeira, S.M., Kirchhoff, L.V., and Donelson, J.E. (2000). Amastin mRNA abundance in *Trypanosoma cruzi* is controlled by a 3′-untranslated region position-dependent cis-element and an untranslated region-binding protein. J. Biol. Chem. *275*, 12051–12060.

Couvreur, B., Wattiez, R., Bollen, A., Falmagne, P., Le Ray, D., and Dujardin, J.C.(2002). Eubacterial HslV and HslU subunits homologs in primordial eukaryotes. Mol. Biol. Evol. *19*, 2110–2117.

Coux, O. (2002). The 26S proteasome. Prog. Mol. Subcell. Biol. *29*, 85–107.

D'Orso, I., and Frasch, A.C. (2001). Functionally different AU- and G-rich cis-elements confer developmentally regulated mRNA stability in *Trypanosoma cruzi* by interaction with specific RNA-binding proteins. J. Biol. Chem. *276*, 15783–15793.

Das, A., and Bellofatto, V. (2003). RNA polymerase II-dependent transcription in trypanosomes is associated with a SNAP complex-like transcription factor. Proc. Natl. Acad. Sci. USA *100*, 80–85.

Das, M., Mukherjee, S.B., and Shaha, C. (2001). Hydrogen peroxide induces apoptosis-like death in *Leishmania donovani* promastigotes. J. Cell. Sci. *114*, 2461–2469.

de Diego, J.L., Katz, J.M., Marshall, P., Gutierrez, B., Manning, J.E., Nussenzweig, V., and Gonzalez, J. (2001). The ubiquitin-proteasome pathway plays an essential role in proteolysis during *Trypanosoma cruzi* remodeling. Biochemistry. *40*, 1053–1062.

de Moor, C.H., and Richter, J.D. (2001). Translational control in vertebrate development. Int. Rev. Cytol. *203*, 567–608.

Deana, A., and Belasco, J.G. (2005). Lost in translation: the influence of ribosomes on bacterial mRNA decay. Genes Dev.*19*, 2526–2533.

Dell, K.R., and Engel, J.N. (1994). Stage-specific regulation of protein phosphorylation in *Leishmania major*. Mol. Biochem. Parasitol. *64*, 283–292.

Di Noia, J.M., D'Orso, I., Sanchez, D.O., and Frasch, A.C. (2000). AU-rich elements in the 3′-untranslated region of a new mucin-type gene family of *Trypanosoma cruzi* confers mRNA instability and modulates translation efficiency. J. Biol. Chem. *275*, 10218–10227.

Diehl, S., Diehl, F., El-Sayed, N.M., Clayton, C., and Hoheisel, J.D. (2002). Analysis of stage-specific gene expression in the bloodstream and the procyclic form of *Trypanosoma brucei* using a genomic DNA-microarray. Mol. Biochem. Parasitol. *123*, 115–123.

DiPaolo, C., Kieft, R., Cross, M., and Sabatini, R. (2005). Regulation of trypanosome DNA glycosylation by a SWI2/SNF2-like protein. Mol. Cell. *17*, 441–451.

Djikeng, A., Shi, H., Tschudi, C., and Ullu, E. (2001). RNA interference in *Trypanosoma brucei*: cloning of small interfering RNAs provides evidence for retroposon-derived 24–26-nucleotide RNAs. Rna *7*, 1522–1530.

Dubessay, P., Blaineau, C., Bastien, P., Tasse, L., Van Dijk, J., Crobu, L., and Pages, M. (2006). Cell cycle-dependent expression regulation by the proteasome pathway and characterization of the nuclear targeting signal of a *Leishmania major* Kin-13 kinesin. Mol. Microbiol. *59*, 1162–1174.

Dumas, C., Chow, C., Muller, M., and Papadopoulou, B. (2006). A novel class of developmentally regulated noncoding RNAs in *Leishmania*. Eukaryot. Cell. *5*, 2033–2046.

Eddy, S.R. (2001). Non-coding RNA genes and the modern RNA world. Nat. Rev. Genet. 2, 919–929.

El-Fakhry, Y., Ouellette, M., and Papadopoulou, B. (2002). A proteomic approach to identify developmentally regulated proteins in *Leishmania infantum*. Proteomics 2, 1007–1017.

El-Sayed, N.M., Myler, P.J., Bartholomeu, D.C., Nilsson, D., Aggarwal, G., Tran, A.N., Ghedin, E., Worthey, E.A., Delcher, A.L., Blandin, G., Westenberger, S.J., Caler, E., *et al.* (2005). The genome sequence of *Trypanosoma cruzi*, etiologic agent of Chagas disease. Science. *309*, 409–415.

Fatica, A., Morlando, M., and Bozzoni, I. (2000). Yeast snoRNA accumulation relies on a cleavage-dependent/polyadenylation-independent 3′-processing apparatus. Embo J. *19*, 6218–6229.

Folgueira, C., Quijada, L., Soto, M., Abanades, D.R., Alonso, C., and Requena, J.M. (2005). The translational efficiencies of the two *Leishmania infantum* HSP70 mRNAs, differing in their 3′-untranslated regions, are affected by shifts in the temperature of growth through different mechanisms. J. Biol. Chem. *280*, 35172–35183.

Foucher, A.L., Papadopoulou, B., and Ouellette, M. (2006). Prefractionation by digitonin extraction increases representation of the cytosolic and intracellular proteome of *Leishmania infantum*. J. Prot. Res. 5, 1741–1750.

Freistadt, M.S., Cross, G.A., and Robertson, H.D. (1988). Discontinuously synthesized mRNA from *Trypanosoma brucei* contains the highly methylated 5′ cap structure, m7GpppA*A*C(2′-O)mU*A. J. Biol. Chem. *263*, 15071–15075.

Furger, A., Schurch, N., Kurath, U., and Roditi, I. (1997). Elements in the 3′ untranslated region of procyclin mRNA regulate expression in insect forms of *Trypanosoma brucei* by modulating RNA stability and translation. Mol. Cell. Biol. *17*, 4372–4380.

Garlapati, S., Dahan, E., and Shapira, M. (1999). Effect of acidic pH on heat shock gene expression in *Leishmania*. Mol. Biochem. Parasitol. *100*, 95–101.

Gebauer, F., Grskovic, M., and Hentze, M.W. (2003). Drosophila sex-lethal inhibits the stable association of the 40S ribosomal subunit with msl-2 mRNA. Mol. Cell. *11*, 1397–1404.

Gibson, W.C., Swinkels, B.W., and Borst, P. (1988). Post-transcriptional control of the differential expression of phosphoglycerate kinase genes in *Trypanosoma brucei*. J. Mol. Biol. *201*, 315–325.

Gilinger, G., and Bellofatto, V. (2001). Trypanosome spliced leader RNA genes contain the first identified RNA polymerase II gene promoter in these organisms. Nucleic Acids Res. *29*, 1556–1564.

Gille, C., Goede, A., Schloetelburg, C., Preissner, R., Kloetzel, P.M., Gobel, U.B., and Frommel, C. (2003). A comprehensive view on proteasomal sequences: implications for the evolution of the proteasome. J. Mol. Biol. *326*, 1437–1448.

Glaser, T.A., Baatz, J.E., Kreishman, G.P., and Mukkada, A.J. (1988). pH homeostasis in *Leishmania donovani* amastigotes and promastigotes. Proc. Natl. Acad. Sci. USA *85*, 7602–7606.

Glaser, T.A., Moody, S.F., Handman, E., Bacic, A., and Spithill, T.W. (1991). An antigenically distinct lipophosphoglycan on amastigotes of *Leishmania major*. Mol. Biochem. Parasitol. *45*, 337–344.

Gonzalez, J., Ramalho-Pinto, F.J., Frevert, U., Ghiso, J., Tomlinson, S., Scharfstein, J., Corey, E.J., and Nussenzweig, V. (1996). Proteasome activity is required for the stage-specific transformation of a protozoan parasite. J. Exp. Med. *184*, 1909–1918.

Gonzalez-Aseguinolaza, G., Almazan, F., Rodriguez, J.F., Marquet, A., and Larraga, V. (1997). Cloning of the gp63 surface protease of *Leishmania infantum*. Differential post-translational modifications correlated with different infective forms. Biochim. Biophys. Acta. *(1361*, 92–102.

Haile, S., Estevez, A.M., and Clayton, C. (2003). A role for the exosome in the in vivo degradation of unstable mRNAs. Rna. *9*, 1491–1501.

Hartmann, C., Hotz, H.R., McAndrew, M., and Clayton, C. (1998). Effect of multiple downstream splice sites on polyadenylation in *Trypanosoma brucei*. Mol. Biochem. Parasitol. *93*, 149–152.

Hermoso, T., and Jaffe, C.L. (1993). Phosphorylation of proteins in virulent promastigotes from *Leishmania major*. Biol. Res. *26*, 267–271.

Holzer, T.R., McMaster, W.R., and Forney, J.D. (2006). Expression profiling by whole-genome interspecies microarray hybridization reveals differential gene expression in procyclic promastigotes, lesion-derived amastigotes, and axenic amastigotes in *Leishmania mexicana*. Mol. Biochem. Parasitol. *146*, 198–218.

Horn, D. (2001). Nuclear gene transcription and chromatin in *Trypanosoma brucei*. Int. J. Parasitol. *31*, 1157–1165.

Hotz, H.R., Hartmann, C., Huober, K., Hug, M., and Clayton, C. (1997). Mechanisms of developmental regulation in *Trypanosoma brucei*: a polypyrimidine tract in the 3′-untranslated region of a surface protein mRNA affects RNA abundance and translation. Nucleic Acids Res. *25*, 3017–3026.

Houseley, J., LaCava, J., and Tollervey, D. (2006). RNA-quality control by the exosome. Nat. Rev. Mol. Cell. Biol. *7*, 529–539.

Hua, S., To, W.Y., Nguyen, T.T., Wong, M.L., and Wang, C.C. (1996). Purification and characterization of proteasomes from *Trypanosoma brucei*. Mol. Biochem. Parasitol. *78*, 33–46.

Huang, J., and van der Ploeg, L.H. (1991a). Maturation of polycistronic pre-mRNA in *Trypanosoma brucei*: analysis of trans splicing and poly(A) addition at nascent RNA transcripts from the hsp70 locus. Mol. Cell. Biol. *11*, 3180–3190.

Huang, J., and Van der Ploeg, L.H. (1991b). Requirement of a polypyrimidine tract for trans-splicing in trypanosomes: discriminating the PARP promoter from the immediately adjacent 3′ splice acceptor site. Embo J. *10*, 3877–3885.

Huang, Y., and Carmichael, G.C. (1996). Role of polyadenylation in nucleocytoplasmic transport of mRNA. Mol. Cell. Biol. *16*, 1534–1542.

Hug, M., Hotz, H.R., Hartmann, C., and Clayton, C. (1994). Hierarchies of RNA-processing signals in a trypanosome surface antigen mRNA precursor. Mol. Cell. Biol. *14*, 7428–7435.

Hummel, H.S., Gillespie, R.D. and Swindle, J. (2000). Mutational analysis of 3′ splice site selection during trans-splicing. J. Biol. Chem. *275*, 35522–35531.

Irmer, H., and Clayton, C. (2001). Degradation of the unstable EP1 mRNA in *Trypanosoma brucei* involves initial destruction of the 3′-untranslated region. Nucleic Acids Res. *29*, 4707–4715.

Ivens, A.C., Peacock, C.S., Worthey, E.A., Murphy, L., Aggarwal, G., Berriman, M., Sisk, E., Rajandream, M.A., Adlem, E., Aert, R., Anupama, A., Apostolou, Z., *et al.* (2005). The genome of the kinetoplastid parasite, *Leishmania major*. Science *309*, 436–442.

Johnson, P.J., Kooter, J.M., and Borst, P. (1987). Inactivation of transcription by UV irradiation of *T. brucei* provides evidence for a multicistronic transcription unit including a VSG gene. Cell *51*, 273–281.

Jones, A., Faldas, A., Foucher, A., Hunt, E., Tait, A., Wastling, J.M., and Turner, C.M. (2006). Visualisation and analysis of proteomic data from the procyclic form of *Trypanosoma brucei*. Proteomics *6*, 259–267.

Kapler, G.M., and Beverley, S.M. (1989). Transcriptional mapping of the amplified region encoding the dihydrofolate reductase-thymidylate synthase of *Leishmania major* reveals a high density of transcripts, including overlapping and antisense RNAs. Mol. Cell. Biol. *9*, 3959–3972.

Kelly, B.L., Nelson, T.N., and McMaster, W.R. (2001). Stage-specific expression in *Leishmania* conferred by 3′ untranslated regions of *L. major* leishmanolysin genes (GP63). Mol. Biochem. Parasitol. *116*, 101–104.

Komar, A.A., and Hatzoglou, M. (2005). Internal ribosome entry sites in cellular mRNAs: mystery of their existence. J. Biol. Chem. *280*, 23425–23428.

Konno, N., Ishii, M., Nagai, A., Watanabe, T., Ogasawara, A., Mikami, T., and Matsumoto, T. (2006). Mechanism of Candida albicans transformation in response to changes of pH. Biol. Pharm. Bull. *29*, 923–926.

Kooter, J.M., De Lange, T., and Borst, P. (1984). Discontinuous synthesis of mRNA in trypanosomes. Embo J. 3, 2387–2392.

Krobitsch, S., Brandau, S., Hoyer, C., Schmetz, C., Hubel, A., and Clos, J. (1998). *Leishmania donovani* heat shock protein 100. Characterization and function in amastigote stage differentiation. J. Biol. Chem. *273*, 6488–6494.

Lamontagne, J., and Papadopoulou, B. (1999). Developmental regulation of spliced leader RNA gene in *Leishmania donovani* amastigotes is mediated by specific polyadenylation. J. Biol. Chem. *274*, 6602–6609.

Langford, C.K., Ewbank, S.A., Hanson, S.S., Ullman, B., and Landfear, S.M. (1992). Molecular characterization of two genes encoding members of the glucose transporter superfamily in the parasitic protozoan *Leishmania donovani*. Mol. Biochem. Parasitol. 55, 51–64.

Larreta, R., Soto, M., Quijada, L., Folgueira, C., Abanades, D.R., Alonso, C., and Requena, J.M. (2004). The expression of HSP83 genes in *Leishmania infantum* is affected by temperature and by stage-differentiation and is regulated at the levels of mRNA stability and translation. BMC Mol Biol 5, 3.

LeBowitz, J.H., Smith, H.Q., Rusche, L., and Beverley, S.M. (1993). Coupling of poly(A) site selection and trans-splicing in *Leishmania*. Genes Dev. *7*, 996–1007.

LeBowitz, J.H. (1994). Transfection experiments with *Leishmania*. Methods Cell. Biol. *45*, 65–78.

Leifso, K., Cohen-Freue, G., Dogra, N., Murray, A., and McMaster, W.R. (2007). Genomic and proteomic expression analysis of *Leishmania* promastigote and amastigote life stages: the *Leishmania* genome is constitutively expressed. Mol. Biochem. Parasitol. *152*, 35–46.

Liang, X.H., Haritan, A., Uliel, S., and Michaeli, S. (2003). trans and cis splicing in trypanosomatids: mechanism, factors, and regulation. Eukaryot. Cell. *2*, 830–840.

Lim, L.P., Lau, N.C., Garrett-Engele, P., Grimson, A., Schelter, J.M., Castle, J., Bartel, D.P., Linsley, P.S., and Johnson, J.M. (2005). Microarray analysis shows that some microRNAs downregulate large numbers of target mRNAs. Nature. *433*, 769–773.

Liniger, M., Bodenmuller, K., Pays, E., Gallati, S., and Roditi, I. (2001). Overlapping sense and antisense transcription units in *Trypanosoma brucei*. Mol. Microbiol. *40*, 869–878.

Lopez-Estrano, C., Tschudi, C., and Ullu, E. (1998). Exonic sequences in the 5′ untranslated region of alpha-tubulin mRNA modulate trans splicing in *Trypanosoma brucei*. Mol. Cell. Biol. *18*, 4620–4628.

Luo, H., and Bellofatto, V. (1997). Characterization of two protein activities that interact at the promoter of the trypanosomatid spliced leader RNA. J. Biol. Chem. *272*, 33344–33352.

Luo, H., Gilinger, G., Mukherjee, D., and Bellofatto, V. (1999). Transcription initiation at the TATA-less spliced leader RNA gene promoter requires at least two DNA-binding proteins and a tripartite architecture that includes an initiator element. J. Biol. Chem. *274*, 31947–31954.

MacFarlane, J., Blaxter, M.L., Bishop, R.P., Miles, M.A., and Kelly, J.M. (1990). Identification and characterisation of a *Leishmania donovani* antigen belonging to the 70-kDa heat-shock protein family. Eur J Biochem *190*, 377–384.

Mair, G., Shi, H., Li, H., Djikeng, A., Aviles, H.O., Bishop, J.R., Falcone, F.H., Gavrilescu, C., Montgomery, J.L., Santori, M.I., Stern, L.S., Wang, Z., Ullu, E., and Tschudi, C. (2000a). A new twist in trypanosome RNA metabolism: cis-splicing of pre-mRNA. Rna. *6*, 163–169.

Mair, G., Ullu, E. and Tschudi, C. (2000b). Cotranscriptional cap 4 formation on the *Trypanosoma brucei* spliced leader RNA. J. Biol. Chem. *275*, 28994–28999.

Manning-Cela, R., Gonzalez, A., and Swindle, J. (2002). Alternative splicing of LYT1 transcripts in *Trypanosoma cruzi*. Infect. Immun. *70*, 4726–4728.

Marchesini, N., and Docampo, R. (2002). A plasma membrane P-type H(+)-ATPase regulates intracellular pH in *Leishmania mexicana amazonensis*. Mol. Biochem. Parasito.l *119*, 225–236.

Marchetti, M.A., Tschudi, C., Silva, E., and Ullu, E. (1998). Physical and transcriptional analysis of the *Trypanosoma brucei* genome reveals a typical eukaryotic arrangement with close interspersion of RNA polymerase II- and III-transcribed genes. Nucleic Acids Res. *26*, 3591–3598.

Martinez-Calvillo, S., Yan, S., Nguyen, D., Fox, M., Stuart, K., and Myler, P.J. (2003). Transcription of *Leishmania major* Friedlin chromosome 1 initiates in both directions within a single region. Mol. Cell. *11*, 1291–1299.

Martinez-Calvillo, S., Nguyen, D., Stuart, K., and Myler, P.J. (2004). Transcription initiation and termination on *Leishmania major* chromosome 3. Eukaryot. Cell. *3*, 506–517.

Matthews, K.R., Tschudi, C., and Ullu, E. (1994). A common pyrimidine-rich motif governs trans-splicing and polyadenylation of tubulin polycistronic pre-mRNA in trypanosomes. Genes Dev. *8*, 491–501.

Mattick, J.S. and Makunin, I.V. (2005). Small regulatory RNAs in mammals. Hum. Mol. Genet. *14 Spec No 1*, R121–132.

Mazumder, B., Seshadri, V., and Fox, P.L. (2003). Translational control by the 3′-UTR: the ends specify the means. Trends Biochem. Sci. *28*, 91–98.

McNicoll, F., Muller, M., Cloutier, S., Boilard, N., Rochette, A., Dube, M., and Papadopoulou, B. (2005). Distinct 3′-untranslated region elements regulate stage-specific mRNA accumulation and translation in *Leishmania*. J. Biol. Chem. *280*, 35238–35246.

McNicoll, F., Drummelsmith, J., Muller, M., Madore, E., Boilard, N., Ouellette, M., and Papadopoulou, B. (2006). A combined proteomic and transcriptomic approach to the study of stage differentiation in *Leishmania infantum*. Proteomics 6, 3567–3581.

Meijer, H.A., and Thomas, A.A. (2002). Control of eukaryotic protein synthesis by upstream open reading frames in the 5′-untranslated region of an mRNA. Biochem. J. *367*, 1–11.

Milhausen, M., Nelson, R.G., Sather, S., Selkirk, M., and Agabian, N. (1984). Identification of a small RNA containing the trypanosome spliced leader: a donor of shared 5′ sequences of trypanosomatid mRNAs? Cell. *38*, 721–729.

Minning, T.A., Bua, J., Garcia, G.A., McGraw, R.A., and Tarleton, R.L. (2003). Microarray profiling of gene expression during trypomastigote to amastigote transition in *Trypanosoma cruzi*. Mol. Biochem. Parasitol. *131*, 55–64.

Mishra, K.K., Holzer, T.R., Moore, L.L., and LeBowitz, J.H. (2003). A negative regulatory element controls mRNA abundance of the *Leishmania mexicana* Paraflagellar rod gene PFR2. Eukaryot. Cell. *2*, 1009–1017.

Monnerat, S., Martinez-Calvillo, S., Worthey, E., Myler, P.J., Stuart, K.D., and Fasel, N. (2004). Genomic organization and gene expression in a chromosomal region of *Leishmania major*. Mol. Biochem. Parasitol. *134*, 233–243.

Morlando, M., Greco, P., Dichtl, B., Fatica, A., Keller, W., and Bozzoni, I. (2002). Functional analysis of yeast snoRNA and snRNA 3′-end formation mediated by uncoupling of cleavage and polyadenylation. Mol. Cell. Biol. *22*, 1379–1389.

Murphy, W.J., Watkins, K.P., and Agabian, N. (1986). Identification of a novel Y branch structure as an intermediate in trypanosome mRNA processing: Evidence for *trans* splicing. Cell. *47*, 517–525.

Murray, H.W., Berman, J.D., Davies, C.R. and Saravia, N.G. (2005). Advances in leishmaniasis. Lancet *366*, 1561–1577.

Myler, P.J., Audleman, L., deVos, T., Hixson, G., Kiser, P., Lemley, C., Magness, C., Rickel, E., Sisk, E., Sunkin, S., Swartzell, S., Westlake, T., *et al.* (1999). *Leishmania major* Friedlin chromosome 1 has an unusual distribution of protein-coding genes. Proc. Natl. Acad. Sci. USA 96, 2902–2906.

Myung, K.S., Beetham, J.K., Wilson, M.E., and Donelson, J.E. (2002). Comparison of the post-transcriptional regulation of the mRNAs for the surface proteins PSA (GP46) and MSP (GP63) of *Leishmania chagasi*. J. Biol. Chem. *277*, 16489–16497.

Navarro, M., Cross, G.A., and Wirtz, E. (1999). *Trypanosoma brucei* variant surface glycoprotein regulation involves coupled activation/inactivation and chromatin remodeling of expression sites. Embo J. *18*, 2265–2272.

Noll, T.M., Desponds, C., Belli, S.I., Glaser, T.A., and Fasel, N.J. (1997). Histone H1 expression varies during the *Leishmania major* life cycle. Mol. Biochem. Parasitol. *84*, 215–227.

Nugent, P.G., Karsani, S.A., Wait, R., Tempero, J., and Smith, D.F. (2004). Proteomic analysis of *Leishmania mexicana* differentiation. Mol. Biochem. Parasitol. *136*, 51–62.

Papadopoulou, B., Roy, G., and Ouellette, M. (1994). Autonomous replication of bacterial DNA plasmid oligomers in *Leishmania*. Mol. Biochem. Parasitol. *65*, 39–49.

Parker, R., and Song, H. (2004). The enzymes and control of eukaryotic mRNA turnover. Nat. Struct. Mol. Biol. *11*, 121–127.

Peacock, C.S., Seeger, K., Harris, D., Murphy, L., Ruiz, J.C., Quail, M.A., Peters, N., Adlem, E., Tivey, A., Aslett, M., Kerhornou, A., Ivens, A., Fraser, A., Rajandream, M.A., Carver, T., Norbertczak, H., Chillingworth, T., Hance, Z., Jagels, K., Moule, S., Ormond, D., Rutter, S., Squares, R., Whitehead, S., Rabbinowitsch, E., Arrowsmith, C., White, B., Thurston, S., Bringaud, F., Baldauf, S.L., Faulconbridge, A., Jeffares, D., Depledge, D.P., Oyola, SO., Hilley, J.D., Brito, L.O., Tosi, L.R., Barrell, B., Cruz, A.K., Mottram, J.C., Smith, D.F., and Berriman, M. (2007). Comparative genomic analysis of three *Leishmania* species that cause diverse human disease. Nat. Genet. *39*, 839–847.

Pellé, R., and Murphy, N.B. (1993). Stage-specific differential polyadenylation of mini-exon derived RNA in African trypanosomes. Mol. Biochem. Parasitol. *59*, 277–286.

Perry, K.L., Watkins, K.P., and Agabian, N. (1987). Trypanosome mRNAs have unusual 'cap 4' structures acquired by addition of a spliced leader. Proc. Natl. Acad. Sci. USA *84*, 8190–8194.

Perumal, K., and Reddy, R. (2002). The 3′ end formation in small RNAs. Gene Expr. *10*, 59–78.

Pillai, R.S. (2005). MicroRNA function: multiple mechanisms for a tiny RNA? Rna *11*, 1753–1761.

Plewes, K.A., Barr, S.D., and Gedamu, L. (2003). Iron superoxide dismutases targeted to the glycosomes of *Leishmania chagasi* are important for survival. Infect. Immun. *71*, 5910–5920.

Purdy, J.E., Donelson, J.E., and Wilson, M.E. (2005). Regulation of genes encoding the major surface protease of *Leishmania chagasi* via mRNA stability. Mol. Biochem. Parasitol. *142*, 88–97.

Quijada, L., Soto, M., Alonso, C., and Requena, J.M. (2000). Identification of a putative regulatory element in the 3′-untranslated region that controls expression of HSP70 in *Leishmania infantum*. Mol. Biochem. Parasitol. *110*, 79–91.

Rivero-Lezcano, O.M., Chicharro, C., and Rivas, L. (1997). Acidic pH stress induces protein tyrosine phosphorylation in *Leishmania pifanoi*. Mol. Biochem. Parasitol. *84*, 123–129.

Robertson, C.D. (1999). The *Leishmania mexicana* proteasome. Mol Biochem Parasitol. *103*, 49–60.

Robinson, K.A. and Beverley, S.M. (2003). Improvements in transfection efficiency and tests of RNA interference (RNAi) approaches in the protozoan parasite *Leishmania*. Mol. Biochem. Parasitol. *128*, 217–228.

Rochette, A., McNicoll, F., Girard, J., Breton, M., Leblanc, E., Bergeron, M.G., and Papadopoulou, B. (2005). Characterization and developmental gene regulation of a large gene family encoding amastin surface proteins in *Leishmania* spp. Mol. Biochem. Parasitol. *140*, 205–220.

Sacks, D., and Kamhawi, S. (2001). Molecular aspects of parasite-vector and vector-host interactions in leishmaniasis. Annu. Rev. Microbiol. *55*, 453–483.

Salotra, P., Ralhan, R. and Sreenivas, G. (2000). Heat-stress induced modulation of protein phosphorylation in virulent promastigotes of *Leishmania donovani*. Int. J. Biochem. Cell. Biol. *32*, 309–316.

Sarkar, D. and Bhaduri, A. (1995). Temperature-induced rapid increase in cytoplasmic free $Ca2^+$ in pathogenic *Leishmania donovani* promastigotes. FEBS Lett. *375*, 83–86.

Saxena, A., Worthey, E.A., Yan, S., Leland, A., Stuart, K.D. and Myler, P.J. (2003). Evaluation of differential gene expression in *Leishmania major* Friedlin procyclics and metacyclics using DNA microarray analysis. Mol. Biochem. Parasitol. *129*, 103–114.

Saxena, A., Lahav, T., Holland, N., Aggarwal, G., Anupama, A., Huang, Y., Volpin, H., Myler, P.J., and Zilberstein, D. (2007). Analysis of the *Leishmania donovani* transcriptome reveals an ordered progression of transient and permanent changes in gene expression during differentiation. Mol. Biochem. Parasitol. *152*, 53–65.

Schimanski, B., Laufer, G., Gontcharova, L., and Gunzl, A. (2004). The *Trypanosoma brucei* spliced leader RNA and rRNA gene promoters have interchangeable TbSNAP50-binding elements. Nucleic Acids Res. *32*, 700–709.

Schimanski, B., Nguyen, T.N., and Gunzl, A. (2005). Characterization of a multisubunit transcription factor complex essential for spliced-leader RNA gene transcription in *Trypanosoma brucei*. Mol. Cell. Biol. *25*, 7303–7313.

Schroder, M., and Kaufman, R.J. (2005). ER stress and the unfolded protein response. Mutat. Res. *569*, 29–63.

Schurch, N., Hehl, A., Vassella, E., Braun, R., and Roditi, I. (1994). Accurate polyadenylation of procyclin mRNAs in *Trypanosoma brucei* is determined by pyrimidine-rich elements in the intergenic regions. Mol. Cell. Biol. *14*, 3668–3675.

Segal, S.P., Dunckley, T., and Parker, R. (2006). Sbp1p affects translational repression and decapping in Saccharomyces cerevisiae. Mol. Cell. Biol. *26*, 5120–5130.

Sereno, D., Roy, G., Lemesre, J.L., Papadopoulou, B., and Ouellette, M. (2001). DNA transformation of *Leishmania infantum* axenic amastigotes and their use in drug screening. Antimicrob. Agents Chemother. *45*, 1168–1173.

Siegel, T.N., Tan, K.S., and Cross, G.A. (2005). Systematic study of sequence motifs for RNA trans splicing in *Trypanosoma brucei*. Mol. Cell. Biol. *25*, 9586–9594.

Sogin, M.L., Hinkle, G., and Leipe, D.D. (1993). Universal tree of life. Nature. 36, 795.

Somanna, A., Mundodi, V., and Gedamu, L. (2002). Functional analysis of cathepsin B-like cysteine proteases from *Leishmania donovani* complex. Evidence for the activation of latent transforming growth factor beta. J. Biol. Chem. *277*, 25305–25312.

Storz, G., Altuvia, S., and Wassarman, K.M. (2005). An abundance of RNA regulators. Annu. Rev. Biochem. *74*, 199–217.

Sturm, N.R., Fleischmann, J., and Campbell, D.A. (1998). Efficient trans-splicing of mutated spliced leader exons in *Leishmania tarentolae*. J. Biol. Chem. *273*, 18689–18692.

Sutton, R.E., and Boothroyd, J.C. (1986). Evidence for trans splicing in trypanosomes. Cell. *47*, 527–535.

Svensson, F., Ceriani, C., Wallstrom, E.L., Kockum, I., Algranati, I.D., Heby, O., and Persson, L. (1997). Cloning of a trypanosomatid gene coding for an ornithine decarboxylase that is metabolically unstable even though it lacks the C-terminal degradation domain. Proc. Natl.Acad. Sci. USA *94*, 397–402.

Tarun, S.Z., Jr., and Sachs, A.B. (1996). Association of the yeast poly(A) tail binding protein with translation initiation factor eIF-4G. Embo J. *15*, 7168–7177.

Teixeira, S.M., Russell, D.G., Kirchhoff, L.V., and Donelson, J.E. (1994). A differentially expressed gene family encoding 'amastin,' a surface protein of *Trypanosoma cruzi* amastigotes. J. Biol. Chem. *269*, 20509–20516.

Turco, S.J., and Descoteaux, A. (1992). The lipophosphoglycan of *Leishmania* parasites. Annu. Rev. Microbiol. *46*, 65–94.

Ullu, E., Matthews, K.R., and Tschudi, C. (1993). Temporal order of RNA-processing reactions in trypanosomes: rapid trans splicing precedes polyadenylation of newly synthesized tubulin transcripts. Mol. Cell. Biol. *13*, 720–725.

Ullu, E., Tschudi, C., and Günzl, A. (1996). Trans-splicing in trypanosomatid protozoa. a.M.P.e. In D. F. Smith (ed.), Oxford University Press, Oxford, pp. 115–133.

Ullu, E., Tschudi, C., and Chakraborty, T. (2004). RNA interference in protozoan parasites. Cell. Microbiol. *6*, 509–519.

Van Hellemond, J.J., and Mottram, J.C. (2000). The CYC3 gene of *trypanosoma brucei* encodes a cyclin with a short half-life. Mol. Biochem. Parasitol. *111*, 275–282.

van Leeuwen, F., Kieft, R., Cross, M., and Borst, P. (1998). Biosynthesis and function of the modified DNA base beta-D-glucosyl-hydroxymethyluracil in *Trypanosoma brucei*. Mol. Cell. Biol.*18*, 5643–5651.

Vassella, E., Braun, R., and Roditi, I. (1994). Control of polyadenylation and alternative splicing of transcripts from adjacent genes in a procyclin expression site: a dual role for polypyrimidine tracts in trypanosomes? Nucleic Acids Res. 22, 1359–1364.

Walker, J., Vasquez, J.J., Gomez, M.A., Drummelsmith, J., Burchmore, R., Girard, I., and Ouellette, M. (2006). Identification of developmentally-regulated proteins in *Leishmania panamensis* by proteome profiling of promastigotes and axenic amastigotes. Mol. Biochem. Parasitol. *147*, 64–73.

Wang, C.C., Bozdech, Z., Liu, C.L., Shipway, A., Backes, B.J., Harris, J.L., and Bogyo, M. (2003). Biochemical analysis of the 20 S proteasome of *Trypanosoma brucei*. J. Biol. Chem. *278*, 15800–15808.

Webb, J.R., Campos-Neto, A., Skeiky, Y.A., and Reed, S.G. (1997). Molecular characterization of the heat-inducible LmST11 protein of *Leishmania major*. Mol. Biochem. Parasitol. *89*, 179–193.

Wek, R.C., Jiang, H.Y., and Anthony, T.G. (2006). Coping with stress: eIF2 kinases and translational control. Biochem. Soc. Trans. *34*, 7–11.

Wiese, M., Kuhn, D. and Grunfelder, C.G. (2003). Protein kinase involved in flagellar-length control. Eukaryot Cell 2, 769–777.

Worhunsky, D.J., Godek, K., Litsch, S., and Schlax, P.J. (2003). Interactions of the non-coding RNA DsrA and RpoS mRNA with the 30 S ribosomal subunit. J. Biol. Chem. *278*, 15815–15824.

Worthey, E.A., Martinez-Calvillo, S., Schnaufer, A., Aggarwal, G., Cawthra, J., Fazelinia, G., Fong, C., Fu, G., Hassebrock, M., Hixson, G., Ivens, A.C., Kiser, P., Marsolini, *et al.* (2003). *Leishmania major* chromosome 3 contains two long convergent polycistronic gene clusters separated by a tRNA gene. Nucleic Acids Res. *31*, 4201–4210.

Wouters, B.G., van den Beucken, T., Magagnin, M.G., Koritzinsky, M., Fels, D., and Koumenis, C. (2005). Control of the hypoxic response through regulation of mRNA translation. Semin. Cell. Dev. Biol. *16*, 487–501.

Wu, Y., El-Fakhry, Y., Sereno, D., Tamar, S., and Papadopoulou, B. (2000). A new developmentally regulated gene family in *Leishmania* amastigotes encoding a homolog of amastin surface proteins. Mol. Biochem. Parasitol. *110*, 345–357.

Yao, Y., Toth, C.R., Huang, L., Wong, M.L., Dias, P., Burlingame, A.L., Coffino, P., and Wang, C.C. (1999). Alpha5 subunit in *Trypanosoma brucei* proteasome can self-assemble to form a cylinder of four stacked heptamer rings. Biochem. J. *344 Pt 2*, 349–358.

Zalfa, F., Giorgi, M., Primerano, B., Moro, A., Di Penta, A., Reis, S., Oostra, B., and Bagni, C. (2003). The fragile X syndrome protein FMRP associates with BC1 RNA and regulates the translation of specific mRNAs at synapses. Cell. *112*, 317–327.

Zamora-Veyl, F.B., Kroemer, M., Zander, D. and Clos, J. (2005). Stage-specific expression of the mitochondrial co-chaperonin of *Leishmania donovani*, CPN10. Kinetoplastid Biol. Dis. *4*, 3.

Zeiner, G.M., Sturm, N.R., and Campbell, D.A. (2003). The *Leishmania tarentolae* spliced leader contains determinants for association with polysomes. J. Biol. Chem. *278*, 38269–38275.

Zhang, W.W., and Matlashewski, G. (1997). Loss of virulence in *Leishmania donovani* deficient in an amastigote-specific protein, A2. Proc. Natl. Acad. Sci. USA *94*, 8807–8811.

Zilberstein, D., Philosoph, H. and Gepstein, A. (1989). Maintenance of cytoplasmic pH and proton motive force in promastigotes of *Leishmania donovani*. Mol. Biochem. Parasitol. *36*, 109–117.

Zilberstein, D., and Gepstein, A. (1993). Regulation of L-proline transport in *Leishmania donovani* by extracellular pH. Mol. Biochem. Parasitol. *61*, 197–205.

Zilberstein, D., and Shapira, M. (1994). The role of pH and temperature in the development of *Leishmania* parasites. Annu. Rev. Microbiol. *48*, 449–470.

Zilka, A., Garlapati, S., Dahan, E., Yaolsky, V., and Shapira, M. (2001). Developmental regulation of HSP83 in *Leishmania*: transcript levels are controlled by the efficiency of 3′ RNA processing and preferential translation is directed by a determinant in the 3′ UTR. J. Biol.Chem. *276*, 47922–47929.

The *Leishmania* Proteome

4

Nicolas Fasel, Nathalie Acestor, Amal El Fadili-Kündig,
Iveth Gonzalez and Slavica Masina

Abstract

Genome projects for the most medically important protozoan parasites have been initiated in the hope that their sequencing will help to understand the biology of the parasites and to identify new targets for urgently needed drugs, or for the development of vaccines. The completion of the genome sequence of *Leishmania major* corresponds to the achievement of one important milestone in the quest for such information. Data mining of the sequences of the related genomes of *Trypanosoma brucei* (*T. brucei*), *T. cruzi* and for other *Leishmania* species such as *L. braziliensis*, *L. infantum* and *L. mexicana* provides additional tools for the identification, comparative analysis and functional definition of proteins. Taken together, this mass of information opens the door for large-scale proteomic studies to dissect not only protein expression, regulation and function, but also biological processes involving complex protein–protein interactions, post-translational regulation events and post-translational modifications.

Considering that gene expression in trypanosomatids is largely regulated at the post-transcriptional level, a systematic analysis of proteins at a high throughput level is an obvious approach. Recent technological developments allowing the rapid identification of proteins by peptide fingerprinting and mass spectrometry have opened the door for such analysis.

The potential size of the *Leishmania* proteome

The most remarkable finding in trypanosomatids is the organization of their transcription units, whereby the predicted genes are grouped into large transcription clusters on the same DNA strand. These clusters can contain 100 genes and can be arranged either 'divergently' or 'convergently' into polycistronic units which are processed to monocistronic mRNA by polyadenylation and *trans*-splicing of a 'mini-exon' or 'splice-leader' (SL) to the primary transcript. As a consequence of this organisation, regulation of gene expression in *Leishmania* takes place mainly at the post-transcriptional level and can occur by controlling maturation, stability and translation of the processed transcripts (Clayton, 2002; Campbell *et al.*, 2003; Palenchar *et al.*, 2006). In this respect, numerous genes coding for proteins with RNA binding motifs have been identified in the genome, supporting the notion that a specific machinery involved in post-transcriptional mechanisms regulates gene expression.

Considering that transcription in trypanosomatids generates very large polycistronic transcripts containing up to 100 distinct genes as deduced from genome sequence analysis (see Chapter 2), precise regulation of individual gene expression cannot efficiently function via RNA polymerase promoter control. Thus, control of gene expression has been shown to occur at multiple levels, such as *trans*-splicing, polyadenylation, mRNA stability, transcript elongation, RNA translation and protein stability (Huang *et al.*, 1991a; Huang *et al.*, 1991b; Kapotas *et al.*, 1993; Graham *et al.*, 1995; Lee, 1998).

Not surprisingly, in the light of the recent genome sequencing data, very few RNA polymerase promoters were identified in trypanosomatids. In particular, no RNA polymerase II promoters were described for *Leishmania* to date (Martinez-Calvillo *et al.*, 2003; Worthey *et al.*, 2003a; Monnerat *et al.*, 2004), with the notable exception of the splice leader RNA promoter (Luo *et al.*, 1999; Mair *et al.*, 2000b; Gilinger *et al.*, 2001). It has been proposed that AT-rich regions could serve as entry sites for RNA polymerase II (Tosato *et al.*, 2001; Belli *et al.*, 2003). RNA polymerase I promoters were described in the rDNA locus of several *Leishmania* species (Gay *et al.*, 1996; Uliana *et al.*, 1996; Yan *et al.*, 1999) likewise in *T. cruzi* (Figueroa-Angulo *et al.*, 2003). Other exceptions include, for example, the RNA polymerase I promoters for the variant surface glycoprotein (VSG) and the procyclic acidic repetitive protein (PARP) genes in *T. brucei* (Rudenko *et al.*, 1992).

In any case, the fact that the main control of gene expression occurs at the post-transcriptional level precluded the identification of genes differentially regulated in the life cycle by RNA profiling, and puts more emphasis on a thorough analysis of differential gene expression at the protein level.

From the 32.8-megabase haploid genome of *L. major* (Ivens *et al.*, 2005), it can be estimated that there are 8272 protein-coding genes, of which 36% have a putative function when a cut off size of 100 amino acids is chosen as a mean value for a given coding region. As for other organisms, this number is clearly an underestimate of the total number of proteins present in *Leishmania* due to the importance of post-translational modifications (PTMs). In the human genome, for example, it is currently estimated that the number of genes ranges from 20,000 to 25,000. Complexity is generated at several levels, mainly attributed to alternative splicing and PTMs. For example, in the human chromosome 21, it was estimated that up to 75% of the primary transcripts are alternatively processed.

In *Trypanosoma*, alternative trans-splicing is thus far considered as a rare event, with only few examples of alternative RNA processing (Manning-Cela *et al.*, 2002). However, it cannot be excluded that RNA processing may participate in the generation of additional complexity of the *Leishmania* proteome. Importantly, the old conviction that *cis*-splicing does not exist in the trypanosomatid kingdom has been proven wrong. Extensive analysis of the -poly(A) polymerase (PAP) gene of *T. brucei* and *T. cruzi* revealed the existence of *cis*-splicing introns in these organisms (Mair *et al.*, 2000a). In addition, these introns are apparently processed by a mechanism similar to the classical *cis*-splicing mechanism, which obeys the GT/AG rule. More recently, characterisation of the *cis*-spliceosomal U1 small nuclear ribonucleoprotein (U1 snRNP) as well as cloning of the gene encoding the U1 small nuclear RNA (snRNA) of *T. brucei* has been described (Palfi *et al.*, 2002). It is thus likely that both *cis*- and *trans*-splicing are ancient acquisitions of the eukaryotic cell, despite the fact that *cis*-splicing has not, as yet, been described in *Leishmania*.

More importantly, it is likely that PTMs of proteins is, and will be, even more of a major factor contributing to the complexity of the *Leishmania* proteome and other trypanosomatid species. As for higher eukaryotes, several proteins in *Leishmania* are cleaved thus eliminating signal sequences, pro-peptides and initiator methionines, or are cleaved, by specific parasitic proteases. The *Leishmania* proteins can also be modified by the addition of simple chemical groups such as acetyl, methyl and phosphoryl groups as well as more complex molecules, such as sugars, lipids and glycolipids. Finally, proteins can be subject to ubiquitination, hyperoxidation transformation of the sulfhydryl group into sulfenic, sulfinic or sulfonic acid or nitrosylation. Today, more than two hundred different types of PTMs are known with others surely yet to be discovered. Some of these PTMs may be specific to higher eukaryotes and it cannot be excluded that others are specific to trypanosomatids or at least to lower eukaryotes. In humans, when combining the complexity generated by alternative splicing together with that produced by PTMs, it appears that the number of different protein molecules expressed by the 20,000 to 25,000 protein-encoding genes is probably close to one million. It is clear that this number is still a matter for discussion and that the total number of proteins in *Leishmania* will be proportionally smaller considering that RNA processing is not a major mechanism to generate alternative mRNAs. In other lower eukaryotes such as yeast, 6280 genes have an estimate number of less than 5600 true open reading frames. It is also unclear to what extent the number of short proteins (less than 100 codons) has been underestimated.

Another mechanism that could influence the total number of proteins possibly synthesized in *Leishmania* and in *Trypanosoma* is antisense transcription of the large poly-cistronic transcription units (Kapler *et al.*, 1989; Liniger *et al.*, 2001; Belli *et al.*, 2003; Monnerat *et al.*, 2004). Although the low-abundance antisense transcripts are polyadenylated, contain open reading frames and could code for polypeptides, protein synthesis from these RNAs has not, as yet been demonstrated. If these RNAs are bona-fide mRNAs that can be synthesized by both strands, it would definitively increase the size of the coding potential of the genome and participate directly in an increased complexity of the proteome. In any case, it is obvious that the number of different protein molecules in *Leishmania* will be less than the number predicted in humans but will still be several factors over the 8300 predicted protein-coding genes.

However, obtaining an estimate of the complexity of the *Leishmania* proteome is presently not a major priority. It is more important to focus our efforts on specific biological questions and to obtain the appropriate tools to answer these questions. Stage-specific expression, virulence factors and specific pathways and organelles (such as the glycosome, the flagellar pocket and the kinetoplast) are among the important aspects of Trypanosomatidae and *Leishmania* biology.

A genome-wide approach to analyse gene expression has been applied to *Leishmania* and in other trypanosomatids using DNA arrays (Saxena *et al.*, 2003; Akopyants *et al.*, 2004; Almeida *et al.*, 2004; Duncan, 2004; Duncan *et al.*, 2004). These data showed large discrepancies between the different studies which could be due the size, number, or the type of probes used (genomic, cDNA, or oligonucleotide). Lately, a differential interspecies analysis was performed for the transcriptome of *L. mexicana* using high density whole genome oligonucleotide microarrays designed from the genome sequence of *Leishmania* (Holzer *et al.*, 2006). This study shows that only 3.5% of all genes could be differentially regulated between promastigote and lesion derived amastigotes. This result is in agreement with the data of Akopyants *et al.*, 2004 and with the expression profiling of *T. brucei* (Diehl *et al.*, 2002; Brems *et al.*, 2005). Taken together, it appears that only a relative small fraction of the *Leishmania* genes are differentially regulated between promastigotes and amastigotes and further confirms that proteomics is a necessary approach to investigate the differential expression at the RNA level during the parasite life cycle and to study differential translational regulation and post-translational modifications.

In conclusion, in addition to certain genes, which have been shown to be regulated at the transcription or mRNA stability level which could account for only a small percentage of detected differentially regulated genes, the systematic analysis of proteins is an important technological component to identify specific proteins in particular stage specific proteins, and to define the importance of PTMs in their function. Thus, it will eliminate the biological restriction due to the mode of regulation and take into account only the final product(s). A systematic proteomic analysis of the different stages of the parasite will likely increase the list of differentially expressed genes. Preliminary results already point in this direction. These gene products are likely to be the most interesting to follow, as well as gene products and pathways which are unique to *Leishmania* and *Trypanosoma* in the perspective of the design of new drugs and new antigens for vaccine development. In addition, protein profiling will help to solve more specific questions such as virulence factors, drug resistance and structural features concerning trypanosomatids.

Practical concerns with analysing the *Leishmania* proteome

Sensu stricto, the proteome is defined as the total number of proteins present in a cell at a defined time in a defined state. Proteome analysis relies mainly on the ability to separate individual proteins from a complex mixture and to identify these proteins by well known methods including low-throughput sequencing using Edman's degradation or higher-throughput proteomic techniques based on mass spectrometry (MS). The latter includes peptide mass fingerprinting on simple instruments or on instruments such as an MS/MS that is capable of more than one round of mass spectrometry.

Most of the basic proteomic technologies are described in reviews and in detailed protocols and thus will be mentioned only briefly here. Two-dimensional gel electrophoresis (2DE) allows identification of the relative masses of proteins and their isoelectric point. Proteins are separated according to net charge in the first dimension (isoelectric focusing, IEF) and by molecular weight in the second dimension (SDS-PAGE) (O'Farrell *et al.*, 1975). Separation by 2DE is extremely efficient in resolving a large number of proteins. Following staining of the gel, proteins are excised and then cleaved by specific proteases (often trypsin). The resulting proteolytic peptides (most of the time tryptic peptides) can then be extracted from the gel slice, have their molecular masses and sequences determined using mass-spectrometry-based sequencing methods, followed by protein identification using database searching. However, we should keep in mind that 2DE gels are extremely labour-intensive, requiring several separations for comparative analyses and are inadequate to resolve several classes of proteins such as acidic or basic proteins, very large or small proteins, extremely hydrophobic and membrane-associated proteins, and low abundance proteins. Despite the introduction of specialized pH gradients allowing for increased resolution (Bjellqvist *et al.*, 1993), a high percentage of proteins within an organism cannot be mapped and identified using the 2DE-MS approach. Alternatively, one dimension gel electrophoresis, or separation by liquid chromatography can be used depending on parameters such as the abundance of a protein following fractionation or prior affinity purification.

In the end, identification always comes down to the comparison of experimentally obtained masses with theoretical masses of proteins present in databases. Thus, the identification of a given protein relies on bioinformatic tools to determine possible sequence matches, functional domains, important amino acid motifs and evolutionary relationships between proteins. Several software packages are available and new developments should allow the representation and comparison of liquid-chromatography-mass spectrometry data. For example, MSight, now available as a free software (www.expasy.org/MSight/), can display and compare LC-MS and LC-MS/MS data as two-dimensional density maps (Palagi *et al.*, 2005). This software is designed to construct and manipulate MS images, allowing the user to have an immediate overview of complex datasets. More importantly, it allows the analyser to exploit the correlation between mass and elution time and to align and compare samples in order to accurately quantify peak intensities and thus determine significant differences (Palagi *et al.*, 2005).

Protein profiling comparison involves the identification and quantitation of protein components in various biological systems. It will always remain difficult to obtain a global comprehensive proteome because of its enormous complexity and dynamic range. It is only by sub-proteomic fractionations that a simplification of the proteome will provide practical tools towards the characterization of the entire proteome. Although limited, stable isotope labelling technology, by both metabolic and chemical methods, has been the most commonly used approach for proteome-wide profiling. Different techniques have been developed such as differential staining of gels with fluorescent dyes, tagging and chemical modification methods such as isotope-coded affinity tags (ICATs) or labelling with 'heavy' amino acids as SILAC (stable isotope labelling with amino acids in cell culture).

Isotope-coded affinity tagging (ICAT) reagents were developed for the quantification and identification of proteins within a complex proteome (Gygi *et al.*, 1999). They usually consist of a thiol reactive group, a biotin affinity tag for isolation and purification of tagged peptides and a linker region consisting of either eight hydrogens (light tag) or eight deuteriums (heavy tag) (Gygi *et al.*, 1999; Moseley, 2001; Adamczyk *et al.*, 2002). Briefly, proteins are labelled either with the light or the heavy ICAT reagents, mixed, digested by trypsin and then separated and analysed by LC/MS/MS. Peptides are quantified based on relative signal intensities of identical peptides labelled with either the heavy or light form of the ICAT reagent. This approach circumvents the need for gel-based separation (Moseley, 2001; Lee *et al.*, 2002; Schmidt *et al.*, 2004). For kinetoplastids, ICAT reagents have been used successfully to compare the developmental stages of *T. cruzi* (Paba *et al.*, 2004). Thus, ICAT labelling can be applied to solving problems such as differential gene expression between life cycle stages, effect of stress or drugs, as well as determination of virulence factors in virulent versus attenuated strains of *Leishmania* parasites.

Isotope-coded affinity tagging reagents have undergone several improvements (Smolka *et al.*, 2001; Smolka *et al.*, 2002; Patel *et al.*, 2004; van Swieten *et al.*, 2005) including the development of fluorescent (visible) ICAT reagents (Bottari *et al.*, 2004; Gu *et al.*, 2004), cleavable ICAT reagents (Hansen *et al.*, 2003; Li *et al.*, 2003; Li *et al.*, 2004; Yi *et al.*, 2005) and incorporating different reactive moieties (Righetti *et al.*, 2003) (Moseley, 2001; Smolka *et al.*, 2002). The major limitation of ICAT reagents is the requirement of a free cysteine on the protein to be labelled. Therefore, complementary approaches must be used to maximize proteome coverage.

A simple approach based on the incorporation of amino acids with substituted isotope nuclei has been applied in various experimental designs and is promised to widen the application for cells in culture. SILAC is a simple and accurate method for mass spectrometric–based quantitative proteomics (Ong *et al.*, 2002). It relies on the metabolic incorporation of a quantitative label. Briefly, two groups of cells are grown in media differing only by the presence of a 'light' or a 'heavy' form of a particular amino acid (e.g. L-leucine versus deuterated L-leucine). Through the use of essential amino acids, the cells are forced to use the particular labelled or unlabelled form of the amino acid. Thus, with each cell division, the cell population replaces at least half of the original form of the amino acid. Unlabelled and labelled samples can be combined prior to lysis of the cells and treated as a single sample in all subsequent the steps and in the final quantitative analysis. The quantitative analysis is based on standard protocols such as peptide mass fingerprinting. Peptides from one sample are distinguished from peptides of the other sample by different masses due to the incorporation of 'heavy' amino acids.

Similarly, specific protocols to study protein modifications have been developed for various organisms in particular for mammalian cells. Specialized methods are already available to study phosporylation (phosphoproteomics) and glycosylation (glycoproteomics). Such methods could also be applied to the analysis of *Leishmania* proteomes. Identification of phosphorylated proteins remains a difficult task despite technological advances in protein purification methods and mass spectrometry. Identification of tyrosine-phosphorylated proteins can be obtained by coupling SILAC to mass spectrometry. This strategy can be used to identify potential kinase substrates without prior knowledge of the signalling pathways and can also be applied to profiling of specific kinases in cells. Due to its sensitivity and general applicability, this approach will be useful for investigating signalling pathways in a global fashion and for using phosphoproteomics for functional annotation of genomes. Although it may not be of major concern for *Leishmania* proteomics, the major limitation of SILAC-labelling is the difficulty of labelling tissue samples, where metabolic incorporation of labelled amino acids is required.

In addition to these simple techniques, low resolution techniques such as circular dichroism, Fourier transform infrared spectroscopy and small angle X-ray scattering can be used to study the secondary structure of proteins, together with X-ray crystallography and nuclear magnetic resonance to characterize the three-dimensional structure of peptides and proteins. To identify and quantify all the proteins found in cells, tandem mass spectrometry combined with reverse phase chromatography or 2-DE is used. Protein–protein interaction studies rely on widely used techniques such as affinity and immuno-affinity chromatography, yeast two-hybrid techniques and fluorescence resonance energy transfer (FRET). Mapping the location of proteins and protein–protein interactions in whole cells during key cell events is a new branch of proteomics gaining increasing importance. Determination of the location of labelled proteins or protein complexes in an intact cell, frequently correlated with images of cells from light-based microscopes can now be performed by X-ray tomography and optical fluorescence microscopy. It is evident that proteomic studies in *Leishmania* will take advantage of all these technological developments.

Several of the basic approaches described here have been used in the first proteomic analyses of *Leishmania* but it is clear that the level of analysis, e.g. the numbers of identified proteins reached for *Leishmania* is far below that reported for other organisms. Furthermore, problems specific to *Leishmania* will have to be solved before we can rely on reproducible reference maps. Several factors influence directly the proteome of a cell. Among the most obvious parameter to consider for *Leishmania*, are the species, the stage of the parasite in the life cycle, the growth medium and its serum content. Additional factors should also be considered such as concentration of media components, external pressure, temperature, pH, oxygen pressure, oxidative state and the parasite density. Furthermore, variations are known to occur between *Leishmania* isolates. It is known that, in different *Leishmania* strains, proteins such as histone H1, have different isoforms (due to variations in a number of repeated motifs) (Belli *et al.*, 1999) which will have different migration patterns in 2DE or different purification properties. Furthermore, infectivity can be different between one isolate and another (Ritter *et al.*, 2004). In addition to variability in the *Leishmania* isolate, species or genetic polymorphisms, the preparation of the sample and the conditions of separation should not be underestimated. *Leishmania* is known to express a high level of glycosylated cell surface antigens such as lipophosphoglycans, proteophosphoglycans, GPI-anchored proteins, glycosinositol-phospholipids and N-glycan glycoproteins. These particular structures can affect directly or indirectly the separation protocols and the resolution on 2DE gels or liquid chromatography. *Leishmania* is also known to express a high level of active proteolytic enzymes mainly in the infectious stages of the life cycle. The presence of specific protease inhibitors in the sample preparation is thus required. The *Leishmania* proteome is also known to have an extensive over-representation of spots in the acidic regions coupled to a low representation of basic proteins. To solve this kind of specific problem, modification of current protocols was necessary (Brobey *et al.*, 2007). Finally, it is known from researchers in the field that independent laboratories can obtain different proteomic patterns using similar samples. Thus, it is likely that, in the next years, we will have to handle several proteomes and it will be necessary to find ways in solving this problem to avoid that in every proteomic experiment we rediscover a *Leishmania* proteome and accumulate the same kind of data.

However, despite the difficulties, some features are quite similar in all the data reported thus far. Among the most often encountered features of proteomic analysis is the presence of proteolytic processing of specific proteins and the difficulty in resolving some spots in 2DE gels. In addition to the problems of proteolytic cleavage, the presence of cell surface complex glycosylated molecules is also an important issue, which is not yet fully resolved.

Applications of proteomics to *Leishmania*

Protein profiling has been applied to different aspects of *Leishmania* biology such as stage specific gene expression, characterization of immunodominant antigens, drug resistance, identification of virulence factors and the editosome. These studies were the first of their kind and it is likely that researchers will make use of alternative protein separation and specific fractionation techniques in conjunction with MS identification to obtain a more detailed picture of the *Leishmania* proteome. Such an approach has been recently reported (Foucher *et al.*, 2006). Differential solubilization of membranes using digitonin was applied to *Leishmania* parasites. This protocol facilitated the resolution and depth of comparative studies by removing abundant proteins such as tubulins and associated proteins of the cytoskeleton (Foucher *et al.*, 2006). Most of the data obtained thus far represent only the tip of the iceberg of the *Leishmania* proteome and additional efforts will be required to obtain the remainder of the *Leishmania* proteome or at least subproteome. Important aspects of *Leishmania* biology have not taken advantage of proteomics. For example, PTMs are known to be important in trypanosomatids. In addition to PTM of membrane proteins, which have been extensively studied, it is expected that signalling pathways depending on phosphorylation and specific kinases will benefit from phosphoproteomics approaches developed in higher eukaryotes. Phosphoproteomics will play an important role in solving, on a wide scale, biochemical pathways in *Leishmania* and *Trypanosoma* including the dissection of pathways involved in differentiation of the parasites between specific stages of the life cycle.

Stage-specific gene expression

Among the different domains of proteomic applications, studying differential gene expression in the various stages of the life cycle of the parasite is the most obvious, though likely to be one of the most challenging issues. Several publications have reported the differential expression of several proteins using 2DE gels and MS identification using different *Leishmania* species as model organisms.

As a start, most of these studies rely on reference maps. These maps correspond to a map of as many proteins as possible from an organism (Cordwell *et al.*, 2001). The map is then used as a reference for experiments aimed at comparing protein profiles of different life stages of *Leishmania* or isolates with varying virulence properties (Acestor *et al.*, 2002; Drummelsmith *et al.*, 2003). Several reference maps have been completed for other pathogens. These maps cover a limited percentage of the total number of open reading frames. For example, in *Haemophilus influenzae*, 33% of the open reading frames are covered (Bumann *et al.*, 2001). The success of reference mapping is a direct result of the completion of sequencing. They must give a reproducible and interpretable image of the proteome of *Leishmania* parasites at a defined time, and strict criteria in the preparation of the sample must be followed. These reference maps may be of two types. They could rely simply on representations of 2DE gels obtained in different conditions, or on mass peptide fingerprints as reported for *T. cruzi* (Atwood *et al.*, 2005). Only the former type of reference mapping has been used, thus far, for *Leishmania*. The gathered pattern is directly influenced by the conditions used and the analysis system. Even so, it is not easy to compare reference maps from one laboratory to maps from another laboratory and as mentioned earlier, there is a high risk that the proteome is rediscovered in every experiment. Furthermore, it has become apparent from proteomic studies in other model organisms that this type of analysis is quite inefficient, in that the number of successfully identified and quantified peptides is about an order of magnitude lower than the number of detectable peptides present in the sample (Li *et al.*, 2004) and that this representation is biased toward higher abundance proteins. In several of the reports concerned with *Leishmania* proteomes, only 2DE in narrow pI ranges (5–7) of soluble lysates were performed, thus representing mainly highly abundant proteins and eliminating a large proportion of possibly differentially regulated proteins. Even so, we can still gather important information from an estimate of the percentage of differentially regulated genes and on potential new target genes.

In the transformation from the promastigote to the intracellular amastigote stage, only a few proteins have been described as having a pivotal functional role. One key function in the life cycle seems to be played by heat shock protein 90 (HSP90) (Clos *et al.*, 2001; Graefe *et al.*, 2002). From the information accumulated over the last years on *Leishmania* parasites, it can be expected that specific classes of proteins would be changed in their expression and could complement the information gathered on HSP90. The metabolism and the morphology of *Leishmania* change when the parasite enters into the macrophage, loses it flagellum and differentiates into a non-motile form surviving at a low pH and at 37°C. Thus, it is not too surprising that most of the differentially regulated proteins fall into groups of functionally related proteins including those implicated in the cytoskeleton and cell membrane, stress response, energy and amino acid metabolism, cell cycle and proliferation (Bente *et al.*, 2003; Walker *et al.*, 2006b). In addition to these classes of proteins, translation initiation and elongation factors as well as ribosomal proteins were identified, some of them being uniquely expressed in the amastigote stage (Bente *et al.*, 2003; Nugent *et al.*, 2004), indicative of an extensive remodelling of the protein synthesis machinery.

A recent study on *L. panamensis* brought additional information on differential expression during development (Walker *et al.*, 2006b). Comparison of protein profiling between *L. panamensis* promastigotes and axenically grown amastigotes allowed, in a sample representative of only 8 to 10% of the 8300 predicted genes, the identification of 75 differentially regulated protein spots in amastigotes, of which, 24 were unique to this stage, the other one being overexpressed by a factor up to 5-fold. Among the identified peptide sequences, the most interesting was the amastigote expressed cysteine synthase which is found only in plants and microorganisms. Other interesting novel proteins expressed in a stage-specific manner were identified after pre-factionation using different concentrations of digitonin (Foucher *et al.*, 2006). This serial extraction protocol was used to analyse subcellular fractions of *Leishmania infantum* promastigotes and amastigotes. This approach increased proteome coverage and could also be used to confirm that the presence of the given protein in a fraction was predictive of its cellular localization. In addition to previously described proteins, several new proteins, many of them being hypothetical proteins, could be attributed either to the promastigote or amastigote cell stage (Foucher *et al.*, 2006). Interestingly, some proteins were found in unexpected fractions. For example, enolase was found not only in the cytosol as predicted but could also have an additional localization, possibly in the glycosome. Although enolase has been shown to be present in the cytosol of *T. brucei* (Hannaert *et al.*, 2003), another localization cannot be formally excluded. Furthermore, enolase was shown to be present at the cell surface in yeast, where it could have a different role suggesting that enolase could have two functions (Lopez-Villar *et al.*, 2006). The 'enolase case' is a good example where the pre-fractionation approach coupled to protein identification not only gives a better representation of the proteome but also opens a new field of investigation.

In each proteomics study reported thus far, proteolytically cleaved polypeptides have been identified in high abundance in certain parasitic stages (Acestor *et al.*, 2002; Bente *et al.*, 2003; Drummelsmith *et al.*, 2003; Nugent *et al.*, 2004; Walker *et al.*, 2006b). For example, low molecular weight β-tubulin and HSP70 spots were present in higher abundance in amastigotes even in the presence of an extensive protease inhibitor cocktail and urea. In contrary, HSP60 is found in both life stages, mainly in a complete form in the amastigote, but as several proteolytic fragments in the promastigote form (Foucher *et al.*, 2006). Although it is tempting to suggest a stage-specific processing of the HSP60 protein, it is questionable if this finding has biological relevance for *Leishmania* survival and differentiation.

Identification of immunodominant antigens

In many previous studies, several *Leishmania* proteins have been purified using specific antisera from patients. Only groups having access to protein sequencing facilities using Edman degradation were able to obtain the correct sequence and hence identify the parasite antigen. Even so, protein sequencing encounters major difficulties due to blocked N-termini. Furthermore, sequences were often only partial with several uncertainties. Considering that the *Leishmania* genome was not complete during these early investigations, the identification of the correct gene by cloning and sequencing was necessary to confirm the identity of the polypeptide and verify the amino acid sequence. These approaches were taking a considerable amount of time and did not exclude the possibility of problems in the identification of the expected gene. A proteomic approach for these kind of studies is more straightforward. The immunologic approach could be extremely efficient in the identification and characterization of other antigens in *L. infantum* and in other *Leishmania* species using sera from promastigote immunized rabbits (Dea-Ayuela *et al.*, 2006), patients, or other mammalian hosts such as dogs harbouring different pathologies. Pointing into this direction, a wider range study on the host immune response of *Leishmania* antigens was performed by proteome serology using sera from patients and their corresponding *L. donovani* parasite isolates (Forgber *et al.*, 2006). This interesting study shows that, although some antigens are shared among different isolates, there are extensive differences and no immunodominant antigens. Although the genetic polymorphism of the major histocompatibility complex of the individuals and its importance in the control of the T cell response could explain these results, it is nevertheless a valid demonstration of the extensive use of the parasite genome information. This study shows also differences in the antigenicity of the parasites from different but neighbouring endemic regions. As discussed by the authors, it suggests that antigenic drifts in the parasites are likely to be essential for immune evasion. This study could serve as a base to identify surrogate markers for resistance or susceptibility, which is essential in the development of vaccines and provides promising tools in the monitoring of the immune response and diagnosis.

Drug resistance

One important problem in infectious diseases is the ease with which several parasites can manipulate gene expression. Similarly to cancer cells, such a capacity has a direct incidence on the resistance of parasites to specific drugs. Drug resistance is an increasing problem in leishmaniasis. In some regions of the world, up to 80% of cases exhibit clinical resistance to certain drugs. It is therefore important to start comparative studies of various isolates harbouring different drug resistance phenotypes. Alternatively, isolates resistant to different drugs could be compared which will be helpful to dissect the complexity of drug resistance pathways.

Several *Leishmania major* isolates with varying resistance to methotrexate were compared by protein profiling (Drummelsmith *et al.*, 2004). Numerous differentially expressed proteins were identified in *L. major* parasites that have been either shocked with, or selected *in vitro* for high level resistance to the model antifolate drug methotrexate. As expected, some were associated with the stress response, whereas others were found to be overexpressed due to genetic linkage to primary resistance mediators present on DNA amplicons. Several proteins not previously associated with resistance were also identified. This study revealed novel roles for methionine adenosyltransferase and S-adenosylmethionine in methotrexate resistance strains.

Methotrexate resistance was correlated to increased cellular S-adenosylmethionine levels. This thorough comparative proteomic analysis highlights the variety of responses required for drug resistance to be achieved. Similarly, preliminary characterization of the serine hydroxymethyltransferase (SHMT) was obtained by digitonin pre-fractionation (Foucher *et al.*, 2006). This study was followed by a more detailed characterization of two SHMT isoforms showing that they were preferentially expressed in the amastigote stage of the life cycle, that RNA of one of the isoform can be modulated by glycine, serine and folate and that overexpression of the other isoform increased resistance to antifolate methotrexate (Gagnon *et al.*, 2006).

It is likely that this kind of analysis will gain importance in the coming years for a better understanding of the drug resistance mechanism and for the comparison of various human isolates corresponding to species with differential resistance to drugs. Comparative analysis of genetically related pairs of antimony sensitive and antimony-resistant *L. donovani* strains isolated from kala-azar patients highlighted a number of differentially regulated proteins. Interestingly, drug-induced cell death was modified in the resistant parasite (Vergnes *et al.*, 2007). Two proteins, HSP83 and the small kinetoplastid calpain related protein were proposed to be implicated in the drug-resistant cell death phenotype (Vergnes *et al.*, 2007).

Identification of virulence factors

For many years, differential infectivity has been ascribed to various parasite isolates. Only in some instances, was it possible to attribute a role for a specific gene to increased or decreased infectivity. In most cases, it was impossible to identify the genes involved in differential infectivity or pathogenicity. Furthermore, most of the analyses were limited in the number of genes or pathways, which could be identified. A proteomic approach was thus applied as a possible solution to these problems.

Parasites of the *Leishmania Viannia* subgenus are the major causative agents of mucocutaneous leishmaniasis (MCL) in the New World. MCL is characterized by parasite dissemination (metastasis) from the original cutaneous lesion to form debilitating secondary lesions in the nasopharyngeal mucosa. A protein profiling-approach was used to identify gene products associated with metastasis using laboratory clones of *L. (V.) guyanensis* with stable phenotypes ranging from highly metastatic (M^+) through to infrequently metastatic (M^+/M^-) and non-metastatic (M^-). These data were confirmed using human clinical isolates (Walker *et al.*, 2006a). The application of 2DE and MS to comparative analyses of protein expression in M^+ and M^- laboratory clones and isolates generated reference proteome maps for *L. guyanensis*, identified landmark proteins and, most importantly, enabled the identification of two differentially expressed proteins, EF-1β and TXNPx, potentially linked to the metastatic phenotype. The reported metabolic functions of EF-1β and TXNPx support their strong potential as authentic dissemination determinants. Both polypeptides have been described as being active in defence against chemical and oxidant stress providing a basis for further elucidation of the importance of resistance to oxidative stress in the pathogenesis underlying MCL. The resistance and the functional role of TXNPx with metastasis was further confirmed (Acestor *et al.*, 2006). In addition to these proteins, several proteins of unknown function were also identified (Walker *et al.*, 2006a). The functional characterization of such proteins will require additional effort and screening of a large number of human isolates using specific antibodies to obtain a better picture of the metastatic process. These findings validate the utility of proteomics as a tool for phenotypic analyses of *Leishmania*. It also opens new ways into the investigation of the different *Leishmania* species responsible for the various forms of leishmaniasis.

Structural studies of the editosome

Comparative studies of specific structures or purified organelles can be performed using a proteomic approach. One particular example of such an approach is the dissection of the editosome in trypanosomatids. Kinetoplastid protists in general, possess a two-unit genome, a nuclear genome and a mitochondrial genome, which in contrast to most other eukaryotes, is located in a single tubular mitochondrion, called the kinetoplast. RNA editing, a post-transcriptional modification of RNAs that involves the insertion/deletion or modification of specific residues in mitochondrial RNAs, was discovered in trypanosomatids, prior to being described in several other organisms including mammals. RNA editing occurs at an incredibly high rate in the kinetoplast of trypanosomatids, and up to hundreds of U-insertions/deletions were observed in a single transcript (Benne *et al.*, 1986).

A great effort has been performed in order to identify all proteins that are part of the editosome, that is, the protein complexes implicated in RNA editing. It appears that RNA editing occurs within ribonucleoprotein particles (RNPs). A 20S RNP was first isolated from mitochondrial extracts, which contains guide RNAs, RNA ligases, editing-site-specific endonuclease, terminal uridylyl transferase (TUTase) and 3′-U exonuclease activities (Seiwert *et al.*, 1994; Kable *et al.*, 1996; Panigrahi *et al.*, 2001a; Panigrahi *et al.*, 2001b). A second 35–40S RNP containing additional proteins was also isolated (Pollard *et al.*, 1992). Of the four enzymatic activities described in these RNPs, the genes encoding a 52 kDa and a 48 kDa RNA ligase are the only ones which were identified (McManus *et al.*, 2001; Schnaufer *et al.*, 2001). Several RNA binding proteins may be implicated in editing and are currently being studied, these include, gBP21, a gRNA binding protein (Muller *et al.*, 2001), TBRGG1, an oligo-U binding protein associated with the 35–40S RNP (Vanhamme *et al.*, 1998), and REAP-1, which binds purine-rich sequences and is also associated with the 35–40S complex (Madison-Antenucci *et al.*, 2001). A DEAD-box RNA helicase (mHel61p) could additionally be important for dissociation of the two RNAs after editing is completed (Missel *et al.*, 1997). Finally, four other putative editing complex proteins have been identified by mass spectrometry. These show no sequence homology to other proteins (Panigrahi *et al.*, 2001b). These data were the basis for additional studies comparing the editosomes of different trypanosomatids demonstrating that protein domains important for catalysis, molecular interaction and structure are generally well conserved within other protein families with some being additionally found in novel combinations in the editosome proteins (Worthey *et al.*, 2003b).

Such combined approaches can be applied to other cell compartments and to interacting proteins present in cellular complexes present in different trypanosomatids.

Post-translational modifications

The activities of most cellular proteins are regulated by post-translational modifications. One of the most important PTM is reversible protein phosphorylation, which decorates more than 30% of the human proteome and regulates signal transduction pathways under normal conditions, as well as in several pathological disorders such as neurodegenerative diseases, diabetes, autoimmune diseases, and several forms of cancer. Detailed phosphorylation studies have not yet been performed at the proteome level in *Leishmania*, although some reports describe the presence of stage-specific protein kinases (PK) (Bente *et al.*, 2003).

Bioinformatics searches (Parsons *et al.*, 2005) for typical and atypical protein kinases in the genomes of three trypanosomatids revealed a total of 176 PKs in *T. brucei*, 190 in *T. cruzi* and 199 in *L. major*, most of which were orthologous across the three species. This represents approximately 30% of the number of PKs in the human host. Trypanosomatids lack receptor-linked tyrosine and tyrosine kinase-like kinases, although they do possess dual-specificity kinases. Trypanosomatids possess a large set of PKs, comprising approximately 2% of each genome, suggesting a key role for phosphorylation in parasite biology. The presence of numerous PKs with significant sequence similarity to known drug targets, as well as a large number of unusual kinases that might represent novel targets, strongly argue for functional analysis of these molecules and of the phosphorylated proteins. It is evident that proteomics studies will take advantage of this valuable information and of specific knock-out and overexpressing transgenic parasites.

In addition to well-known modifications such as glycosylation and glycolipid addition, other PTMs will likely play essential roles in *Leishmania* biology. When infecting the macrophage, *Leishmania* parasites are submitted to an oxidative stress. This stress can induce hyper-oxidation of proteins and modify their pI. For example, it is known that hyperoxidation leads to the formation of sulfenic, sulfinic and sulfonic acids at a reduced cysteine.

New protein modifications have been described in human cells such as glutathionylated proteins in oxidatively stressed human T lymphocytes (Fratelli *et al.*, 2002; Fratelli *et al.*, 2003; Fratelli *et al.*, 2004). They were identified by redox proteomics. Among the glutathionylated proteins, specific enzymes were identified such as enolase or peroxiredoxin, which have been shown to be differentially expressed in the life cycle of the parasite. In the trypanosomatid oxidative detoxification system, glutathione is replaced by trypanothione. Will we find trypanothionylated enolase or peroxiredoxin? The question remains open.

Interestingly, in the search for virulence factors in MCL, several similar proteins had very similar molecular masses but different pIs (Walker *et al.*, 2006a). Several mechanisms modifying the charge of the proteins could be responsible for such a shift in pI. It is obvious that many questions are pending and that we are facing challenging issues mainly in the identification of differentially regulated proteins present at low abundance.

Specific organelles: the glycosome

Glycosomes of trypanosomatids belong to the microbody family of organelles. In trypanosomatids, these organelles are not only involved in carbohydrate metabolism but are also predicted to carry out gluconeogenesis, reactions of the hexose-monophosphate pathway, purine salvage and pyrimidine biosynthesis, β-oxidation of fatty acids, fatty acid elongation and the biosynthesis of ether lipids. In addition, they seem to catalyse several reactions of isoprenoid synthesis and are involved in oxidant stress protection. Several of the implicated enzymes have already been identified in *T. brucei* (Opperdoes, 1987).

Bioinformatics prediction of the trypanosomatid glycosomal proteome has been recently reported (Opperdoes *et al.*, 2006). The authors based their approach on the presence of either a C-terminal or an N-terminal peroxisomal targeting sequence. From those proteins with known function it appears that the predicted glycosomal proteomes are very similar in all three organisms with respect to enzyme content. They are predicted to also carry out gluconeogenesis, reactions of the hexose-monophosphate pathway, purine salvage and pyrimidine biosynthesis, β-oxidation of fatty acids, fatty acid elongation and the biosynthesis of ether lipids. In addition, they seem to catalyse several reactions of isoprenoid synthesis and are involved in oxidant stress protection. This study provides the groundwork for proteomic studies in the different stages of the life cycle of the parasite.

Perspectives

Bioinformatics and data management

In the coming years, it will be necessary to annotate all the *Leishmania* proteins known in *L. major* and in the most medically relevant species such as *L. infantum*, *L. donovani* and species of the *Leishmania Viannia* subgenus. This task should also be done for other infectious protozoa such as *T. brucei* and *T. cruzi* and also for non-infectious trypanosomatids such as *Crithidia fasciculata*. This information could then be integrated into a trypanosomatid orthologues data base similarly to that reported for the glycosome (Opperdoes *et al.*, 2006).

Such a database could also contain annotation of known polymorphisms at the protein sequence level and of all known post-translational modifications in *Leishmania* proteins. As they are difficult to predict, most post-translational modifications (PTMs) described thus far come from experimental data and from bioinformatics searches such as the one reported on the kinome (Parsons *et al.*, 2005). Following strict rules, such data can be transferred 'by similarity' to orthologous proteins or to other members of the same protein family.

High-throughput projects have been developed in order to identify human proteins subject to specific PTMs (www.hupo.org). Thus far, mainly phosphorylations, glycosylations, sumoylation and ubiquitination have been investigated. The data produced in humans should be continuously integrated into the trypanosomatid database. Tight links to structural information should be provided such as links to the PDB/RCSB 3D-structure database, which includes many features useful to structural biologists, such as literature references concerning X-ray and NMR papers (www.expasy.org/sw3d; www.wwpdb.org; www.rcsb.org). *Leishmania* structural biology could thus benefit from other available databases. Less than 10% of all human proteins have been characterized at the level of their 3D-structure. This percentage will rapidly expand in the next years. It will provide homology-derived models not only for all human proteins for which such an approach is scientifically relevant but also likely for *Leishmania* proteins.

Technological development

New technological approaches developed for mammalian and model organisms such as yeast and plants could help in solving parasite specific technical problems. Mass spectrometry-based quantitative proteomics has become an important component of biological and clinical research. New methods based on the selection and chemical synthesis of isotopically labelled reference peptides will help in uniquely identifying a particular protein and the addition of a panel of such peptides to the sample mixture consisting of tryptic peptides from the proteome in question (Pan *et al.*, 2005). Recently, the capability to isotopically label reference peptides has been extended from two labels to multiple labels, allowing for the simultaneous quantification of multiplex samples. These quantitative approaches gained progressively more popularity with the significant improvement of mass spectrometry performance.

In this respect, a new platform was set up consisting of a peptide separation module for the generation of ordered peptide arrays from the combined peptide sample on the sample plate of a MALDI mass spectrometer, a high throughput MALDI-TOF/TOF mass spectrometer, and a suite of software tools for the selective analysis of the targeted peptides and the interpretation of the results (Lie *et al.*, 2005). This method was applied to the analysis of the human blood serum proteome (Pan *et al.*, 2005). It demonstrates the feasibility of using mass spectrometry-based proteomics as a high throughput screening technology for the detection and quantification of targeted proteins in a complex system.

Clinical proteomics

A major focus in proteomics is the capability to conduct comparative analyses of biofluids from healthy and disease-affected patients for the identification of highly specific biomarkers and/or the development of MS-based diagnostic platforms. Much effort has gone into optimizing the biofluid proteome coverage that can be obtained using these technologies, leaving proteomics poised to make an important impact in disease diagnostics in the future. Such approaches have been used in cancer studies, raising new possibilites in diagnostic and treatment (Brusic *et al.*, 2007). For example, in cancer patient sera, tumour-derived proteins are present at levels 10^6–10^9-fold lower than albumin. Correspondingly, new platforms have been developed in order to pre-fractionate the samples (e.g. SELDI-TOF-MS). Furthermore, human tumour tissue is associated with sample contamination by non-tumour cells and serum proteins. Finally, in most instances, only a moderate number of tumours is analysed, questioning the representativity of clinical materials. It will therefore be of great importance that biomarkers identified by proteomics are validated in representative materials by other techniques, such as immunohistochemistry. In leishmaniasis, the biomarker approach can be applied to human biofluids such as the blood serum of infected patients, whereby sera of healthy versus symptomatic and asymptomatic patients may be examined, being also aware that patients could suffer from other infectious diseases which could complicate the interpretation of the results. This kind of study could bring surprising results. For example, a unique human biomarker, peroxiredoxin II, normally an intracellular protein implicated in detoxification, which had previously never been detected in human blood was identified in patients serum upon infection with the SARS virus. This finding could have significance for the development of new diagnostic methods for detecting SARS infections and for designing new therapies for SARS patients (Chen *et al.*, 2004). In addition, based on recent advances in mass spectrometry such as surface-enhanced laser desorption-ionization time of flight (SELDI-TOF), proteomic fingerprinting has been used for the diagnosis of human African trypanosomiasis (Papadopoulos *et al.*, 2004; Agranoff *et al.*, 2005). In *Leishmania*, a similar approach may be applied. For example, the identified biomarkers could serve as surrogate markers for resistance or susceptibility, and in the development of drug therapy and vaccines. These markers could come from the host or from parasite proteins and could potentially be used to follow the outcome of the disease and act as diagnostic tools. *Leishmania* animal models together with human sera collected over the years in clinical trials would be valuable material to start such studies.

Chemical proteomics

Proteomics employs a number of newly emerging tools. One such tool is chemical proteomics. This is a branch of proteomics devoted to the exploration of protein function using both *in vitro* and *in vivo* chemical probes. Chemical proteomics aims to define protein function and mechanism at the level of protein-ligand interactions and promises to be an important advance in the characterization of enzymatic activities. The approach is based on the design of activity-based probes (ABP) and activity-based protein profiling methods used in conjunction with libraries of synthetic small molecules to screen for pharmacological tools to disrupt/alter basic biological processes (Sadaghiani *et al.*, 2007).

There is often no direct correlation between the abundance of an enzyme and its activity. ABP can help to solve the limitation of profiling enzymatic activities (Cravatt *et al.*, 2000; Greenbaum *et al.*, 2002). ABPs can be designed for a distinct subset of enzymatic targets and thus facilitate the identification as well as activity-based protein profiling of enzymes in complex proteomes. ABPs consist of three elements, namely, a reactive group, a linker and a tag region (Jeffery *et al.*, 2003). The reactive group serves as the key functional element that directs covalent, activity-dependent modification of a target enzyme or enzyme family. The tag is used to visualize the resulting modified target enzymes. Several different ABPs have been developed to target serine hydrolases (Liu *et al.*, 1999; Kidd *et al.*, 2001), caspase family cysteine proteases (Thornberry *et al.*, 1994; Faleiro *et al.*, 1997; Winssinger *et al.*, 2002), papain family cysteine proteases (Bogyo *et al.*, 1997a; Greenbaum *et al.*, 2000), metalloproteases (Chan *et al.*, 2004; Saghatelian *et al.*, 2004), phosphatases (Lo *et al.*, 2002), the proteasome (Bogyo *et al.*, 1997b; Kessler *et al.*, 2001; Nazif *et al.*, 2001) and the ubiquitin-specific proteases (Borodovsky *et al.*, 2002; Jeffery and Bogyo, 2003; Hemelaar *et al.*, 2004a; Hemelaar *et al.*, 2004b; Ovaa *et al.*, 2004; Speers *et al.*, 2004; Borodovsky *et al.*, 2005).

Activity-based probes set the stage for more classical genetic and biochemical studies. These reagents can be applied to study for example several key enzyme families in *Leishmania*. However, they could be limited in their application to *Leishmania* by the fact that selective affinity labels have not been identified for a large number of proteins/protein families. With the increase of drug resistance in *Leishmania* species, ABPs could be instrumental to the design of new drugs. For example, ABPs have been used to characterize the major cysteine proteases of *T. brucei* (Caffrey *et al.*, 2001). At least one of these target proteases, trypanopain-Tb, is a potential target for the development of novel drug therapies. Additionally, a series of peptide vinyl sulphone ABPs were used to extensively characterize the proteasome of *T. brucei* (Wang *et al.*, 2003). Even if homologous enzymes are present in the parasite and in the host, differences in the substrate and inhibitor specificities of the parasite enzyme relative to the human enzyme could help and lead to the generation of novel anti-parasitic agents.

Conclusions

Proteome analysis promises to be essential for the identification of stage-specific genes and pathways, virulence factors, resistance mechanisms and serum biomarkers associated with leishmaniases. In addition, proteome analysis technologies offer the opportunity to examine protein expression profiles, protein–protein interactions and the activation of specific signalling pathways, which could be valuable for the design of new drugs. It is evident that the *Leishmania* research field will benefit highly from these technological developments, which will provide new solutions to prevent, or treat, leishmaniasis.

References

Acestor, N., Masina, S., Ives, A., Walker, J., Saravia, N.G., and Fasel, N. (2006). Resistance to oxidative stress is associated with metastasis in mucocutaneous leishmaniasis. J. Infect. Dis. *194*, 1160–1167.

Acestor, N., Masina, S., Walker, J., Saravia, N.G., Fasel, N., and Quadroni, M. (2002). Establishing two-dimensional gels for the analysis of *Leishmania* proteomes. Proteomics *2*, 877–879.

Adamczyk, M., Gebler, J.C., and Wu, J. (2002). Identification of phosphopeptides by chemical modification with an isotopic tag and ion trap mass spectrometry. Rapid Commun. Mass. Spectrom. *16*, 999–1001.

Agranoff, D., Stich, A., Abel, P., and Krishna, S. (2005). Proteomic fingerprinting for the diagnosis of human African trypanosomiasis. Trends Parasitol. *21*, 154–157.

Akopyants, N.S., Matlib, R.S., Bukanova, E.N., Smeds, M.R., Brownstein, B.H., Stormo, G.D., and Beverley, S.M. (2004). Expression profiling using random genomic DNA microarrays identifies differentially expressed genes associated with three major developmental stages of the protozoan parasite *Leishmania major*. Mol. Biochem. Parasitol. *136*, 71–86.

Almeida, R., Gilmartin, B.J., McCann, S.H., Norrish, A., Ivens, A.C., Lawson, D., Levick, M.P., Smith, D.F., Dyall, S.D., Vetrie, D., Freeman, T.C., Coulson, R.M., Sampaio, I., Schneider, H., and Blackwell, J.M. (2004). Expression profiling of the *Leishmania* life cycle: cDNA arrays identify developmentally regulated genes present but not annotated in the genome. Mol. Biochem. Parasitol. *136*, 87–100.

Atwood, J.A., 3rd, Weatherly, D.B., Minning, T.A., Bundy, B., Cavola, C., Opperdoes, F.R., Orlando, R., and Tarleton, R.L. (2005). The *Trypanosoma cruzi* proteome. Science. *309*, 473–476.

Belli, S., Formenton, A., Noll, T., Ivens, A., Jacquet, R., Desponds, C., Hofer, D., and Fasel, N. (1999). *Leishmania major*: histone H1 gene expression from the sw3 locus. Exp. Parasitol. *91*, 151–160.

Belli, S.I., Monnerat, S., Schaff, C., Masina, S., Noll, T., Myler, P.J., Stuart, K., and Fasel, N. (2003). Sense and antisense transcripts in the histone H1 (HIS-1) locus of *Leishmania major*. Int. J. Parasitol. *33*, 965–975.

Benne, R., Van den Burg, J., Brakenhoff, J.P., Sloof, P., Van Boom, J.H., and Tromp, M.C. (1986). Major transcript of the frameshifted coxII gene from trypanosome mitochondria contains four nucleotides that are not encoded in the DNA. Cell. *46*, 819–826.

Bente, M., Harder, S., Wiesgigl, M., Heukeshoven, J., Gelhaus, C., Krause, E., Clos, J., and Bruchhaus, I. (2003). Developmentally induced changes of the proteome in the protozoan parasite *Leishmania donovani*. Proteomics *3*, 1811–1829.

Bjellqvist, B., Hughes, G.J., Pasquali, C., Paquet, N., Ravier, F., Sanchez, J.C., Frutiger, S., and Hochstrasser, D. (1993). The focusing positions of polypeptides in immobilized pH gradients can be predicted from their amino acid sequences. Electrophoresis *14*, 1023–1031.

Bogyo, M., Gaczynska, M., and Ploegh, H.L. (1997a). Proteasome inhibitors and antigen presentation. Biopolymers *43*, 269–280.

Bogyo, M., McMaster, J.S., Gaczynska, M., Tortorella, D., Goldberg, A.L., and Ploegh, H. (1997b). Covalent modification of the active site threonine of proteasomal beta subunits and the Escherichia coli homolog HslV by a new class of inhibitors. Proc. Natl. Acad. Sci. USA *94*, 6629–6634.

Borodovsky, A., Ovaa, H., Kolli, N., Gan-Erdene, T., Wilkinson, K.D., Ploegh, H.L., and Kessler, B.M. (2002). Chemistry-based functional proteomics reveals novel members of the deubiquitinating enzyme family. Chem. Biol. *9*, 1149–1159.

Borodovsky, A., Ovaa, H., Meester, W.J., Venanzi, E.S., Bogyo, M.S., Hekking, B.G., Ploegh, H.L., Kessler, B.M., and Overkleeft, H.S. (2005). Small-molecule inhibitors and probes for ubiquitin- and ubiquitin-like-specific proteases. Chembiochemistry *6*, 287–291.

Bottari, P., Aebersold, R., Turecek, F., and Gelb, M.H. (2004). Design and synthesis of visible isotope-coded affinity tags for the absolute quantification of specific proteins in complex mixtures. Bioconjug. Chem. *15*, 380–388.

Brems, S., Guilbride, D.L., Gundlesdodjir-Planck, D., Busold, C., Luu, V.D., Schanne, M., Hoheisel, J., and Clayton, C. (2005). The transcriptomes of *Trypanosoma brucei* Lister 427 and TREU927 bloodstream and procyclic trypomastigotes. Mol Biochem. Parasitol. *139*, 163–172.

Brobey, R.K., and Soong, L. (2007). Establishing a liquid-phase IEF in combination with 2-DE for the analysis of *Leishmania* proteins. Proteomics *7*, 116–120.

Brusic, V., Marina, O., Wu, C.J., and Reinherz, E.L. (2007). Proteome informatics for cancer research: from molecules to clinic. Proteomics *7*, 976–991.

Bumann, D., Meyer, T.F., and Jungblut, P.R. (2001). Proteome analysis of the common human pathogen Helicobacter pylori. Proteomics *1*, 473–479.

Caffrey, C.R., Hansell, E., Lucas, K.D., Brinen, L.S., Alvarez Hernandez, A., Cheng, J., Gwaltney, S.L., 2nd, Roush, W.R., Stierhof, Y.D., Bogyo, M., Steverding, D., and McKerrow, J.H. (2001). Active site mapping, biochemical properties and subcellular localization of rhodesain, the major cysteine protease of *Trypanosoma brucei* rhodesiense. Mol. Biochem. Parasitol. *118*, 61–73.

Campbell, D.A., Thomas, S., and Sturm, N.R. (2003). Transcription in kinetoplastid protozoa: why be normal? Microbes Infect. *5*, 1231–1240.

Chan, E.W., Chattopadhaya, S., Panicker, R.C., Huang, X., and Yao, S.Q. (2004). Developing photoactive affinity probes for proteomic profiling: hydroxamate-based probes for metalloproteases. J. Am. Chem. Soc. *126*, 14435–14446.

Chen, J.H., Chang, Y.W., Yao, C.W., Chiueh, T.S., Huang, S.C.,Chien, K.Y., Chen, A., Chang, F.Y., Wong, C.H.,Chen, Y.J. (2004). Plasma proteome of severe acute respiratory syndrome analyzed by two-dimensional gel electrophoresis and mass spectrometry. Proc. Natl. Acad. Sci. USA *101*, 17039–17044.

Clayton, C.E. (2002). Life without transcriptional control? From fly to man and back again. Embo J. *21*, 1881–1888.

Clos, J., Klaholz, L., Kroemer, M., Krobitsch, S., and Lindquist, S. (2001). Heat shock protein 100 and the amastigote stage-specific A2 proteins of *Leishmania donovani*. Med. Microbiol. Immunol (Berl). *190*, 47–50.

Cordwell, S.J., Nouwens, A.S., and Walsh, B.J. (2001). Comparative proteomics of bacterial pathogens. Proteomics *1*, 461–472.

Cravatt, B.F., and Sorensen, E.J. (2000). Chemical strategies for the global analysis of protein function. Curr. Opin. Chem. Biol. *4*, 663–668.

Dea-Ayuela, M.A., Rama-Iniguez, S., and Bolas-Fernandez, F. (2006). Proteomic analysis of antigens from *Leishmania infantum* promastigotes. Proteomics *6*, 4187–4194.

Diehl, S., Diehl, F., El-Sayed, N.M., Clayton, C., and Hoheisel, J.D. (2002). Analysis of stage-specific gene expression in the bloodstream and the procyclic form of *Trypanosoma brucei* using a genomic DNA-microarray. Mol. Biochem. Parasitol. *123*, 115–123.

Drummelsmith, J., Brochu, V., Girard, I., Messier, N., and Ouellette, M. (2003). Proteome mapping of the protozoan parasite *Leishmania* and application to the study of drug targets and resistance mechanisms. Mol. Cell. Proteomics *2*, 146–155.

Drummelsmith, J., Girard, I., Trudel, N., and Ouellette, M. (2004). Differential protein expression analysis of *Leishmania major* reveals novel roles for methionine adenosyltransferase and S-adenosylmethionine in methotrexate resistance. J. Biol. Chem. *279*, 33273–33280.

Duncan, R. (2004). DNA microarray analysis of protozoan parasite gene expression: outcomes correlate with mechanisms of regulation. Trends Parasitol. *20*, 211–215.

Duncan, R.C., Salotra, P., Goyal, N., Akopyants, N.S., Beverley, S.M., and Nakhasi, H.L. (2004). The application of gene expression microarray technology to kinetoplastid research. Curr. Mol. Med. *4*, 611–621.

Faleiro, L., Kobayashi, R., Fearnhead, H., and Lazebnik, Y. (1997). Multiple species of CPP32 and Mch2 are the major active caspases present in apoptotic cells. Embo J. *16*, 2271–2281.

Figueroa-Angulo, E., Martinez-Calvillo, S., Lopez-Villasenor, I., and Hernandez, R. (2003). Evidence supporting a major promoter in the *Trypanosoma cruzi* rRNA gene. FEMS Microbiol. Lett. *225*, 221–225.

Forgber, M., Basu, R., Roychoudhury, K., Theinert, S., Roy, S., Sundar, S., and Walden, P. (2006). Mapping the Antigenicity of the Parasites in *Leishmania donovani* Infection by Proteome Serology. PLoS ONE. *1*, e40.

Foucher, A.L., Papadopoulou, B., and Ouellette, M. (2006). Prefractionation by digitonin extraction increases representation of the cytosolic and intracellular proteome of *Leishmania infantum*. J. Proteome Res. *5*, 1741–1750.

Fratelli, M., Demol, H., Puype, M., Casagrande, S., Eberini, I., Salmona, M., Bonetto, V., Mengozzi, M.,

Duffieux, F., Miclet, E., Bachi, A., Vandekerckhove, J., Gianazza, E., and Ghezzi, P. (2002). Identification by redox proteomics of glutathionylated proteins in oxidatively stressed human T lymphocytes. Proc. Natl. Acad. Sci. USA *99*, 3505–3510.

Fratelli, M., Demol, H., Puype, M., Casagrande, S., Villa, P., Eberini, I., Vandekerckhove, J., Gianazza, E., and Ghezzi, P. (2003). Identification of proteins undergoing glutathionylation in oxidatively stressed hepatocytes and hepatoma cells. Proteomics *3*, 1154–1161.

Fratelli, M., Gianazza, E., and Ghezzi, P. (2004). Redox proteomics: identification and functional role of glutathionylated proteins. Expert Rev. Proteomics *1*, 365–376.

Gagnon, D., Foucher, A., Girard, I., and Ouellette, M. (2006). Stage specific gene expression and cellular localization of two isoforms of the serine hydroxymethyltransferase in the protozoan parasite *Leishmania*. Mol. Biochem. Parasitol. *150*, 63–71.

Gay, L.S., Wilson, M.E., and Donelson, J.E. (1996). The promoter for the ribosomal RNA genes of *Leishmania chagasi*. Mol. Biochem. Parasitol. *77*, 193–200.

Gilinger, G., and Bellofatto, V. (2001). Trypanosome spliced leader RNA genes contain the first identified RNA polymerase II gene promoter in these organisms. Nucleic Acids Res. *29*, 1556–1564.

Graefe, S.E., Wiesgigl, M., Gaworski, I., Macdonald, A., and Clos, J. (2002). Inhibition of HSP90 in *Trypanosoma cruzi* induces a stress response but no stage differentiation. Eukaryot. Cell. *1*, 936–943.

Graham, S.V., and Barry, J.D. (1995). Transcriptional regulation of metacyclic variant surface glycoprotein gene expression during the life cycle of *Trypanosoma brucei*. Mol. Cell. Biol. *15*, 5945–5956.

Greenbaum, D., Medzihradszky, K.F., Burlingame, A., and Bogyo, M. (2000). Epoxide electrophiles as activity-dependent cysteine protease profiling and discovery tools. Chem. Biol. *7*, 569–581.

Greenbaum, D.C., Arnold, W.D., Lu, F., Hayrapetian, L., Baruch, A., Krumrine, J., Toba, S., Chehade, K., Bromme, D., Kuntz, I.D., and Bogyo, M. (2002). Small molecule affinity fingerprinting. A tool for enzyme family subclassification, target identification, and inhibitor design. Chem. Biol. *9*, 1085–1094.

Gu, S., Liu, Z., Pan, S., Jiang, Z., Lu, H., Amit, O., Bradbury, E.M., Hu, C.A., and Chen, X. (2004). Global investigation of p53-induced apoptosis through quantitative proteomic profiling using comparative amino acid-coded tagging. Mol. Cell. Proteomics *3*, 998–1008.

Gygi, S.P., Rist, B., Gerber, S.A., Turecek, F., Gelb, M.H., and Aebersold, R. (1999). Quantitative analysis of complex protein mixtures using isotope-coded affinity tags. Nat. Biotechnol. *17*, 994–999.

Hannaert, V., Albert, M.A., Rigden, D.J., da Silva Giotto, M.T., Thiemann, O., Garratt, R.C., Van Roy, J., Opperdoes, F.R., and Michels, P.A. (2003). Kinetic characterization, structure modelling studies and crystallization of *Trypanosoma brucei* enolase. Eur. J. Biochem. *270*, 3205–3213.

Hansen, K.C., Schmitt-Ulms, G., Chalkley, R.J., Hirsch, J., Baldwin, M.A., and Burlingame, A.L. (2003). Mass spectrometric analysis of protein mixtures at low levels using cleavable 13C-isotope-coded affinity tag and multidimensional chromatography. Mol. Cell. Proteomics *2*, 299–314.

Hemelaar, J., Borodovsky, A., Kessler, B.M., Reverter, D., Cook, J., Kolli, N., Gan-Erdene, T., Wilkinson, K.D., Gill, G., Lima, C.D., Ploegh, H.L., and Ovaa, H. (2004a). Specific and covalent targeting of conjugating and deconjugating enzymes of ubiquitin-like proteins. Mol. Cell. Biol. *24*, 84–95.

Hemelaar, J., Galardy, P.J., Borodovsky, A., Kessler, B.M., Ploegh, H.L., and Ovaa, H. (2004b). Chemistry-based functional proteomics: mechanism-based activity-profiling tools for ubiquitin and ubiquitin-like specific proteases. J. Proteome Res. *3*, 268–276.

Holzer, T.R., McMaster, W.R., and Forney, J.D. (2006). Expression profiling by whole-genome interspecies microarray hybridization reveals differential gene expression in procyclic promastigotes, lesion-derived amastigotes, and axenic amastigotes in *Leishmania mexicana*. Mol. Biochem. Parasitol. *146*, 198–218.

Huang, J., and van der Ploeg, L.H. (1991a). Maturation of polycistronic pre-mRNA in *Trypanosoma brucei*: analysis of trans splicing and poly(A) addition at nascent RNA transcripts from the hsp70 locus. Mol. Cell. Biol. *11*, 3180–3190.

Huang, J., and Van der Ploeg, L.H. (1991b). Requirement of a polypyrimidine tract for trans-splicing in trypanosomes: discriminating the PARP promoter from the immediately adjacent 3′ splice acceptor site. Embo J. *10*, 3877–3885.

Ivens, A.C., Peacock, C.S., Worthey, E.A., Murphy, L., Aggarwal, G., Berriman, M., Sisk, E., Rajandream, M.A., Adlem, E., Aert, R., *et al.* (2005). The genome of the kinetoplastid parasite, *Leishmania major*. Science. *309*, 436–442.

Jeffery, D.A., and Bogyo, M. (2003). Chemical proteomics and its application to drug discovery. Curr. Opin. Biotechnol. *14*, 87–95.

Kable, M.L., Seiwert, S.D., Heidmann, S., and Stuart, K. (1996). RNA editing: a mechanism for gRNA-specified uridylate insertion into precursor mRNA. Science. *273*, 1189–1195.

Kapler, G.M., and Beverley, S.M. (1989). Transcriptional mapping of the amplified region encoding the dihydrofolate reductase-thymidylate synthase of *Leishmania major* reveals a high density of transcripts, including overlapping and antisense RNAs. Mol. Cell. Biol. *9*, 3959–3972.

Kapotas, N., and Bellofatto, V. (1993). Differential response to RNA trans-splicing signals within the phosphoglycerate kinase gene cluster in *Trypanosoma brucei*. Nucleic Acids Res. *21*, 4067–4072.

Kessler, B.M., Tortorella, D., Altun, M., Kisselev, A.F., Fiebiger, E., Hekking, B.G., Ploegh, H.L., and Overkleeft, H.S. (2001). Extended peptide-based inhibitors efficiently target the proteasome and reveal overlapping specificities of the catalytic beta-subunits. Chem. Biol. *8*, 913–929.

Kidd, D., Liu, Y., and Cravatt, B.F. (2001). Profiling serine hydrolase activities in complex proteomes. Biochemistry. *40*, 4005–4015.

Lee, H., Griffin, T.J., Gygi, S.P., Rist, B., and Aebersold, R. (2002). Development of a multiplexed microcapillary liquid chromatography system for high-throughput proteome analysis. Anal. Chem. *74*, 4353–4360.

Lee, M.G. (1998). The 3′ untranslated region of the hsp 70 genes maintains the level of steady state mRNA in *Trypanosoma brucei* upon heat shock. Nucleic Acids Res. *26*, 4025–4033.

Li, J., Steen, H., and Gygi, S.P. (2003). Protein profiling with cleavable isotope-coded affinity tag (cICAT) reagents: the yeast salinity stress response. Mol. Cell. Proteomics *2*, 1198–1204.

Li, X.J., Pedrioli, P.G., Eng, J., Martin, D., Yi, E.C., Lee, H., and Aebersold, R. (2004). A tool to visualize and evaluate data obtained by liquid chromatography-electrospray ionization-mass spectrometry. Anal. Chem. *76*, 3856–3860.

Li, X.J., Yi, E.C., Kemp, C.J., Zhang, H., and Aebersold, R. (2005). A software suite for the generation and comparison of peptide arrays from sets of data collected by liquid chromatography-mass spectrometry. Mol. Cell. Proteomics *4*, 1328–1340.

Liniger, M., Bodenmuller, K., Pays, E., Gallati, S., and Roditi, I. (2001). Overlapping sense and antisense transcription units in *Trypanosoma brucei*. Mol. Microbiol. *40*, 869–878.

Liu, Y., Patricelli, M.P., and Cravatt, B.F. (1999). Activity-based protein profiling: the serine hydrolases. Proc. Natl. Acad. Sci USA *96*, 14694–14699.

Lo, L.C., Pang, T.L., Kuo, C.H., Chiang, Y.L., Wang, H.Y., and Lin, J.J. (2002). Design and synthesis of class-selective activity probes for protein tyrosine phosphatases. J. Proteome Res. *1*, 35–40.

Lopez-Villar, E., Monteoliva, L., Larsen, M.R., Sachon, E., Shabaz, M., Pardo, M., Pla, J., Gil, C., Roepstorff, P., and Nombela, C. (2006). Genetic and proteomic evidences support the localization of yeast enolase in the cell surface. Proteomics *6 Suppl. 1*, S107–118.

Luo, H., Gilinger, G., Mukherjee, D., and Bellofatto, V. (1999). Transcription initiation at the TATA-less spliced leader RNA gene promoter requires at least two DNA-binding proteins and a tripartite architecture that includes an initiator element. J. Biol. Chem. *274*, 31947–31954.

Madison-Antenucci, S., and Hajduk, S.L. (2001). RNA editing-associated protein 1 is an RNA binding protein with specificity for preedited mRNA. Mol. Cell. *7*, 879–886.

Mair, G., Shi, H., Li, H., Djikeng, A., Aviles, H.O., Bishop, J.R., Falcone, F.H., Gavrilescu, C., Montgomery, J.L., Santori, M.I., Stern, L.S., Wang, Z., Ullu, E., and Tschudi, C. (2000a). A new twist in trypanosome RNA metabolism: cis-splicing of pre-mRNA. Rna *6*, 163–169.

Mair, G., Ullu, E., and Tschudi, C. (2000b). Cotranscriptional cap 4 formation on the *Trypanosoma brucei* spliced leader RNA. J. Biol. Chem. *275*, 28994–28999.

Manning-Cela, R., Gonzalez, A., and Swindle, J. (2002). Alternative splicing of LYT1 transcripts in *Trypanosoma cruzi*. Infect. Immun. *70*, 4726–4728.

Martinez-Calvillo, S., Nguyen, D., Stuart, K., and Myler, P.J. (2004). Transcription initiation and termination on *Leishmania major* chromosome 3. Eukaryot. Cell. *3*, 506–517.

Martinez-Calvillo, S., Yan, S., Nguyen, D., Fox, M., Stuart, K., and Myler, P.J. (2003). Transcription of *Leishmania major* Friedlin chromosome 1 initiates in both directions within a single region. Mol. Cell. *11*, 1291–1299.

McManus, M.T., Shimamura, M., Grams, J., and Hajduk, S.L. (2001). Identification of candidate mitochondrial RNA editing ligases from *Trypanosoma brucei*. Rna. *7*, 167–175.

Missel, A., Souza, A.E., Norskau, G., and Goringer, H.U. (1997). Disruption of a gene encoding a novel mitochondrial DEAD-box protein in *Trypanosoma brucei* affects edited mRNAs. Mol. Cell. Biol. *17*, 4895–4903.

Monnerat, S., Martinez-Calvillo, S., Worthey, E., Myler, P.J., Stuart, K.D., and Fasel, N. (2004). Genomic organization and gene expression in a chromosomal region of *Leishmania major*. Mol. Biochem. Parasitol. *134*, 233–243.

Moseley, M.A. (2001). Current trends in differential expression proteomics: isotopically coded tags. Trends Biotechnol. *19*, S10–16.

Muller, U.F., Lambert, L., and Goringer, H.U. (2001). Annealing of RNA editing substrates facilitated by guide RNA-binding protein gBP21. Embo J. *20*, 1394–1404.

Nazif, T., and Bogyo, M. (2001). Global analysis of proteasomal substrate specificity using positional-scanning libraries of covalent inhibitors. Proc. Natl. Acad. Sci. USA *98*, 2967–2972.

Nugent, P.G., Karsani, S.A., Wait, R., Tempero, J., and Smith, D.F. (2004). Proteomic analysis of *Leishmania mexicana* differentiation. Mol. Biochem. Parasitol. *136*, 51–62.

O'Farrell, T.P., and Dilley, J.V. (1975). A comparison of radiation response, cyanide toxicity and sulfur transferase activity in native North American rodents. Comp. Biochem. Physiol B. *50*, 443–447.

Ong, S.E., Blagoev, B., Kratchmarova, I., Kristensen, D.B., Steen, H., Pandey, A., and Mann, M. (2002). Stable isotope labeling by amino acids in cell culture, SILAC, as a simple and accurate approach to expression proteomics. Mol. Cell. Proteomics *1*, 376–386.

Opperdoes, F.R. (1987). Compartmentation of carbohydrate metabolism in trypanosomes. Annu. Rev. Microbiol. *41*, 127–151.

Opperdoes, F.R., and Szikora, J.P. (2006). In silico prediction of the glycosomal enzymes of *Leishmania major* and trypanosomes. Mol. Biochem. Parasitol. *147*, 193–206.

Ovaa, H., Kessler, B.M., Rolen, U., Galardy, P.J., Ploegh, H.L., and Masucci, M.G. (2004). Activity-based ubiquitin-specific protease (USP) profiling of virus-infected and malignant human cells. Proc. Natl. Acad. Sci. USA *101*, 2253–2258.

Paba, J., Santana, J.M., Teixeira, A.R., Fontes, W., Sousa, M.V., and Ricart, C.A. (2004). Proteomic analysis of the human pathogen *Trypanosoma cruzi*. Proteomics *4*, 1052–1059.

Palagi, P.M., Walther, D., Quadroni, M., Catherinet, S., Burgess, J., Zimmermann-Ivol, C.G., Sanchez J.C., Binz, P.A., Hochstrasser, D.F., and Appel R.D. (2005). MSight: an image analysis software for liquid chromatography-mass spectrometry. Proteomics 5, 2381–2384.

Palenchar, J.B., and Bellofatto, V. (2006). Gene transcription in trypanosomes. Mol. Biochem. Parasitol. *146*, 135–141.

Palfi, Z., Lane, W.S., and Bindereif, A. (2002). Biochemical and functional characterization of the cis-spliceosomal U1 small nuclear RNP from *Trypanosoma brucei*. Mol. Biochem. Parasitol. *121*, 233–243.

Pan, S., Zhang, H., Rush, J., Eng, J., Zhang, N., Patterson, D., Comb M.J., and Aebersold, R. (2005). High throughput proteome screening for biomarker detection. Mol. Cell. Proteomics *4*, 182–190.

Panigrahi, A.K., Gygi, S.P., Ernst, N.L., Igo, R.P., Jr., Palazzo, S.S., Schnaufer, A., Weston, D.S., Carmean, N., Salavati,*et al.* (2001a). Association of two novel proteins, TbMP52 and TbMP48, with the *Trypanosoma brucei* RNA editing complex. Mol. Cell. Biol. *21*, 380–389.

Panigrahi, A.K., Schnaufer, A., Carmean, N., Igo, R.P., Jr., Gygi, S.P., Ernst, N.L., Palazzo, S.S., Weston, D.S., Aebersold, R., Salavati, R., and Stuart, K.D. (2001b). Four related proteins of the *Trypanosoma brucei* RNA editing complex. Mol. Cell. Biol. *21*, 6833–6840.

Papadopoulos, M.C., Abel, P.M., Agranoff, D., Stich, A., Tarelli, E., Bell, B.A., Planche, T., Loosemore, A., Saadoun, S., Wilkins, P., and Krishna, S. (2004). A novel and accurate diagnostic test for human African trypanosomiasis. Lancet. *363*, 1358–1363.

Parsons, M., Worthey, E.A., Ward, P.N., and Mottram, J.C. (2005). Comparative analysis of the kinomes of three pathogenic trypanosomatids: *Leishmania major*, *Trypanosoma brucei* and *Trypanosoma cruzi*. BMC Genomics *6*, 127.

Patel, A., and Perrin, D.M. (2004). Efficient synthesis of isotopically pure isotope-coded affinity tagging reagents. Bioconjug. Chem. *15*, 224–230.

Pollard, V.W., Harris, M.E., and Hajduk, S.L. (1992). Native mRNA editing complexes from *Trypanosoma brucei* mitochondria. Embo J. *11*, 4429–4438.

Righetti, P.G., Castagna, A., Antonucci, F., Piubelli, C., Cecconi, D., Campostrini, N., Zanusso, G., and Monaco, S. (2003). The proteome: anno Domini (2002). Clin. Chem. Lab. Med. *41*, 425–438.

Ritter, U., Mattner, J., Rocha, J.S., Bogdan, C., and Korner, H. (2004). The control of *Leishmania (Leishmania) major* by TNF in vivo is dependent on the parasite strain. Microbes Infect. *6*, 559–565.

Rudenko, G., Lee, M.G., and Van der Ploeg, L.H. (1992). The PARP and VSG genes of *Trypanosoma brucei* do not resemble RNA polymerase II transcription units in sensitivity to Sarkosyl in nuclear run-on assays. Nucleic Acids Res. *20*, 303–306.

Sadaghiani, A.M., Verhelst, S., and Bogyo, M. (2007). Tagging and detection strategies for activity-based proteomics. Curr. Opin. Chem. Biol. *11*, 20-28.

Saghatelian, A., Jessani, N., Joseph, A., Humphrey, M., and Cravatt, B.F. (2004). Activity-based probes for the proteomic profiling of metalloproteases. Proc. Natl. Acad. Sci. USA *101*, 10000–10005.

Saxena, A., Worthey, E.A., Yan, S., Leland, A., Stuart, K.D., and Myler, P.J. (2003). Evaluation of differential gene expression in *Leishmania major* Friedlin procyclics and metacyclics using DNA microarray analysis. Mol. Biochem. Parasitol. *129*, 103–114.

Schmidt, F., Donahoe, S., Hagens, K., Mattow, J., Schaible, U.E., Kaufmann, S.H., Aebersold, R., and Jungblut, P.R. (2004). Complementary analysis of the *Mycobacterium tuberculosis* proteome by two-dimensional electrophoresis and isotope-coded affinity tag technology. Mol. Cell. Proteomics 3, 24–42.

Schnaufer, A., Panigrahi, A.K., Panicucci, B., Igo, R.P., Jr., Wirtz, E., Salavati, R., and Stuart, K. (2001). An RNA ligase essential for RNA editing and survival of the bloodstream form of *Trypanosoma brucei*. Science. *291*, 2159–2162.

Seiwert, S.D., and Stuart, K. (1994). RNA editing: transfer of genetic information from gRNA to precursor mRNA in vitro. Science. *266*, 114–117.

Smolka, M., Zhou, H., and Aebersold, R. (2002). Quantitative protein profiling using two-dimensional gel electrophoresis, isotope-coded affinity tag labeling, and mass spectrometry. Mol. Cell. Proteomics *1*, 19–29.

Smolka, M.B., Zhou, H., Purkayastha, S., and Aebersold, R. (2001). Optimization of the isotope-coded affinity tag-labeling procedure for quantitative proteome analysis. Anal. Biochem. *297*, 25–31.

Speers, A.E., and Cravatt, B.F. (2004). Chemical strategies for activity-based proteomics. Chembiochemistry 5, 41–47.

Thornberry, N.A., Peterson, E.P., Zhao, J.J., Howard, A.D., Griffin, P.R., and Chapman, K.T. (1994). Inactivation of interleukin-1 beta converting enzyme by peptide (acyloxy)methyl ketones. Biochemistry. 33, 3934–3940.

Tosato, V., Ciarloni, L., Ivens, A.C., Rajandream, M.A., Barrell, B.G., and Bruschi, C.V. (2001). Secondary DNA structure analysis of the coding strand switch regions of five *Leishmania major* Friedlin chromosomes. Curr. Genet. *40*, 186–194.

Uliana, S.R., Fischer, W., Stempliuk, V.A., and Floeter-Winter, L.M. (1996). Structural and functional characterization of the *Leishmania amazonensis* ribosomal RNA promoter. Mol. Biochem. Parasitol. *76*, 245–255.

van Swieten, P.F., Leeuwenburgh, M.A., Kessler, B.M., and Overkleeft, H.S. (2005). Bioorthogonal organic chemistry in living cells: novel strategies for labeling biomolecules. Org. Biomol. Chem. 3, 20–27.

Vanhamme, L., Perez-Morga, D., Marchal, C., Speijer, D., Lambert, L., Geuskens, M., Alexandre, S., Ismaili, N., Goringer, U., Benne, R., and Pays, E. (1998). *Trypanosoma brucei* TBRGG1, a mitochondrial oligo(U)-binding protein that co-localizes with an in vitro RNA editing activity. J. Biol. Chem. *273*, 21825–21833.

Vergnes, B., Gourbal, B., Girard, I., Sundar, S., Drummelsmith, J., and Ouellette, M. (2007). A proteomics screen implicates HSP83 and a small kinetoplastid calpain-related protein in drug resistance in *Leishmania donovani* clinical field isolates by modulating drug-induced programmed cell death. Mol. Cell. Proteomics 6, 88–101.

Walker, J., Acestor, N., Gongora, R., Quadroni, M., Segura, I., Fasel, N., and Saravia, N.G. (2006a). Comparative protein profiling identifies elongation factor-1beta and tryparedoxin peroxidase as factors associated with metastasis in *Leishmania guyanensis*. Mol. Biochem. Parasitol. *145*, 254–264.

Walker, J., Vasquez, J.J., Gomez, M.A., Drummelsmith, J., Burchmore, R., Girard, I., and Ouellette, M. (2006b). Identification of developmentally-regulated proteins in *Leishmania panamensis* by proteome profiling of promastigotes and axenic amastigotes. Mol. Biochem. Parasitol. *147*, 64–73.

Wang, C.C., Bozdech, Z., Liu, C.L., Shipway, A., Backes, B.J., Harris, J.L., and Bogyo, M. (2003). Biochemical analysis of the 20 S proteasome of *Trypanosoma brucei*. J. Biol. Chem. *278*, 15800–15808.

Winssinger, N., Ficarro, S., Schultz, P.G., and Harris, J.L. (2002). Profiling protein function with small molecule microarrays. Proc. Natl. Acad.Sci. USA *99*, 11139–11144.

Worthey, E.A., Martinez-Calvillo, S., Schnaufer, A., Aggarwal, G., Cawthra, J., Fazelinia, G., Fong, C., Fu, G., Hassebrock, M., Hixson, G., *et al.* (2003a). *Leishmania major* chromosome 3 contains two long convergent polycistronic gene clusters separated by a tRNA gene. Nucleic Acids Res. *31*, 4201–4210.

Worthey, E.A., Schnaufer, A., Mian, I.S., Stuart, K., and Salavati, R. (2003b). Comparative analysis of editosome proteins in trypanosomatids. Nucleic Acids Res. *31*, 6392–6408.

Yan, S., Lodes, M.J., Fox, M., Myler, P.J., and Stuart, K. (1999). Characterization of the *Leishmania donovani* ribosomal RNA promoter. Mol. Biochem. Parasitol. *103*, 197–210.

Yi, E.C., Li, X.J., Cooke, K., Lee, H., Raught, B., Page, A., Aneliunas, V., Hieter, P., Goodlett, D.R., and Aebersold, R. (2005). Increased quantitative proteome coverage with (13)C/(12)C-based, acid-cleavable isotope-coded affinity tag reagent and modified data acquisition scheme. Proteomics 5, 380–387.

Analysis of the *Leishmania* Metabolome

5

Malcolm J. McConville, David P. De Souza, Eleanor C. Saunders, James Pyke, Thomas Naderer, Miriam A. Ellis, M. Fleur Sernee, Julie E. Ralton and Vladimir A. Likić

Abstract

The sequencing of the *L. major* genome has provided unprecedented insights into the metabolic capacity of *Leishmania* parasites. However, we still have a relatively limited understanding of which metabolic pathways are essential for parasite survival in the sandfly and mammalian host or how species-specific differences in *Leishmania* metabolism contribute to differences in tissue tropism and disease patterns. The *L. major* genome also contains a large number of genes with no homologues in other eukaryotes. Metabolite profiling, or metabolomics, is an emerging field in functional genomics that is both complimentary to transcript and protein profiling approaches, and a powerful tool in its own right for defining metabolic processes and annotating gene function. In this review, we provide an overview of *Leishmania* metabolism and describe new analytical approaches that are being used to directly characterize the *Leishmania* metabolome. Finally, we highlight major metabolic pathways in *Leishmania* that appear to be essential for parasite survival and/or differentiation in the mammalian host.

Introduction

Leishmania parasites develop as extracellular promastigotes in the digestive tract of the sandfly vector and as obligate intracellular amastigotes within the phagolysosomes of mammalian macrophages. Each of these major developmental stages may progress through a number of morphological and/or physiological states that exhibit markedly different growth rates (Sacks and Kamhawi, 2001). For example, some species of *Leishmania* appear to progress through at least two separate, consecutive growth cycles in the sandfly midgut (procyclic, leptomonad), each of which give rise to a non-dividing stage (nectomonad, metacyclic) (Rogers *et al.*, 2002; Gossage *et al.*, 2003). The non-proliferating metacyclic promastigotes share many characteristics with cultured stationary phase promastigotes and appear to be well adapted for survival in the mammalian host (Sacks and Kamhawi, 2001). It is likely that nutrient levels play a key role in regulating the timing, prevalence and persistence of each of these stages. Whether intracellular amastigotes also alternate between different physiological or metabolic states has not been directly addressed. Indeed there is little information on the growth rates of *Leishmania* amastigotes in the mammalian host. Recent studies suggest that amastigotes expand rapidly during an initial silent phase of infection in susceptible mice, but that parasite numbers plateau following the appearance of lesions (Belkaid *et al.*, 2000; Engwerda *et al.*, 2004). Amastigotes in the latter phase may enter a metabolically quiescent latent phase and persist for decades in genetically resistant individuals. The presence of persistent *Leishmania* parasites is indicated by the re-emergence of disease in immunocompromised individuals years after resolution of the initial infection. While amastigotes appear to primarily infect macrophages, they can also colonize other host cells, including fibroblasts (Bogdan *et al.*, 2000), dendritic cells (Moll *et al.*, 1995) and neutrophils (van Zandbergen *et al.*, 2004), each of which are likely to contain biochemically distinct phagosomes. Finally, different species of *Leishmania* often target different tissues leading to markedly different disease manifestations (McMahon-Pratt and Alexander, 2004). Even between species, such as *L. major* and *L. mexicana*, that cause very similar diseases, amastigotes stages can induce markedly different phagosomes (from one parasite per vacuole to large communual vacuoles) and depend on different virulence factors and metabolic pathways for survival in macrophages (Turco *et al.*, 2001; Stafford *et al.*, 2002). The current literature thus indicates enormous species and stage-specific differences in the metabolic state of *Leishmania* in both the insect vector and mammalian host.

Studies on the readily cultured promastigote stages over the last three decades have provided a general overview of the metabolic capacity of these stages and their specific nutrient requirements. They have also led to the identification of several metabolic pathways in these evolutionarily divergent eukaryotes that differ substantially from those in the host cells. Importantly, it is now possible to explicitly test whether these metabolic processes contribute to the virulence of these parasites *in vivo* following targeted gene deletions or the overexpression of genes of interest (Beverley, 2003). However, in contrast to the wealth of information available on cultured promastigote stages, comparatively little is known about the metabolism of intracellular amastigotes *in vivo*. Most biochemical studies on the amastigote stages have been limited to amastigotes freshly isolated from macrophages or lesions. Differentiation protocols for obtaining axenic amastigotes have also been developed for some species (Bates, 1994; Zilberstein and Shapira, 1994; Saar *et al.*, 1998; Barak *et al.*, 2005). Axenic amastigotes are characteristically obtained by subjecting promastigotes to mild heat shock and acidic pH in serum-rich medium. While these forms exhibit many similarities to lesion- or macrophage-derived amastigotes (Saar *et al.*, 1998), the culture conditions used are unlikely to mimic the nutrient composition of the macrophage phagosome or induce the same metabolic responses of amastigotes *in vivo*. Remarkably, and in contrast to the experience with many other microbial pathogens, transcript and protein profiling approaches of different developmental stages of *Leishmania* have provided few insights into the metabolism of intracellular amastigotes (Duncan *et al.*, 2004). The vast majority of the genes encoding metabolic enzymes (>95%) are constitutively expressed in all developmental stages (Akopyants *et al.*, 2004; Almeida *et al.*, 2004; Holzer *et al.*, 2006; Leifso *et al.*, 2007). This reflects, in part, the use of polycistronic transcription in trypanosomatids which in turn has abrogated the need for the network of transcription factors that regulate gene transcription in most other organisms (Clayton, 2002). While it is important to emphasize that some metabolic processes are partly regulated by stage-specific changes in the stability and translation efficiency of mRNA (Clayton, 2002), it is clear that the regulation of *Leishmania* metabolism is largely controlled by post-translational mechanisms, including the modification of proteins, compartmentalization, protein oligomerization, and metabolic controls (ter Kuile, 1999).

While traditional biochemical and genetic approaches generally focus on individual metabolic pathways, direct analysis of the cellular metabolome is increasingly being used to obtain a more global view of metabolic processes and networks (Goodacre *et al.*, 2004). Metabolite profiling and fingerprinting approaches, collectively called metabolomics, involve the non-targeted, holistic analysis of changes in the complete set of metabolites in a cell (the metabolome) in response to environmental and cellular changes (Fernie *et al.*, 2004; Goodacre *et al.*, 2004). Metabolomics represent an emerging arm of functional genomics which has been made feasible by recent developments in both analytical instrumentation and bioinformatics (Weckwerth and Morgenthal, 2005). While it has been extensively applied to the analysis of metabolism in plants (Urbanczyk-Wochniak *et al.*, 2003; Cook *et al.*, 2004; Tohge *et al.*, 2005; Weckwerth *et al.*, 2004), and some animal and microbial systems (Raamsdonk *et al.*, 2001; Nicholson *et al.*, 2002), there are relatively few examples where it has been used to analyse the metabolism of microbial pathogens. Metabolomic approaches are likely to be particularly suitable for detecting stage-specific or environmentally induced changes in metabolism in *Leishmania* and other tryposomatids that are not mediated by changes in mRNA or protein expression. Moreover, additional tools that provide an unbiased coverage of cellular metabolism are urgently needed to define the function of genes in the trypanosomatid genomes that share no homology to genes from other organisms (Raamsdonk *et al.*, 2001).

In this chapter, we provide an overview of *Leishmania* metabolism and highlight a number of metabolic pathways that differ substantially from those in animal cells. We also describe the use of metabolite profiling and fingerprinting approaches to define the stage-specific changes in the *Leishmania* metabolome. Finally, we summarize metabolic processes in *Leishmania* that are required for survival of promastigote and/or amastigote stages in mammalian macrophages.

Overview of *Leishmania* metabolism

The following sections provide an overview of distinctive aspects of *Leishmania* metabolism that have been identified both experimentally and/or by analysis of the *L. major* genome. Many of these pathways are compartmentalized, as summarized in Fig. 5.1. Aspects of *Leishmania* metabolism have been covered in other recent reviews (Opperdoes and Michels, 2001; Hannaert *et al.*, 2003; Bringaud *et al.*, 2006; Opperdoes and Coombs, 2007).

Carbohydrate metabolism

All stages of *Leishmania* express the enzymes for glycolysis and the pentose phosphate pathway, which are used for energy generation and/or production of reducing equivalents and precursors for RNA and DNA synthesis, respectively (Fig. 5.2). Many of these enzymes are partially (Hannaert *et al.*, 1992; Nyame *et al.*, 1994) or completely sequestered within the glycosome, which is thus the major organelle for carbohydrate metabolism (Hannaert *et al.*, 2003). *Leishmania* are predicted to utilize a wide variety of hexoses and pentoses that are interconverted with glucose for use in these pathways (Opperdoes and Coombs, 2007). Disaccharides, such as sucrose, can also be catabolized and used as major carbon source by promastigote stages (Naderer, T, Sernee, MF. and McConville, MJ, unpublished). The initial glycolytic reactions present in glycosomes result in the net consumption of ATP and production of NAD^+. As the glycosome membrane is thought to be relatively impermeable to most metabolites, glycosomal pools of ATP and $NADH^+$ must be regenerated within the glycosome. Glycosomal ATP synthesis may be catalysed by 3-phosphoglycerate kinase (dually localized in the glycoscome and cytosol) (Concepcion *et al.*, 2001), a phosphoenolpyruvate carboxykinase and/or a novel pyrophosphatase-dependent pyruvate phosphate dikinase (Hannaert *et al.*, 2003). Similarly, NAD^+ can be regenerated by conversion of oxaloacetate to malate and succinate by a NADH-dependent malate dehydrogenase and fumarate reductase, respectively (Fig. 5.1). Interestingly, the gluconeogenic enzyme, fructose-1,6-bisphosphatase, is also localized to glycosomes and, together with glycolytic enzymes, is constitutively expressed in all developmental stages (Naderer *et al.*, 2006). The co-expression and co-localization of gluconeogenic and glycolytic enzymes would allow these parasites to rapidly adapt to different carbon sources, but raises the questions as to how these pathways are regulated without a futile cycle. The pentose phosphate pathway is also active in *Leishmania* promastigotes and all but one of the genes (the transaldolase) predict that the encoded proteins are glycosomal proteins (Maugeri *et al.*, 2003). This pathway provides the major source of NADPH for anabolic reactions and regeneration of cellular antioxidant systems. Indeed, a *L. infantum* mutant, adapted to nitric oxide stress, expressed high levels of the pentose phosphate shunt enzyme, 6-phosphogluconate dehydrogenase (Holzmuller *et al.*, 2005). It is thought that the localization of glycolytic, gluconeogenic and pentose phosphate pathways to the glycosome may facilitate the regulation of these pathways. This is true for the first step in the glycolytic pathway, catalysed by hexose kinase. Unlike other eukaryotic hexose kinases, the trypanosomatid hexose kinases are not allosterically regulated by ATP and relocation of this enzyme to the cytosol results in run-away phosphorylation of hexose pools and cell death when glucose is the major carbon source (Kessler and Parsons, 2005). However, many other glycosomal enzymes can be relocated to the cytosol with minimal perturbation of metabolic flux

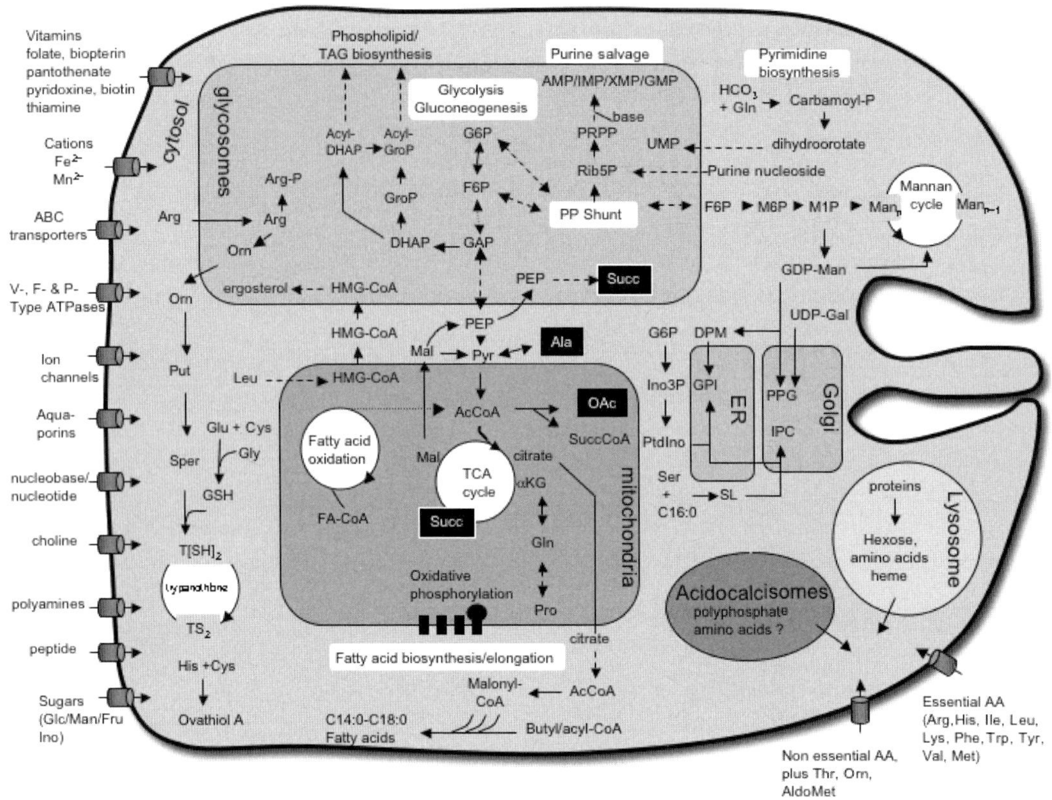

Figure 5.1 Overview of the *Leishmania* metabolome. Major metabolic pathways discussed in this Chapter. Metabolic steps are represented by arrows. Solid arrows indicate direct steps in a pathway, dashed arrows indicate that known multiple steps in a pathway are not shown. The subcellular localization of specific pathways is based on the subcellular localization of one or more enzymes in the pathway and/or the presence of targeting presequences in those enzymes. It should be noted that many glycosomal enzymes are often partially localized in the cytosol.

Abbreviations used: AA, amino acid (standard three-letter code used for specific amino acids); AcCoA, acetyl-CoA; AcylDHAP, acyldihydroxyacetonephosphate; Acyl-GroP, acyl-glycerol-3-phosphate; Ala, alanine; AldoMet, S-adenosylmethionine; AMP, adenosine monophosphate, Arg, arginine; Arg-P, arginine-phosphate; Asp, aspartate; C16:0, palmitateCys, cysteine; DHAP, dihydroxyacetonephosphate; DPM, dolichol-phosphate-mannose; F6P, fructose-6-phosphate; FA-CoA, fatty acyl-CoA; Fum, fumarate; GAP, glyceraldehyde-3-phosphate; G6P, glucose-6-phosphate; GDP-Man, GDP-mannose; Glu, glutamate; Gly, glycine; GPI, glycosylphosphatidylinositols; GroP, glycerol-3-phosphate; GSH, glutathione; His, histidine; HMG-CoA, 3-hydroxy-3-methylglutaryl CoA; IMP, inosine monophosphate; Ino, *myo*-inositol; Ino3P, inositol-3-phosphate; IPC, inositolphosphoceramide; αKG, α-ketoglutarate; LPG, lipophosphoglycan; Mal, malate; M1P, mannose-1-phosphate; M6P, mannose-6-phosphate; OAc, acetate; Orn, ornithine; PEP, phosphoenolpyruvate; PPG, proteophosphoglycan; PRPP, phosphoribose pyrophosphate; PPG, proteophosphoglycan; PP*i*, pyrophosphate; Pro, proline; PtdIno, phosphatidylinositol; Pyr, pyruvate; Put, putrescine; Rib5P, ribose-5-phosphate; Sper, spermidine; SL, sphingosine; Succ, succinate; $T[SH]_2$ and TS_2, trypanothione (reduced and non-reduced); UDP-Gal, UDP-galactopyranose.

(Roberts *et al.*, 2004; Zarella-Boitz *et al.*, 2004) indicating that most glycosomal metabolites are in equilibrium with cytosolic pools. Essentially nothing is known about transport of metabolites into or out of the glycosome, and no transporter proteins have been localized to the glycosomal membrane.

TCA cycle and mitochondrial respiration

Mitochondrial respiration is likely to be important for all developmental stages of *Leishmania* (Van Hellemond and Tielens, 1997). When glucose is plentiful, the major products of the glycolytic pathway, phosphoenolpyruvate and

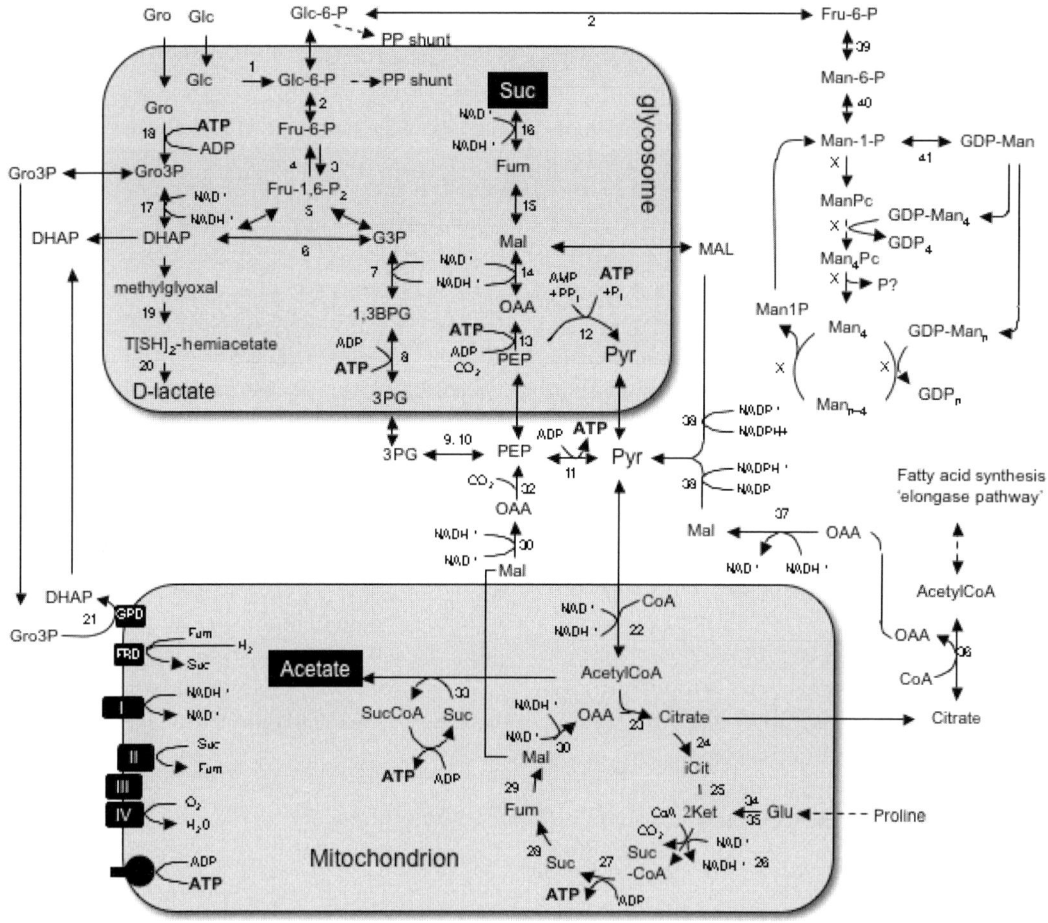

Figure 5.2 Central carbon metabolism in *Leishmania*. The presence and subcellular localization of glycolytic pathway, pentose phosphate shunt and TCA cycle enzymes are based on experimental observations and analysis of the *L. major* genome. The pathway of mannan biosynthesis has recently been deduced (Sernee *et al.*, 2006). Aspects of this scheme are also based on studies undertaken on the insect (procyclic) stage of *T. brucei* (Van Hellemond *et al.*, 1998; Coustou *et al.*, 2003; Lamour *et al.*, 2005).

Abbreviations used: 1,3BPG, 1,3-bisphosphoglycerate; 2KG, α-ketoglutarate; 3PG, 3-phosphoglycerate; DHAP, dihydroxyacetonephosphate; Fru-6-P, fructose-6-phosphate; Fru-1,6-P_2, fructose-1,6-bisphosphate; Fum, fumerate; Glc-6-P, glucose-6-phosphate; G3P, glyceraldehyde-3-phosphate; Gro, glycerol; Gro3P, glycerol-3-phosphate; Glu, glutamate; iCit, isocitrate; Mal, malate; Man-6-P, mannose-6-phosphate; Man-1-P, mannose-1-phosphate; ManPc, mannose-cyclic-phosphate; Man_n, mannan; PEP, phosphoenolpyruvate; Pyr, pyruvate; OAA, oxaloacetate; Suc, succinate; SucCoA, succinyl-CoA; PP shunt, pentose phosphate shunt; I-IV, complexes of the respiratory chain; Enzymes; 1, hexose knase; 2, glucose-6-phosphate isomerase; 3, phosphofructokinase; 4, fructose-1,6-bisphosphatase; 5, aldolase; 6, triose-phosphate isomerase; 7, glyceroaldehyde-3-phosphate dehydrogenase; 8, phosphoglycerate kinase; 9, phosphoglycerate mutase, 10, enolase; 11, pyruvate kinase; 12, pyruvate phosphate dikinase; 13, phosphoenolpyruvate carboxykinase; 14, malate dehydrogenase; 15, fumarate hydratase; 16, NADH-dependent fumarate reductase; 17, glycerol-3-phosphate dehydrogenase; 18, glycerol kinase; 19, glyoxylase-I; 20, glycoxylase-II; 21, FAD-dependent glycerol-3-phosphate dehydrogenase; 22, pyruvate dehydrogenase complex; 23, citrate synthase; 24, aconitase; 25, isocitrate dehydrogenase; 26, 2-ketoglutarate dehydrogenase; 27, succinyl-CoA synthetase; 28, succinate dehydrogenase; 29, fumarate hydratase; 30, malate dehydrogenase; 31, PEP carboxykinase; 33, acetate: succinate CoA transferase and/or acetyl-CoA synthase 34, glutamate aminotransferase; glutamate dehydrogenase; 36, citrate lyase; 37, malate dehydrogenase; 38, malic enzyme; 39, phosphomannoisomerase; 40, phosphomannomutase; 41, GDP-Man pyrophosphorylase; X, unknown enzymes.

pyruvate are, respectively, catabolized to succinate in the glycosome or transported to mitochondria and oxidized to CO_2 or acetate (Fig. 5.2). The generation of CO_2 reflects the oxidation of acetyl-CoA (generated by the mitochondrial pyruvate dehydrogenase complex) in the conventional TCA cycle, while the generation of acetate reflects the breakdown of acetyl-CoA by acetyl-CoA synthetase, or the combined actions of acetate:succinate CoA transferase and succinyl-CoA ligase (Van Hellemond *et al.*, 1998; ter Kuile, 1999; van Weelden *et al.*, 2005) (Fig. 5.2). Dependence on TCA cycle activity may increase markedly in the intracellular amastigote stages that are likely exposed to low glucose, but relatively high levels of amino acids (and lipids) (Naderer *et al.*, 2006). The expression of the mitochondrial isoform of malate dehydrogenase is highly upregulated (>100-fold) in *L. mexicana* amastigotes compared to promastigotes (Leroux *et al.*, 2006). Interestingly, the cytoplasmic and glycosomal isoforms of this enzyme are also highly upregulated in amastigotes (Leroux *et al.*, 2006). Cytoplasmic malate dehydrogenase, may be part of a pathway that converts mitochondria-derived citric acid to pyruvate with generation of NADPH, essential for sustaining the oxidative defences of intracellular amastigotes (Fig. 5.2).

Leishmania mitochondria contain the components of a functional respiratory chain, encoded by both nuclear and mitochondrial (kinetoplast) maxicircle genes. Analysis of the *L. major* genome indicate the presence of NADH dehydrogenase (complex I), succinate dehydrogenase (complex II), Coenzyme Q-cytochrome c reductase (complex III), and the proton pumping cytochrome c oxidase (complex IV) (Santhamma and Bhaduri, 1995; Bermudez *et al.*, 1997). While inhibitors of complex I (such as rotenone) had no effect on the respiration of *L. donovani* promastigotes (Santhamma and Bhaduri, 1995), inhibitors of complex II-IV (i.e. malonate, thenoyltrifluoroacetone, antimycin and cyanide) are effective, although sometimes partial inhibitors of respiration, and induce a reversible state of metabolic arrest in a number of *Leishmania* species (Van Hellemond and Tielens, 1997). *Leishmania* also contain a fumarate reductase that catalyses the reverse reaction of succinate dehydrogenase, and may donate electrons through complex II (Santhamma and Bhaduri, 1995; Chen *et al.*, 2001). Animal cells lack fumarate reductase and chalcone inhibitors of this enzyme have strong anti-*Leishmania* activity *in vitro* and *in vivo* (Chen *et al.*, 2001). The electrochemical gradient generated by the respiratory chain is used by an oligomycin- and DCCD-sensitive mitochondrial ATP-synthase. Other mitochondrial dehydrogenases that contribute to the reduction of ubiquinone are the FAD^+-dependent glycerol-3-phosphate dehydrogenase (catalysing the oxidation of glycosome-derived glycerol-3-phosphate to dihydroxyacetonephosphate), and the membrane-bound proline dehydrogenase (Guerra *et al.*, 2006). By contributing to the redox balance of the glycosome, FAD^+-dependent glycerol-3-phosphate dehydrogenase may increase the efficiency of glycolysis (by avoiding the need for glycosomal succinate fermentation) and allow all developmental stages to utilize glycerol (derived from phospholipid and triacyl glycerol breakdown) as a carbon source (Guerra *et al.*, 2006).

Energy stores

Leishmania accumulate a unique type of carbohydrate reserve material composed of linear chains of β-1,2-linked mannose (Previato *et al.*, 1984; Ralton *et al.*, 2003). These oligosaccharides replace conventional carbohydrate reserves of other eukaryotes such as glycogen or starch. Mannan accumulates in stationary phase promastigotes, when carbon sources are in excess of requirements for growth (Keegan and Blum, 1992; Ralton *et al.*, 2003). Mannan also accumulates when promastigotes are subjected to heat shock and in intracellular amastigotes, indicating that carbon sources are not limiting in the macrophage phagosome (Ralton *et al.*, 2003). Mannan biosynthesis occurs in the cytoplasm and the major steps have recently been elucidated (Sernee *et al.*, 2006) (Fig. 5.2). *De novo* synthesis of mannan requires the initial synthesis of a novel mannose-cyclic-phosphate primer. This sugar is initially elongated with >4 mannose residues before the cyclic phosphate residue is removed. Neutral mannan oligomers are further elongated with β-1,2-linked mannose residues (donated from GDP-Man) to form the mixed population of oligomers 4–50 residues long (Fig. 5.2). Mannan accumulates in the cytosol (up to 10 mM) when glucose is in excess of normal energy requirements of the parasite, suggesting that glycosomal pools of glucose-6-phosphate or fructose-6-phosphate are in rapid equilibrium with cytoplasmic pools of these sugars. The predominant localization of glucose-6-phosphate isomerase in the cytosol of *L. mexicana* promastigotes (Nyame *et al.*, 1994), indicates that glucose-6-phosphate may be the major sugar phosphate to be exported from the glycosome. Mannan is catabolized when carbon sources are limiting, although none of the enzymes involved in its degradation have been defined (Ralton *et al.*, 2003). Several other trypanosomatids synthesize mannan, including *Crithidia fasciculata* and *Herpetomonas samuelpessoai* (Mendonca-Previato *et al.*, 1979), although neither *T. brucei* nor *T. cruzi* appears to synthesize mannan.

As in other protists and metazoans, longer term storage of reduced carbon in *Leishmania* involves lipids. Fatty acids synthesized *de novo* or acquired from the extracellular milieu can be incorporated into triacylglycerols (TAG) that in turn are packaged into lipid bodies. In yeast and other eukaryotes, lipid body biogenesis is thought to involve the deposition of TAG within the ER membrane bilayer. The *L. major* genome contains a number of TAG lipases that may be involved in mobilizing these bodies during extended nutrient starvation (T. Naderer and M.J. McConville, unpublished data). Released fatty acids can be catabolized by β-oxidation in both the glycosomes and in the mitochondria. While a number of different fatty acyl-CoA synthetases and acyl-CoA dehydrogenases with different chain-length specificity have been identified, most of the genes encoding β-oxidation enzymes contain mitochondrial targeting signals, and the role of glycosomal fatty acid oxidation is currently undefined (Berriman *et al.*, 2005). The major products of TAG breakdown are glycerol and acetyl-CoA. The glycerol could potentially constitute an important carbon source for intracellular amastigotes, while acetyl-CoA could feed into the TCA cycle. However, *Leishmania* appear to lack the enzymes of the glyoxylate shunt (despite initial reports that the activities might exist (Simon *et al.*, 1978)) that are required for net conversion of acetyl-CoA to sugars via the TCA cycle and gluconeogenesis (Berriman *et al.*, 2005; Naderer *et al.*, 2006). It is therefore unlikely that *Leishmania* promastigotes or amastigotes can utilize fatty acid as their sole or major carbon source.

Lipid biosynthesis

In most eukaryotes and prokaryotes, bulk fatty acid biosynthesis is mediated by either type I and/or type II fatty acid synthases. The genomes of trypanosomatids, including *Leishmania*, encode a type II fatty acid synthase machinery, which in *T. brucei*, is localized to the mitochondria (Morita *et al.*, 2000; Stephens *et al.*, 2007). Metabolic labelling experiments with *T. brucei* indicate that this pathway is active and important in providing lipids and lipid-precursors (including the C8 octanoate precursor for mitochondrial lipoic acid synthesis) that are likely to be essential for mitochondrial function (Stephens *et al.*, 2007). However, this pathway may not be the major pathway for *de novo* fatty acid biosynthesis in trypanosomatids. Englund and colleagues have recently shown that a family of β-ketoacyl-CoA synthases, or elongases (ELO) may drive bulk fatty acid biosynthesis in *T. brucei*, *Leishmania* and *T. cruzi* (Lee *et al.*, 2006). The elongases are located on ER membranes, and utilize malonyl-CoA as the two carbon donor and CoA as the carrier and activating thiol (Lee *et al.*, 2006). The typanosomatid elongase system is modular, with each ELO enzyme exhibiting selectivity for elongating different chain length precursors. The shortest acyl-CoA substrate in this pathway is butyryl-CoA. The pathway and carbon source required for butyryl-CoA synthesis has not been defined. The localization of this pathway on the cytoplasmic face of microsomes implies that mitochondrial citric acid is exported to the cytoplasm and converted to oxaloacetate and acetyl-CoA, with the latter metabolite then being carboxylated to malonyl-CoA.

Fatty acids synthesized *de novo* by the elongase system or scavenged from the host are incorporated into the full repertoire of glycerol-phospholipids. The initial steps in glycerolipid synthesis involve the transfer of fatty acyl chains from acyl-CoA donors to either glycerol-3-phosphate or DHAP. While acyl-glycerol-3-phosphate is expected to be the precursor for glycerolipid biosynthesis, targeted deletion of the *L. major* glycerol-3-phosphate acyltransferase had little effect on phospholipid biosynthesis or parasite growth (Zufferey and Mamoun, 2005). In contrast, deletion of the *L. major* glycosomal DHAP-acyltransferase had a profound affect on the capacity of promastigote to survive in stationary growth phase or to infect macrophages and mice (Zufferey and Ben Mamoun, 2006). Taken together with the analysis of an ether-lipid mutant (Zufferey *et al.*, 2003), it appears that acyl-DHAP is the primary precursor for both ester- and ether-glycerolipid biosynthesis and that the conversion of acyl-DHAP to acyl-glycerol-3-phosphate and alkyl-glycerol-3-phosphate, respectively, represents an important branch point in leishmanial lipid biosynthesis.

Leishmania can also synthesize sphingolipids *de novo*, a pathway that is initiated by serine palmitoyltransferase. Remarkably, this enzyme is only expressed in the promastigote stages (one of the few examples of tight stage-specific expression of a metabolic enzyme), and amastigote stages appear to scavenge all their sphingolipid requirements from the host (Zhang *et al.*, 2003; Denny *et al.*, 2004). Sphingolipids are converted to ceramide which is further modified with inositol to form inositolphosphoceramide (IPC), the major sphingolipid of these parasites (Singh *et al.*, 1988; Ralton and McConville, 1998). A putative IPC-synthase has recently been identified in *L. mexicana* (Denny *et al.*, 2006) and like the fungal counterparts, is likely to be localized in the Golgi apparatus. Pathways of phospholipid, isoprenoid and sterol biosynthesis are also similar to those that occur in yeast, with two significant differences. First, the major aminophospholipids of *Leishmania* are enriched in alkylacylglycerol lipids (Singh *et al.*, 1988; Lux *et al.*, 2000). Second, the major precursor for isoprenoid and sterol biosynthesis, hydroxymethylglutaryl-CoA (HMG-CoA) is derived from the mitochondrial breakdown of leucine, rather than acetate (Ginger *et al.*, 1999; Ginger *et al.*, 2001). As the enzymes involved in isoprenoid synthesis are localized to the glycosomes, HMG-CoA must presumably be transported between the mitochondria and glycosomes (Fig. 5.1).

Amino acid metabolism

Leishmania promastigotes and amastigotes can use amino acids as their major or sole carbon source in the absence of carbohydrates (Burchmore *et al.*, 2003). Recent studies suggest that amino acid uptake and catabolism is important for amastigote proliferation in macrophage phagolysosomes. First, *Leishmania* amastigotes exhibit high rates of amino acid uptake and express specific amino acid permeases, concommitant with a decrease in hexose uptake (Hart and Coombs, 1982; Mazareb *et al.*, 1999; Akerman *et al.*, 2004; Geraldo *et al.*, 2005). Second, targeted deletion of *L. major* fructose-1,6-bisphosphatase, a key enzyme in the gluconeogenic pathway, prevented amastigote proliferatation in macrophages and the development of lesions in mice (Naderer *et al.*, 2006). Third, the lytic capacity of amastigote lysosomes is considerably higher than in promastigote stages, providing an internal source of amino acids, (Mottram *et al.*, 1996). Finally, all stages of *Leishmania* have the capacity to catabolize most amino acids (Hart and Coombs, 1982). Pathways delineated experimentally or predicted from the *L. major* genome (Opperdoes and Coombs, 2007) include the following. Glutamate can be deaminated to α-ketoglutarate by both a mitochondrial NAD-dependent glutamate dehydrogenase and a putative cytosolic NADPH-dependent isozyme. While the mitochondrial α-ketoglutarate can be fed into the TCA cycle, the fate of the cytosolic pool is unclear. Proline is catabolized by the sequential actions of a mitochondrial proline dehydrogenase and a pyrroline-5-carboxylate dehydrogenase, to glutamate. This pathway generates $FADH_2$, NADH and TCA cycle intermediates and proline is preferentially utilized by some trypanosomatids (i.e. *T. brucei* procyclics) when exogenous glucose levels are low (Lamour *et al.*, 2005). Alanine, serine and cysteine can all be converted to pyruvate. Threonine can either be converted to serine by the THF-dependent serine hydroxymethyltransferase, prior to its conversion to pyruvate or catabolized via the Ser/Thr dehydratase and the ketobutyrate pathway to succinyl-CoA (Opperdoes and Coombs, 2007). Methionine, valine and isoleucine may also be degraded to succinyl-CoA and succinate, and used to fuel the TCA cycle or the conversion of acetyl-CoA (by acetate:succinyl-CoA tranferase) to acetate and ATP (Berriman *et al.*, 2005). Aspartate feeds directly into the TCA cycle following conversion to oxaloacetate by a mitochondrial aspartate aminotransferase or to fumarate via a functional purine-nucleotide cycle. Glycine can be split into CO_2 and formic acid by the glycine cleavage system, or converted to serine and pyruvate by the THF-coupled serine hydroxymethyltransferase, and Thr/Ser dehydratase (Opperdoes and Coombs, 2007). *L. major* lacks the enzymes for breakdown of aromatic amino acids, tryptophan, lysine and histidine (Berriman *et al.*, 2005).

The amino acid biosynthetic capacity of *L. major* is similar to metazoans. These parasites lack enzymes for *de novo* synthesis of arginine, histidine, isoleucine, leucine, lysine, phenylalanine, tryptophan, tyrosine and valine, which thus need to be scavenged from the environment, but can synthesize all non-essential amino acids (Steiger and Steiger, 1976; Steiger and Steiger, 1977). While many of the genes required for *de novo* biosynthesis of threonine and methionine are present, the enzymes required for synthesis of homoserine (a common precursor in both pathways) are missing (Opperdoes and Coombs, 2007). Consequently, *Leishmania* are auxotrophic for methionine (Vickers *et al.*, 2006). Methionine is converted to S-adenosylmethionine (AdoMet), the methyl donor for all transmethylation reactions except that involved in the regeneration of methionine from homocysteine (Reguera *et al.*, 2007). Supplementation of *L. major* promastigote cultures with homocysteine partially relieved the methionine auxotrophy, indicating that the AdoMet cycle is active in this stage (Vickers *et al.*, 2006). However, methioine synthesis from homocysteine is not required for amastigote infectivity, indicating that the macrophage phagolysosome contains sufficient levels of methionine for growth (Vickers *et al.*, 2006). AdoMet can also be internalized by *Leishmania* to supplement this *de novo* pathway.

L. major is predicted to contain all the enzymes in the urea cycle except arginine-succinate lyase (Berriman *et al.*, 2005) and is therefore auxotrophic for arginine. This amino acid is taken up by a high-affinity, plasma membrane permease (Shaked-Mishan *et al.*, 2006) and is converted in the glycosome to either phosphoarginine (an important high-energy phosphagen (Colasante *et al.*, 2006)) or to the non-proteinogenic amino acid, ornithine (Roberts *et al.*, 2004). Ornithine is the precursor for polyamines that are required for normal growth as well as the biosynthesis of trypanothione, a major thiol in trypanosomatids. The conversion of ornithine to putrescine and spermidine is catalysed by the cytosolic enzymes; ornithine decarboxylase and spermidine synthase (Jiang *et al.*, 1999; Roberts *et al.*, 2004) (Fig. 5.1). Unexpectedly, given the importance of these polyamines, *Leishmania* apparently lack pathways for converting arginine to putrescine via agmatinase, for synthesizing ornithine from glutamate or for synthesizing spermine (Jiang *et al.*, 1999; Roberts *et al.*, 2004). They are also unable to degrade spermine to spermidine or spermidine to putrescine. Based on the biochemical phenotype of *Leishmania* mutants deficient in ornithine decarboxylase, S-adenosylmethionine decarboxylase (required for the synthesis of S-adenosylmethioninamine) and arginase, it is likely that spermidine is the essential polyamine (Jiang *et al.*, 1999; Roberts *et al.*, 2002; Roberts *et al.*, 2004) (Fig. 5.1). *Leishmania* are capable of supplementing their minimal polyamine biosynthetic capabilities by scavenging a range of polyamines from the extracellular milieu. Specific transporter activities for putrescine and spermidine have been detected (Basselin *et al.*, 2000) and a gene encoding a polyamine transporter has been recently identified in *L. donovani* (Hasne and Ullman, 2005). *Leishmania* amastigotes may be able to obtain all their spermidine requirements from the macrophage phagolysosome as the *L. major* Δarg mutant is able to survive in macrophages despite being a strict polyamine auxotroph (Roberts *et al.*, 2004; Reguera *et al.*, 2005). Furthermore, specific inhibition of the macrophage arginase attenuated growth of *L. major* amastigotes in macrophages *in vitro* and *in vivo* (Kropf *et al.*, 2005), indicating that the availability of host polyamines regulate growth in intracellular parasite stages.

Purine salvage and pyrimidine synthesis

Leishmania lack the enzymes required for *de novo* synthesis of inosine monophosphate from phosphoribosyl pyrophosphate and are thus auxotrophic for purines. All promastigote stages express plasma membrane and secreted nucleotidase/nucleases and polytopic membrane transporters that mediate the degradation of extracellular nucleotides and nucleic acids and the uptake of purine (and pyrimidine) nucleosides (Vasudevan *et al.*, 1998; Sopwith *et al.*, 2002; Landfear *et al.*, 2004; Joshi and Dwyer, 2007). The *L. donovani* promastigote NT1 (NT1.1/NT1.2) and NT2 transporters are responsible for uptake of adenosine /pyrimidine nucleosides and guanosine/inosine, respectively (Carter *et al.*, 2001; Landfear *et al.*, 2004). Amastigote stages may express additional or alternative transporters for the uptake of purine nucleosides (adenosine, guanosine and inosine) (Ghosh and Mukherjee, 2000), purine nucleobases (adenine, guanine and xanthine) (Al-Salabi and de Koning, 2005) and the pyrimidine base, uracil (Papageorgiou *et al.*, 2005). The K_m for the nucleoside transporters are 100- to 1000-fold lower than mammalian nucleoside transporters, and at least some of these activites are post-translationally upregulated when exogenous substrates are low, facilitating effective uptake in the macrophage phagosomes (Seyfang and Landfear, 1999; Carter *et al.*, 2001). Internalized purine bases are converted to their respective nucleotides by three distinct phosphoribosyl-transferases (PRTs) (Boitz and Ullman, 2006a,b). While the *L. donovani* adenine-PRT is localized in the cytosol, the hypoxanthine-guanine- and xanthine-PRTs are primarily localized in the glycosomes (Zarella-Boitz *et al.*, 2004). However, glycosomal localization was not essential for xanthine-PRT function (Zarella-Boitz *et al.*, 2004) and the reason for the compartmentalization of some, but not all of the PRT remains unclear. *L. donovani* promastigotes lacking both hypoxanthine-guanine- and adenine-PRTs grow normally when xanthine is provided indicating that *Leishmania* contain all the enzymes needed for interconverting purine nucleotides, including a mitochondrial purine-nucleotide cycle for the interconversion of IMP and AMP. In addition to expressing pyrimidine transporters, *Leishmania* have the capacity for *de novo* synthesis of pyrimidines from glutamate, bicarbonate and aspartate. While the initial steps in pyrimidine biosynthesis occur in the cytoplasm, the last two steps in this pathway are predicted to occur in the glycosome.

Vitamins and cofactors

Leishmania stages are auxotrophic for several vitamins that must be scavenged from the host environment. The *L. major* genome lacks enzymes for *de novo* synthesis of folic acid, biopterin, pantothenate, pyridoxine, biotin, thiamine, riboflavin, nicotinic acid and cyanocobalamin, all of which are typically included in defined growth media (Steiger and Steiger, 1976; Merlen *et al.*, 1999). Putative genes for a *de novo* synthesis of ascorbic acid are predicted from the *L. major* genome and a FAD-dependent oxidoreductase that might catalyse the final step in the pathway partially characterized in *T. brucei* (Wilkinson *et al.*, 2005). However, direct evidence for this pathway or a requirement for exogenous ascorbic acid has not yet been demonstrated. Specific transporters for the uptake and reduction of both biopterin and folate to their active intracellular forms (BH_4 and tetrahydrofolate) have been identified (Ouellette *et al.*, 2002). Interestingly, both the *Leishmania* biopterin transporter, BT1, and pteridine reductase (PTR1) can transport and reduce folates, respectively, but not antifolates such as methotrexate (that inhibits dihydrofolate reductase), and are frequently upregulated in methotrexate-resistant parasites (Kundig *et al.*, 1999; Lemley *et al.*, 1999). The function of biopterin in *Leishmania* metabolism remains enigmatic. Biopterins can partially substitute the requirement of folate and there is evidence that biopterin can be converted to 5′-methyltetrahydrofolate and 10′-formyl-tetrahydrofolate *in vivo* (Beck and Ullman, 1991), although *Leishmania* lack the expected enzymes activities for this pathway (Ouellette *et al.*, 2002). Interestingly, surface BT1 expression and internal BH_4 levels decrease when *L. major* promastigotes reach stationary phase, coinciding with metacyclogenesis. Moreover, *L. major* mutants lacking either BT1 or PTR1 transform more efficiently to metacylic promastigotes and were more infective in mice (Cunningham and Beverley, 2001; Cunningham *et al.*, 2001). Thus a decrease in intracellular levels of BH_4 provides a signal or modulates a metabolic pathway that is required for metacyclogenesis and virulence in the mammalian host. Folates on the other hand serve as cofactors in a variety of one-carbon transfer reactions, and are minimally required for biosynthesis of thymidylate in *Leishmania*, as well as interconversion of serine and glycine and methionine biosynthesis (Ouellette *et al.*, 2002). *Leishmania* also exhibit an absolute requirement for haem (Steiger and Steiger, 1977). Haem is normally synthesized from succinyl-CoA, but *L. major* lack the first two enzymes in this pathway, despite having the last three enzymes, all of which appear to have been derived by lateral transfer from bacteria (Berriman *et al.*, 2005). Uptake of haemoglobin (the major *in vivo* source of haem) may be mediated by endocytosis, through a 46 kDa haemoglobin receptor that shares sequence similarity to the parasite hexose kinase (Krishnamurthy *et al.*, 2005).

Trypanothione and ovothiol biosynthesis

Trypanosomatids synthesize four low molecular weight thiols; glutathione, glutathionylspermidine, trypanothione (N^1,N^8-bis(glutathionyl) spermidine) and ovothiol A (N-methyl-4-mercaptohistidine), that together comprise the unique thiol-redox system of these parasites. While the cellular levels of trypanothione remain fairly constant in different species and developmental stages (7–9 mM), the cellular level of ovothiol A can vary by more than 200-fold. For example, ovothiol levels in *L. donovani* promastigotes and amastigotes range from ~10 mM to below levels of detection, respectively (Ariyanayagam and Fairlamb, 2001). In contrast, comparable levels of ovothiol (5–10 mM) occur in promastigote and amastigote stages of *L. major* (Ariyanayagam and Fairlamb, 2001). The pathway of trypanothione biosynthesis and the properties of the enzymes involved in maintaining trypanothione in a reduced state have been extensively reviewed (Steenkamp, 2002). Trypanothione is synthesized by two ATP-dependent reactions from the tripeptide glutathione and spermidine. Interestingly, glutathionylspermidine synthase can also catalyse the reverse reaction to regenerate spermidine and glutathione, under conditions of polyamine deficiency. In addition to playing a pivotal role in the antioxidant defence system of these parasites, trypanothione also serves as a direct or indirect (via tryparedoxin) donor of reducing equivalents for some enzymes, such as ribonucleotide (Dormeyer *et al.*, 2001), as well as the functioning of the glyoxylase detoxification system (Ariza *et al.*, 2006). Methylglyoxal is a toxic 2-oxoaldehyde generated during glycolysis that can form advanced glycation products with proteins and nucleic acids. Trypanothione forms a hemiacetal adduct with methylglyoxylate in the glycosome, generating the preferred substrate for two enzymes; lactoylglutathione lyase and hydroxyacylglutathione hydrolase. The products of this detoxification pathway are D-lactate and trypanothione, and the former can subsequently be converted to pyruvate (Ariza *et al.*, 2006). Unlike trypanothione, ovothiol A occurs in a number of other protists and in marine invertebrates, where it is thought to function as a bulk antioxidant and a one-electron donor. Ovothiol A biosynthesis has been studied in *Crithidia fasciculata*; histidine and cysteine are converted to S-(4-histidyl)-cysteine-sulfoxide, which is subsequently lysed to 4-mercatohistidine and pyruvate. 4-Mercaptohistidine is then methylated by S-adenosylmethionine to generate ovothiol A (Vogt *et al.*, 2001; Steenkamp, 2002).

ER and Golgi glycosylation pathways

The major surface macromolecules of *Leishmania* promastigotes are glycosyl-phosphatidylinositol (GPI)-anchored lipophosphoglycan (LPG), GPI-anchored glycoproteins and proteophosphoglycans (PPGs), and a family of free GPI glycolipids, termed glycoinositolphospholipids (GIPLs) (Naderer *et al.*, 2004). All the steps involved in the biosynthesis of the protein anchors, including the transfer of GlcNAc to PI, the de-N-acetylation of GlcNAc-PI and subsequent modification with three mannose residues (donated from dolichol-phosphate-mannose) and ethanolamine-phosphate (donated from PE), take place in the ER (McConville *et al.*, 2002). Mature GPI proteins can be reversibly modified with an additional inositol-linked fatty acid, although this is removed prior to addition of the GPI to newly synthesized proteins in the lumen of the ER (Hilley *et al.*, 2000).

The GPI anchor of LPG is assembled on the same subpool of alkylacyl-PI as the protein anchors and the first three steps of these pathways are the same (Ralton and McConville, 1998). However, the Man_1GlcN-PI species is modified with a second mannose residue in α1–3, rather than α1–6 linkage, donated directly from GDP-Man rather than dolichol-phosphate mannose (Ilgoutz *et al.*, 1999). The Man_2GlcN-PI anchor precursor is subsequently transported to the Golgi apparatus where it is modified with three galactose residues donated from UDP-galactofuranose (UDP-Gal_f) and UDP-galactopyranose (UDP-Gal) (Turco *et al.*, 2001). The UDP-Gal mutase responsible for the synthesis of UDP-Gal_f donor has recently been identified and shown to belong to a new class of eukaryotic enzymes (Beverley *et al.*, 2005; Kleczka *et al.*, 2007). The mature LPG anchor precursor is subsequently modified with glucose-1-phosphate (on a core mannose) and long chains of repeating Galβ1-Manα-P phosphodisaccharides, assembled by sequential transfer of mannose-1-phosphate and galactose from GDP-Man and UDP-Gal, respectively. These phosphoglycan chains can be elaborated with single monosaccharide or long oligosaccharide side chains, and are generally capped with 2–4 αMan residues (McConville *et al.*, 1995). The side chain composition of LPG (and related proteophosphoglycans) is likely regulated by the species-specific expression of large families of Golgi glycosyltransferases (Dobson *et al.*, 2003; Dobson *et al.*, 2003), the activity of Golgi sugar nucleotide transporters (Ma *et al.*, 1997) and rate of transport through the Golgi apparatus. In contrast to the protein and LPG anchors, the GIPLs are assembled on PI species with a much more heterogeneous lipid composition (Ralton and McConville, 1998). They also have more polymorphic glycan head-groups. Depending on the species and developmental stage, the GIPLs can have the same glycan sequence as the protein anchors, the LPG anchors, a combination of both anchors, or be elaborated with distinct side chains not found on either class of anchors (McConville *et al.*, 2002). The rate of biosynthesis of anchor and free GPIs varies markedly in different developmental stages, being highest in rapidly dividing promastigotes and lowest in stationary phase promastigotes and lesion-derived amastigotes (Ralton *et al.*, 2003). These stage-specific differences suggest that GPI biosynthesis closely matches parasite growth rate and requirement of new membrane components. Interestingly, amastigote stages of *Leishmania* express very low or undetectable levels of LPG and also down-regulate the surface expression of GPI-anchored proteins, such as gp63 (McConville and Blackwell, 1991). In contrast, they continue to express very high levels of free GPIs that consequently make up the major surface components of intracellular stages (McConville and Blackwell, 1991). As all growth stages of *Leishmania* appear to constitutively express the enzymes involved in phosphoglycan biosynthesis, it is likely that the stage-specific expression of LPG biosynthesis is regulated at the level of synthesis of different GPI precursors.

Many cell surface and secretory proteins of *Leishmania* are modified with elaborate phosphoglycan chains, as well as conventional N-linked glycans (Ilg *et al.*, 1994; Ilg, 2000). The protein-linked phosphoglycans are structurally related to those found on LPG, although significant differences in chain length and degree of branching can occur (Ilg *et al.*, 1994; Ilg *et al.*, 1996). Protein phosphoglycosylation (PPG) occurs in the Golgi apparatus and is required for the assembly of large oligomeric proteophosphoglycan complexes that are tethered to the plasma membrane via a GPI anchor (mPPG) or secreted into the extracellular milieu (PPG) (Ilg, 2000). Protein phosphoglycosylation is initiated by transfer of Manα-1-P to serine-rich repetitive domains in the polypeptide backbone (Moss *et al.*, 1999). The Manα1-P-serine epitope primes the synthesis of short mannan side chains composed of α1,2-linked Man or phosphodisaccharide repeat units. The latter can be modified with monosaccharide/oligosaccharide side chains (as in LPG) or with additional phosphodisaccharide chains (not found in LPG). The structures of the protein-linked phosphoglycans vary in different developmental stages, possibly reflecting stage-specific regulation of Golgi glycosyltransferases and/or the rate of transport of PPGs through the secretory pathway.

Phosphate metabolism

In common with other protists, and some prokaryotes, trypanosomatids accumulate pyrophosphate and polyphosphates in specialized acidic vacuoles, termed acidocalcisomes (Docampo and Moreno, 2001). Polyphosphates are linear polymers of orthophosphate residues linked by high energy phosphoanhydride bonds. Trypanosomatid polyphosphates have comparatively short chains (2–5 residues), but may reach concentrations of 3–8 M and form microcrystalline aggregates in acidocalcisomes (Moreno *et al.*, 2000; Moreno *et al.*, 2002). The role(s) of polyphosphate in trypanosomatid metabolism is unclear. They may constitute an alternative energy to ATP, sequester cations (Ca^{2+} or basic amino acids) in acidocalcisomes, or protect against osmotic stresses. There is also evidence that polyphosphate may be required for the transition to stationary growth (Docampo and Moreno, 2001). Polyphosphates are likely synthesized by an undefined polyphosphate kinase and are degraded by an exopolyphosphatase. The *L. major* exopolyphosphatase is located in both acidocalcisomes and in the cytosol (Rodrigues *et al.*, 2002). Pyrophosphate generated by exophosphatases in the acidocalcisomes may subsequently be used by several pyrophosphate-dependent enzymes in the cytosol and glycosomes (Hannaert *et al.*, 2003). Cytoplasmic pools of pyrophosphate also fuel the V-H^+-pyrophosphatase on the acidocalcisome membrane that is needed to acidify the lumen of this organelle (Hill *et al.*, 2000). *Leishmania* promastigotes also contain high levels of phosphoarginine, which may be generated by a glycosomal arginine kinase (Moreno *et al.*, 2000). Phosphoarginine could constitute a phosphate store and/or buffer the luminal levels of ATP/ADP in the glycosome. In support of this function, overexpression of arginine kinase in *T. cruzi* confers some protection against nutritional stress (Pereira *et al.*, 2003).

Direct analysis of the *Leishmania* metabolome

Despite the insights provided by traditional biochemical and genetic approaches and the annotation of the *L. major* genome, it is likely that some metabolic processes have yet to be identified, and a global view of metabolism in *Leishmania* is lacking. Even at the most basic level, little is known about how central carbon metabolism and different pathways of biosynthesis and catabolism are coordinately regulated during parasite differentiation. A full understanding of parasite metabolism also requires a detailed analysis of metabolic networks, in order to predict the consequences of stage-specific changes in enzyme expression or pharmacological inhibition of a particular pathway.

Metabolite profiling or fingerprinting provides an alternative approach for analysing metabolic processes and is increasingly being applied to a wide variety of different biological systems. While metabolite profiling aims to both quantitate and identify as many metabolites as possible, metabolite fingerprinting aims to quantitate as many signals as possible without identification of the metabolite, allowing greater coverage of the metabolome and high throughput capability (Goodacre *et al.*, 2004). Recent advancements in analytical chemistry and bioinformatics have been critical in facilitating the development of metabolomic approaches. For example, the coupling of gas chromatography (GC), liquid chromatography (LC) or capillary electrophoresis (CE) with time-of-flight (TOF) and Fourier transform ion cyclotron resonance (FTICR) mass spectrometers have made it feasible to simultaneously separate and detect large numbers of small molecules in a high throughput manner (Halket *et al.*, 2005). Parallel improvements in signal processing and the development of new algorithms and statistical approaches have also been critical for processing and interpreting the large data sets generated by these non-biased holistic approaches.

Analysis of the *Leishmania* metabolome

Several issues need to be considered during the design of metabolomic experiments on microorganisms. First, if comparisons are to be made between different species, strains, growth states or mutants, it is important to ensure that cells are maintained under the same carefully controlled growth conditions. Reproducible growth in batch cultures can be achieved using standard or fully defined medium and the same initial inoculum. However, different species of *Leishmania* often display a preference for different media and variation in the growth rate of strains needs to be taken into account. The use of continuous or chemostat culture systems offer further control over growth conditions, including the capacity to grow parasites under constant conditions, at fixed rates, and with a growth-limiting factor chosen by the investigator (Ter Kuile and Opperdoes, 1991; ter Kuile, 1999).

Second, it is imperative that metabolism of cells should be stopped (quenched) as rapidly as possible during harvesting in order to obtain a true snapshot of the metabolome. Intracellular concentrations of metabolites can change more rapidly than mRNA and protein (e.g. ATP has a half life of less than 0.1 s), particularly when cells are concentrated during harvesting. Methods for rapidly quenching metabolism include; (i) rapid filtration through an ultrafiltration membrane; (ii) dilution of cells in perchloric acid and (iii) dilution of cells in 60% methanol at –20°C and rapid centrifugation (Saez and Lagunas, 1976; Theobald *et al.*, 1993). We have found that rapid chilling of culture media to 0°C, followed by rapid centrifugation of cells provides a highly reproducible method that is suitable for processing large number of replicate samples (D. De Souza, E. Saunders, J. Pyke and M. McConville, unpublished data).

Third, consideration should be given to the number of replicate samples taken. Analysis of replicate samples greatly increases the power of subsequent statistical analyses and the ability of the investigator to distinguish between biological and analytical variability. The number of replicates required will depend on the analytical systems used, but most analyses will typically include four or more independent replicate samples.

Fourth, metabolite extraction and fractionation protocols can represent a significant source of variability in metabolomic studies. The method used to extract metabolites should be optimized to maximize extraction of major metabolite classes with minimum fractionation steps. In addition to increasing variability, fractionation procedures reduce the representation of the metabolome. Several extraction protocols have been utilized for microbial metabolomics studies. These include, (i) boiling in ethanol-buffered solution (Gonzalez *et al.*, 1997), (ii) extraction in 60% perchloric acid, followed by removal of acid with potassium perchlorate (Lowry *et al.*, 1971; Rainey and MacKenzie, 1991; Gupta *et al.*, 1999) (iii) chloroform extraction at –45°C (Ruijter *et al.*, 1996) and (iv) extraction in chloroform:methanol:water (1:3:1 v/v) at 0°C, followed by aqueous biphasic partitioning (De Souza, D., Saunders, E., Pyke, J., McConville M., manuscript in preparation).

Microbial metabolomes can be analysed using a variety of different procedures, ranging from thin-layer chromatography (TLC) to nuclear magnetic resonance spectroscopy (NMR) and advanced "hyphenated" techniques such as GC-MS, LC-MS, CE-MS and LC-diode array/fluorescence detection. The latter methods combine chromatographic procedures for separating metabolites based on their physical and chemical properties, with (mass) spectral-based identification of each metabolite (Want *et al.*, 2005). GC-MS is probably the best developed of these methods, providing high separation efficiency, sensitivity, reproducibility (in retention times and quantitation) and dynamic range (Halket *et al.*, 2005; Want *et al.*, 2005). Furthermore, samples analysed by GC-quadrupole-MS or GC-TOF-MS are ionized under the same conditions (i.e. by electron impact at 70 eV), allowing direct comparison of mass spectra obtained with different instruments. Comprehensive mass spectral reference libraries for hundreds of thousands of compounds are available from commercial suppliers and some metabolomics laboratories. While many volatile and medium polar compounds can be analysed directly by GC-MS, derivatization is generally required for detection of polar metabolites. In the past decade, derivatization by oximation and subsequent silylation, usually trimethylsilylation or *tert*-butyldimethylsilylation, is commonly used. Oximation of keto groups improves the GC properties of many metabolites and prevents enolization reactions that increase the number of peaks per analyte (Halket *et al.*, 2005). The utility of GC-MS for metabolite profiling has been significantly increased with the commercial development of GC-TOF-MS and GC-GC-MS instruments. TOF-MS instruments have higher scan rates than quadrupole-MS detectors, and relative ion intensities remain constant over a peak (Wagner *et al.*, 2003). Both properties facilitate the identification of peaks in a completely unbiased manner and deconvolution of coeluting compounds in complex chromatograms (Wagner *et al.*, 2003). Fast scanning TOF-MS can also be used with ultrafast GC or GC-GC to improve the throughput capacity, sensitivity and/or resolution of complex mixtures (Halket *et al.*, 2005). The major limitations of GC-MS, with regard to metabolomic analyses, include a rather limited mass range (generally under 1000 Da) and the need for derivatization which introduces variability and complicates characterization of new metabolites. At present, conventional GC-MS approaches allow quantitation of 100–300 metabolites, while GC-TOF-MS has been used to identify up to 1000 metabolites (Cook *et al.*, 2004; Weckwerth *et al.*, 2004).

Notwithstanding the superior resolution and robustness of GC-MS, LC-based approaches are essential for the detection of highly involatile or heat-sensitive compounds, such as nucleotides, oligosaccharides, polyphosphates, thiols, CoA esters and complex lipids. LC-coupled to amperometric, fluorescence, diode array and MS detectors have been used to identify oligosaccharides (van Weelden *et al.*, 2003), polyamines and thiols (Shim and Fairlamb, 1988; Roberts *et al.*, 2001) glycolytic and TCA cycle intermediates (van Weelden *et al.*, 2003) and lipids (Roberts *et al.*, 2001; Zhang *et al.*, 2003) in trypanosomatid parasites. LC approaches have the advantage that derivitization is generally not necessary. However, a major problem with quantitative LC-MS approaches is ion suppression, where compound ionization is suppressed by the presence of co-eluting analytes, jeopardizing accurate quantitative analyses. Moreover, many metabolites are unstable at high temperature or at non-neutral pHs, restricting the range of LC conditions that can be employed. To date, only a limited number of holistic LC methods have been reported using a range of different approaches (Tolstikov and Fiehn, 2002; Lafaye *et al.*, 2005). A number of recent developments may overcome some of these limitations and help with the analysis of polar or complex secondary metabolites. Hybrid MS detectors such as the linear ion-trap combined with FT-ICR-MS detector can be used to directly analyse complex metabolite mixtures with very high-resolution MS detection, allowing determination of the elemental composition of individual metabolites (Hirai *et al.*, 2004; Tohge *et al.*, 2005). Direct infusion atmospheric pressure ionization (API) coupled with FT-ICR-MS has already been used to identify ~6,000 metabolites in extracts of ripening tomatoes (Tohge *et al.*, 2005). New LC-based laboratory-on-a-chip or micro total analytical systems (μTAS) coupled to MS also have the potential to provide superior separation efficiency and reproducibility than current systems. Finally, strategies for quantitating metabolites by LC-MS following the spiking of samples with ^{13}C- or ^{15}N-labelled metabolites are being developed to circumvent the problem of ion suppression (Lafaye *et al.*, 2005). Databases established for metabolite identification include the National Institutes of Standards and Technology (NIST) commercial database (http://www.nist.gov/srd/nist1.htm), SRI International BioCyc (http://biocyc.org), Atomic Reconstruction of Metabolism project (http://www.metabolome.jp), Wiley Registry of Mass Spectral Data (http://www.wileyregistry.com), the GOLM open-access database at the Max Planck Institute of Molecular Plant Physiology (http://www.mpimp-golm.mpg.de/mms-library/), and METLIN (http://metlin.scripps.edu/). We have recently established a curated metabolic database (LeishCyc) for *Leishmania* (http://bioinformatics.bio21.unimelb.edu.au/leishcyc.html) that contains all experimentally determined and predicted metabolic pathways, enzymes and metabolites in these parasites.

NMR is another technique that is both complimentary to the MS methods and widely used for metabolite profiling. The major advantages of NMR approaches include the use of underivitized samples, ease of detection of highly polar metabolites and the ability to define the labelling pattern of metabolites following exposure of organisms to labelled substrates such as ^{13}C-glucose (Moreno *et al.*, 2000; Krishnan *et al.*, 2005). However, NMR approaches are much less sensitive than MS methods and typical analyses may include <20 metabolites (Krishnan *et al.*, 2005). This is clearly a major shortcoming when biological material is limited. Even when samples are non-limiting, as for example, when parasites are cultured *in vitro*, the requirement for more material increases the challenge of rapidly quenching metabolism during cell harvesting and metabolite extraction.

Data processing and bioinformatics

Statistical analysis of metabolomic data sets are often required to identify metabolites or signal patterns which differentiate between samples, to identify correlations within or between data sets, and to classify samples of unknown origin. Robust statistical analyses are possible because it is relatively easy to generate large numbers of replicate samples (typically >8) with current metabolite profiling platforms. Computer processing of metabolomics data involves at least three distinct stages: signal processing, post-processing, and statistical analysis. Signal processing typically involves noise smoothing, baseline correction, peak detection, peak quantitation and various data-based normalizations. In addition, in mass-spectrometry-based methods overlapping peaks may be deconvoluted based on the time profile of individual ion-chromatograms, and mass-spectra are often matched against reference libraries for the purpose of metabolite identification. In the post-processing phase an attempt is made to correlate peaks from different experiments (either within the sets of replicate experiments, or between different sets of replicate experiments). The goal of the post-processing phase is to compile the experimental information into a data matrix suitable for subsequent statistical analysis (the data matrix is a two-dimensional matrix whose rows correspond to experiments and columns correspond to individual analytes). Repeated cycles of statistical analysis and data mining may follow with the goal of consolidation of the experimental information with the experimental set-up, known biology of the system, and generation of new hypotheses. Commonly used multivariate data analysis tools include both unsupervised (i.e. principal component analysis (PCA) and hierarchical cluster analysis) and supervised (i.e. discriminant analysis) approaches. One aim of multivariate analysis is data simplification or dimension reduction while retaining as much as possible of the information present in the original data. For example, PCA reduces the dimensionality of the original data by finding a new set of orthogonal axes, which are formed as a linear combination of original variables, and point in the direction of maximal variability in data. The first principal component explains the greatest amount of variance in the original data set, the second principal component explains the largest variance independent of the first component, and so on. It is often useful to project the metabolomic data set onto the space defined by the first two principal components. Such a projection results in a two-dimensional plot allowing the visualization of inherent data structure and data outliers (Fig. 5.3).

The secreted and cellular metabolome of *Leishmania* spp.

Several investigators have identified stage-specific changes in the metabolome of *Leishmania* using NMR. In the first of these studies, Rainey and MacKenzie analysed the major cellular and secreted metabolites of *L. pifanoi* promastigotes and axenic amastigotes cultivated in the presence of ^{13}C-glucose (Rainey and MacKenzie, 1991). These semi-quantitative analyses revealed significant differences in the way in which promastigote and axenic amastigote stages metabolize glucose under both aerobic and anaerobic conditions. *L. pifanoi* promastigotes accumulated and secreted large amounts of succinate and acetate, consistent with these compounds being major end-products of glycosomal and mitochondrial oxidation of glucose, respectively (Fig. 5.2). In contrast, amastigote stages accumulated high levels of alanine and secreted both acetate and alanine into the medium. These differences are interesting given that both stages were exposed to the same nutritional environment. More recently, Gupta *et al.* used ^{1}H-NMR to analyse the polar metabolites of *L. donovani* promastigotes, axenic amastigote and true lesion amastigotes (Gupta *et al.*, 1999). In addition to the major products of central carbon metabolism (succinate, acetate and amino acids), high level of acetoacetate, α-glycerophosphoryl choline, β-hydroxybutyrate and betaine were present in cellular extracts of these stages. However, significant quantitative differences were observed in major metabolite levels, particularly between cultured promastigotes and lesion amastigotes, with axenic amastigotes having

A

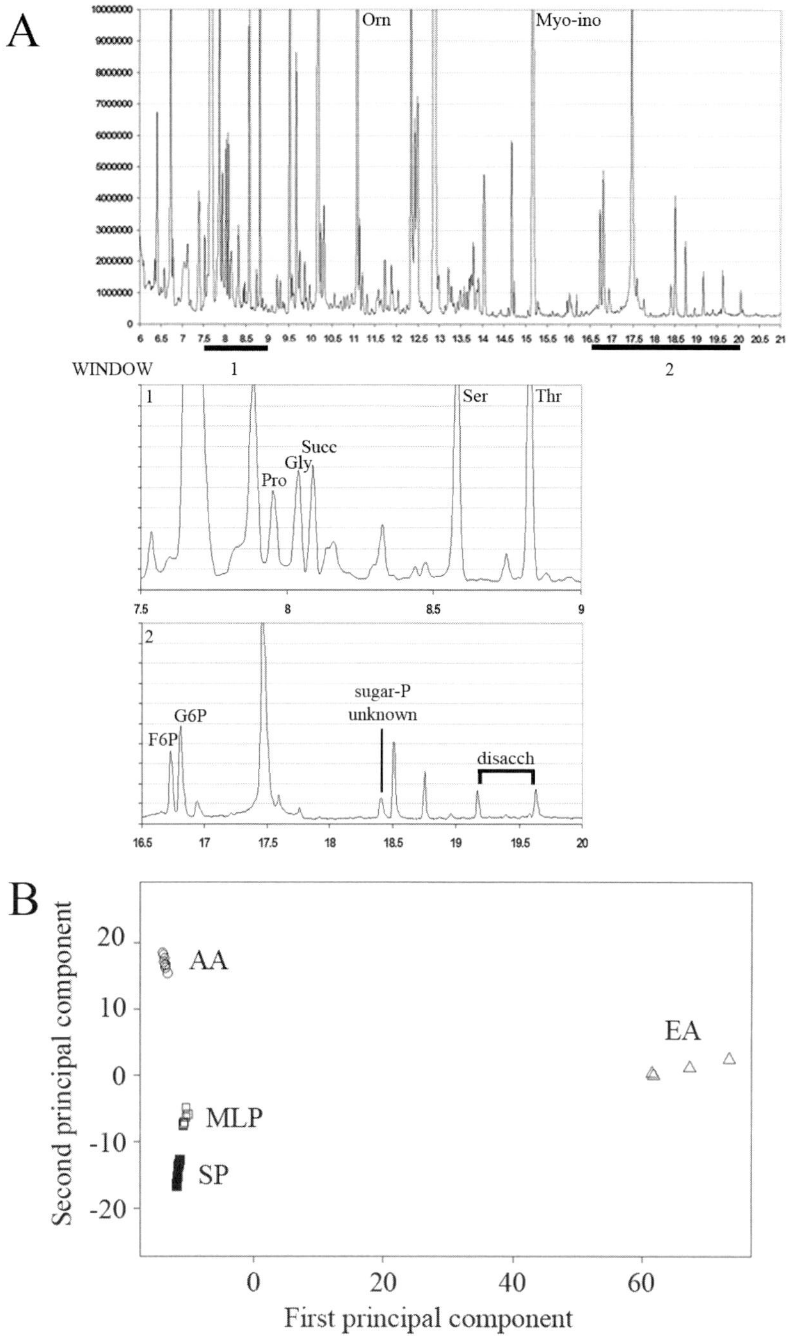

Figure 5.3 GC-MS chromatogram of *L. mexicana* promastigote polar metabolite extract**.** Panel A. Polar metabolites from *L. mexicana* promastigotes were derivitized by oximation and silylation and analysed by GC-MS. Windows 1 and 2 show detail of the 7.5–9 min and 16.5–20 min time periods underlined in upper panel containing amino acids and sugar phosphate/disaccharides, respectively. Pro, proline; Gly, glycine; Succ, succinate; Ser, serine; Thr, threonine; F6P, fructose-6-phosphate, G6P, glucose-6-phosphate; disacch, the disaccharide peaks Manα1–2Man and sucrose. Panel B. Principal Component Analysis (PCA) of the polar metabolites of *L. mexicana* promastigote (harvested in mid-log phase (MLP) and stationary growth phase (SP)), axenic amastigotes (differentiated from promastigotes *in vitro;* AA) and intracellular amastigote (isolated from the J774 macrophage-like cell line; EA). The replicate extracts of different developmental stages cluster closely together and are clearly separated from each other. Note that the intracellular amastigotes are clearly distinct from the cultivated stages.

intermediate profiles despite being morphologically similar to lesion amastigotes (Gupta *et al.*, 1999). Axenic amastigotes may thus maintain many aspects of promastigote-like metabolism when cultivated in rich medium.

We have recently used GC-MS and GC-TOF-MS to analyse stage-specific changes in *L. mexicana* metabolism in log and stationary phase promastigotes, axenic amastigotes and intracellular amastigotes (isolated from J774 macrophages or mouse bone marrow-derived macrophages) (D. De Souza, E. Saunders, V. Likic and M.J. McConville, unpublished data). After quenching metabolism, metabolites were extracted in cold organic solvents and polar and apolar metabolites separated by phase partitioning. Following derivatization (methoximation and trimethylsilylation), the polar metabolite extracts were analysed by GC-MS. More than 150 peaks were robustly integrated and 50 metabolites identified, including amino acids, sugar phosphates, organic acids and fatty acids (Fig. 5.3A) and a number of previously unidentified disaccharides. These analyses revealed marked quantitative changes in the metabolomes of the different promastigote stages and intracellular amastigotes (Fig. 5.3B). Significant changes (> 5-fold) were observed in the levels of 20% of the metabolites, when profiles from log and stationary phase *L. mexicana* promastigotes were compared with those from axenic and intracellular amastigotes. While all four stages could be clearly differentiated from each other, a continuum in levels of individual metabolites was often observed from log to stationary promastigotes, and from non-dividing amastigotes to intracellular amastigotes. For example, several metabolites were present at very low levels in the promastigote stages but increased dramatically in amastigote stages (i.e. citrate, *myo*-inositol, β-alanine, hexadecanoic acid and mannan). The accumulation of TCA cycle intermediates such as citrate and several putative and confirmed storage carbohydrates in the latter stages suggests that both axenic amastigotes and intracellular amastigotes are not limited by carbon sources, but may enter a slow or quiescent growth phase allowing the channelling of carbon into storage materials. In support of this conclusion, the steady-state level of ornithine, which is required to sustain polyamine biosynthesis and cell proliferation, is down-regulated by >100-fold in stationary phase promastigotes and intracellular amastigotes. Whether the slow growth phenotype of intracellular amastigotes is a programmed adaptation to prevent overgrowth of the phagosome, or the absence of an essential nutrient is unclear. Finally, thin layer chromatographic and GC-MS analysis of the lipid fraction of isolated amastigotes provides ample evidence that intracellular stages can scavenge a wide range of lipids from the phagosomes (McConville and Blackwell, 1991; Winter *et al.*, 1994; E. Saunders, D. De Souza and M. McConville, unpublished data). In summary, these analyses demonstrate that all developmental stages of *L. mexicana* can be clearly defined by GC-MS metabolite profiling indicating marked differences in physiological state. They also highlight the presence of novel or unanticipated metabolic pathways, as well as the contribution of metabolite scavenging mechanisms in the intracellular stages.

It is interesting to contrast the magnitude of the changes in the metabolome with changes observed in DNA-microarray and proteomic analyses. While some variation is observed between the latter studies, less than 5% of mRNAs changed by more than two-fold when different promastigote (log and stationary phase) and axenic amastigote stages were compared. Of the transcripts that did change, less than 10% were directly associated with metabolism (Akopyants *et al.*, 2004; Almeida *et al.*, 2004; Holzer *et al.*, 2006). A similar pattern has emerged from proteomic analysis (Nugent *et al.*, 2004; Leifso *et al.*, 2007) and analyses of enzyme activities in trypanosomatids grown at different growth rates in chemostats (ter Kuile, 1999; ter Kuile and Westerhoff, 2001). Overall, it is clear that changes in the metabolome of different developmental stages is far greater than would be predicted from detectable changes in mRNA or protein. Given the stage-specific variability in the GC-MS profiles, it was surprising to find to a remarkable constancy in metabolite levels within a particular stage. The same developmental stages from an isolate analysed before and after passage through a mouse clustered tightly together and well separated from other developmental stages (D. De Souza, E. Saunders, V. Likić, and M. McConville, unpublished data). Maintenance of constant steady state levels of metabolites is commonly observed in metazoan tissues, even under conditions where metabolic fluxes change dramatically. It has been proposed that metabolite stability could reflect cooperativity in allosteric enzymes exhibiting feedback inhibition (Fell, 2005). Alternatively, it may reflect coordinated changes in the activity of all enzymes within a metabolic network, such that an increase in flux is paralleled by an equivalent increase in enzyme levels /activities – the so-called multi-site modulation (Fell, 2005). At present there is little evidence that the latter occurs in *Leishmania*, highlighting the potential importance of post-translational modifications and allosteric controls in controlling metabolic flux in these parasites.

Metabolic pathways required for virulence in the mammalian host

The up-regulation of a particular metabolic process in pathogenic *Leishmania* stages does not necessarily indicate that it is essential for virulence. The demonstration that disruption of enzyme function results in loss of virulence in the mammalian host is a key step in validating a pathway as a drug target. There is also considerable interest in identifying *Leishmania* mutants that persist within the mammalian host without inducing pathology, as possible live attenuated vaccines (Titus *et al.*, 1995; Uzonna *et al.*, 2004). *Leishmania* mutants that grow at the same rate as wild type parasites in culture, but exhibit a growth defect in macrophages or mice (growth *in vivo* mutant; *giv*) appear to fall into three categories (Table 5.1). The most common class of mutants identified to date exhibit a delayed-lesion phenotype in susceptible mice, indicating a defect in the initial stages of infection, but not in amastigote survival (P^{giv}/A^{WT}). These mutants could be deficient in processes that are required for differentiation of procyclic promastigotes to infective metacyclic promastigotes, invasion and survival of metacyclic promastigotes in macrophages or differentiation of metacyclic promastigotes to amastigotes. The second class of mutants are poorly equipped to survive in macrophages and mice as either promastigotes or amastigotes (P^{giv}/A^{giv}). This class may also include mutants that are unable to differentiate into amastigotes. These mutants fail to establish an infection in mice or are cleared after a relatively short period. The third class of *Leishmania* mutants are able to persist long term in susceptible mice, without inducing pathology (P/A^{Per}). Examples of this class have only recently been identified (Table 5.1).

Leishmania mutants displaying a delayed lesion phenotype (P^{giv}:A^{WT})

A number of *L. major* mutants with defects in surface glycocalyx biosynthesis survive poorly as promastigotes, but grow normally as amastigotes, in macrophages. The first of these mutants to be generated was the *L. major Δlpg1* mutant that is unable to synthesize the glycan core of LPG, but expresses all other surface and secreted glycoconjugates (Spath *et al.*, 2000). *L. major Δlpg1* promastigotes invade macrophages with the same kinetics as wild type promastigotes, but are subsequently cleared by the host cell. However, some surviving parasites differentiate to amastigotes and proliferate normally within macrophages. This mutant induces lesions after a delay of several weeks and isolated amastigotes exhibit the same virulence phenotype as wild type parasites. The virulence phenotype of the *L. major Δlpg1* mutant is largely consistent with previous studies suggesting that LPG protects promastigotes from complement-mediated lysis, is involved in parasite uptake and internalization, and modulates macrophage responses to internalized promastigotes (McConville *et al.*, 2002; Naderer *et al.*, 2004). Interestingly, analysis of the *L. major Δlpg1*–macrophage interaction indicated that LPG was not essential or responsible for many of these processes (Spath *et al.*, 2003). For example, while LPG-minus promastigotes were more sensitive to complement lysis than wild type *L. major* promastigotes, expression of LPG was not required for promastigote internalization by macrophages, inactivation of macrophage signalling pathways or inhibition of macrophage IL-12 secretion (Spath *et al.*, 2003). Furthermore, the continued infectivity of *L. major Δlpg1* amastigotes was anticipated from earlier biochemical studies showing that LPG levels are massively downregulated in this stage (Moody *et al.*, 1993). Remarkably, deletion of *lpg1* in *L. mexicana* had no effect on promastigote

Table 5.1 Metabolic pathways required for survival of *Leishmania* promastigotes or amastigotes in macrophages, but not for growth in rich medium

Pathway	Gene	Defect	Type	Reference
Folate/biopterin metabolism	*lmj-dhfr-ts*	Thymidine auxot'y	P^{giv}, A^{per}	(Titus *et al.*, 1995)
	lmj-bt1	Biopterin uptake	P^{vir}, A?	(Cunningham *et al.*, 2001)
	lmj-ptr1	Biopterin reductase	P^{vir}, A?	(Cunningham *et al.*, 2001)
Trypanothione metabolism	*ld-tr*	Trypanothione reductase	P^{giv}, A^{giv}	(Dumas *et al.*, 1997)
Mannose metabolism	*lmx-pmm*	All glycoproteins/ glycolipids/mannan	P^{giv} A^{giv}	(Garami *et al.*, 2001)
	lmx-gdpmp	All glycoprotein/ Glycolipids/mannan	P^{giv} A^{giv}	(Garami and Ilg, 2001)
	lmx-pmi	Man auxotrophy	P^{giv} A^{giv}	(Garami and Ilg, 2001)
	lmx-dpms	LPG, GIPL, GPI-protein	P^{giv} A^{giv}	(Garami *et al.*, 2001)
LPG	*lmj-lpg1*	LPG anchor	P^{giv}, A^{WT}	(Spath *et al.*, 2000)
	lmj-ugm	LPG, polar GIPL	P^{giv}, A^{WT}	(Kleczka *et al.*, 2007)
	lmx-lpg1	LPG	P^{WT}, A^{WT}	(Ilg 2000)
Phosphoglycan	*lmj-lpg2*	LPG, PPG	P^{giv}, A^{per}	(Spath *et al.*, 2003)
	lmx-lpg2	LPG, PPG	P^{WT}, A^{WT}	(Ilg *et al.*, 2001)
Ether lipids	*lmj-ads*	All ether lipids	P^{giv}, A^{WT}	(Zufferey *et al.*, 2003)
Sphingolipids	*lmj-spt*	Sphingolipid, IPC	P^{giv}, A^{WT}	(Zhang *et al.*, 2003; Denny *et al.*, 2004)
Glycerolipids	*lmj-dat*	Acyl/alkylglycerolipids	P^{giv}, A^{giv}	(Zufferey and Ben Mamoun, 2006)
Glucose uptake	*lmx-gt1/-gt2/-gt3*	CHO metabolism	P^{giv}, A^{giv}	(Burchmore *et al.*, 2003)
Inositol synthesis	*lmx-ino1*	Inositol auxotrophy	P^{giv}, A^{giv}	(Ilg 2002)
Gluconeogenesis	*lmj-fbp*	Gluconeogenesis	P^{WT}, A^{per}	(Naderer *et al.*, 2004)
Protein turnover	*lmx-cpB*	Protein degradation	P^{giv}, A^{per}	(Mottram *et al.*, 1996)
	lmx-atg4	Autophagy	P^{giv}, A?	(Besteiro *et al.*, 2006)
Signalling	*lmx-mapkk4*	Protein phosphorylation	P^{giv}, A^{giv}	(Kuhn and Wiese, 2005)
	lmx-lmpk	Protein phosphorylation	P^{wt} A^{giv}	(Wiese and Gorcke, 2001)
Heat shock	*lmj-hsp100*	Heat shock response/ differentiation	P^{giv}, A^{giv}	(Hubel *et al.*, 1997)
Fe^{3+} uptake	*lam-lit1*	Intracellular iron levels	P^{giv}, A^{per}	(Huynh *et al.*, 2006)

Classification of *Leishmania* mutants with reduced growth phenotype *in vivo* (*giv*).

Promastigote growth in macrophages and/or mice the same (P^{WT}), reduced (P^{giv}) or enhanced (P^{vir}) compared with wild type promastigotes. Amastigote growth in macrophages and/or mice the same (A^{WT}) or reduced (A^{giv}) compared to wild type amastigotes. *Leishmania* mutants that fail to induce lesions but persist in infected mice (A^{per}) Lmj = *L. major*; Lmx = *L. mexicana;* Ld = *L. donovani.* Enzymes: ads, alkyldihydroxyacetone synthase; atg4, autophagy gene; bt, biopterin transporter; cpB, cysteine protease; dat, dihydroxyacetone phosphate acyltransferase; dhfr-ts, dihydrofolate reductase-thymidine synthase; dpms, dolichol phosphate mannose synthase; fbp, fructose-1,6,-bisphosphatase; gdpmp, GDP-Man phosphorylase; gt, glucose transporter; hsp100, heat shock protein 100; ino1, inositol-3-phosphate synthase; lit1, ion transporter; lpg1, putative galactofuranosyltransferase; lpg2, GDP-Man transporter (Golgi apparatus); mpk, mitogen activated protein (MAP) kinase; mapkk4, MAP kinase, kinase 4; pmm, phosphomannose isomerase; pmi, phosphomannose mutase; ptr, pteridine reductase; ugm, UDP-Gal mutase; spt, serine:palmitoyl transferase; tr, trypanothione reductase.

infectivity in macrophages or mice (Ilg, 2000), despite the fact that both *L. major* and *L. mexicana* promastigotes synthesize similar levels of LPG. However, more detailed profiling of stage-specific expression of LPG in these species revealed interesting differences. For example, while *L. major* and *L. donovani* metacyclic promastigotes accumulate structurally modified forms of LPG that form large complexes on the cell surface (McConville *et al.*, 1992; Sacks *et al.*, 1995), *L. mexicana* promastigotes express very low levels of LPG when they enter stationary growth phase (Ilg *et al.*, 1992; Ralton *et al.*, 2003). *L. mexicana* stationary phase promastigotes thus appear to remodel their surface glycocalyx whilst still in the sandfly vector, possibly accounting for the dispensability of LPG in macrophage invasion and survival.

The *L. major Δads* mutant also exhibits a delayed lesion phenotype in mice. This mutant lacks the enzyme alkyl-dihydroxyacetonephosphate-synthase (ADS), that catalyses the first committed step in ether lipid biosynthesis. Consequently, *Δads* promastigotes lack LPG, GIPLs and plasmanyl-phospholipids (Zufferey *et al.*, 2003). Like the *L. major Δlpg1* mutant, *L. major Δads* promastigotes infect macrophages but are subsequently eliminated, while in mice, surviving amastigote induce a delayed lesion phenotype (Zufferey *et al.*, 2003). This phenotype was surprising given that GIPLs and ether-phospholipids are expressed at high levels in amastigote stages (Zufferey *et al.*, 2003) and several studies had suggested that parasite-derived ether lipids inhibit signal transduction pathways in macrophages (Turco, 1999). Interestingly, macrophage signalling was still impaired following infection with the *Δads* mutant, indicating that other molecules can fulfil these functions or that the GIPLs/plasmanyl-phospholipids are not normally involved *in vivo* (Turco, 1999). In considering possible substitutes, *Leishmania* amastigotes have been shown to scavenge a range of lipids, including glycosphingolipids, from the host phagolysosome membrane which may bypass the need for endogenous glycolipids (McConville and Blackwell, 1991; Schneider *et al.*, 1993; Winter *et al.*, 1994).

Scavenging of host lipids could also account for the relatively mild loss of virulence phenotype of several other *Leishmania* surface-coat and plasma membrane lipid mutants (Garami *et al.*, 2001). Analysis of the *L. major Δspt* sphingolipid mutant provides the strongest evidence that intracellular amastigotes depend on lipid scavenging from the host cell for important membrane lipids. This mutant lacks the enzyme, serine palmitoyltransferase (SPT), and is unable to synthesize sphingosine or the major cellular sphingolipid, inositol-phospho-ceramide (IPC) (Zhang *et al.*, 2003). *L. major Δspt* promastigotes grow normally in culture, but exhibit a number of defects in membrane transport, flagellum morphology and biogenesis of acidocalcisomes upon entering stationary growth (Zhang *et al.*, 2003; Denny *et al.*, 2004; Tull *et al.*, 2004). Stationary phase *Δspt* promastigotes are poorly virulent in macrophages and mice, while surviving amastigotes are as virulent as wild type amastigotes and synthesized parasite-specific sphingolipids, such as IPC (Zhang *et al.*, 2005). The sphingolipid base for these lipids is presumably derived from the breakdown of complex host glycosphingolipids or by uptake of simple host sphingolipids. Intriguingly, wild type amastigotes lack SPT, one of the few examples of stage-specific expression of a metabolic enzyme, suggesting that sphingolipid scavenging is essential for intracellular survival (Zhang *et al.*, 2003; Denny *et al.*, 2004).

Leishmania promastigotes preferentially utilize glucose over other carbon sources (Hart and Coombs, 1982) and appear to be dependent on glucose uptake during the initial stages of mammalian infection. Promastigote stages of a *L. mexicana* mutant lacking the three functional glucose transporters, *LMGT1*, *LMGT2* and *LMGT3*, (*Δgt*) are highly sensitive to heat shock and are unable to survive in macrophages (Burchmore *et al.*, 2003). Because this mutant is rapidly cleared by macrophages, it has been difficult to determine whether glucose uptake is also essential for intracellular survival of the amastigote stages. Not withstanding the observation that axenic *L. mexicana* amastigotes appear to require glucose for survival in rich medium, lesion-derived amastigotes take up glucose at very low rates (Burchmore and Hart, 1995) and also appear to utilize other carbon sources in the macrophage (Naderer *et al.*, 2006). It is possible that promastigotes stages are heavily dependent on scavenging glucose during the initial stages of infection, but may switch to utilizing other carbon sources following differentiation to amastigotes stages.

Leishmania mutants that fail to differentiate to metacylic promastigotes might be expected to have a $P^{giv}A^{WT}$ phenotype. Conversely, mutants with an increased propensity to differentiate to metacyclic promastigotes should exhibit increased virulence *in vivo*. This is the case for a *L. major* mutant lacking the enzyme pteridine reductase 1 (PTR1), which is involved in reducing exogenous folate and biopterin to tetrahydrofolate and BH_4, respectively. BH_4 levels normally decrease when *Leishmania* promastigotes reach stationary phase and differentiate to metacyclic promastigotes (Cunningham and Beverley, 2001). Deletion of either PTR1 or the biopterin transporter, BT1, further decreased BH_4 levels and increased the proportion of promastigotes that differentiated to metacyclic promastigotes and the infectivity of these parasites in mice (Cunningham *et al.*, 2001). These findings suggest that *Leishmania* may have evolved regulatory mechanisms for limiting their own growth and pathogenesis in the mammalian host, thereby increasing the opportunity for transmission to the next host.

Leishmania mutants attenuated in both promastigote and amastigote stages (P^{giv}/A^{giv})

One of the first genes to be deleted in *Leishmania* was *dhfr-ts*, encoding dihydrofolate reductase-thymidylate synthase. The *L. major Δdhfr-ts* mutant is auxotrophic for thymidine and is unable to replicate within macrophages following differentiation into amastigotes. The *Δdhfr-ts* mutant survived in BALB/c mice for up to 2 months and were subsequently cleared unless the infected mice were intravenously supplied with a source of thymidine (Titus *et al.*, 1995). Because of its low infectivity, this mutant was tested for its potential as a live vaccine. Partial and strong protection was obtained when the *Δdhfr-ts* mutant was used to immunize highly susceptible BALB/c and more resistant CBA mice, respectively. However, challenges were always associated with some pathology and the eventual clearance of this mutant by the host may prevent the development of long term immunity (Titus *et al.*, 1995).

The maintenance of intracellular redox balance of *Leishmania* amastigotes is likely to be important for survival in the oxidizing environment of the macrophage phagosome. This is supported by genetic ablation of the gene encoding trypanothione reductase (TR). Targeted deletion of both genes of trypanothione reductase was not achievable, suggesting that this enzyme is essential for normal growth (Dumas *et al.*, 1997). However, a *L. donovani* heterozygote knockout cell line retaining one functional copy of the trypanothione reductase gene grew normally in culture and was able to differentiate to amastigotes *in vitro*. This mutant is rapidly cleared by human macrophages expressing a strong oxidative burst, but not by the murine macrophage-like J774 cell line that is unable to produce reactive oxygen metabolites. A similar phenotype was observed in a *L. major Δtr/+* heterozygote mutant (Dumas *et al.*, 1997) and by expression of dominant negative TR mutant protein that effectively downregulated TR activity (Tovar *et al.*, 1998). Reduced levels of trypanothione reductase thus compromise the ability of *Leishmania* amastigotes to survive in activated macrophages.

Promastigote metacyclogenesis and promastigote–amastigote differentiation is associated with morphological changes in lysosome structure (Mullin *et al.*, 2001) and increased expression of lysosomal cysteine proteases (Rosenthal, 1999). Targeted deletion of the *L. mexicana* cysteine proteinase B array (encoded by a tandem array of 19 genes) greatly reduced promastigote infectivity in murine peritoneal macrophages, while amastigote growth in macrophages was less effected (Mottram *et al.*, 1996; Frame *et al.*, 2000). However, both promastigote and amastigote stages of the *Δlmcpb* mutant induce lesions only after a prolonged delay. Similarly, decreased lesion development can be achieved by treating *L. mexicana*-infected mice with cysteine protease inhibitors (Selzer *et al.*, 1999; Frame *et al.*, 2000). Lysosomal proteases may contribute to the virulence of *Leishmania* through several processes. Firstly, the turnover of endogenous proteins via the secretory pathway (Mullin *et al.*, 2001) or via the autophagic pathway (Besteiro *et al.*, 2006) may be required for remodelling of intracellular organelles, metabolic pathways and cell morphology during parasite differentiation

(McConville *et al.*, 2002). For example, genetic deletion of proteins involved in autophagy in *L. mexicana*, such as the agt4 homologue, reduced the capacity of procyclic promastigotes to differentiate to metacyclic promastigotes and survive in macrophages (Besteiro *et al.*, 2006). Secondly, autophagy and lysosomal degradation may also provide an important source of amino acids and other nutrients to sustain growth during early stages of infection. Indeed the *L. mexicana Δagt4* mutant is unable to withstand nutrient starvation to the same extent as wild type parasites (Besteiro *et al.*, 2006). Thirdly, intracellular amastigotes also endocytose host proteins that accumulate in the lumen of the phagolysosome providing an additional carbon source for this stage (Schaible *et al.*, 1999). The potential importance of endocytosed proteins as a nutrient source is suggested by the finding that a *L. mexicana* mutant defective in the early steps of endocytosis are unable to survive in macrophages or susceptible mice (Gokool, 2003). Finally, lysosomal proteases secreted by amastigotes may inhibit signal transduction pathways, such as those involving the NF-κB transcription factor and IL-3 secretion, in the macrophage host cell (Cameron *et al.*, 2004). How the parasite proteases access the host cytosol remains unknown.

All *Leishmania* glycoconjugates contain mannose, and mannose metabolism is essential for parasite infectivity in both the insect and mammalian host, but not for growth in rich medium (Garami and Ilg, 2001; Garami *et al.*, 2001). *L. mexicana* mutants lacking key enzymes in mannose activation (i.e. phosphomannose mutase (PMM) and GDP-Man pyrophosphorylase (GDPMP)) are highly sensitive to mild heat shock that normally induce promastigote-amastigote differentiation, and nutrient starvation conditions (Garami and Ilg, 2001; Garami *et al.*, 2001; Ralton *et al.*, 2003). This phenotype is distinct from the *L. mexicana* mutants defective in the synthesis of surface glycoconjugates (such as the *Δdpms* and *DIG1* mutants; Table 5.1), indicating that other mannose-containing molecules may be important for the parasite stress response and virulence (Naderer and McConville, 2002). For example, the *L. mexicana Δpmm* and *Δgdpmp* mutants only synthesize highly truncated N-linked glycans, that may lead to increased protein misfolding in the endoplasmic reticulum and sensitivity to heat stress, as well as loss of virulence (Kink and Chang, 1987). The *L. mexicana Δpmm* and *Δgmp* mutants are also unable to synthesize intracellular mannan, that accumulates in response to heat shock and pharmacological inhibitors of heat shock protein function (Ralton *et al.*, 2003). In this respect, mannan may have dual function as both a metabolic reserve material and as a compatible solute, protecting cytoplasmic proteins from misfolding and aggregation during cellular stress (Ralton *et al.*, 2003).

Deletion of the gene encoding *L. major* dihydroxyacetonephosphate acyltransferase (DAT) had a profound affect on the viability of non-dividing promastigotes and resulted in complete loss of virulence in BALB/c mice (Zufferey and Ben Mamoun, 2006). LmDAT initiates one arm of glycerolipid biosynthesis and downstream products of this pathway can be incorporated into bulk acyl-glycerolipids or the alkyl-glycerolipids. Interestingly, deletion of *L. major* glycerol-3-phosphate acyltransferase (that catalyses the first reaction in a parallel and partially redundant pathway leading to acyl-glycerolipids) had little affect on parasite growth or lipid metabolism (Zufferey and Mamoun, 2005). Given that ether lipids are not required for normal promastigote growth (as indicated by analysis of the *L. major Δads* mutant) it is possible that some *de novo* synthesis of acyl-glycerol lipids via the LmDAT pathway is essential for the viability of slow-growing parasite stages. This would be surprising given the capacity of *Leishmania* amastigotes to scavenge lipids from the host. Alternatively, pertubation of dihydroxyacetone metabolism by disruption of this pathway may have other, as yet, undefined affects on parasite metabolism.

Very little is known about signal transduction pathways in *Leishmania* that are required for differentiation and metabolic adaptation to the different host environments. The *L. major* genome contains a large number of protein kinases that might be involved in signal transduction. Targeted deletion of a *L. mexicana* gene encoding a putative MAP kinase kinase, termed *lmxPK4*, had no effect on promastigote growth *in vitro*, but had profound effects on promastigote differentiation in macrophages and the development of lesions in mice (Kuhn and Wiese,

2005). Interestingly, second site bypass mutants frequently arose, indicating that LmxPK4 function can be bypassed by other proteins. As the LmxPK4 protein was only expressed in promastigote stages, despite constitutive expression of LmxPK4 mRNA, this kinase may be required for promastigote-amastigote differentiation. A second MAP kinase homologue, termed LMPK, was identified by complementation of a mutant that was initially generated to test the function of two acid phosphatase genes (Wiese and Gorcke, 2001). The *Δlmpk* mutant can differentiate into amastigotes *in vitro* and in macrophages. However, *in vitro* differentiated amastigotes do not proliferate and intracellular amastigotes are rapidly cleared. As LMPK is expressed in both promastigote and amastigote stages, this protein may be required for amastigote proliferation.

Leishmania mutants that can persist in susceptible animals without causing pathology

While deletion of the *L. mexicana LPG2* gene had no affect on parasite virulence, deletion of the same gene in *L. major* resulted in loss of virulence in macrophages (Spath *et al.*, 2003). This mutant cannot synthesize LPG or PPGs and is highly susceptible to the macrophage oxidative burst. Remarkably, the *L. major Δlpg2* mutant persisted for many months in susceptible mice without causing detectable pathology (Spath *et al.*, 2003). Given that these parasites fail to survive in macrophages *in vitro*, it is likely that they persist within other host cells (Bogdan *et al.*, 2000). Persistent parasites were also isolated from infected mice lacking inducible nitric oxide synthase, indicating that this phenotype is an intrinsic property of the parasite mutant rather than the consequence of an active host immune responses. A second site revertant of the *L. major Δlpg2* mutant was recovered that induced lesions in mice and was able to survive as amastigotes in macrophages (Spath *et al.*, 2004). The *Δlpg2*-revertant still lacked the capacity to make phosphoglycans and promastigotes were severely attenuated in macrophages. *L. major* promastigotes thus appear to be highly dependent on phosphoglycan biosynthesis for survival in macrophages, while amastigotes appear to have the capacity to bypass this requirement under selective pressure. Interestingly, deletion of the *LPG2* gene in *L. mexicana* had no detectable affect on promastigote or amastigote infectivity or the capacity of these parasites to cause lesions (Ilg *et al.*, 2001). It is possible that the metabolic by-pass in *L. major Δlpg2*-revertant parasites is constitutively active in both promastigote and amastigote stages of *L. mexicana*.

It is likely that other *Leishmania* mutants that have previously been classified as avirulent in mice, are in fact able to persist for long periods of time. For example, *L. major* and *L. amazonensis* mutants deficient in gluconeogenesis (*Δfbp*) (Naderer *et al.*, 2006) and iron uptake (*Δlit*) (Huynh *et al.*, 2006), respectively, are able to invade macrophages as promastigotes and differentiate to amastigotes, but then grow very slowly, if at all, in *ex vivo* infected macrophages. While these mutants fail to induce significant lesions in BALB/c mice, significant numbers of parasites (approximately 10-fold less than wild type) can be detected in draining lymph nodes. These 'persistent' parasites retained the same biochemical phenotype of the original mutant, suggesting that they are able to proliferate in macrophages and/or other host cells that are more permissive than those used in the *in vitro* infection experiments. The detection and characterization of mutants with this phenotype is clearly of interest in defining metabolic traits that allow parasites to induce active disease, versus maintaining a persistent 'latent' infection.

Concluding remarks

Traditional biochemical pathway analysis of cultivatable stages of *Leishmania* has provided an overview of *Leishmania* metabolism and identified a number of areas where parasite metabolism differs markedly from the mammalian host. Reverse and forward genetic tools have been used to characterize metabolic genes in *Leishmania*, and to investigate the consequences of perturbing specific metabolic pathways on virulence in the sandfly and mammalian hosts. However, the next few years should witness an even greater expansion in our understanding of parasite metabolism, propelled by the availability of the TriTryp genomes (eventually covering the genomes of many different trypanosomatid species and strains) and the development of new analytical tools for directly analysing the metabolome of these parasites. The latter are likely to assume a much greater importance when analysing trypanosomatid metabolism because of the presence of polycistronic gene expression and the absence of complex networks of regulatory transcription factors in these parasites. Indeed, there is increasing evidence that many stage-specific changes in metabolism are not reflected by concomitant changes in the expression of mRNA or protein of key enzymes. These observations suggest that many metabolic processes in *Leishmania* are regulated post-translationally (via protein modifications, compartmentalization and/or oligomerization) or by metabolic control. Direct analysis of the metabolome is thus likely to be essential for identifying major changes in metabolism that are associated with parasite differentiation and adaptation to varying environmental stimuli. It is anticipated that metabolite profiling will be useful in defining (i) molecular processes that underpin promastigote and amastigote differentiation; (ii) species-specific differences in metabolism that contribute to the distinct tissue- and cell-type tropism of some *Leishmania* spp; (iii) the metabolic state of parasites during periods of active and latent disease; (iv) parasite responses to drug treatment in order to detect compensatory responses and potential drug-resistance mechanisms. Finally, metabolomic profiling is likely to be powerful for annotating gene function. This is an area that is relevant to all microbial pathogens and a key to the future identification of new drug targets. Collectively, these applications provide a powerful incentive for developing the technologies and bioinformatics tools that underpin metabolomic and other profiling approaches.

Acknowledgements

The published and unpublished work from this laboratory reported in this paper was supported by the Australian National Health and Medical Research Council and the Howard Hughes Medical Institute. We thank Drs Fred Opperdoes and Graham Coombs for communicating manuscripts before publication.

References

Akerman, M., Shaked-Mishan, P., Mazareb, S., Volpin, H. and Zilberstein, D. (2004). Novel motifs in amino acid permease genes from *Leishmania*. Biochem. Biophys. Res. Commun. *325*, 353–366.

Akopyants, N.S., Matlib, R.S., Bukanova, E.N., Smeds, M.R., Brownstein, B.H., Stormo, G.D. and Beverley, S.M. (2004). Expression profiling using random genomic DNA microarrays identifies differentially expressed genes associated with three major developmental stages of the protozoan parasite *Leishmania major*. Mol. Biochem. Parasitol. *136*, 71–86.

Al-Salabi, M.I. and de Koning, H.P. (2005). Purine nucleobase transport in amastigotes of *Leishmania mexicana*, involvement in allopurinol uptake. Antimicrob. Agents Chemother. *49*, 3682–3689.

Almeida, R., Gilmartin, B.J., McCann, S.H., Norrish, A., Ivens, A.C., Lawson, D., Levick, M.P., Smith, D.F., Dyall, S.D., Vetrie, D., Freeman, T.C., Coulson, R.M., Sampaio, I., Schneider, H. and Blackwell, J.M. (2004). Expression profiling of the *Leishmania* life cycle, cDNA arrays identify developmentally regulated genes present but not annotated in the genome. Mol. Biochem. Parasitol. *136*, 87–100.

Ariyanayagam, M.R. and Fairlamb, A.H. (2001). Ovothiol and trypanothione as antioxidants in trypanosomatids. Mol. Biochem. Parasitol. *115*, 189–198.

Ariza, A., Vickers, T.J., Greig, N., Armour, K.A., Dixon, M.J., Eggleston, I.M., Fairlamb, A.H. and Bond, C.S. (2006). Specificity of the trypanothione-dependent *Leishmania major* glyoxalase I, structure and biochemical comparison with the human enzyme. Mol. Microbiol. 59, 1239–1248.

Barak, E., Amin-Spector, S., Gerliak, E., Goyard, S., Holland, N. and Zilberstein, D. (2005). Differentiation of *Leishmania donovani* in host-free system, analysis of signal perception and response. Mol. Biochem. Parasitol. *141*, 99–108.

Basselin, M., Coombs, G.H. and Barrett, M.P. (2000). Putrescine and spermidine transport in *Leishmania*. Mol. Biochem. Parasitol. *109*, 37–46.

Bates, P.A. (1994). The developmental biology of *Leishmania* promastigotes. Exp. Parasitol. *79*, 215–218.

Beck, J.T. and Ullman, B. (1991). Biopterin conversion to reduced folates by *Leishmania donovani* promastigotes. Mol. Biochem. Parasitol. *49*, 21–28.

Belkaid, Y., Mendez, S., Lira, R., Kadambi, N., Milon, G. and Sacks, D. (2000). A natural model of *Leishmania major* infection reveals a prolonged 'silent' phase of parasite amplification in the skin before the onset of lesion formation and immunity. J. Immunol. *165*, 969–977.

Bermudez, R., Dagger, F., D'Aquino, J.A., Benaim, G. and Dawidowicz, K. (1997). Characterization of mitochondrial electron-transfer in *Leishmania mexicana*. Mol. Biochem. Parasitol. *90*, 43–54.

Berriman, M., Ghedin, E., Hertz-Fowler, C., Blandin, G., Renauld, H., Bartholomeu, D.C., Lennard, N.J., Caler, E., Hamlin, N.E., Haas, B., Bohme, U., Hannick, L., Aslett, M.A., Shallom, J., Marcello, L., Hou, L., Wickstead, B., Alsmark, U.C., Arrowsmith, C., Atkin, R.J., Barron, A.J., Bringaud, F., Brooks, K., Carrington, M., Cherevach, I., Chillingworth, T.J., Churcher, C., Clark, L.N., Corton, C.H., Cronin, A., Davies, R.M., Doggett, J., Djikeng, A., Feldblyum,

T., Field, M.C., Fraser, A., Goodhead, I., Hance, Z., Harper, D., Harris, B.R., Hauser, H., Hostetler, J., Ivens, A., Jagels, K., Johnson, D., Johnson, J., Jones, K., Kerhornou, A.X., Koo, H., Larke, N., Landfear, S., Larkin, C., Leech, V., Line, A., Lord, A., Macleod, A., Mooney, P.J., Moule, S., Martin, D.M., Morgan, G.W., Mungall, K., Norbertczak, H., Ormond, D., Pai, G., Peacock, C.S., Peterson, J., Quail, M.A., Rabbinowitsch, E., Rajandream, M.A., Reitter, C., Salzberg, S.L., Sanders, M., Schobel, S., Sharp, S., Simmonds, M., Simpson, A.J., Tallon, L., Turner, C.M., Tait, A., Tivey, A.R., Van Aken, S., Walker, D., Wanless, D., Wang, S., White, B., White, O., Whitehead, S., Woodward, J., Wortman, J., Adams, M.D., Embley, T.M., Gull, K., Ullu, E., Barry, J.D., Fairlamb, A.H., Opperdoes, F., Barrell, B.G., Donelson, J.E., Hall, N., Fraser, C.M., Melville, S.E. and El-Sayed, N.M. (2005). The genome of the African trypanosome *Trypanosoma brucei*. Science *309*, 416–422.

Besteiro, S., Williams, R.A., Morrison, L.S., Coombs, G.H. and Mottram, J.C. (2006). Endosome sorting and autophagy are essential for differentiation and virulence of *Leishmania major*. J. Biol. Chem. *281*, 11384–11396.

Beverley, S.M. (2003). Protozomics, trypanosomatid parasite genetics comes of age. Nat. Rev. Genet. *4*, 11–19.

Beverley, S.M., Owens, K.L., Showalter, M., Griffith, C.L., Doering, T.L., Jones, V.C. and McNeil, M.R. (2005). Eukaryotic UDP-galactopyranose mutase (GLF gene) in microbial and metazoal pathogens. Eukaryot Cell *4*, 1147–1154.

Bogdan, C., Donhauser, N., Doring, R., Rollinghoff, M., Diefenbach, A. and Rittig, M.G. (2000). Fibroblasts as host cells in latent leishmaniosis. J. Exp. Med. *191*, 2121–2130.

Boitz, J.M. and Ullman, B. (2006a). A conditional mutant deficient in hypoxanthine-guanine phosphoribosyltransferase and xanthine phosphoribosyltransferase validates the purine salvage pathway of *Leishmania donovani*. J. Biol. Chem. *281*, 16084–16089.

Boitz, J.M. and Ullman, B. (2006b). *Leishmania donovani* singly deficient in HGPRT, APRT or XPRT are viable in vitro and within mammalian macrophages. Mol. Biochem. Parasitol. *148*, 24–30.

Bringaud, F., Riviere, L. and Coustou, V. (2006). Energy metabolism of trypanosomatids, adaptation to available carbon sources. Mol. Biochem. Parasitol. *149*, 1–9.

Burchmore, R.J. and Hart, D.T. (1995). Glucose transport in amastigotes and promastigotes of *Leishmania mexicana* mexicana. Mol. Biochem. Parasitol. *74*, 77–86.

Burchmore, R.J., Rodriguez-Contreras, D., McBride, K., Merkel, P., Barrett, M.P., Modi, G., Sacks, D. and Landfear, S.M. (2003). Genetic characterization of glucose transporter function in *Leishmania mexicana*. Proc. Natl. Acad. Sci. USA *100*, 3901–3906.

Cameron, P., McGachy, A., Anderson, M., Paul, A., Coombs, G.H., Mottram, J.C., Alexander, J. and Plevin, R. (2004). Inhibition of lipopolysaccharide-induced macrophage IL-12 production by *Leishmania mexicana* amastigotes, the role of cysteine peptidases and the NF-kappaB signaling pathway. J. Immunol. *173*, 3297–3304.

Carter, N.S., Landfear, S.M. and Ullman, B. (2001). Nucleoside transporters of parasitic protozoa. Trends Parasitol. *17*, 142–145.

Chen, M., Zhai, L., istensen, S.B., Theander, T.G. and Kharazmi, A. (2001). Inhibition of fumarate reductase in *Leishmania major* and *L. donovani* by chalcones. Antimicrob. Agents Chemother. *45*, 2023–2029.

Clayton, C.E. (2002). Life without transcriptional control? From fly to man and back again. Embo J. *21*, 1881–1888.

Colasante, C., Ellis, M., Ruppert, T. and Voncken, F. (2006). Comparative proteomics of glycosomes from bloodstream form and procyclic culture form *Trypanosoma brucei* brucei. Proteomics *6*, 3275–3293.

Concepcion, J.L., Adje, C.A., Quinones, W., Chevalier, N., Dubourdieu, M. and Michels, P.A. (2001). The expression and intracellular distribution of phosphoglycerate kinase isoenzymes in *Trypanosoma cruzi*. Mol. Biochem. Parasitol. *118*, 111–121.

Cook, D., Fowler, S., Fiehn, O. and Thomashow, M.F. (2004). A prominent role for the CBF cold response pathway in configuring the low-temperature metabolome of Arabidopsis. Proc. Natl. Acad. Sci. USA. *101*, 15243–15248.

Coustou, V., Besteiro, S., Biran, M., Diolez, P., Bouchaud, V., Voisin, P., Michels, P.A., Canioni, P., Baltz, T. and Bringaud, F. (2003). ATP generation in the *Trypanosoma brucei* procyclic form, cytosolic substrate level is essential, but not oxidative phosphorylation. J. Biol. Chem. *278*, 49625–49635.

Cunningham, M.L. and Beverley, S.M. (2001). Pteridine salvage throughout the *Leishmania* infectious cycle, implications for antifolate chemotherapy. Mol. Biochem. Parasitol. *113*, 199–213.

Cunningham, M.L., Titus, R.G., Turco, S.J. and Beverley, S.M. (2001). Regulation of differentiation to the infective stage of the protozoan parasite *Leishmania major* by tetrahydrobiopterin. Science *292*, 285–287.

Denny, P.W., Goulding, D., Ferguson, M.A. and Smith, D.F. (2004). Sphingolipid-free *Leishmania* are defective in membrane trafficking, differentiation and infectivity. Mol. Microbiol. 52, 313–327.

Denny, P.W., Shams-Eldin, H., Price, H.P., Smith, D.F. and Schwarz, R.T. (2006). The protozoan inositol phosphorylceramide synthase, a novel drug target that defines a new class of sphingolipid synthase. J. Biol. Chem. *281*, 28200–28209.

Dobson, D.E., Mengeling, B.J., Cilmi, S., Hickerson, S., Turco, S.J. and Beverley, S.M. (2003). Identification of genes encoding arabinosyltransferases (SCA) mediating developmental modifications of lipophosphoglycan required for sandfly transmission of *Leishmania major*. J. Biol. Chem. *278*, 28840–28848.

Dobson, D.E., Scholtes, L.D., Valdez, K.E., Sullivan, D.R., Mengeling, B.J., Cilmi, S., Turco, S.J. and Beverley, S.M. (2003). Functional identification of galactosyltransferases (SCGs) required for species-specific modifications of the lipophosphoglycan adhesin controlling *Leishmania major*-sandfly interactions. J. Biol. Chem. *278*, 15523–15531.

Docampo, R. and Moreno, S.N. (2001). The acidocalcisome. Mol. Biochem. Parasitol. *114*, 151–159.

Dormeyer, M., Reckenfelderbaumer, N., Ludemann, H. and Krauth-Siegel, R.L. (2001). Trypanothione-dependent synthesis of deoxyribonucleotides by *Trypanosoma brucei* ribonucleotide reductase. J. Biol. Chem. *276*, 10602–10606.

Dumas, C., Ouellette, M., Tovar, J., Cunningham, M.L., Fairlamb, A.H., Tamar, S., Olivier, M. and Papadopoulou, B. (1997). Disruption of the trypanothione reductase gene of *Leishmania* decreases its ability to survive oxidative stress in macrophages. EMBO J. *16*, 2590–2598.

Duncan, R.C., Salotra, P., Goyal, N., Akopyants, N.S., Beverley, S.M. and Nakhasi, H.L. (2004). The application of gene expression microarray technology to kinetoplastid research. Curr. Mol. Med. *4*, 611–621.

Engwerda, C.R., Ato, M. and Kaye, P.M. (2004). Macrophages, pathology and parasite persistence in experimental visceral leishmaniasis. Trends Parasitol. *20*, 524–530.

Fell, D.A. (2005). Enzymes, metabolites and fluxes. J. Exp. Bot 56, 267–272.

Fernie, A.R., Trethewey, R.N., Krotzky, A.J. and Willmitzer, L. (2004). Metabolite profiling, from diagnostics to systems biology. Nat. Rev. Mol. Cell Biol. 5, 763–769.

Frame, M.J., Mottram, J.C. and Coombs, G.H. (2000). Analysis of the roles of cysteine proteinases of *Leishmania mexicana* in the host–parasite interaction. Parasitology 121 (Pt 4), 367–377.

Garami, A. and Ilg, T. (2001). Disruption of mannose activation in *Leishmania mexicana*, GDP-mannose pyrophosphorylase is required for virulence, but not for viability. EMBO J. *20*, 3657–3666.

Garami, A. and Ilg, T. (2001). The role of phosphomannose isomerase in *Leishmania mexicana* glycoconjugate synthesis and virulence. J. Biol. Chem. *276*, 6566–6575.

Garami, A., Mehlert, A. and Ilg, T. (2001). Glycosylation defects and virulence phenotypes of *Leishmania mexicana* phosphomannomutase and dolicholphosphate-mannose synthase gene deletion mutants. Mol. Cell Biol. *21*, 8168–8183.

Geraldo, M.V., Silber, A.M., Pereira, C.A. and Uliana, S.R. (2005). Characterisation of a developmentally regulated amino acid transporter gene from *Leishmania amazonensis*. FEMS Microbiol. Lett. *242*, 275–280.

Ghosh, M. and Mukherjee, T. (2000). Stage-specific development of a novel adenosine transporter in *Leishmania donovani* amastigotes. Mol. Biochem. Parasitol. *108*, 93–99.

Ginger, M.L., Chance, M.L. and Goad, L.J. (1999). Elucidation of carbon sources used for the biosynthesis of fatty acids and sterols in the trypanosomatid *Leishmania mexicana*. Biochem. J. 342 (Pt 2), 397–405.

Ginger, M.L., Chance, M.L., Sadler, I.H. and Goad, L.J. (2001). The biosynthetic incorporation of the intact leucine skeleton into sterol by the trypanosomatid *Leishmania mexicana*. J. Biol. Chem. *276*, 11674–11682.

Gokool, S. (2003). Sigma 1- and mu 1-Adaptin homologues of *Leishmania mexicana* are required for parasite survival in the infected host. J. Biol. Chem. *278*, 29400–29409.

Gonzalez, B., Francois, J. and Renaud, M. (1997). A rapid and reliable method for metabolite extraction in yeast using boiling buffered ethanol. Yeast *13*, 1347–1355.

Goodacre, R., Vaidyanathan, S., Dunn, W.B., Harrigan, G.G. and Kell, D.B. (2004). Metabolomics by numbers, acquiring and understanding global metabolite data. Trends Biotechnol. *22*, 245–252.

Gossage, S.M., Rogers, M.E. and Bates, P.A. (2003). Two separate growth phases during the development of *Leishmania* in sandflies, implications for understanding the life cycle. Int. J. Parasitol. 33, 1027–1034.

Guerra, D.G., Decottignies, A., Bakker, B.M. and Michels, P.A. (2006). The mitochondrial FAD-dependent glycerol-3-phosphate dehydrogenase of Trypanosomatidae and the glycosomal redox balance of insect stages of *Trypanosoma brucei* and *Leishmania* spp. Mol. Biochem. Parasitol. *149*, 155–169.

Gupta, N., Goyal, N., Singha, U.K., Bhakuni, V., Roy, R. and Rastogi, A.K. (1999). Characterization of intracellular metabolites of axenic amastigotes of *Leishmania donovani* by 1H NMR spectroscopy. Acta Trop 73, 121–133.

Halket, J.M., Waterman, D., Przyborowska, A.M., Patel, R.K., Fraser, P.D. and Bramley, P.M. (2005). Chemical derivatization and mass spectral libraries in metabolic profiling by GC/MS and LC/MS/MS. J. Exp. Bot 56, 219–243.

Hannaert, V., Blaauw, M., Kohl, L., Allert, S., Opperdoes, F.R. and Michels, P.A. (1992). Molecular analysis of the cytosolic and glycosomal glyceraldehyde-3-phosphate dehydrogenase in *Leishmania mexicana*. Mol. Biochem. Parasitol. 55, 115–126.

Hannaert, V., Bringaud, F., Opperdoes, F.R. and Michels, P.A. (2003). Evolution of energy metabolism and its compartmentation in Kinetoplastida. Kinetoplastid Biol. Dis. 2, 11.

Hart, D.T. and Coombs, G.H. (1982). *Leishmania mexicana*, energy metabolism of amastigotes and promastigotes. Exp. Parasitol. 54, 397–409.

Hasne, M.P. and Ullman, B. (2005). Identification and characterization of a polyamine permease from the protozoan parasite *Leishmania major*. J. Biol. Chem. *280*, 15188–15194.

Hill, J.E., Scott, D.A., Luo, S. and Docampo, R. (2000). Cloning and functional expression of a gene encoding a vacuolar-type proton-translocating pyrophosphatase from *Trypanosoma cruzi*. Biochem. J. *351*, 281–288.

Hilley, J.D., Zawadzki, J.L., McConville, M.J., Coombs, G.H. and Mottram, J.C. (2000). *Leishmania mexicana* mutants lacking glycosylphosphatidylinositol (GPI),protein transamidase provide insights into the biosynthesis and functions of GPI-anchored proteins. Mol. Biol. Cell *11*, 1183–1195.

Hirai, M.Y., Yano, M., Goodenowe, D.B., Kanaya, S., Kimura, T., Awazuhara, M., Arita, M., Fujiwara, T. and Saito, K. (2004). Integration of transcriptomics and metabolomics for understanding of global responses to nutritional stresses in Arabidopsis thaliana. Proc. Natl. Acad. Sci. USA. *101*, 10205–10210.

Holzer, T.R., McMaster, W.R. and Forney, J.D. (2006). Expression profiling by whole-genome interspecies microarray hybridization reveals differential gene expression in procyclic promastigotes, lesion-derived amastigotes, and axenic amastigotes in *Leishmania mexicana*. Mol. Biochem. Parasitol. *146*, 198–218.

Holzmuller, P., Hide, M., Sereno, D. and Lemesre, J.L. (2005). *Leishmania infantum* amastigotes resistant to nitric oxide cytotoxicity, Impact on in vitro parasite developmental cycle and metabolic enzyme activities. Infect. Genet. Evol.

Hubel, A., Krobitsch, S., Horauf, A. and Clos, J. (1997). *Leishmania major* Hsp100 is required chiefly in the mammalian stage of the parasite. Mol. Cell Biol. *17*, 5987–5995.

Huynh, C., Sacks, D.L. and Andrews, N.W. (2006). A *Leishmania amazonensis* ZIP family iron transporter is essential for parasite replication within macrophage phagolysosomes. J. Exp. Med. *203*, 2363–2375.

Ilg, T. (2000). Lipophosphoglycan is not required for infection of macrophages or mice by *Leishmania mexicana*. Embo J. *19*, 1953–1962.

Ilg, T. (2000). Proteophosphoglycans of *Leishmania*. Parasitol. Today *16*, 489–497.

Ilg, T. (2002). Generation of myo-inositol-auxotrophic *Leishmania mexicana* mutants by targeted replacement of the myo-inositol-1-phosphate synthase gene. Mol. Biochem. Parasitol. *120*, 151–156.

Ilg, T., Demar, M. and Harbecke, D. (2001). Phosphoglycan repeat-deficient *Leishmania mexicana* parasites remain infectious to macrophages and mice. J. Biol. Chem. *276*, 4988–4997.

Ilg, T., Etges, R., Overath, P., McConville, M.J., Thomas-Oates, J., Thomas, J., Homans, S.W. and Ferguson, M.A. (1992). Structure of *Leishmania mexicana* lipophosphoglycan. J. Biol. Chem. *267*, 6834–6840.

Ilg, T., Overath, P., Ferguson, M.A., Rutherford, T., Campbell, D.G. and McConville, M.J. (1994). O- and N-glycosylation of the *Leishmania mexicana*-secreted acid phosphatase. Characterization of a new class of phosphoserine-linked glycans. J. Biol. Chem. *269*, 24073–24081.

Ilg, T., Stierhof, Y.D., Craik, D., Simpson, R., Handman, E. and Bacic, A. (1996). Purification and structural characterization of a filamentous, mucin-like proteophosphoglycan secreted by *Leishmania* parasites. J. Biol. Chem. *271*, 21583–21596.

Ilgoutz, S.C., Zawadzki, J.L., Ralton, J.E. and McConville, M.J. (1999). Evidence that free GPI glycolipids are essential for growth of *Leishmania mexicana*. Embo J. *18*, 2746–2755.

Jiang, Y., Roberts, S.C., Jardim, A., Carter, N.S., Shih, S., Ariyanayagam, M., Fairlamb, A.H. and Ullman, B. (1999). Ornithine decarboxylase gene deletion mutants of *Leishmania donovani*. J. Biol. Chem. *274*, 3781–3788.

Joshi, M.B. and Dwyer, D.M. (2007). Molecular and functional analyses of a novel, class-I secretory nuclease from the human pathogen, *Leishmania donovani*. J. Biol. Chem.

Keegan, F.P. and Blum, J.J. (1992). Utilization of a carbohydrate reserve comprised primarily of mannose by *Leishmania donovani*. Mol. Biochem. Parasitol. *53*, 193–200.

Kessler, P.S. and Parsons, M. (2005). Probing the role of compartmentation of glycolysis in procyclic form *Trypanosoma brucei*, RNA interference studies of PEX14, hexokinase, and phosphofructokinase. J. Biol. Chem. *280*, 9030–9036.

Kink, J.A. and Chang, K.P. (1987). Tunicamycin-resistant *Leishmania mexicana* amazonensis, expression of virulence associated with an increased activity of N-acetylglucosaminyltransferase and amplification of its presumptive gene. Proc. Natl. Acad. Sci. USA *84*, 1253–1257.

Kleczka, B., Lamerz, A.C., van Zandbergen, G., Wenzel, A., Gerardy-Schahn, R., Wiese, M. and Routier, F.H. (2007). Targeted gene deletion of *Leishmania major* UDP-galactopyranose mutase leads to attenuated virulence. J. Biol. Chem.

Krishnamurthy, G., Vikram, R., Singh, S.B., Patel, N., Agarwal, S., Mukhopadhyay, G., Basu, S.K. and Mukhopadhyay, A. (2005). Hemoglobin receptor in *Leishmania* is a hexokinase located in the flagellar pocket. J. Biol. Chem. *280*, 5884–5891.

Krishnan, P., Kruger, N.J. and Ratcliffe, R.G. (2005). Metabolite fingerprinting and profiling in plants using NMR. J. Exp. Bot *56*, 255–265.

Kropf, P., Fuentes, J.M., Fahnrich, E., Arpa, L., Herath, S., Weber, V., Soler, G., Celada, A., Modolell, M. and Muller, I. (2005). Arginase and polyamine synthesis are key factors in the regulation of experimental leishmaniasis in vivo. FASEB J. *19*, 1000–1002.

Kuhn, D. and Wiese, M. (2005). LmxPK4, a mitogen-activated protein kinase kinase homologue of *Leishmania mexicana* with a potential role in parasite differentiation. Mol. Microbiol. *56*, 1169–1182.

Kundig, C., Leblanc, E., Papadopoulou, B. and Ouellette, M. (1999). Role of the locus and of the resistance gene on gene amplification frequency in methotrexate-resistant *Leishmania tarentolae*. Nucleic Acids Res. *27*, 3653–3659.

Lafaye, A., Junot, C., Pereira, Y., Lagniel, G., Tabet, J.C., Ezan, E. and Labarre, J. (2005). Combined proteome and metabolite-profiling analyses reveal surprising insights into yeast sulfur metabolism. J. Biol. Chem. *280*, 24723–24730.

Lafaye, A., Labarre, J., Tabet, J.C., Ezan, E. and Junot, C. (2005). Liquid chromatography-mass spectrometry and 15N metabolic labeling for quantitative metabolic profiling. Anal. Chem. *77*, 2026–2033.

Lamour, N., Riviere, L., Coustou, V., Coombs, G.H., Barrett, M.P. and Bringaud, F. (2005). Proline metabolism in procyclic *Trypanosoma brucei* is down-regulated in the presence of glucose. J. Biol. Chem. *280*, 11902–11910.

Landfear, S.M., Ullman, B., Carter, N.S. and Sanchez, M.A. (2004). Nucleoside and nucleobase transporters in parasitic protozoa. Eukaryot. Cell *3*, 245–254.

Lee, S.H., Stephens, J.L., Paul, K.S. and Englund, P.T. (2006). Fatty acid synthesis by elongases in trypanosomes. Cell *126*, 691–699.

Leifso, K., Cohen-Freue, G., Dogra, N., Murray, A. and McMaster, W.R. (2007). Genomic and proteomic expression analysis of *Leishmania* promastigote and amastigote life stages, The *Leishmania* genome is constitutively expressed. Mol. Biochem. Parasitol. *152*, 35–46.

Lemley, C., Yan, S., Dole, V.S., Madhubala, R., Cunningham, M.L., Beverley, S.M., Myler, P.J. and Stuart, K.D. (1999). The *Leishmania donovani* LD1 locus gene ORFG encodes a biopterin transporter (BT1). Mol. Biochem. Parasitol. *104*, 93–105.

Leroux, A., Fleming-Canepa, X., Aranda, A., Maugeri, D., Cazzulo, J.J., Sanchez, M.A. and Nowicki, C. (2006). Functional characterization and subcellular localization of the three malate dehydrogenase isozymes in *Leishmania* spp. Mol. Biochem. Parasitol. *149*, 74–85.

Lowry, O.H., Carter, J., Ward, J.B. and Glaser, L. (1971). The effect of carbon and nitrogen sources on the level of metabolic intermediates in Escherichia coli. J. Biol. Chem. *246*, 6511–6521.

Lux, H., Heise, N., Klenner, T., Hart, D. and Opperdoes, F.R. (2000). Ether–lipid (alkyl-phospholipid) metabolism and the mechanism of action of ether – lipid analogues in *Leishmania*. Mol. Biochem. Parasitol. *111*, 1–14.

Ma, D., Russell, D.G., Beverley, S.M. and Turco, S.J. (1997). Golgi GDP-mannose uptake requires *Leishmania* LPG2. A member of a eukaryotic family of putative nucleotide-sugar transporters. J. Biol. Chem. *272*, 3799–3805.

Maugeri, D.A., Cazzulo, J.J., Burchmore, R.J., Barrett, M.P. and Ogbunude, P.O. (2003). Pentose phosphate metabolism in *Leishmania mexicana*. Mol. Biochem. Parasitol. *130*, 117–125.

Mazareb, S., Fu, Z.Y. and Zilberstein, D. (1999). Developmental regulation of proline transport in *Leishmania donovani*. Exp. Parasitol. *91*, 341–348.

McConville, M.J. and Blackwell, J.M. (1991). Developmental changes in the glycosylated phosphatidylinositols of *Leishmania donovani*. Characterization of the promastigote and amastigote glycolipids. J. Biol. Chem. *266*, 15170–15179.

McConville, M.J., Mullin, K.A., Ilgoutz, S.C. and Teasdale, R.D. (2002). Secretory pathway of trypanosomatid parasites. Microbiol. Mol. Biol. Rev. *66*, 122–154.

McConville, M.J., Schnur, L.F., Jaffe, C. and Schneider, P. (1995). Structure of *Leishmania* lipophosphoglycan, inter- and intra-specific polymorphism in Old World species. Biochem. J. 310 (Pt 3), 807–818.

McConville, M.J., Turco, S.J., Ferguson, M.A. and Sacks, D.L. (1992). Developmental modification of lipophosphoglycan during the differentiation of *Leishmania major* promastigotes to an infectious stage. Embo J. *11*, 3593–3600.

McMahon-Pratt, D. and Alexander, J. (2004). Does the *Leishmania major* paradigm of pathogenesis and protection hold for New World cutaneous leishmaniases or the visceral disease? Immunol. Rev. *201*, 206–224.

Mendonca-Previato, L., Gorin, P.A. and Previato, J.O. (1979). Investigations on polysaccharide components of cells of Herpetomonas samuelpessoai grown on various media. Biochemistry *18*, 149–154.

Merlen, T., Sereno, D., Brajon, N., Rostand, F. and Lemesre, J.L. (1999). *Leishmania* spp, completely defined medium without serum and macromolecules (CDM/LP) for the continuous in vitro cultivation of infective promastigote forms. Am. J. Trop. Med. Hyg. *60*, 41–50.

Moll, H., Flohe, S. and Rollinghoff, M. (1995). Dendritic cells in *Leishmania major*-immune mice harbor persistent parasites and mediate an antigen-specific T cell immune response. Eur. J. Immunol. *25*, 693–699.

Moody, S.F., Handman, E., McConville, M.J. and Bacic, A. (1993). The structure of *Leishmania major* amastigote lipophosphoglycan. J. Biol. Chem. *268*, 18457–18466.

Moreno, B., Rodrigues, C.O., Bailey, B.N., Urbina, J.A., Moreno, S.N., Docampo, R. and Oldfield, E. (2002). Magic-angle spinning (31)P NMR spectroscopy of condensed phosphates in parasitic protozoa, visualizing the invisible. FEBS Lett. *523*, 207–212.

Moreno, B., Urbina, J.A., Oldfield, E., Bailey, B.N., Rodrigues, C.O. and Docampo, R. (2000). 31P NMR spectroscopy of *Trypanosoma brucei, trypanosoma cruzi*, and *Leishmania major*. Evidence for high levels of condensed inorganic phosphates. J. Biol. Chem. *275*, 28356–28362.

Morita, Y.S., Paul, K.S. and Englund, P.T. (2000). Specialized fatty acid synthesis in African trypanosomes, myristate for GPI anchors. Science *288*, 140–143.

Moss, J.M., Reid, G.E., Mullin, K.A., Zawadzki, J.L., Simpson, R.J. and McConville, M.J. (1999). Characterization of a novel GDP-mannose, Serine-protein mannose-1-phosphotransferase from *Leishmania mexicana*. J. Biol. Chem. *274*, 6678–6688.

Mottram, J.C., Souza, A.E., Hutchison, J.E., Carter, R., Frame, M.J. and Coombs, G.H. (1996). Evidence from disruption of the lmcpb gene array of *Leishmania mexicana* that cysteine proteinases are virulence factors. Proc. Natl. Acad. Sci. USA 93, 6008–6013.

Mullin, K.A., Foth, B.J., Ilgoutz, S.C., Callaghan, J.M., Zawadzki, J.L., McFadden, G.I. and McConville, M.J. (2001). Regulated degradation of an endoplasmic reticulum membrane protein in a tubular lysosome in *Leishmania mexicana*. Mol. Biol. Cell *12*, 2364–2377.

Naderer, T., Ellis, M.A., Sernee, M.F., De Souza, D.P., Curtis, J., Handman, E. and McConville, M.J. (2006). Virulence of *Leishmania major* in macrophages and mice requires the gluconeogenic enzyme fructose-1, 6-bisphosphatase. Proc. Natl. Acad. Sci. USA. *103*, 5502–5507.

Naderer, T. and McConville, M.J. (2002). Characterization of a *Leishmania mexicana* mutant defective in synthesis of free and protein-linked GPI glycolipids. Mol. Biochem. Parasitol. *125*, 147–161.

Naderer, T., Vince, J.E. and McConville, M.J. (2004). Surface determinants of *Leishmania* parasites and their role in infectivity in the mammalian host. Curr. Mol. Med. *4*, 649–665.

Nicholson, J.K., Connelly, J., Lindon, J.C. and Holmes, E. (2002). Metabonomics, a platform for studying drug toxicity and gene function. Nat. Rev. Drug Discov *1*, 153–161.

Nugent, P.G., Karsani, S.A., Wait, R., Tempero, J. and Smith, D.F. (2004). Proteomic analysis of *Leishmania mexicana* differentiation. Mol. Biochem. Parasitol. *136*, 51–62.

Nyame, K., Do-Thi, C.D., Opperdoes, F.R. and Michels, P.A. (1994). Subcellular distribution and characterization of glucosephosphate isomerase in *Leishmania mexicana* mexicana. Mol. Biochem. Parasitol. *67*, 269–279.

Opperdoes, F. and Coombs, G.H. (2007). Metabolismof *Leishmania*; proven and predicted. Trends in Parasitology in press.

Opperdoes, F.R. and Michels, P.A. (2001). Enzymes of carbohydrate metabolism as potential drug targets. Int. J. Parasitol. *31*, 482–490.

Ouellette, M., Drummelsmith, J., El-Fadili, A., Kundig, C., Richard, D. and Roy, G. (2002). Pterin transport and metabolism in *Leishmania* and related trypanosomatid parasites. Int. J. Parasitol. *32*, 385–398.

Papageorgiou, I.G., Yakob, L., Al Salabi, M.I., Diallinas, G., Soteriadou, K.P. and De Koning, H.P. (2005). Identification of the first pyrimidine nucleobase transporter in *Leishmania*, similarities with the *Trypanosoma brucei* U1 transporter and anti-leishmanial activity of uracil analogues. Parasitology *130*, 275–283.

Pereira, C.A., Alonso, G.D., Ivaldi, S., Silber, A.M., Alves, M.J., Torres, H.N. and Flawia, M.M. (2003). Arginine kinase overexpression improves *Trypanosoma cruzi* survival capability. FEBS Lett. *554*, 201–205.

Previato, J.O., Xavier, M.T., Brazil, R.P., Gorin, P.A. and Mendonca-Previato, L. (1984). Formation of (1 – – 2)-linked beta-D-mannopyranan by *Leishmania mexicana* amazonensis, relationship with certain *Crithidia* and *Herpetomonas* species. J. Parasitol. *70*, 449–450.

Raamsdonk, L.M., Teusink, B., Broadhurst, D., Zhang, N., Hayes, A., Walsh, M.C., Berden, J.A., Brindle, K.M., Kell, D.B., Rowland, J.J., Westerhoff, H.V., van Dam, K. and Oliver, S.G. (2001). A functional genomics strategy that uses metabolome data to reveal the phenotype of silent mutations. Nat. Biotechnol. *19*, 45–50.

Rainey, P.M. and MacKenzie, N.E. (1991). A carbon-13 nuclear magnetic resonance analysis of the products of glucose metabolism in *Leishmania pifanoi* amastigotes and promastigotes. Mol. Biochem. Parasitol. *45*, 307–315.

Ralton, J.E. and McConville, M.J. (1998). Delineation of three pathways of glycosylphosphatidylinositol biosynthesis in *Leishmania mexicana*. Precursors from different pathways are assembled on distinct pools of phosphatidylinositol and undergo fatty acid remodeling. J. Biol. Chem. *273*, 4245–4257.

Ralton, J.E., Naderer, T., Piraino, H.L., Bashtannyk, T.A., Callaghan, J.M. and McConville, M.J. (2003). Evidence that intracellular beta1–2 mannan is a virulence factor in *Leishmania* parasites. J. Biol. Chem. *278*, 40757–40763.

Reguera, R.M., Redondo, C.M., Perez-Pertejo, Y. and Balana-Fouce, R. (2007). S-Adenosylmethionine in protozoan parasites, Functions, synthesis and regulation. Mol. Biochem. Parasitol. *152*, 1–10.

Reguera, R.M., Tekwani, B.L. and Balana-Fouce, R. (2005). Polyamine transport in parasites, a potential target for new antiparasitic drug development. Comp. Biochem. Physiol. C Toxicol. Pharmacol. *140*, 151–164.

Roberts, S.C., Jiang, Y., Jardim, A., Carter, N.S., Heby, O. and Ullman, B. (2001). Genetic analysis of spermidine synthase from *Leishmania donovani*. Mol. Biochem. Parasitol. *115*, 217–226.

Roberts, S.C., Scott, J., Gasteier, J.E., Jiang, Y., Brooks, B., Jardim, A., Carter, N.S., Heby, O. and Ullman, B. (2002). S-adenosylmethionine decarboxylase from *Leishmania donovani*. Molecular, genetic, and biochemical characterization of null mutants and overproducers. J. Biol. Chem. *277*, 5902–5909.

Roberts, S.C., Tancer, M.J., Polinsky, M.R., Gibson, K.M., Heby, O. and Ullman, B. (2004). Arginase plays a pivotal role in polyamine precursor metabolism in *Leishmania*. Characterization of gene deletion mutants. J. Biol. Chem. *279*, 23668–23678.

Rodrigues, C.O., Ruiz, F.A., Vieira, M., Hill, J.E. and Docampo, R. (2002). An acidocalcisomal exopolyphosphatase from *Leishmania major* with high affinity for short chain polyphosphate. J. Biol. Chem. *277*, 50899–50906.

Rogers, M.E., Chance, M.L. and Bates, P.A. (2002). The role of promastigote secretory gel in the origin and transmission of the infective stage of *Leishmania mexicana* by the sandfly *Lutzomyia longipalpis*. Parasitology *124*, 495–507.

Rosenthal, P.J. (1999). Proteases of protozoan parasites. Adv. Parasitol. *43*, 105–159.

Ruijter, G.J., Panneman, H., van den Broeck, H.C., Bennett, J.M. and Visser, J. (1996). Characterisation of the *Aspergillus nidulans* frA1 mutant, hexose phosphorylation and apparent lack of involvement of hexokinase in glucose repression. FEMS Microbiol. Lett. *139*, 223–228.

Saar, Y., Ransford, A., Waldman, E., Mazareb, S., Amin-Spector, S., Plumblee, J., Turco, S.J. and Zilberstein, D. (1998). Characterization of developmentally-regulated activities in axenic amastigotes of *Leishmania donovani*. Mol. Biochem. Parasitol. *95*, 9–20.

Sacks, D. and Kamhawi, S. (2001). Molecular aspects of parasite-vector and vector-host interactions in leishmaniasis. Annu. Rev. Microbiol. *55*, 453–483.

Sacks, D.L., Pimenta, P.F., McConville, M.J., Schneider, P. and Turco, S.J. (1995). Stage-specific binding of *Leishmania donovani* to the sandfly vector midgut is regulated by conformational changes in the abundant surface lipophosphoglycan. J. Exp. Med. *181*, 685–697.

Saez, M.J. and Lagunas, R. (1976). Determination of intermediary metabolites in yeast. Critical examination of the effect of sampling conditions and recommendations for obtaining true levels. Mol. Cell Biochem. *13*, 73–78.

Santhamma, K.R. and Bhaduri, A. (1995). Characterization of the respiratory chain of *Leishmania donovani* promastigotes. Mol. Biochem. Parasitol. *75*, 43–53.

Schaible, U.E., Schlesinger, P.H., Steinberg, T.H., Mangel, W.F., Kobayashi, T. and Russell, D.G. (1999).

Parasitophorous vacuoles of *Leishmania mexicana* acquire macromolecules from the host cell cytosol via two independent routes. J. Cell Sci. *112*, 681–693.

Schneider, P., Rosat, J.P., Ransijn, A., Ferguson, M.A. and McConville, M.J. (1993). Characterization of glycoinositol phospholipids in the amastigote stage of the protozoan parasite *Leishmania major*. Biochem. J. 295 (Pt 2), 555–564.

Selzer, P.M., Pingel, S., Hsieh, I., Ugele, B., Chan, V.J., Engel, J.C., Bogyo, M., Russell, D.G., Sakanari, J.A. and McKerrow, J.H. (1999). Cysteine protease inhibitors as chemotherapy, lessons from a parasite target. Proc. Natl. Acad. Sci. USA. *96*, 11015–11022.

Sernee, M.F., Ralton, J.E., Dinev, Z., Khairallah, G.N., O'Hair, R.A., Williams, S.J. and McConville, M.J. (2006). *Leishmania* beta-1, 2-mannan is assembled on a mannose-cyclic phosphate primer. Proc. Natl. Acad. Sci. USA. *103*, 9458–9463.

Seyfang, A. and Landfear, S.M. (1999). Substrate depletion upregulates uptake of myo-inositol, glucose and adenosine in *Leishmania*. Mol. Biochem. Parasitol. *104*, 121–130.

Shaked-Mishan, P., Suter-Grotemeyer, M., Yoel-Almagor, T., Holland, N., Zilberstein, D. and Rentsch, D. (2006). A novel high-affinity arginine transporter from the human parasitic protozoan *Leishmania donovani*. Mol. Microbiol. *60*, 30–38.

Shim, H. and Fairlamb, A.H. (1988). Levels of polyamines, glutathione and glutathione-spermidine conjugates during growth of the insect trypanosomatid Crithidia fasciculata. J. Gen. Microbiol. *134*, 807–817.

Simon, M.W., Martin, E. and Mukkada, A.J. (1978). Evidence for a functional glyoxylate cycle in the leishmaniae. J. Bacteriol. *135*, 895–899.

Singh, B.N., Costello, C.E., Beach, D.H. and Holz, G.G., Jr. (1988). Di-O-alkylglycerol, mono-O-alkylglycerol and ceramide inositol phosphates of *Leishmania mexicana* mexicana promastigotes. Biochem. Biophys. Res. Commun. *157*, 1239–1246.

Sopwith, W.F., Debrabant, A., Yamage, M., Dwyer, D.M. and Bates, P.A. (2002). Developmentally regulated expression of a cell surface class I nuclease in *Leishmania mexicana*. Int. J. Parasitol. *32*, 449–459.

Spath, G.F., Epstein, L., Leader, B., Singer, S.M., Avila, H.A., Turco, S.J. and Beverley, S.M. (2000). Lipophosphoglycan is a virulence factor distinct from related glycoconjugates in the protozoan parasite *Leishmania major*. Proc. Natl. Acad. Sci. USA *97*, 9258–9263.

Spath, G.F., Garraway, L.A., Turco, S.J. and Beverley, S.M. (2003). The role(s) of lipophosphoglycan (LPG) in the establishment of *Leishmania major* infections in mammalian hosts. Proc. Natl. Acad. Sci. USA *100*, 9536–9541.

Spath, G.F., Lye, L.F., Segawa, H., Sacks, D.L., Turco, S.J. and Beverley, S.M. (2003). Persistence without pathology in phosphoglycan-deficient *Leishmania major*. Science *301*, 1241–1243.

Spath, G.F., Lye, L.F., Segawa, H., Turco, S.J. and Beverley, S.M. (2004). Identification of a compensatory mutant (lpg2-REV) of *Leishmania major* able to survive as amastigotes within macrophages without LPG2-dependent glycoconjugates and its significance to virulence and immunization strategies. Infect. Immun. *72*, 3622–3627.

Stafford, J.L., Neumann, N.F. and Belosevic, M. (2002). Macrophage-mediated innate host defense against protozoan parasites. Crit. Rev. Microbiol. *28*, 187–248.

Steenkamp, D.J. (2002). Thiol metabolism of the trypanosomatids as potential drug targets. IUBMB Life *53*, 243–248.

Steiger, R.F. and Steiger, E. (1976). A defined medium for cultivating *Leishmania donovani* and *L. braziliensis*. J. Parasitol. *62*, 1010–1011.

Steiger, R.F. and Steiger, E. (1977). Cultivation of *Leishmania donovani* and *Leishmania braziliensis* in defined media, nutritional requirements. J. Protozool. *24*, 437–441.

Stephens, J.L., Lee, S.H., Paul, K.S. and Englund, P.T. (2007). Mitochondrial Fatty Acid Synthesis in *Trypanosoma brucei*. J. Biol. Chem. *282*, 4427–4436.

ter Kuile, B.H. (1999). Regulation and adaptation of glucose metabolism of the parasitic protist *Leishmania donovani* at the enzyme and mRNA levels. J. Bacteriol. *181*, 4863–4872.

Ter Kuile, B.H. and Opperdoes, F.R. (1991). Chemostat cultures of *Leishmania donovani* promastigotes and *Trypanosoma brucei* procyclic trypomastigotes. Mol. Biochem. Parasitol. *45*, 171–173.

ter Kuile, B.H. and Westerhoff, H.V. (2001). Transcriptome meets metabolome, hierarchical and metabolic regulation of the glycolytic pathway. FEBS Lett. *500*, 169–171.

Theobald, U., Mailinger, W., Reuss, M. and Rizzi, M. (1993). In vivo analysis of glucose-induced fast changes in yeast adenine nucleotide pool applying a rapid sampling technique. Anal. Biochem. *214*, 31–37.

Titus, R.G., Gueiros-Filho, F.J., de Freitas, L.A. and Beverley, S.M. (1995). Development of a safe live *Leishmania* vaccine line by gene replacement. Proc. Natl. Acad. Sci. USA. *92*, 10267–10271.

Tohge, T., Nishiyama, Y., Hirai, M.Y., Yano, M., Nakajima, J., Awazuhara, M., Inoue, E., Takahashi, H., Goodenowe, D.B., Kitayama, M., Noji, M., Yamazaki, M. and Saito, K. (2005). Functional genomics by integrated analysis of metabolome and transcriptome of Arabidopsis plants overexpressing an MYB transcription factor. Plant J. *42*, 218–235.

Tolstikov, V.V. and Fiehn, O. (2002). Analysis of highly polar compounds of plant origin, combination of hydrophilic interaction chromatography and electrospray ion trap mass spectrometry. Anal. Biochem. *301*, 298–307.

Tovar, J., Cunningham, M.L., Smith, A.C., Croft, S.L. and Fairlamb, A.H. (1998). Downregulation of *Leishmania donovani* trypanothione reductase by heterologous expression of a trans-dominant mutant homologue, effect on parasite intracellular survival. Proc. Natl. Acad. Sci. USA 95, 5311–5316.

Tull, D., Vince, J.E., Callaghan, J.M., Naderer, T., Spurck, T., McFadden, G.I., Currie, G., Ferguson, K., Bacic, A. and McConville, M.J. (2004). SMP-1, a member of a new family of small myristoylated proteins in kinetoplastid parasites, is targeted to the flagellum membrane in Leishmania. Mol. Biol. Cell *15*, 4775–4786.

Turco, S.J. (1999). Adversarial relationship between the leishmania lipophosphoglycan and protein kinase C of host macrophages. Parasite Immunol. *21*, 597–600.

Turco, S.J., Spath, G.F. and Beverley, S.M. (2001). Is lipophosphoglycan a virulence factor? A surprising diversity between *Leishmania* species. Trends Parasitol. *17*, 223–226.

Urbanczyk-Wochniak, E., Luedemann, A., Kopka, J., Selbig, J., Roessner-Tunali, U., Willmitzer, L. and Fernie, A.R. (2003). Parallel analysis of transcript and metabolic profiles, a new approach in systems biology. EMBO Rep. *4*, 989–993.

Uzonna, J.E., Spath, G.F., Beverley, S.M. and Scott, P. (2004). Vaccination with phosphoglycan-deficient *Leishmania major* protects highly susceptible mice from virulent challenge without inducing a strong Th1 response. J. Immunol. *172*, 3793–3797.

Van Hellemond, J.J., Opperdoes, F.R. and Tielens, A.G. (1998). Trypanosomatidae produce acetate via a mitochondrial acetate,succinate CoA transferase. Proc. Natl. Acad. Sci. USA *95*, 3036–3041.

Van Hellemond, J.J. and Tielens, A.G. (1997). Inhibition of the respiratory chain results in a reversible metabolic arrest in *Leishmania* promastigotes. Mol. Biochem. Parasitol. *85*, 135–138.

van Weelden, S.W., Fast, B., Vogt, A., van der Meer, P., Saas, J., van Hellemond, J.J., Tielens, A.G. and Boshart, M. (2003). Procyclic *Trypanosoma brucei* do not use Krebs cycle activity for energy generation. J. Biol. Chem. *278*, 12854–12863.

van Weelden, S.W., van Hellemond, J.J., Opperdoes, F.R. and Tielens, A.G. (2005). New functions for parts of the Krebs cycle in procyclic *Trypanosoma brucei*, a cycle not operating as a cycle. J. Biol. Chem. *280*, 12451–12460.

van Zandbergen, G., Klinger, M., Mueller, A., Dannenberg, S., Gebert, A., Solbach, W. and Laskay, T. (2004). Cutting edge, neutrophil granulocyte serves as a vector for *Leishmania* entry into macrophages. J. Immunol. *173*, 6521–6525.

Vasudevan, G., Carter, N.S., Drew, M.E., Beverley, S.M., Sanchez, M.A., Seyfang, A., Ullman, B. and Landfear, S.M. (1998). Cloning of *Leishmania* nucleoside transporter genes by rescue of a transport-deficient mutant. Proc. Natl. Acad. Sci. USA. *95*, 9873–9878.

Vickers, T.J., Orsomando, G., de la Garza, R.D., Scott, D.A., Kang, S.O., Hanson, A.D. and Beverley, S.M. (2006). Biochemical and genetic analysis of methylenetetrahydrofolate reductase in leishmania metabolism and virulence. J. Biol. Chem. *281*, 38150–38158.

Vogt, R.N., Spies, H.S. and Steenkamp, D.J. (2001). The biosynthesis of ovothiol A (N-methyl-4-mercaptohistidine). Identification of S-(4′-L-histidyl)-L-cysteine sulfoxide as an intermediate and the products of the sulfoxide lyase reaction. Eur. J. Biochem. *268*, 5229–5241.

Wagner, C., Sefkow, M. and Kopka, J. (2003). Construction and application of a mass spectral and retention time index database generated from plant GC/EI-TOF-MS metabolite profiles. Phytochemistry *62*, 887–900.

Want, E.J., Cravatt, B.F. and Siuzdak, G. (2005). The expanding role of mass spectrometry in metabolite profiling and characterization. Chembiochem *6*, 1941–1951.

Weckwerth, W., Loureiro, M.E., Wenzel, K. and Fiehn, O. (2004). Differential metabolic networks unravel the effects of silent plant phenotypes. Proc. Natl. Acad. Sci. U S A. *101*, 7809–7814.

Weckwerth, W. and Morgenthal, K. (2005). Metabolomics, from pattern recognition to biological interpretation. Drug Discovery Today,Targets. *10*, 1551–1558.

Wiese, M. and Gorcke, I. (2001). Homologues of LMPK, a mitogen-activated protein kinase from *Leishmania mexicana*, in different *Leishmania* species. Med. Microbiol. Immunol. (Berl) *190*, 19–22.

Wilkinson, S.R., Prathalingam, S.R., Taylor, M.C., Horn, D. and Kelly, J.M. (2005). Vitamin C biosynthesis in trypanosomes, a role for the glycosome. Proc. Natl. Acad. Sci. USA. *102*, 11645–11650.

Winter, G., Fuchs, M., McConville, M.J., Stierhof, Y.D. and Overath, P. (1994). Surface antigens of *Leishmania mexicana* amastigotes, characterization of glycoinositol phospholipids and a macrophage-derived glycosphingolipid. J. Cell Sci. 107 (Pt 9), 2471–2482.

Zarella-Boitz, J.M., Rager, N., Jardim, A. and Ullman, B. (2004). Subcellular localization of adenine and xanthine phosphoribosyltransferases in *Leishmania donovani*. Mol. Biochem. Parasitol. *134*, 43–51.

Zhang, K., Hsu, F.F., Scott, D.A., Docampo, R., Turk, J. and Beverley, S.M. (2005). *Leishmania* salvage and remodelling of host sphingolipids in amastigote survival and acidocalcisome biogenesis. Mol. Microbiol. *55*, 1566–1578.

Zhang, K., Showalter, M., Revollo, J., Hsu, F.F., Turk, J. and Beverley, S.M. (2003). Sphingolipids are essential for differentiation but not growth in Leishmania. Embo J. *22*, 6016–6026.

Zilberstein, D. and Shapira, M. (1994). The role of pH and temperature in the development of *Leishmania* parasites. Annu. Rev. Microbiol. *48*, 449–470.

Zufferey, R., Allen, S., Barron, T., Sullivan, D.R., Denny, P.W., Almeida, I.C., Smith, D.F., Turco, S.J., Ferguson, M.A. and Beverley, S.M. (2003). Ether phospholipids and glycosylinositolphospholipids are not required for amastigote virulence or for inhibition of macrophage activation by *Leishmania major*. J. Biol. Chem. *278*, 44708–44718.

Zufferey, R. and Ben Mamoun, C. (2006). *Leishmania major* expresses a single dihydroxyacetone phosphate acyltransferase localized in the glycosome, important for rapid growth and survival at high cell density and essential for virulence. J. Biol. Chem. *281*, 7952–7959.

Zufferey, R. and Mamoun, C.B. (2005). The initial step of glycerolipid metabolism in *Leishmania major* promastigotes involves a single glycerol-3-phosphate acyltransferase enzyme important for the synthesis of triacylglycerol but not essential for virulence. Mol. Microbiol. *56*, 800–810.

Physiological and Biochemical Aspects of *Leishmania* Development

6

Dan Zilberstein

Abstract

Leishmania is an obligate intracellular parasitic protozoon that cycles between the midgut of sandflies (the vector) and phagolysosomes of mammalian macrophages (the host). Host invasion involves attachment to macrophages, phagocytosis and development inside phagolysosomes. During the latter phase, promastigotes (the extracellular form) differentiate into amastigotes (the intracellular form), thus adapting to live in the hydrolytic environment of the lysosome. This chapter summarizes current knowledge about how promastigotes differentiate into amastigotes when surrounded by a phagolysosome milieu. *In vivo* studies indicate that once inside macrophages, promastigotes start to differentiate into amastigotes only when exposed to an acidic environment, which occurs after infected phagosomes fuse with late endosomes. To study in detail the molecular mechanisms underlying *Leishmania* differentiation, a model system has been developed where the parasites differentiate in axenic culture. Host-free differentiation is achieved by shifting cultured promastigotes to a lysosome-like environment and has enabled many insights into parasite life inside phagolysosomes. Here I describe how amastigotes were axenized and used for the study of *Leishmania* intracellular development.

Introduction

Leishmania belong to a large group of unicellular parasitic protozoa that cycle between insect vectors and mammalian hosts, each of which represent distinct and harsh environments. The extracellular forms of these organisms reside and proliferate as non-infective procyclic promastigotes in the midgut of sandflies, a hydrolytic environment of relatively basic pH with a mean temperature of 26°C. They become infective by differentiating into non-proliferating metacyclic promastigotes that subsequently migrate along the gut to the fly's mouth (Chang *et al.*, 1985). Upon host infection, promastigotes are phagocytosed by macrophages, where they differentiate into amastigotes inside the phagolysosomes (Chang and Dwyer, 1976). Phagolysosomes are also hydrolytic but very acidic (pH 4.5–5.5) with a temperature ~10°C higher than in the vector, e.g. 33–37°C, the precise temperature dependent on whether the infection is cutaneous or visceral (Antoine *et al.*, 1998). The mechanisms *Leishmania* parasites employ to cope with concomitant large changes in temperature (~10°C) and hydrogen concentration (more than two orders of magnitude) during infection has intrigued cell physiologists for many years (Zilberstein and Shapira, 1994; Burchmore and Barrett, 2001).

Numerous studies have been carried out to investigate various aspects of *Leishmania* parasitism and the host immune response, using experimental models that include mice, hamsters, macrophage cell lines and macrophages freshly isolated from animals. However, it is difficult to measure nutrient uptake by either promastigotes or amastigotes when they are inside sandflies or macrophages. Therefore, for those researchers interested in the physiology of *Leishmania* adaptation and development, use of the aforementioned models involves separating the parasites from their hosts or vectors. The time it takes to purify amastigotes from mice spleen, liver or foot pad and the use of neutral pH solutions during purification raises questions concerning the validity of data generated by such experiments. Our experience is that as soon as amastigotes encounter neutral pH they start differentiating back into promastigotes (Saar *et al.*, 1998). Thus, there is demand for model *Leishmania* systems where both promastigotes and amastigotes can be monitored more easily and authentically. This chapter describes the development of axenic culturing and discusses how it has improved our understanding of *Leishmania* parasitism.

Promastigotes are easy to axenize

Axenic liquid cultures were developed in the 1960s and 1970s based on the concept that providing synthetically an environment and nutrients that resemble insect midgut should allow promastigote growth (Neal and Miles, 1963; Lemma and Schiller, 1964; Steiger and Steiger, 1977; Berens and Marr, 1978). Similarly, agar-based solid media have also been developed for culturing promastigotes (Iovannisci and Ullman, 1983). Axenic promastigotes resemble the morphology of insect-derived promastigotes and, depending on the number of passages, are infective. These cultures have been used widely, enabling many discoveries that have advanced our knowledge of *Leishmania* biology and, in some cases, even biology in general. For example, axenic cultures facilitated identification of lipophosphoglycan (LPG; Turco, 1988) that covers promastigotes and was found to play important roles in parasite virulence (Spath *et al.*, 2000), and the biological processes of RNA editing (Simpson *et al.*, 2004) and solute transport (Landfear, 2000) were better characterized thanks to this model system. Tools for reverse genetics such as stable transfection, gene replacement and expression of foreign transposons were all developed using axenic promastigotes (Beverley, 2003). Moreover, using axenic promastigotes at various time points along their growth curve to infect mice or cultured macrophages led to the discovery of metacyclogenesis, the process whereby non-pathogenic procyclics differentiate into infective metacyclics (Da Silva and Sacks, 1987), which occurs naturally in the sandflies' midgut and had been inaccessible to scrutiny. Importantly, axenic cultures have proven stable, consistent and therefore useful for studying various aspects of promastigote biology. They have become so routine that at times, it is forgotten that axenic cultures represent *in vitro* promastigotes, which if cultured for too long become virulence-attenuated and most likely different from those that grow inside the vector. In order to retain promastigote virulence the parasites must be recycled periodically in animals.

Studies of axenic promastigote lines have become such a focus for the field of molecular parasitology that many scientists mistakenly believe it is reasonable to extrapolate information about the entire parasite life cycle from this model. In particular, research into drug development and mechanisms of drug resistance, as well as into biochemical aspects of parasitism, has used axenic promastigote lines to infer properties of the life cycle. For example, for years the concept was that *Leishmania* parasites are insensitive to pentavalent antimony, the anti-leishmanial drug of choice. It was hypothesized that intracellular amastigotes are killed because macrophages reduce it to the toxic Sb^{III} (Shaked-Mishan *et al.*, 2001). Subsequent development of axenic amastigote cultures demonstrated that this concept was wrong. Amastigotes take up Sb^{V} and reduce it using a stage-specific enzymatic mechanism (Ephros *et al.*, 1999; Shaked-Mishan *et al.*, 2001; Zhou *et al.*, 2004).

It became clear in the late 1980s that development of axenic amastigotes was necessary to enable insight into parasites life inside host macrophages.

Leishmania development inside macrophage phagolysosomes

Macrophage-invading promastigotes differentiate into amastigotes when they reach intracellular acidic pH

It became apparent from studies in the 1970s using cultured and animal-derived macrophages that *Leishmania* differentiation from the extracellular infectious form to the intracellular replicative phenotype occurs within the phagolysosomes of macrophages (Alexander and Vickerman, 1975; Chang and Dwyer, 1976; Berman *et al.*, 1979).

The *in vitro* studies were corroborated by the *in vivo* demonstration that when Balb/c mice are infected by *Leishmania* containing colloidal gold, the gold particles co-localize with amastigotes in the lysosomes of foot pad macrophages (Berman *et al.*, 1981). This and other research suggests that phagocytosis is the basic mechanism underlying *Leishmania* entry into the host, although some active involvement of the parasites cannot be excluded (Antoine *et al.*, 1998).

Indeed, exactly how parasites enter macrophages and the details of the invasion pathway are still not fully understood. During normal phagocytosis it takes about 30 minutes for lysosome markers to localize in a phagosome, whereas it takes more then twice that time for infective promastigotes to appear in phagosomes (Antoine *et al.*, 1998). Lipophosphoglycan (LPG), a major component of the promastigote surface membrane, was found to play a key role in the infective pathway as it restricts transiently the fusion of nascent promastigote-containing phagosomes with late endocytic compartments (Desjardins and Descoteaux, 1997). Specifically, two hours after formation, nearly all phagosomes containing latex beads acquire BSA-gold from the endocytic pathway, but only half of those harbouring promastigotes do so (Desjardins and Descoteaux, 1997). A series of genetic and biochemical studies have confirmed that LPG transiently blocks phagosome maturation (Turco and Descoteaux, 1992; Desjardins and Descoteaux, 1997; Holm *et al.*, 2001; Turco *et al.*, 2001). For example, within an hour of infecting J774 macrophages, almost all phagosomes formed around *L. donovani* lpg2 mutant promastigotes accumulate the lysosomal markers Rab7 or LAMP-1, whereas only half of the vacuoles bearing wild-type promastigotes acquire these late endosomal markers (Scianimanico *et al.*, 1999). In general, mutant promastigotes that lack LPG are delivered more quickly to lysosomes and, consequently, are killed by macrophages (Scianimanico *et al.*, 1999; Spath *et al.*, 2000; Turco *et al.*, 2001; Spath *et al.*, 2003). However, two recent studies have implied that LPG might not be the only factor that regulates *Leishmania* entry into the host. Firstly, *L. major* metacyclic promastigotes, a form containing a thick LPG layer, and amastigotes that lack LPG demonstrate similar rates of host entry and of phagolysosome formation (Courret *et al.*, 2002) and, secondly, LPG null mutants of *L. mexicana* are as effective as wild type in macrophage invasion (Ilg, 2000). The demonstration that virulent *L. chagasi* are localized in caveolae during phagocytosis by host macrophages, has prompted speculation that these specialized membrane domains play a role in intracellular parasite survival by targeting parasites to a phagocytosis pathway where lysosome fusion is delayed (Rodriguez *et al.*, 2006). Thus, cholesterol-containing macrophage membrane domains, such as caveolae, along with LPG likely influence the parasite invasion pathway.

It is well accepted that parasite-containing phagosome biogenesis is attenuated, a stage in the invasion pathway referred to as 'pregnant pause' that allows promastigotes to prepare for the imminent environmental transition. It has been suggested that during the pause infectious promastigotes establish vacuoles that are separate from the endosomal pathway, and only then differentiate into the replicative form adapted to proliferate inside acidic phagolysosomes (Swanson and Fernandez-Moreira, 2002). In line with this, post-pregnant-pause phagolysosomes exhibit an acidic luminal pH and contain various active acidic hydrolases (Antoine *et al.*, 1990; Alexander and Russell, 1992; Lang *et al.*, 1994). Thus, it seems reasonable to assume that whereas the pre-pregnant-pause invasion pathway is dependent on the presence of metacyclic-specific parasite surface LPG, everything that happens after the pause is independent of LPG. Based on this model for the invasion pathway, my colleagues and I proposed that in order to promote promastigote-to-amastigote differentiation in axenic culture the post-pregnant-pause conditions must be employed (Saar *et al.*, 1998).

Over the years, many attempts have been made using various pH-specific probes to establish what are the post-pregnant-pause pH conditions experienced by the parasite, i.e. the pH of *Leishmania* infected phagolysosomes in macrophages. Rivas and Chang (Rivas and Chang, 1983) determined a pH value of approximately 5 for *L. mexicana*-infected phagolysosomes. Later, using more advanced methods, the pH of *L. amasonensis*-infected phagolysosomes was determined to range between 4.7 and 5.3 (Antoine *et al.*, 1990). Infected vacuoles were noted to be more acidic than parasite-free lysosomes that exhibited pH 5.2–5.5. Phagolysosomes containing Old World and New World *Leishmania* species were observed to display similar pH, even though parasite content per vacuole is different. Vacuoles containing *L. major* are acidified rapidly and reach pH 5 less than 30 minutes after infection (Sturgill-Koszycki *et al.*, 1994). Taken together, regardless of *Leishmania* species, the pH of parasite-infected phagolysosomes appears to be 5–5.5 and the vacuole contains hydrolases and proteases typical of lysosomes (Burchmore and Barrett, 2001).

Several groups have measured the rate of promastigote-to-amastigote differentiation inside the parasitophorous vacuoles. The first estimate was that it takes 2–5 days for phagocytosed metacyclic promastigotes to differentiate fully into amastigotes (Antoine *et al.*, 1998). More detailed study indicated that morphological transformation occurs after the pregnant pause, i.e. about 12 hours after the initiation of phagocytosis, taking a total of 5 days (Courret *et al.*, 2001).

Only metacyclic promastigotes can invade macrophages successfully

In the mid 1980s David Sacks and colleagues observed that the virulence level of axenic *L. major* promastigotes depends on their growth phase; logarithmic cells are not virulent, whereas stationary cells readily infect macrophages (Franke *et al.*, 1985). Subsequently, they showed that logarithmic cells resemble non-infective procyclic promastigotes whereas stationary cells resemble virulent metacyclic promastigotes (Sacks *et al.*, 1985). Biochemical characterization revealed that this phenomenon correlates with the length and complexity of LPG in the parasite outer membrane (Sacks, 1992). Metacyclics are covered by a thick layer of modified LPG, in contrast procyclics are coated only by a thin layer of this molecule. The same study found that amastigotes do not synthesize LPG and in fact, shed this molecule during differentiation. Recent research using axenic amastigotes indicates that parasites start to shed LPG as early as 2 hours after the initiation of differentiation (Barak, E., Turco, S. and Zilberstein, D. unpublished). The timing of LPG shedding supports the notion that LPG loss is involved in the release from pregnant pause during macrophage infection (Swanson and Fernandez-Moreira, 2002).

For more than a decade it was agreed that successful macrophage invasion is dependent solely on metacyclogenesis. However, a recent study overturned this view with the discovery that apoptotic cells are required to be present in a mixture with live metacyclic promastigotes in order for invasion to be successful. The apoptotic cells appear to be necessary because they stop PMN cells from producing inflammatory agents like TGF-beta (Van *et al.*, 2006). Thus, metacyclic promastigotes bearing a thick LPG together with apoptotic cells enable parasites to survive host defence mechanisms. The point where the parasite no longer needs to be averting host defence mechanisms is likely marked by the pregnant pause, when the parasites prepare themselves for differentiation by shedding LPG and starting to express amastigote genes. Based on this view of the invasion pathway, my colleagues and I have hypothesized that parasites not needing to deal with host defence mechanisms (i.e. in host-free culture) can differentiate into amastigotes without going through metacyclogenesis. This premise is supported by the fact that it is possible to induce differentiation of logarithmic phase promastigotes in host-free culture (see below).

Axenization of *Leishmania* amastigotes in host-free culture

Leishmania respond to the acidic host environment

As mentioned above, in their cycling between vector and host, *Leishmania* encounter significant changes in temperature and pH, of ~10°C and two pH units, respectively. Unsurprisingly, the parasites have developed mechanisms that enable them to sense and adapt rapidly to these changes (Burchmore and Barrett, 2001). When culturing promastigotes, if the pH of the growth medium is shifted from 7 to 4.5–5.5 changes are observed in proline transport activity; the pH optimum for transport changes from 7 to 6 and V^{max} decreases by one order of magnitude (Zilberstein and Gepstein, 1993). The optimum pH of proline transport activity returns to 7 a few hours after the pH of the growth medium is restored to neutral pH. Notably, the lower optimum pH for proline transport activity observed in the growth medium shifting experiments is similar to that determined for amastigotes (Glaser and Mukkada, 1992). This correlation raised the notion that pH shifting the growth medium from 7 to 4.5 without changing the growth temperature is not only being sensed but represents, at least partially, a differentiation trigger. This hypothesis was supported further by the observation that pH shifting induces up-regulated expression of an amastigote-specific antigen on the parasite cell surface. Expression of the amastigote-specific antigen was found to be suppressed rapidly upon returning the growth medium pH to 7 (Zilberstein *et al.*, 1991). In particular, it was noted that the expression level of this antigen is higher in cells resembling morphologically amastigotes than in promastigote-shaped cells. The ability of cultured parasites to sense and respond developmentally to pH changes was demonstrated further by reports that acidic pH alone can induce procyclic-to-metacyclic promastigote differentiation of various *Leishmania* species, both New and Old World (Bates and Tetley, 1993; Zakai *et al.*, 1998). Specifically, when parasites reach stationary growth phase at 26°C a small proportion (up to 25%) differentiate into metacyclics (see above), but if the culturing conditions are pH 5.5 almost 100% differentiate into metacyclics (Zakai *et al.*, 1998).

To sense and act upon changes in extracellular pH there must exist pH sensing and signal transduction pathway(s) in promastigotes. Neither surface nor internal pH sensors have been identified. However, for a few *Leishmania* species, acidic pH-sensitive elements that regulate translation have been characterized to exist in the 3′UTR of certain mRNAs. For example, these *cis*-acting elements are encoded by a number of stage-specific genes, including amastin, A2 and a few enzymes (Boucher *et al.*, 2002; Rochette *et al.*, 2005; McNicoll *et al.*, 2005). Deletion of the 3′UTR element results in pH insensitivity and constitutive gene expression.

In order to sense changes in proton concentrations it is imperative that both promastigotes and amastigotes obtain mechanisms that strongly regulate intracellular pH homeostasis. Indeed, promastigotes and amastigotes are able to maintain constantly neutral intracellular pH despite being exposed to extracellular pHs ranging from 4.5 to almost 9 (Glaser *et al.*, 1988; Zilberstein *et al.*, 1989). In both promastigotes and amastigotes a P-type ATPase in the plasma membrane represents the primary proton pump (Zilberstein and Dwyer, 1988; Liveanu *et al.*, 1991). As well as creating a proton electrochemical gradient across the plasma membrane, which provides the energy required to drive the transport of solutes against their concentration gradients (Zilberstein and Dwyer, 1985; Glaser and Mukkada, 1992; Drew *et al.*, 1995; Klamo *et al.*, 1996; Stein *et al.*, 2003), we have proposed that the primary proton pump works in concert with secondary proton carriers to regulate intracellular pH homeostasis (Zilberstein and Shapira, 1994). Similar mechanisms for maintaining pH homeostasis have been shown to operate in bacteria that, like *Leishmania*, encounter disparate environments with different pHs (Padan *et al.*, 1981; Padan *et al.*, 2005). To date, a gene coding for a putative P-type H^+-ATPase has been cloned (Jiang *et al.*, 1994; Mukherjee *et al.*, 2001), but it is yet to be demonstrated whether this gene functions as the primary proton pump of *Leishmania*.

The roles of heat and acidic pH in inducing differentiation

Following reports in the 1970s that amastigotes proliferate inside acidic phagolysosomes, investigators have sought conditions that mimic the host and thereby induce differentiation of promastigotes into amastigotes and allow stable growth of amastigotes in a host-free environment. In 1984 Pan (Pan, 1984) reported that raising the growth temperature of *L. mexicana* to 32°C induces promastigotes to transform into amastigote-shaped cells. This pioneering work was followed by several studies that replicated the same phenomenon for other New World *Leishmania* species (Shapira *et al.*, 1988; Pan and McMahon-Pratt, 1988; Eperon and McMahon-Pratt, 1989a; Eperon and McMahon-Pratt, 1989b; Lebowitz *et al.*, 1990; Rainey *et al.*, 1991; Traub-Cseko *et al.*, 1993; Duboise *et al.*, 1994; Leon *et al.*, 1994), with select amastigote-specific genes shown to be expressed in a few cases (Ghosh and Honigberg,

1976; Hendricks and Child, 1980). However, these 'amastigotes' do not survive long at high temperature and neutral pH and are inconsistent in their characteristics. Thus, these early attempts failed to establish long-term axenic amastigotes. Old World species are even more sensitive than New World species to growth at high temperature, as exposing *L. donovani* amastigotes to 37°C at pH 7 induces growth arrest within a generation. These arrested promastigotes die after a few days (Barak *et al.*, 2005). Interestingly, replacing the growth medium with serum stabilizes these arrested cells at 37°C and transforms them into a stable culture of axenic amastigote-shaped cells (Joshi *et al.*, 1993).

As mentioned earlier in this chapter, exposing promastigotes to lysosome-like pH without raising the temperature induces expression of amastigote-specific genes (Zilberstein *et al.*, 1991; Mazareb *et al.*, 1999), but does not trigger morphological differentiation. Therefore, it was recognized that although each signal alone can trigger to some extent gene expression and morphological changes, only the combination of high temperature with acidic growth conditions is likely to enable complete differentiation of promastigotes into amastigotes for both Old and New World *Leishmania* species (Bates, 1993; Zilberstein and Shapira, 1994). Indeed, using this approach several laboratories have succeeded in developing methods for *in vitro* parasite differentiation, and for culturing amastigotes axenically (Bates, 1993; Basselin *et al.*, 1996; Chattopadhayay *et al.*, 1996; Gupta *et al.*, 1996; Hodgkinson *et al.*, 1996; Pral and Alfieri, 1996; Callahan *et al.*, 1997; Sereno and Lemesre, 1997c; Saar *et al.*, 1998; Gutierrez *et al.*, 2001; Teixeira *et al.*, 2002; Goyard *et al.*, 2003; Debrabant *et al.*, 2004).

Differentiation methods for New World *Leishmania* species

Bates and colleagues (Bates *et al.*, 1992; Bates, 1994) were the first to induce in host-free culture the differentiation of *L. mexicana* promastigotes into amastigotes using growth conditions that combined acidic pH with high temperature. Initially, they purified amastigotes from mice, which they transformed into procyclic promastigotes by culturing at pH 7, 25°C, in Schneider's *Drosophila* medium. Subsequently, these late log phase procyclics were grown at pH 5.5, and within 6 days most of the parasite population differentiated into metacyclic promastigotes (Bates and Tetley, 1993). Finally, the growth temperature was increased to 32°C, which induced differentiation into amastigote-shaped cells. The resultant *L. mexicana* axenic amastigotes resemble animal-derived amastigotes, in that they exhibit amastigote morphology and express amastigote-specific proteinases and nucleases. Using this method, a complete developmental cycle of *L. mexicana* was established (Bates, 1994). The significance of metacyclogenesis to successful differentiation of New World *Leishmania* species is indicated by the finding that stable culturing of *L. amazonensis* axenic amastigotes occurs only if at least 73% of the original promastigote culture are metacyclics (Cysne-Finkelstein *et al.*, 1998). This method, with a few modifications, has been applied by several other laboratories to induce promastigote-to-amastigote differentiation of other New World *Leishmania* species (Bijovsky, 1994; Callahan *et al.*, 1997; Burchmore and Landfear, 1998).

Although pH values below 6 were used to initiate differentiation in the studies described above, several laboratories found that they could induce New World promastigotes to differentiate at higher pHs. For example, *L. braziliensis* promastigotes differentiate into amastigotes if exposed briefly to pH 6.3 before increasing gradually the temperature to 34°C (Pral and Alfieri, 1996; Balanco *et al.*, 1998). Lemesre and colleagues (Basselin *et al.*, 1996; Sereno and Lemesre, 1997b) used pH 6.5 as the pH trigger for *L. mexicana* differentiation, and designed a growth medium optimized for axenizing amastigotes (Lemesre, 1994; Sereno and Lemesre, 1997c). This latter axenic amastigote system has been utilized to study *Leishmania* responses to drugs as well as host–parasite interactions (Lemesre *et al.*, 1997; Sereno *et al.*, 1997; Sereno and Lemesre, 1997a; Sereno and Lemesre, 1997b).

Differentiation methods for Old World *Leishmania* species

For Old World *Leishmania* species, unlike New World species, long-term non-virulent cultures of promastigotes are as useful precursors for *in vitro* differentiation into amastigotes as parasites isolated freshly from animals (Saar *et al.*, 1998; Debrabant *et al.*, 2004). Saar *et al.*, (Saar *et al.*, 1998) developed an *in vitro* differentiation method for *L. donovani* based on the sequence of environmental changes *Leishmania* encounter during the process of host invasion. Briefly, late log phase promastigotes are incubated for 24 hours at 37°C at pH 7 in a 5% CO_2 environment (in medium containing 25% fetal calf serum). Subsequently, the pH of the growth medium is reduced to 5.5 and incubation continued until the parasites differentiate into amastigotes. Successful differentiation of Old World species has also been achieved by applying the method developed for *L. mexicana*, i.e. promastigotes are first exposed to pH 5.5 and then shifted to 37°C (Debrabant *et al.*, 2004). Axenic amastigotes have been shown to resemble animal-derived amastigotes according to several criteria: (1) they exhibit amastigote morphology; (2) they express the amastigote-specific A2 protein family; (3) they stop synthesizing lipophosphoglycan and this molecule is not present on the parasite surface (see also Mengeling *et al.*, 1997); (4) they express amastigote-type soluble acid phosphatase; (5) they possess amastigote-like proline transport activity; (6) they express low levels of 3′ nucleotidases; and (7) they exhibit amastigote-like DNA and protein synthesis activities (compare with Mukkada *et al.*, 1985).

Importantly, the axenic Old World amastigotes generated by either method infect macrophages and mice as efficiently as animal-derived amastigotes (Debrabant *et al.*, 2004; Barak *et al.*, 2005); Darlyuk *et al.*, unpublished). Thus, axenic promastigote-to-amastigote differentiation is equally effective in retaining virulence of cultured *Leishmania* as recycling promastigotes in animal models.

Further analysis of the role of each differentiation signal revealed that during the period that the parasites are exposed to higher temperature, the promastigotes undergo growth arrest, while the subsequent shift to pH 5.5 releases them from this arrest and triggers morphological transformation into amastigotes (Barak *et al.*, 2005). These observations led to the idea that a single shift of log phase promastigotes to both 37°C and pH 5.5 should also trigger differentiation. Indeed, promastigotes shifted in one step to a higher temperature and acidic pH undergo G1 arrest and then differentiate in the same manner as when a two-step shift method is employed (Shaked-Mishan *et al.*, 2001; Barak *et al.*, 2005).

The Saar *et al.* method has been used successfully by several other groups to differentiate Old World species (Mengeling *et al.*, 1997; Bente *et al.*, 2003; Morales *et al.*, 2007). Goyard *et al.*, (Goyard *et al.*, 2003) applied the Debrabant method to study the molecular genetics of LPG in amastigotes. Lemesre and colleagues applied a patented differentiation method to examine *L. infantum* development (Lemesre, 1994; Sereno *et al.*, 1998; Cibrelus *et al.*, 1999). This same patented method was exploited for studying pentamidine transport in *L. donovani* (Basselin *et al.*, 1996) and the mode of action of antimony (Sereno *et al.*, 1998). Laboratories interested in gene expression have used this axenic amastigote model system to demonstrate that SLRNA expression is regulated developmentally (Lamontagne and Papadopoulou, 1999). Chattopadhayay *et al.*, (Chattopadhayay *et al.*, 1996) attempted to use pH 6 to induce *in vitro* differentiation of *L. donovani* promastigotes isolated from Indian kala-azar patients. Although morphological changes were reported, no biochemical evidence was provided to confirm that these conditions induce complete parasite differentiation.

In summary, from the first culture of axenic promastigotes it took *Leishmania* researchers more than 20 years to develop the culture methods for axenic differentiation and stable maintenance of amastigotes. Importantly, these amastigote culture methods have opened up new horizons for the field of leishmaniasis, as they enable insights into the molecular biology and physiology of the disease causing parasite form, the amastigote. To date, more than 100 research articles have been published that use the axenic amastigote model systems. Notably, axenic differentiation is not a straightforward method and there are some species that do not differentiate in host-free culture, such as *L. major* and many clinically isolated parasites (the author, unpublished observations).

Mechanism of *Leishmania* differentiation

Differentiation initiation

Time-course analyses of *L. donovani* differentiation have shown that it is a regulated process involving ordered changes in morphology and protein and gene expression (Barak *et al.*, 2005; Saxena *et al.*, 2007; Rosenzweig *et al.*, 2007). Based on these studies it was proposed that *L. donovani* differentiation is divided into four phases: (1) 0–4 hours, when promastigotes receive and process the differentiation signal; (2) 5–9 hours, the time they cease movement and aggregate; (3) 10–24 hours, when promastigotes change morphology to amastigote-shaped cells and (4) 25–120 hours, when they complete maturation into amastigotes. The transition between the first and second phases occurs synchronously, while cells arrest at G1 (Barak *et al.*, 2005). Early changes such as upregulated expression of the amastigote-specific A2 gene family and initiation of LPG shedding can be detected 1 and 2 hours after the differentiation signal, respectively. In contrast, late changes, such as the switch to amastigote-type proline transport and increased susceptibility to pentavalent antimony occur only 2 and 3 days post signal, respectively (Ephros *et al.*, 1999; Mazareb *et al.*, 1999).

It is not yet clear whether any host molecules are involved in directing and/or promoting the differentiation of promastigotes into amastigotes. It was found recently that C-reactive protein (CRP), a serum protein that is released into the blood serum in response to infection, activates *in vitro* differentiation of both *L. mexicana* and *L. donovani* (Bee *et al.*, 2001; Mbuchi *et al.*, 2006). When CRP is added to a culture of promastigotes it triggers a transient change in parasite morphology that lasts a few hours, but then the parasites return to their original shape. Thus, even though CRP is capable of inducing morphological changes, by itself it is not sufficient to promote complete differentiation. CRP binds to LPG and mediates the binding of parasites to macrophages, consequently increasing their uptake by human macrophages (Culley *et al.*, 1996; Bodman-Smith *et al.*, 2002). Further studies are required to understand how CRP stimulates parasite differentiation in host-free culture; we speculate that additional host molecules cooperate with CRP to promote differentiation.

Since heat is an important differentiation trigger, it is expected that heat shock proteins play a role in this process. *Leishmania* express constitutively high levels of HSP70 and 83 (each constituting almost 1% of their total soluble protein), indicating that these organisms are ready to respond to temperature shifts throughout their life cycle (Zilberstein and Shapira, 1994; Wiesgigl and Clos, 2001b). Exposing promastigotes to either heat or acidic pH has little effect on HSP83 abundance, but induces an elevated protein turnover rate as well as increased mRNA stability (Garlapati *et al.*, 1999; S. Amin-Spector *et al.*, unpublished). Interestingly, without changing temperature or pH, inhibiting *L. donovani* HSP 83 activity, using the specific inhibitor geldanamycin, induces differentiation of *L. donovani* promastigotes into amastigote-like cells (Wiesgigl and Clos, 2001a). This finding implies that HSP83 negatively regulates the initiation of differentiation and led to the hypothesis that following exposure to the differentiation signal HSP83 activity is reduced, which results in high levels of misfolded proteins that are likely involved in the differentiation process. Indeed, modalities other than heat that modulate protein folding, such as ethanol and synthetic analogues of proline, can be used in combination with acidic pH, instead of heat, to induce parasite differentiation (Zagari *et al.*, 1990; Trotter *et al.*, 2002; Barak *et al.*, 2005).

In general, cells cannot tolerate proteins that are misfolded, therefore, they either refold them using chaperones or label them with ubiquitin for degradation in proteasomes (Glickman and Ciechanover, 2002). A recent study of axenically cultured *L. amazonensis* found that shortly after exposing promastigotes to the differentiation signal, cellular proteins are degraded rapidly (Alves *et al.*, 2005). During the early stages of differentiation of these cells it was noted that the activity of several proteinases is upregulated transiently, activity that likely contributes to the observed protein instability. Preliminary studies in our laboratory support that there is significant accumulation of ubiquitylated protein in the cytosol during the first two phases of *L. donovani* differentiation (E. Barak and D. Zilberstein, unpublished). Taken together with the observation that HSP83 regulates negatively differentiation, it is likely that heat-induced protein misfolding play a role in parasite cellular signalling for promastigote to amastigote differentiation.

Leishmania HSP100, a member of the ClpB protein family, is expressed at a higher level in amastigotes and has been demonstrated to be essential for parasite development inside the host, but not for differentiation in axenic culture (Krobitsch and Clos, 1999). In light of the HSP83 data discussed above, it is notable that both HSP70 and HSP100 function as unfoldases in many cells. However, to date no evidence has been provided to validate whether they exhibit unfoldase activity in *Leishmania* (Bukau *et al.*, 2006).

The first phase of *Leishmania* differentiation is dedicated to signal perception and processing. However, the signal transduction pathway that mediates the differentiation signal has not yet been identified. The fact that *Leishmania* genomes lack genes encoding for G proteins narrows the number of candidate signalling pathways (Ivens *et al.*, 2005), even though G protein-like proteins have been identified (Cassel *et al.*, 1991). Martin Wiese and his research group study the role in *L. mexicana* development of mitogen activated protein kinase (MAPK) family members, proteins involved classically in the developmental signalling pathways of diverse organisms. They report that LmxMPK4 is essential for *L. mexicana* proliferation inside macrophages and mice, likely because parasites are unable to differentiate intracellularly (Kuhn and Wiese, 2005). Other members of the MAPK family appear to be responsible for flagella biogenesis and thus, are more important for promastigotes than amastigotes (Bengs *et al.*, 2005; Erdmann *et al.*, 2006). Nascimento *et al.*, have taken an innovative approach to delineating the differentiation signalling pathway. They expressed the human protein tyrosine phosphatase 1B (PTP-1B) in promastigotes and found that the transgenic cells differentiate spontaneously into amastigotes without requiring any temperature or pH changes. Likewise, when promastigotes are treated with sub-lethal concentrations of G1433, a tyrosine kinase inhibitor, spontaneous differentiation is also observed. Cloning of the *L. donovani* homologue of hPTP-1B and its manipulation confirmed that this protein is indeed involved in directing promastigotes to differentiate into amastigotes (Nascimento *et al.*, 2006). Although these studies do not point to the specific signal transduction pathway that initiates differentiation, they indicate that protein phosphorylation/dephosphorylation must play an important role in triggering the differentiation process. Further support for this conclusion comes from earlier experiments using protein kinase inhibitors (Becker and Jaffe, 1997).

Differentiation requires protein retooling for life inside the host

During the third and fourth phases of *Leishmania* differentiation parasites lose the flagellum, reduce in size, change their shape, and undergo major shifts in their metabolism. For example, metabolic pathways such as glycolysis are more active in promastigotes, while fatty acid oxidation is more active in amastigotes (Opperdoes and Coombs, 2007). Also, gluconeogenesis is essential for the virulence of amastigotes and their proliferation inside macrophages, but promastigotes can survive without this pathway (Naderer *et al.*, 2006). Proteomic analyses performed in promastigotes and amastigotes support that there are many differences between the protein expression profiles of these two life stages (Bente *et al.*, 2003; Nugent *et al.*, 2004; Walker *et al.*, 2006; Leifso *et al.*, 2006).

Thus, it is clear that the promastigote-to-amastigote transition involves ordered changes in protein expression to prepare the parasite for life in the host. Moreover, as discussed above, various studies support that degradation-based remodelling of the proteome occurs during this transition. Recent research indicates that an autophagic protein remodelling pathway exists in *Leishmania*, a pathway described previously only in yeast and mammals. The autophagic pathway occurs in a vesicular compartment called the autophagosome, which is formed by a membranous structure that engulfs the cytoplasm and organelles that are to be degraded (Reggiori and Klionsky, 2005). The genesis of autophagosomal structures requires the activity of two protein conjugation systems. The first system involves the ubiquitin-like protein ATG8, which must be processed proteolytically by a family of peptidases, known as ATG4, before it can act as a key component in the autophagosomal membrane. The second system involves the ubiquitin-like protein ATG12, which is attached covalently to ATG5 (Yorimitsu and Klionsky, 2005). The autophagosome delivers the internalized material to the lysosomal compartment for degradation. Both ATG4 and ATG8 have been cloned and characterized in *L. major* and *L. mexicana* and validated to be essential for autophagy and protein remodelling in these organisms. Importantly, experiments using mutants that lack these genes indicate that autophagy is not just involved, but is essential for the differentiation of promastigotes into amastigotes, both *in vitro* and *in vivo* (Besteiro *et al.*, 2006; Williams *et al.*, 2006). This conclusion is supported also by a study using mutant peptidases that interfere with endocytosis in a dominant negative manner; when endocytosis is disabled the parasites cannot perform autophagic remodelling and consequently, do not differentiate and fail to proliferate as amastigotes inside macrophages (Besteiro *et al.*, 2006).

Notably, time-course analysis of *L. mexicana* differentiation indicates that autophagy occurs predominantly between 18 and 72 hours post-signal, the period when promastigotes undergo morphological changes (Williams *et al.*, 2006). Therefore, it is predicted that major changes in protein abundance should occur specifically during this time frame. Our laboratory set out to validate this prediction, employing a novel proteomics approach, namely an isobaric tagging methodology (iTRAQ; Melanson *et al.*, 2006). This method utilizes amine-reactive isobaric tags to label all peptides in a particular protein digest. An advantage of this method is that 4 protein digest samples can each be tagged differently, allowing direct and parallel comparison of up to four samples. Samples are combined at an equal ratio and subjected to LC-MS/MS. After fragmentation a given peptide produces (up to four) distinct signature ions differing by an m/z value from 114 to 117; the relative intensities of these signals represent the relative abundance in each sample of the analysed peptide.

iTRAQ analysis identified and quantified across seven time points during differentiation 1713 proteins, which represents 21% of the parasite proteome (Rosenzweig *et al.*, 2007). The data reveal a delayed increase in the abundance of gluconeogenesis enzymes, coinciding with a decrease in glycolytic capacity. At the same time, β-oxidation, amino acid catabolism, tricarboxylic acid cycle, mitochondrial respiration chain and oxidative phosphorylation capacities are all up-regulated. Summarily, the differentiating parasite shifts from glucose to fatty acids and amino acids as its main energy source. Furthermore, glycerol and amino acids are used as precursors for sugar synthesis, to compensate for the lack of exogenous sugars. In addition, we observed that during the third stage of differentiation, at the time when promastigotes transform morphologically into amastigotes (around 15 hours post-signal), a highly synchronized decrease occurs in the abundance of translation machinery proteins, including ribosomal protein subunits, translation factors and tRNA synthetases. This synchronous decrease is associated with an estimated 10-fold decrease in cell volume and a 2.5-fold decrease in growth rate, when comparing amastigotes to promastigotes. The fact that such global changes occur during the third differentiation stage, suggests strongly that they are part of the parasite differentiation program and underlie adapting to a generally lower protein synthesis rate. Conspicuously, as predicted, the dramatic changes in the abundance of translation machinery proteins echo the period of maximal autophagy activity, and we consider these proteomic changes to reflect the parasite retooling for life in the host phagolysosome.

Future perspectives

The ability to axenize amastigotes augurs significant advances in our understanding of *Leishmania* virulence, and of intracellular parasitism in general. Nevertheless, the axenic culture conditions in current use remain considerably different from the genuine lysosome milieu. Thus, even though cultured amastigotes express the expected stage-specific genes (see Differentiation methods for New World *Leishmania* species, above), they also express many others that indicate responses to their *in vitro* environment (Holzer *et al.*, 2006; B. Papadopoulou, personal communication). This is perhaps not surprising, since most *Leishmania* researchers use the same medium to grow both promastigotes and amastigotes, with only a few modifications in pH and serum content. Recent observations highlight the metabolic differences between promastigotes and amastigotes (see Differentiation requires protein retooling for life inside the host, above) and thus support that these parasite life stages require more specialized growth conditions. Further study must be carried out to identify axenic culture conditions that mimic better the lysosomal environment. For example, inclusion in the culture medium of candidate host molecules that influence differentiation should be investigated.

In order to advance our understanding of *Leishmania* differentiation, the parasite molecule(s) that are responsible for perceiving and transducing the host signal(s) must be characterized. The identity of this parasite signalling pathway will enable new insights into host–parasite interactions. Time course transcriptome and proteome analyses indicate that major changes in mRNA and/or protein abundance are downstream of the initiation of differentiation. This points to post-translational modifications, i.e. phosphorylation, methylation or acetylation, as the best candidates for events that trigger differentiation.

Other intriguing open questions relate to the molecular mechanisms underlying morphological transformation into amastigotes, i.e. what drives the third and fourth phases of differentiation? Time course analyses suggest strongly that these are the periods when most of the changes occur in gene and protein expression. Future studies using axenic cultures should address how parasites shorten their flagella, how they round, how they metabolically retool for life inside the lysosome, how they change their surface coat, and many other questions.

The era of studying axenic amastigotes is just beginning. I am sure that time will prove that this is the best model system for investigating and characterizing *Leishmania* intracellular development and parasitism.

Acknowledgements

I thank my team Tamar Lahav, Doron Rosenzweig and Drs. Tamar Listovski and Neta Holland for fruitful discussions on various aspects of this chapter. The author is supported by grant 490/02 from The Israel Science Foundation founded by The Israel Academy of Sciences and Humanities, grant 2003237 from the United States-Israel Binational Science Foundation, grant 5905 from the Chief Scientist of the Israel Ministry of Health and by the UNDP/World Bank/WHO Special Programme for Research and Training in Tropical Diseases.

References

Alexander, J. and Russell, D.G. (1992). The Interaction of *Leishmania* species with macrophages. Adv. Parasitol. *31*, 175–233.

Alexander, J. and Vickerman, K. (1975). Fusion of host cell secondary lysosomes with the parasitophorous vacuoles of *Leishmania mexicana*-infected macrophages. J. Protozool. *22*, 502–508.

Alves, C.R., Corte-Real, S., Bourguignon, S.C., Chaves, C.S., and Saraiva, E.M. (2005). *Leishmania amazonensis*: early proteinase activities during promastigote-amastigote differentiation in vitro. Exp. Parasitol. *109*, 38–48.

Antoine, J.C., Prina, E., Jouanna, C., and Bongrand, P. (1990). Parasitophorous vacuoles of *Leishmania amazonensis* – infected macrophages maintain an acidic pH. Infect. Immun. *58*, 779–787.

Antoine, J.C., Prina, E., Lang, T., and Courret, N. (1998). The biogenesis and properties of the parasitophorous vacuoles that harbour *Leishmania* in murine macrophages. Trends Microbiol. *6*, 392–401.

Balanco, J.M., Pral, E.M., Da Silva, S., Bijovsky, A.T., Mortara, R.A., and Alfieri, S.C. (1998). Axenic cultivation and partial characterization of *Leishmania braziliensis* amastigote-like stages. Parasitology *116* (*Pt 2*), 103–113.

Barak, E., Amin-Spector, S., Gerliak, E., Goyard, S., Holland, N., and Zilberstein, D. (2005). Differentiation of *Leishmania donovani* in host-free system: analysis of signal perception and response. Mol. Biochem. Parasitol. *141*, 99–108.

Basselin, M., Lawrence, F., and Robert-gero, M. (1996). Pentamidine uptake in *Leishmania donovani* and *Leishmania amazonensis* promastigotes and axenic amastigotes. Biochem. J. *315*, 631–634.

Bates, P.A. (1993). Axenic culture of *Leishmania* amastigotes. Parasitol. Today 9, 143–146.

Bates, P.A. (1994). Complete developmental cycle of *Leishmania mexicana* in axenic culture. Parasitology *108*, 1–9.

Bates, P.A., Robertson, C.D., Tetley, L., and Coombs, G.H. (1992). Axenic cultivation and characterization of *Leishmania mexicana* amastigotes-like forms. Parasitology *105*, 193–202.

Bates, P.A. and Tetley, T. (1993). *Leishmania mexicana*: induction of metactclogenesis bycultivation of promastigotes at acidic pH. Exp. Parasitol. *76*, 412–423.

Becker, S. and Jaffe, C.L. (1997). Effect of protein kinase inhibitors on the growth, morphology, and infectivity of *Leishmania* promastigotes. Parasitol. Res. *83*, 273–280.

Bee, A., Culley, F.J., Alkhalife, I.S., Bodman-Smith, K.B., Raynes, J.G., and Bates, P.A. (2001). Transformation of *Leishmania mexicana* metacyclic promastigotes to amastigote-like forms mediated by binding of human C-reactive protein. Parasitology *122*, 521–529.

Bengs, F., Scholz, A., Kuhn, D., and Wiese, M. (2005). LmxMPK9, a mitogen-activated protein kinase homologue affects flagellar length in *Leishmania mexicana*. Mol. Microbiol. *55*, 1606–1615.

Bente, M., Harder, S., Wiesgigl, M., Heukeshoven, J., Gelhaus, C., Krause, E.*et al.* (2003). Developmentally induced changes of the proteome in the protozoan parasite *Leishmania donovani*. Proteomics *3*, 1811–1829.

Berens, R.L. and Marr, J.J. (1978). An easily prepared defined medium for cultivation of *Leishmania donovani* promastigotes. J. Parasitol. *64*, 160.

Berman, J.D., Fioretti, T.B., and Dwyer, D.M. (1981). In vivo and in vitro localization of *Leishmania* within macrophage phagolysosomes: use of colloidal gold as a lysosomal label. J. Protozool. *28*, 239–242.

Berman, J.D., Wyler, D.J., and Dwyer, D.M. (1979). Multiplication of *Leishmania* in human macrophages in vitro. Infect. Immun. *26*, 375–379.

Besteiro, S., Williams, R.A., Morrison, L.S., Coombs, G.H., and Mottram, J.C. (2006). Endosome sorting and autophagy are essential for differentiation and virulence of *Leishmania major*. J. Biol. Chem. *281*, 11384–11392.

Beverley, S.M. (2003). Protozomics: trypanosomatid parasite genetics comes of age. Nat. Rev. Genet. *4*, 11–19.

Bijovsky, A.T. (1994). *Leishmania mexicana*: the influence of slightly elevated temperature on the ultrastructure of axenic amastigote-like forms. Parasitol. Res. *80*, 696–698.

Bodman-Smith, K.B., Mbuchi, M., Culley, F.J., Bates, P.A., and Raynes, J.G. (2002). C-reactive protein-mediated phagocytosis of *Leishmania donovani* promastigotes does not alter parasite survival or macrophage responses. Parasite Immunol. *24*, 447–454.

Boucher, N., Wu, Y., Dumas, C., Dube, M., Sereno, D., Breton, M.*et al.* (2002). A common mechanism of stage-regulated gene expression in *Leishmania* mediated by a conserved 3′-untranslated region element. J. Biol. Chem. *277*, 19511–19520.

Bukau, B., Weissman, J., and Horwich, A. (2006). Molecular chaperones and protein quality control. Cell *125*, 443–451.

Burchmore, R.J. and Barrett, M.P. (2001). Life in vacuole. Int. J. Parasitol. *31*, 1311–1320.

Burchmore, R.J. and Landfear, S.M. (1998). Differential regulation of multiple glucose transporter genes in *Leishmania mexicana*. J. Biol. Chem. *273*, 29118–29126.

Callahan, H.L., Portal, A.C., Devereaux, R., and Grogl, M. (1997). An axenic amastigote system for drug screening. Antimicrob. Agents Chemother. *41*, 818–822.

Cassel, D., Shoubi, S., Glusman, G., Cukierman, E., Rotman, M., and Zilberstein, D. (1991). *Leishmania donovani*: characterization of a 38-kDa membrane protein that cross-reacts with the mammalian G-protein transducin. Exp. Parasitol. *72*, 411–417.

Chang, K.P. and dwyer, D.M. (1976). Multiplication of a human parasite (*Leishmania donovani*) in phagolysosomes of hamster macrophages *in vivo*. Science *193*, 678–680.

Chang, K.P., Fong, D., and Bray, R.S. (1985). Biology of *Leishmania* and leishmaniasis. In Leishmaniasis, K.P.Chang and D.Bray, eds. (Amsterdam: Elsevier), pp. 1–30.

Chattopadhayay, R., Kaur, S., Ganguly, N.K., and Mahajan, R.C. (1996). Antigenic differences between axenic amastigotes and promastigotes of *Leishmania donovani*. Indian J Med. Res. *104*, 349–354.

Cibrelus, P., Precigout, E., Sereno, D., Carcy, B., Lemesre, J.L., and Gorenflot, A. (1999). Secreted antigens of the amastigote and promastigote forms of *Leishmania infantum* inducing a humoral response in humans and dogs. Parasite *6*, 121–129.

Courret, N., Frehel, C., Gouhier, N., Pouchelet, M., Prina, E., Roux, P.*et al.* (2002). Biogenesis of *Leishmania*-harbouring parasitophorous vacuoles following phagocytosis of the metacyclic promastigote or amastigote stages of the parasites. J. Cell Sci. *115*, 2303–2316.

Courret, N., Frehel, C., Prina, E., Lang, T., and Antoine, J.C. (2001). Kinetics of the intracellular differentiation of *Leishmania amazonensis* and internalization of host MHC molecules by the intermediate parasite stages. Parasitology *122*, 263–279.

Culley, F.J., Harris, R.A., Kaye, P.M., McAdam K.P.W.J., and Raynes, J.G. (1996). C-reactive protein binds to a novel ligand on *Leishmania donovani* and increases uptake into human macrophages. The Journal of Immunology *156*, 4691–4696.

Cysne-Finkelstein, L., Temporal, R.M., Alves, F.A., and Leon, L.L. (1998). *Leishmania amazonensis*: long-term cultivation of axenic amastigotes is associated to metacyclogenesis of promastigotes. Exp. Parasitol. *89*, 58–62.

Da Silva, R. and Sacks, D.L. (1987). Metacyclogenesis is a major determinant of *Leishmania* promastigote virulence and attenuation. Infect Immun. *55*, 2802–2806.

Debrabant, A., Joshi, M.B., Pimenta, P.F., and Dwyer, D.M. (2004). Generation of *Leishmania donovani* axenic amastigotes: their growth and biological characteristics. Int. J. Parasitol. *34*, 205–217.

Desjardins, M. and Descoteaux, A. (1997). Inhibition of phagolysosomal biogenesis by the *Leishmania* lipophosphoglycan. J. Exp. Med *185*, 2061–2068.

Drew, M.E., Langford, C.K., Klamo, E.M., Russell, D.G., Kavanaugh, M.P., and Landfear, S.M. (1995). Functional expression of a myo-inositol/H+ symporter from *Leishmania donovani*. Mol. Cell Biol. *15*, 5508–5515.

Duboise, S.M., Vannier-Santos, M.A., Costa-Pinto, D., Rivas, L., Pan, A.A.T. Raub-cseko, Y. *et al.* (1994). The biosynthesis, processing, and immunolocalization of *Leishmania pifanoi* amastigote cysteine proteinases. Mol. Biochem. Parasitol. *68*, 119–132.

Eperon, S. and McMahon-Pratt, D. (1989a). Extracellular cultivation and morphological characterization of amastigote-like forms of *Leishmania panamensis* and *L. braziliensis*. J. Protozool. *36*, 502–510.

Eperon, S. and McMahon-Pratt, D. (1989b). Extracellular amastigote-like forms of *Leishmania panamensis* and *L. braziliensis*. J. Protozool. *36*, 510–518.

Ephros, M., Bitnun, A., Shaked, P., Waldman, E., and Zilberstein, D. (1999). Stage-specific activity of pentavalent antimony against *Leishmania donovani* axenic amastigotes. Antimicrob. Agents Chemother. *43*, 278–282.

Erdmann, M., Scholz, A., Melzer, I.M., Schmetz, C., and Wiese, M. (2006). Interacting protein kinases involved in the regulation of flagellar length. Mol. Biol. Cell.

Franke, E.D., McGreevy, P.B., Katz, S.P., and Sacks, D.L. (1985). Growth cycle-dependent generation of complement-resistant *Leishmania* promastigotes. J. Immunol. *134*, 2713–2718.

Garlapati, S., Dahan, E., and Shapira, M. (1999). Effect of acidic pH on heat shock gene expression in *Leishmania*. Mol. Biochem. Parasitol. *100*, 95–101.

Ghosh, D.K. and Honigberg, B.M. (1976). Activities of glucose-6-phosphate, 6-phosphogluconate, and isocitrate dehydrogenases from *Leishmania donovani* cultivated at 25 and 37°C. J. Protozool. *23*, 450–455.

Glaser, T.A., Baatz, J.E., Kreishman, G.P., and Mukkada, A.J. (1988). pH homeostasis in *Leishmania donovani* amastigotes and promastigotes. Proc. Natl. Acad. Sci. USA *85*, 7602–7606.

Glaser, T.A. and Mukkada, A.J. (1992). Proline transport in *Leishmania donovani* amastigotes: dependence on pH gradients and membrane potential. Mol. Biochem. Parasitol. *51*, 1–8.

Glickman, M.H. and Ciechanover, A. (2002). The ubiquitin-proteasome proteolytic pathway: destruction for the sake of construction. Physiol Rev. *82*, 373–428.

Goyard, S., Segawa, H., Gordon, J., Showalter, M., Duncan, R., Turco, S.J. *et al.* (2003). An in vitro system for developmental and genetic studies of *Leishmania donovani* phosphoglycans. Mol. Biochem. Parasitol. *130*, 31–42.

Gupta, N., Goyal, N., Kumar, R., Agrawal, A.K., Seth, P.K., and Rastogi, A.K. (1996). Membrane characterization of amastigote-like forms of *Leishmania donovani*. Trop. Med. Int. Health *1*, 495–502.

Gutierrez, J.A., Puentes, F., Moreno, A., Patarroyo, M.E., and Murillo, L.A. (2001). Identification of a differentially expressed mRNA in axenic *Leishmania panamensis* amastigotes. Mem. Inst. Oswaldo Cruz *96*, 835–838.

Hendricks, L.D. and Child, G.E. (1980). Present knowledge of the *in vitro* cultivation of *Leishmania*. In: The *in vitro* cultivation of the pathogens of tropical diseases, (Basel: Schwabe and Compny), pp. 151–172.

Hodgkinson, V.H., Soong, L., Duboise, S.M., and McMahon-Pratt, D. (1996). *Leishmania amazonensis*: cultivation and characterization of axenic amastigote-like organisms. Exp. Parasitol. *83*, 94–105.

Holm, A., Tejle, K., Magnusson, K.E., Descoteaux, A., and Rasmusson, B. (2001). *Leishmania donovani* lipophosphoglycan causes periphagosomal actin accumulation: correlation with impaired translocation of PKCalpha and defective phagosome maturation. Cell Microbiol. 3, 439–447.

Holzer, T.R., McMaster, W.R., and Forney, J.D. (2006). Expression profiling by whole-genome interspecies microarray hybridization reveals differential gene expression in procyclic promastigotes, lesion-derived amastigotes, and axenic amastigotes in *Leishmania mexicana*. Mol. Biochem. Parasitol. *146*, 198–218.

Ilg, T. (2000). Lipophosphoglycan is not required for infection of macrophages or mice by *Leishmania mexicana*. EMBO J *19*, 1953–1962.

Iovannisci, D.M. and Ullman, B. (1983). High efficiency plating method for *Leishmania* promastigotes in semidefined or completely-defined medium. J. Parasitol. *69*, 633–636.

Ivens, A.C., Peacock, C.S., Worthey, E.A., Murphy, L., Aggarwal, G., Berriman, M.*et al.* (2005). The genome of the kinetoplastid parasite, *Leishmania major*. Science *309*, 436–442.

Jiang, S., Anderson, S.A., Winget, G.D., and Mukkada, A.J. (1994). Plasma membrane K^+/H^+-ATPase from *Leishmania donovani*. J. Cell Physiol *159*, 60–66.

Joshi, M., Dwyer, D.M., and Nakhasi, H.L. (1993). Cloning and characterization of expressed genes from in vitro-grown 'amastigotes' of *Leishmania donovani*. Mol. Biochem. Parasitol. 58, 345–354.

Klamo, E.M., Drew, M.E., Landfear, S.M., and Kavanaugh, M.P. (1996). Kinetics and stoichiometry of a proton/myo-inositol cotransporter. J. Biol. Chem. *271*, 14937–14943.

Krobitsch, S. and Clos, J. (1999). A novel role for 100 kD heat shock proteins in the parasite *Leishmania donovani*. Cell Stress. Chaperones. *4*, 191–198.

Kuhn, D. and Wiese, M. (2005). LmxPK4, a mitogen-activated protein kinase kinase homologue of *Leishmania mexicana* with a potential role in parasite differentiation. Mol. Microbiol. 56, 1169–1182.

Lamontagne, J. and Papadopoulou, B. (1999). Developmental regulation of spliced leader RNA gene in *Leishmania donovani* amastigotes is mediated by specific polyadenylation. J. Biol. Chem. *274*, 6602–6609.

Landfear, S.M. (2000). Genetics and biochemistry of *Leishmania* membrane transporters. Curr. Opin. Microbiol. 3, 417–421.

Lang, T., Hellio, R., Kaye, P.M., and Antoine, J.C. (1994). *Leishmania donovani*-infected macrophages: characterization of the parasitophorous vacuole and potential role of this organelle in antigen presentation. J. Cell Sci. *107 (Pt 8)*, 2137–2150.

Lebowitz, J.H., Coburn, C.M., Macmahon-Pratt, D., and Beverley, S.M. (1990). Development of a stable *Leishmania* expression vector and application to the study of parasite surface antigen genes. Proc. Natl. Acad. Sci. USA *87*, 9736–9740.

Leifso, K., Cohen-Freue, G., Dogra, N., Murray, A., and McMaster, W.R. (2006). Genomic and proteomic expression analysis of *Leishmania* promastigote and amastigote life stages: The *Leishmania* genome is constitutively expressed. Mol. Biochem. Parasitol.

Lemesre, J. l. Methods for the culture *in vitro* of different stages of tissue parasites. [WO94/26899]. (1994). Ref Type: Patent

Lemesre, J.L., Sereno, D., Daulouede, S., Veyret, B., Brajon, N., and Vincendeau, P. (1997). *Leishmania* spp.: nitric oxide-mediated metabolic inhibition of promastigote and axenically grown amastigote forms. Exp. Parasitol. *86*, 58–68.

Lemma, A. and Schiller, E.L. (1964). Extracellular cultivation of the leishmanial bodies of species belonging to the protozoan genus *Leishmania*. Exp. Parasitol. *15*, 503–513.

Leon, L.L., Temporal, R.M., Soares, M.J., and Grimaldi Junior, G. (1994). Proteinase activities during temperature-induced stage differentiation of species complexes of *Leishmania*. Acta Trop. *56*, 289–298.

Liveanu, V., Webster, P., and Zilberstein, D. (1991). Localization of the plasma membrane and mitochondrial H^+-ATPase in *Leishmania donovani* promastigotes. Eur. J. Cell Biol. *54*, 95–101.

Mazareb, S., Fu, Z.Y., and Zilberstein, D. (1999). Developmental regulation of proline transport in *Leishmania donovani*. Exp. Parasitol. *91*, 341–348.

Mbuchi, M., Bates, P.A., Ilg, T., Coe, J.E., and Raynes, J.G. (2006). C-reactive protein initiates transformation of *Leishmania donovani* and *L. mexicana* through binding to lipophosphoglycan. Mol. Biochem. Parasitol. *146*, 259–264.

McNicoll, F., Muller, M., Cloutier, S., Boilard, N., Rochette, A., Dube, M.*et al.* (2005). Distinct 3′untranslated region elements regulate stage-specific mRNA accumulation and translation in *Leishmania*. J. Biol. Chem. *280*, 35238–35246.

Melanson, J.E., Avery, S.L., and Pinto, D.M. (2006). High-coverage quantitative proteomics using amine-specific isotopic labeling. Proteomics *6*, 4466–4474.

Mengeling, B.J., Zilberstein, D., and Turco, S.J. (1997). Biosynthesis of *Leishmania* lipophosphoglycan: solubilization and partial characterization of the initiating mannosylphosphoryltransferase. Glycobiology *7*, 847–853.

Morales, M.A., Renaud, O., Faigle, W., Shorte, S.L., and Spath, G.F. (2007). Overexpression of *Leishmania major* MAP kinases reveals stage-specific induction of phosphotransferase activity. Int. J. Parasitol.

Mukherjee, T., Mandal, D., and Bhaduri, A. (2001). *Leishmania* plasma membrane Mg2+-ATPase is a H^+/K^+-antiporter involved in glucose symport. Studies with sealed ghosts and vesicles of opposite polarity. J. Biol. Chem. *276*, 5563–5569.

Mukkada, A.J., Meade, J.C., Glaser, T.A., and Bonventre, P.F. (1985). Enhanced metabolism of *Leishmania donovani* amastigotes at acid pH: an adaptation for intracellular growth. Science *229*, 1099–1101.

Naderer, T., Ellis, M.A., Sernee, M.F., De Souza, D.P., Curtis, J., Handman, E.*et al.* (2006). Virulence of *Leishmania major* in macrophages and mice requires the gluconeogenic enzyme fructose-1, 6-bisphosphatase. Proc. Natl. Acad. Sci. USA *103*, 5502–5507.

Nascimento, M., Zhang, W.W., Ghosh, A., Houston, D.R., Berghuis, A.M., Olivier, M.*et al.* (2006). Identification and characterization of a protein tyrosine phosphatase in *Leishmania*; involvement in virulence. J. Biol. Chem. *281*, 36257–36268.

Neal, R.A. and Miles, R.A. (1963). Heated blood agar medium for the growth of *Trypanosoma cruzi* and some species of *Leishmania*. Nature *198*, 210–211.

Nugent, P.G., Karsani, S.A., Wait, R., Tempero, J., and Smith, D.F. (2004). Proteomic analysis of *Leishmania mexicana* differentiation. Mol. Biochem. Parasitol. *136*, 51–62.

Opperdoes, F.R. and Coombs, G.H. (2007). Metabolism of *Leishmania*: proven and predicted. Trends Parasitol.

Padan, E., Bibi, E., Ito, M., and Krulwich, T.A. (2005). Alkaline pH homeostasis in bacteria: new insights. Biochim. Biophys. Acta *1717*, 67–88.

Padan, E., Zilberstein, D., and Schuldiner, S. (1981). pH homeostasis in bacteria. Biochim. Biophys. Acta *650*, 151–166.

Pan, A.A. (1984). *Leishmania mexicana*: serial cultivation of intracellular stages in a cell-free medium. Exp. Parasitol. *58*, 72–80.

Pan, A.A. and McMahon-Pratt, D. (1988). Monoclonal antibodies specific for the amastigote stage of *Leishmania pifanoi*. I. Characterization of antigens associated with stage- and species-specific determinants. J. Immunol. *140*, 2406–2414.

Pral, E.M. and Alfieri, S.C. (1996). Uptake of Z-Tyr[125I]-AlaCHN2, an irreversible cysteine proteinase inhibitor, by lesion amastigotes, axenic amastigotes and promastigotes of *Leishmania mexicana*. Braz. J. Med. Biol. Res. *29*, 987–994.

Rainey, P.M., Spithill, T.W., McMahon-Pratt, D., and Pan, A.A. (1991). Biochemical and molecular characterization of *Leishmania pifanoi* amastigotes in continuous axenic culture. Mol. Biochem. Parasitol. *49*, 111–118.

Reggiori, F. and Klionsky, D.J. (2005). Autophagosomes: biogenesis from scratch? Curr. Opin. Cell Biol. *17*, 415–422.

Rivas, L. and Chang, K.P. (1983). Intraparasitophorous Vacuolar pH of *Leishmania mexicana* Infected Macrophages. Biological Bulletin *165*, 536–537.

Rochette, A., McNicoll, F., Girard, J., Breton, M., Leblanc, E., Bergeron, M.G. *et al.* (2005). Characterization and developmental gene regulation of a large gene family encoding amastin surface proteins in *Leishmania* spp. Mol. Biochem. Parasitol. *140*, 205–220.

Rodriguez, N.E., Gaur, U., and Wilson, M.E. (2006). Role of caveolae in *Leishmania chagasi* phagocytosis and intracellular survival in macrophages. Cell Microbiol. *8*, 1106–1120.

Rosenzweig, D., Smith, D., Opperdoes, F.R., Stern, S., Olafson, R.W., and Zilberstein, D. (2007) Retooling *Leishmania* metabolism: from sandfly gut to human macrophage. FASEB J., in press.

Saar, Y., Ransford, A., Waldman, E., Mazareb, S., Amin-Spector, S., Plumblee, J. *et al.* (1998). Characterization of developmentally-regulated activities in axenic amastigotes of *Leishmania donovani*. Mol. Biochem. Parasitol. *95*, 9–20.

Sacks, D.L. (1992). The structure and function of the surface lipophosphoglycan on different developmental stages of *Leishmania* promastigotes. Infect. Agents Dis. *1*, 200–206.

Sacks, D.L., Hieny, S., and Sher, A. (1985). Identification of cell surface carbohydrate and antigenic changes between noninfective and infective developmental stages of *Leishmania major* promastigotes. J. Immunol. *135*, 564–569.

Saxena, A., Lahav, T., Holland, N., Anupama, A., Huang, G., Volpin, H. *et al.* (2007). Analysis of the *Leishmania donovani* transcriptome reveals an ordered progression of transient and permanent changes in gene expression during differentiation. Mol. Biochem. Parasitol. *152*, 53–65.

Scianimanico, S., Desrosiers, M., Dermine, J.F., Meresse, S., Descoteaux, A., and Desjardins, M. (1999). Impaired recruitment of the small GTPase rab7 correlates with the inhibition of phagosome maturation by *Leishmania donovani* promastigotes. Cell Microbiol. *1*, 19–32.

Sereno, D., Cavaleyra, M., Zemzoumi, K., Maquaire, S., Ouaissi, A., and Lemesre, J.L. (1998). Axenically growth amastigotes of *Leishmania infantum* used as an *in vitro* model to investigate the pentavalent antimony mode of action. Antimicrob. Agents Chemother. *42*, 3097–3102.

Sereno, D. and Lemesre, J.L. (1997a). Axenically cultured amastigote forms as an *in vitro* model for investigation of antileishmanial agents. Antimicrob. Agents Chemother. *41*, 972–976.

Sereno, D. and Lemesre, J.L. (1997b). *In vitro* life cycle of pentamidine-resistant amastigotes: stability of the chemoresistant phenotypes is dependent on the level of resistance induced. Antimicrob. Agents Chemother. *41*, 1898–1903.

Sereno, D. and Lemesre, J.L. (1997c). Use of enzymatic micromethod to quantify amastigote stage of *Leishmania amazonensis in vitro*. Parasitol. Res. *83*, 401–403.

Sereno, D., Michon, P., Brajon, N., and Lemesre, J.L. (1997). Phenotypic characterization of *Leishmania mexicana* pentamidine-resistant promastigotes. Modulation of the resistance during *in-vitro* developmental life cycle. C. R. Acad. Sci. III. *320*, 981–987.

Shaked-Mishan, P., Ulrich.N., Ephros, M., and Zilberstein, D. (2001). Novel intracellular Sb^{V} reducing activity correlates with antimony susceptibility in *Leishmania donovani*. J. Biol. Chem. 276, 3971–3976.

Shapira, M., McEwen, J.G., and Jaffe, C.L. (1988). Temperature effects on molecular processes which lead to stage differentiation in *Leishmania*. EMBO J 7, 2895–2901.

Simpson, L., Aphasizhev, R., Gao, G., and Kang, X. (2004). Mitochondrial proteins and complexes in *Leishmania* and *Trypanosoma* involved in U-insertion/deletion RNA editing. RNA. *10*, 159–170.

Spath, G.F., Epstein, L., Leader, B., Singer, S.M., Avila, H.A., Turco, S.J. *et al.* (2000). Lipophosphoglycan is a virulence factor distinct from related glycoconjugates in the protozoan parasite *Leishmania major*. Proc. Natl. Acad. Sci. USA *97*, 9258–9263.

Spath, G.F., Garraway, L.A., Turco, S.J., and Beverley, S.M. (2003). The role(s) of lipophosphoglycan (LPG) in the establishment of *Leishmania major* infections in mammalian hosts. Proc. Natl. Acad. Sci. USA *100*, 9536–9541.

Steiger, R.F. and Steiger, E. (1977). Cultivation of *Leishmania donovani* and *Leishmania braziliensis* in defined media: nutritional requirements. J. Protozool. *24*, 437–441.

Stein, A., Vasudevan, G., Carter, N.S., Ullman, B., Landfear, S.M., and Kavanaugh, M. (2003). Equilibrative nucleoside transporter family members from *Leishmania donovani* are electrogenic proton symporters. J. Biol. Chem. *278*, 35127–35134.

Sturgill-Koszycki, S., Schlesinger, P.H., Chakraborty, P., Haddix, P.L., Collins, H.L., Fok, A.K., *et al.* (1994). Lack of acidification in Mycobacterium phagosomes produced by exclusion of the vesicular proton-ATPase. Science *263*, 678–681.

Swanson, M.S. and Fernandez-Moreira, E. (2002). A microbial strategy to multiply in macrophages: the pregnant pause. Traffic 3, 170–177.

Teixeira, M.C., De Jesus, S.R., Sampaio, R.B., Pontes-De-Carvalho, L., and Dos-Santos, W.L. (2002). A simple and reproducible method to obtain large numbers of axenic amastigotes of different *Leishmania* species. Parasitol. Res. *88*, 963–968.

Traub-Cseko, Y.M., Duboise, M., Boukai, L.K., and McMahon-Pratt, D. (1993). Identification of two distinct cysteine proteinase genes of *Leishmania pifanoi* axenic amastigotes using the polymerase chain reaction. Mol. Biochem. Parasitol. *57*, 101–115.

Trotter, E.W., Kao, C.M., Berenfeld, L., Botstein, D., Petsko, G.A., and Gray, J.V. (2002). Misfolded proteins are competent to mediate a subset of the responses to heat shock in *Saccharomyces cerevisiae*. J. Biol. Chem. *277*, 44817–44825.

Turco, S.J. (1988). The lipophosphoglycan of *Leishmania*. Parasitol. Today *4*, 255–257.

Turco, S.J. and Descoteaux, A. (1992). The lipophosphoglycan of *Leishmania* parasites. Annu. Rev. Microbiol. *46*, 65–94.

Turco, S.J., Spath, G.F., and Beverley, S.M. (2001). Is lipophosphoglycan a virulence factor? A surprising diversity between *Leishmania* species. Trends Parasitol. *17*, 223–226.

Walker, J., Vasquez, J.J., Gomez, M.A., Drummelsmith, J., Burchmore, R., Girard, I., *et al.* (2006). Identification of developmentally-regulated proteins in *Leishmania panamensis* by proteome profiling of promastigotes and axenic amastigotes. Mol. Biochem. Parasitol.

Wiesgigl, M. and Clos, J. (2001b). The heat shock protein 90 of *Leishmania donovani*. Med. Microbiol. Immunol. (Berl.) *190*, 27–31.

Wiesgigl, M. and Clos, J. (2001a). Heat shock protein 90 homeostasis controls stage differentiation in *Leishmania donovani*. Mol. Biol. Cell *12*, 3307–3316.

Williams, R.A., Tetley, L., Mottram, J.C., and Coombs, G.H. (2006). Cysteine peptidases CPA and CPB are vital for autophagy and differentiation in *Leishmania mexicana*. Mol. Microbiol. *61*, 655–674.

Yorimitsu, T. and Klionsky, D.J. (2005). Autophagy: molecular machinery for self-eating. Cell Death Differ. *12* Suppl. 2, 1542–1552.

Zagari, A., Nemethy, G., and Scheraga, H.A. (1990). The effect of the L-azetidine-2-carboxylic acid residue on protein conformation. I. Conformations of the residue and of dipeptides. Biopolymers *30*, 951–959.

Zakai, H.A., Chance, M.L., and Bates, P.A. (1998). *In vitro* stimulation of metacyclogenesis in *Leishmania braziliensis*, *L. donovani*, *L. major* and *L. mexicana*. Parasitology *116*, 305–309.

van Zandenbergen, G., Bollinger, A., Wenzel, A., Kamhawi, S., Voll, R., Klinger, M., *et al.* (2006). *Leishmania* disease development depends on the presence of apoptotic promastigotes in the virulent inoculum. Proc. Natl. Acad. Sci. USA. *103*, 13837–13842.

Zhou, Y., Messier, N., Ouellette, M., Rosen, B.P., and Mukhopadhyay, R. (2004). *Leishmania major* LmACR2 is a pentavalent antimony reductase that confers sensitivity to the drug Pentostam. J. Biol. Chem.

Zilberstein, D. and Dwyer, D.M. (1985). Protonmotive force-driven active transport of D-glucose and L-proline in the protozoan parasite *Leishmania donovani*. Proc. Natl. Acad. Sci. USA *82*, 1716–1720.

Zilberstein, D. and Dwyer, D.M. (1988). Identification of a surface membrane proton-translocating ATPase in promastigotes of the parasitic protozoan *Leishmania donovani*. Biochem. J. *256*, 13–21.

Zilberstein, D. and Gepstein, A. (1993). Regulation of L-proline transport in *Leishmania donovani* by extracellular pH. Mol. Biochem. Parasitol. *61*, 197–206.

Zilberstein, D. and Shapira, M. (1994). The role of pH and temperature in the development of *Leishmania* parasites. Annu. Rev. Microbiol. *48*, 449–470.

Zilberstein, D., Philosoph, H., and Gepstein, A. (1989). Maintenance of cytoplasmic pH and proton motive force in promastigotes of *Leishmania donovani*. Mol. Biochem. Parasitol. *36*, 109–118.

Zilberstein, D., Blumenfeld, N., Liveanu, V., Gepstein, A., and Jaffe, C.L. (1991). Growth at acidic pH induces an amastigote stage-specific protein in *Leishmania* promastigotes. Mol. Biochem. Parasitol. *45*, 175–178.

The Metabolic Repertoire of *Leishmania* and Implications for Drug Discovery

7

Fred R. Opperdoes and Paul A.M. Michels

Abstract

The development of new anti-*Leishmania* compounds and the study of the mode of action of existing drugs has long been hampered by the difficulties of obtaining sufficient numbers of intracellular amastigotes or to culture in the laboratory well-defined parasites representative of intracellular amastigote stages. Most of the early studies have been carried out on promastigotes, of which the metabolism is not necessarily identical to that of amastigotes. Also research has been complicated by the fact that amastigotes hide away inside the phagosome of the host cell, the macrophage, which imposes additional diffusion barriers and the possibility of metabolizing or degrading drugs before they reach the target. Only a limited number of drugs are available for the treatment of leishmaniasis, and hardly anything is known about their mode of action. The recent completion of the *Leishmania major* genome sequencing project now provides a more complete insight in the metabolic capacities of this and related parasites. Important differences between *Leishmania* and trypanosomes have been identified, and these differences could be related to the special adaptations required for the viability of *Leishmania* inside the macrophage's phagosomal compartment and for its virulence. Differences that exist between parasite and host could be exploited as potential targets for drugs with high selectivity. Despite the fact that glycolysis is not an essential pathway in the intracellular amastigote, its overall functionality is of vital importance, since many of its enzymes are involved in the reverse process of gluconeogenesis. Also enzymes responsible for mannan synthesis from either hexoses or gluconeogenic substrates are validated drug targets. Glucose transporters and fructose-bisphosphatase, as an exclusive gluconeogenic enzyme, have been validated as drug targets. By virtue of the distinctive structural properties exhibited by the enzymes shared by the glycolytic and gluconeogenic pathways and the fact that the majority of them are ensconced inside glycosomes, organelles which require a complete machinery for their biosynthesis and for which the individual peroxin proteins differ considerably from their mammalian counterparts, both glycolytic enzymes and peroxins are interesting drug targets. The methylglyoxal bypass, although not yet validated as a target, certainly deserves further study. Folate transporters have not yet been exploited and three haem biosynthetic enzymes which are all conserved within *Leismania* and for which there is no obvious function, deserve further investigation. Other interesting drug targets are NADP-dependent fumarate reductase, acetate:succinate CoA transferase, dihydroorotate dehydrogenase and arginase. Finally genomic information has provided a possible explanation for the mode of action of the newly introduced anti-leishmanial drug miltefosine.

Introduction

In this chapter we consider the impact that the completion of the *Leismania major* genome sequencing project has had upon our understanding of its metabolism and the possibilities that this knowledge provides for the development of new drugs against this medically relevant trypanosomatid. The different environments, in both the phagosomal compartment of the mammalian macrophage and its phlebotomine vector, that are occupied by *Leishmania major* means that a correct interpretation of the metabolic strategies used by the parasite in each of the hosts is particularly difficult. In comparison to several other protozoan parasites for which complete genome information is available, such as the sleeping sickness parasite *Trypanosoma brucei* (Berriman *et al.*, 2005; Ginger *et al.*, 2007) and the malarial parasite *Plasmodium falciparum* (Gardner *et al.*, 2002), little is known about the actual metabolism of the pathogenic stage, the amastigote. Notwithstanding this fact, the availability of genome sequences, coupled to the use of modern molecular approaches to study gene function, have facilitated a significant reappraisal in our fundamental understanding of the metabolism in this organism.

Significant metabolic differences exist between *Leishmania* promastigotes and amastigotes. In the female sandfly, promastigotes are adapted for life within the different parts of the fly's intestinal tract or the proboscis when they migrate there as metacyclics. Sandflies live on a complex diet. In addition to blood, they feed on nectar and aphid honeydew and thus promastigotes should be able to digest not only a variety of sugars but also amino acids. The intracellular amastigote thrives within a phagosome and phagolysosomes of mononuclear cells in a mammalian host. The infectious metacyclic promastigote in the proboscis of the sandfly must not only be capable of reaching this niche but, after its transformation to amastigote, should also be metabolically adapted to this more hostile environment. Apart from the fact that these compartments are acidic, little is known about the conditions prevalent in the phagosome, and this limits our ability to mimic these conditions experimentally. Nevertheless, some inferences could be made from the nature of the hydrolytic enzymes present in the phagosome and transporters in the phagosomal membrane (Burchmore and Barrett, 2001). The extensive information on the genes of *L. major* should give us a reasonably complete idea of the metabolic arsenal that is available to this parasite. However, at present it is far from clear in which stage of the life cycle the many gene products are operational and this hinders our understanding greatly.

Experimental studies of metabolism in *Leishmania* have focused mainly on the easily cultured promastigotes corresponding to the parasites found in the sandfly midgut. Other important lifecycle stages, such as metacyclic promastigotes and amastigote forms, are more difficult to obtain and cannot be isolated easily in sufficient quantities for detailed biochemical studies. Metabolic studies have mainly concentrated on *L. donovani* and *L. mexicana*, with the underlying assumption that metabolic differences between the various species would be minimal. The study of amastigote metabolism originally has been complicated by the presence of the host cell. Although axenic amastigote model systems have more recently also become available, we have, until recently, had far less insight into the parasite's metabolic strategies that are important once it is in the mammalian host. Despite the detailed metabolic maps that can be drawn from the genome-derived sequences it is in most cases impossible to distinguish between typical promastigote and amastigote adaptations. Thus, the *in silico* identification of potential drug targets should be complemented with further global transcriptomic and, in view of the post-transcriptional regulation of expression, particularly proteomic analyses (McNicoll *et al.*, 2006) in order to complete our understanding of the adaptations and important metabolic switches that take place when the parasite invades the macrophage. It will be essential to experimentally demonstrate the functionality of a drug target in the pathogenic stage of the parasite. This should be followed by a (conditional) disruption of both alleles of the corresponding gene in order to demonstrate that the identified target plays an essential role in the survival, infectivity and/or virulence of the parasite. Our discussion focuses mainly on the identification of potential drug targets in *L. major*. When additional information about the target in the amastigote, or information from other trypanosomes, is available, these are discussed in a comparative context.

Drugs available for treatment of leishmaniasis

For the treatment of leishmaniasis the currently used drugs are limited to six. The most widely used first-line drugs available for all forms of leishmaniasis are pentavalent antimony preparations such as pentostam (sodium stibogluconate) and glucantime (*N*-methylglucamine antimonate). These drugs are administered parenterally for 10 to 30 days. Both formulations are equally effective but also equally toxic when efficacy is expressed in relation to the amount of antimony administered. The pentavalent antimony needs to be reduced to the trivalent state by either host or parasite metabolism in order to exert its activity (see Chapter 6). Although antimony-based drugs have been used since 1947, their mode of action is still far from understood; they may disrupt the parasite's energy production or interfere with trypanothione metabolism. Unfortunately, in many parts of the world, the parasite has become resistant to antimony and failures and relapses occur in all forms of leishmaniasis and constitute approximately 10–25% of cases. If antimony-based drugs are not effective, the second line compounds of pentamidine (lomidine) and amphotericin B (fungizone) are used, which have been introduced in 1940 and 1959, respectively. The mode of action of pentamidine is not clear, although there is some indication that it may act on the parasite's mitochondrion (Basselin *et al.*, 2002; Muhkerjee *et al.*, 2006). Amphotericin is a polyene antimycotic drug, which is believed to interact with membrane sterols, such as ergosterol present in *Leishmania*'s plasma membrane, to produce an aggregate that forms a transmembrane channel resulting in the loss of intracellular solutes and ions. Ambisome (introduced in 1994) is a formulation of amphotericin B in liposomes. Owing to the high capacity of cells of the reticuloendothelial system for phagocytosis the drug is specifically targeted and taken up by the host cells of the *Leishmania* parasite. This will increase efficacy and reduce toxicity of the drug. An important drawback of this formulation is its high cost. A new first-line drug for the use of visceral leishmaniasis was introduced in 2002. It is the ether-lipid analogue miltefosine (hexadecylphosphocholine, Impavido). This lysophospholipid, originally used for the treatment of certain types of cancer, has been shown to interfere with the synthesis of phospholipids and sterols in *Leishmania* (Lux *et al.*, 2000; Rakotomanga *et al.*, 2007). The advantage of this drug is that it is given orally and is very effective. Miltefosine was originally tested for visceral leishmaniasis in India. The cure rate by miltefosine is 95%. Studies in Ethiopia showed that it is also effective in Africa and clinical trials in Colombia demonstrated a high efficacy for cutaneous leishmaniasis as well. It is now registered in many countries and is the first orally administered breakthrough therapy for both visceral and cutaneous leishmaniasis. A side effect is gastrointestinal disturbance. However, miltefosine resistance in laboratory strains of *Leishmania* has already been reported. Several other drugs, in particular the antifungal azoles itraconazole, ketoconazole, and fluconazole, have been on limited clinical trials. These drugs are supposed to act via inhibition of cytochrome P450, essential in the synthesis of ergosterol. So far, the results were equivocal. Allopurinol, a drug in use for the treatment of gout, probably functions as an alternative substrate for *Leishmania*'s hypoxanthine-guanine phosphoribosyl transferase, an enzyme located inside glycosomes. The allopurinol riboside produced is then incorporated into RNA causing inhibition of protein synthesis in the parasite. It has been used experimentally for human leishmaniasis, with variable degrees of success, and is also on trial for the treatment of Chagas' disease. This drug is now widely used for the treatment of leishmaniasis in dogs.

Despite the fact that several drugs are available for the treatment of leishmaniasis, new and better drugs are urgently required. Most available drugs are costly, require long treatment regimes and are becoming more and more ineffective.

The process of drug development

Only a very limited number of drugs have been developed for the treatment of leishmaniasis over the last 60 years and the efficacy of their use has been hampered by adverse effects or resistance of the parasite (see above). The greatest benefit of the genome sequencing project would be the ultimate development of new, effective and safe medicines for the treatment of all manifestations of leishmaniasis through the identification of one or more ideal drug targets. It has been the purpose of this review to present a comprehensive analysis of the major metabolic pathways of *L. major* with, where possible, an emphasis on amastigote-specific features that may be essential for the parasite's adaptation to the host. An important assumption made in this review is that metabolic differences between *L. major* and other species of *Leishmania* are minimal or non-existent. A preliminary comparison with the incomplete genomic information already available for *L. infantum* and *L. brasiliensis* seems to confirm this idea (Opperdoes and Coombs, 2007).

Obvious drug targets are those pathways that are absent from the host, or individual reactions that are catalysed by enzymes with structural properties that are distinct from host enzymes. In this respect, enzymes that are the results of events of lateral gene transfer from bacteria to a trypanosomatid ancestor are very interesting candidates. Unique pathways or enzymes should be validated first as drug targets by demonstrating, either by inhibition studies and through reverse genetic approaches, that they fulfil an essential role in the viability or virulence of the amastigote. Here the '*Leishmania* researcher' has a disadvantage with respect to his colleagues working with *T. brucei*. RNA interference cannot be used and thus one has to take the laborious road of the creation of (conditional) gene disruption mutants. It should also be kept in mind that only results obtained with macrophages infected with amastigotes can be considered as relevant. Enzyme inhibitors to be selected as leads should be tailored in such a way that they will be able to reach a target hidden away inside the phagosome of the macrophage. Promising lead compounds should be effective not only in axenic amastigotes but also in *in vitro* studies with amastigotes that multiply in infected macrophages, before they should be used in an appropriate animal model. Finally, formulations should be developed that allow them to be administered *per os* in order to guarantee that such a drug will be available at low cost to the poor populations who are most in need of a new anti-leishmanial drug. Each of these steps towards the development of a new anti-leishmanial drug poses a formidable task with limited chances of success. Because of the low economic returns to be expected for private industry from a drug developed for the poorest populations of the world, the interest for the development of such drugs has always been limited and will certainly not increase in the future. With this chapter we try to facilitate one of the first steps in this process, the identification of new drug targets.

Outline of promastigote and amastigote metabolism

Biochemical analyses on *Leishmania* promastigotes have shown that these stages can use both glucose and amino acids, such as proline, as energy sources. The catabolism of these substrates appears to involve both glycolysis, compartmentalized in peroxisome-like organelles called glycosomes, and mitochondrial metabolism with an active tricarboxylic acid (TCA) cycle and linked electron transport chain. There is evidence that other sugars than glucose could also be used. This would especially be important in the case of midgut stages where plant sugars should be abundant. The presence of a glyoxylate cycle once reported, seems to be absent which would mean that fatty acids may not serve as the sole substrates for gluconeogenesis. In addition to carbon dioxide, other end-products of promastigote metabolism include succinate and smaller amounts of acetate, pyruvate, D-lactate, alanine, ammonia or urea. Considerably more fragmentary is our knowledge of the energy metabolism of amastigotes. The main reason for this is that these intracellular stages have been less available for study. The limited number of studies carried out show that *L. mexicana* amastigotes isolated from *in vivo* lesions have an increased beta-oxidation of fatty acids and a reduced need for proline and glucose consumption (Hart and Coombs, 1982). The full complement of TCA-cycle enzymes and respiratory chain are present. Glycosomes are less abundant in amastigotes than in promastigotes (Coombs *et al.*, 1986; Tetley and Vickerman, 1991), but have a bigger arsenal of enzymes such as malate dehydrogenase and phosphoenolpyruvate carboxykinase (Mottram and Coombs, 1985a, 1985b). Results of recent studies carried out on *in vitro* cultured amastigotes largely agree with the findings with lesion-derived amastigotes, thus opening the opportunity for more extensive analyses of the energy metabolism of the infectious stages of these parasites.

Enzymes of carbohydrate catabolism and gluconeogenesis as drug target

From previous studies, it is well known that the *Leishmania* promastigote relies mainly on glycolysis and amino-acid metabolism for energy generation. Glucose and other hexose sugars, such as fructose and mannose, are converted to carbon dioxide, alanine, succinate and acetate which are excreted as the major end-products (Bringaud *et al.*, 2006). The identification in the *L. major* genome of several enzymes involved in the metabolism of C5 sugars, such as ribokinase, ribulokinase and xylulokinase, suggests that this trypanosomatid is equally able to catabolize sugars, other than the hexoses, present in nectar and aphid honeydew, components of the sandfly diet. The genes for ribulokinase and xylulokinase seem to have been acquired by lateral gene transfer from a bacterium, since they have homologues only in prokaryotes. Promastigotes are also capable of hydrolyzing disaccharides through the secretion of disaccharide splitting enzymes (Blum and Opperdoes, 1994) and in agreement with this at least three bacterial-type genes annotated as sucrase (invertase) or beta-fructosidases have been found in the *L. major* genome. *Leishmania* also accumulates an intracellular storage product, β-mannan, a mannose-rich polysaccharide (Keegan and Blum, 1992, 1993; Ralton *et al.*, 2003), thus some of these enzymes could be involved in the degradation of this polysaccharide as well. Owing to the unique nature of these enzymes and because mannan degradation seems to be important for amastigote virulence, they would be interesting drug targets, as discussed below.

The activities of many enzymes of carbohydrate metabolism (Table 7.1) in both *L. mexicana* cultured promastigotes and amastigotes isolated from lesions in mice have been described (Coombs *et al.*, 1982; Mottram and Coombs, 1985a, 1985b). These activities involve several glycolytic enzymes, fructose-1,6-bisphosphatase, a key enzyme of gluconeogenesis, malate dehydrogenase and phosphoenolpyruvate carboxykinase, and glucose-6-phosphate dehydrogenase, the first committed enzyme of the pentose-phosphate pathway. The specific activities of some glycolytic

Table 7.1 Enzymes of carbohydrate metabolism of *L. major*

Enzyme name	Pathway	GeneDB accession
2,3-Bisphosphoglycerate-independent phosphoglycerate mutase	Glycolysis gluconeogenesis	LmjF36.6650
2,3-Bisphosphoglycerate-independent phosphoglycerate mutase	Glycolysis gluconeogenesis	LmjF28.2220
2-Oxoglutarate dehydrogenase e1 component, mitochondrial precursor, putative	Krebs	LmjF36.3470
2-Oxoglutarate dehydrogenase, e3 component, lipoamidedehydrogenase-like	Krebs	LmjF31.2640
2-Oxoglutarate dehydrogenase, e3 component, Lipoamidedehydrogenase-like	Krebs carbohydrate	LmjF31.2640
Aconitase, probable	Krebs	LmjF18.0510
ACPM, acyl carrier protein, putative	Fatty acid synthesis | NADH dehydrogenase | respiratory chain	LmjF27.0290
Adenylate kinase	Glycolysis	LmjF36.1360
Adenylate kinase, putative|*Leishmania major*	Glycolysis	LmjF21.1250
Alanine amino transferase putative, hypothetical 55.0 kDa protein, probable	Glycolysis amino-acid metabolism pyruvate alanine	LmjF12.0630

Table 7.1 Continued

Enzyme name	Pathway	GeneDB accession
Aquaglyceroporin	Glycolysis glycerol fatty acid	LmjF31.0020
ATPase alpha subunit	ATPase|respiratory chain|complex V | mitochondrial	LmjF05.0510
ATPase beta subunit, putative	ATPase|respiratory chain|complex V | mitochondrial	LmjF25.1180
ATPase epsilon subunit, possible	ATPase|respiratory chain | complex V| mitochondrial	LmjF36.0620
beta-Fructosidase-like	Glycolysis carbohydrate	LmjF23.0870
Citrate lyase beta chain / citryl-coa lyase subunit-like	Krebs | fatty acid synthesis	LmjF24.1440
Citrate synthase	Krebs | fatty acid synthesis	LmjF18.0680
Citrate synthase	Krebs	LmjF18.0680
COX11 | cytochrome c oxidase assembly protein, putative	Respiratory chain | mitochondrial	LmjF3.0100
Cytochrome c oxidase subunit 5, probable	Respiratory chain | mitochondrial	LmjF26.1710
Cytochrome c oxidase subunit iv	Respiratory chain | mitochondrial	LmjF12.0670
Cytochrome c oxidase subunit vi	Respiratory chain | mitochondrial	LmjF21.1710
Cytochrome c, probable	Respiratory chain | mitochondrial	LmjF16.1320
Cytochrome c1 precursor-like	Respiratory chain complex III | mitochondrial	LmjF36.3090
Cytochrome c1 precursor-like	Respiratory chain complex III | mitochondrial	LmjF13.0850
Cytochrome c1, putative	Respiratory chain complex III | mitochondrial	LmjF07.0060
Electron transfer flavoprotein-ubiquinone oxidoreductase, putative	Respiratory chain | mitochondrial	LmjF07.0600
Enolase	Glycolysis gluconeogenesis	LmjF14.1160
Fructose-1,6-bisphosphate aldolase	Carbohydrate glycolysis Calvin gluconeogenesis	LmjF36.1260
Fumarate hydratase class I, anaerobic	Krebs carbohydrate	LmjF29.1960
Fumarate hydratase (glycosomal)	Krebs carbohydrate	LmjF24.0320
Galactokinase	Carbohydrate galactose glycolysis	LmjF35.2740
Glucokinase	Carbohydrate |glycolysis	LmjF36.2320
Glucosamine-6-phosphate deaminase	Glycolysis	LmjF32.3260
Glucose transporter, putative, probably lmgt1	Glycolysis	LmjF36.6300
Glucose-6-phosphate isomerase	Carbohydrate glycolysis gluconeogenesis	LmjF12.0530
Glyceraldehyde-3-phosphate dehydrogenase	Carbohydrate glycolysis Calvin gluconeogenesis	LmjF30.2970
Glyceraldehyde-3-phosphate dehydrogenase	Carbohydrate glycolysis Calvin gluconeogenesis	LmjF30.2970

Table 7.1 Continued

Enzyme name	Pathway	GeneDB accession
Glycerol kinase	Carbohydrate glycolysis	LmjF35.3080
Glycerol-3-phosphate dehydrogenase, FAD dependent mitochondrial	Carbohydrate glycolysis respiratory chain| mitochondrial	LmjF28.0240
Glycerol-3-phosphate dehydrogenase (NAD)	Carbohydrate glycolysis	LmjF10.0510
Glycerol-3-phosphate dehydrogenase, possible mitochondrial	Carbohydrate glycolysis respiratory chain| mitochondrial	LmjF20.0430
Hexokinase	Glycolysis carbohydrate	LmjF21.0240
Isocitrate dehydrogenase	Krebs	LmjF33.2550
Isocitrate dehydrogenase [NADP], mitochondrial precursor, putative	Krebs	LmjF10.0290
Isocitrate dehydrogenase [NADP], mitochondrial precursor, putative	Krebs cycle	LmjF10.0290
Malate dehydrogenase mitochondrial MDH	Mitochondrial | Krebs | TCA cycle	LmjF34.0140
Malate synthase	Krebs carbohydrate glyoxylate	Absent
Mannose-1-phosphate guanyltransferase| GDP-D-mannose pyrophosphorylase	Glycosylation | ascorbate biosynthesis plants 4	LmjF23.0110
Monocarboxylate transporter-like	Glycolysis Krebs	LmjF05.0480
NADH-dependent fumarate reductase, FRDg	Krebs carbohydrate	LmjF35.1180
NADH-dependent fumarate reductase, putative	Krebs carbohydrate	LmjF35.1180
NADH-ubiquinone oxidoreductase 20 kDa subunit, mitochondrial precursor, putative	Respiratory chain| mitochondrial	LmjF27.0740
NDI1, NADH dehydrogenase, putative (rotenone insensitive subunit)	Respiratory chain NADH | mitochondrial	LmjF36.5380
NI8M, hypothetical protein, conserved	NADH dehydrogenase | respiratory chain	LmjF32.3170
NUAM, NADH-ubiquinone oxidoreductase 75 kDa subunit, mitochondrial precursor (EC 1.6.5.3), probable	Respiratory chain NADH | mitochondrial	LmjF18.1480
NUBM, NADH dehydrogenase	Respiratory chain NADH | mitochondrial	LmjF05.0980
NUBM, NADH dehydrogenase, putative	Respiratory chain NADH | mitochondrial	LmjF05.0980
NUEM, NADH dehydrogenase, possible	Respiratory chain NADH | mitochondrial	LmjF18.0080
NUHM, NADH-ubiquinone oxidoreductase 24 kDa subunit, Mitochondrial precursor-like protein	Respiratory chain NADH | mitochondrial	LmjF17.0270
NUKM, NADH-ubiquinone oxidoreductase 20 kDa subunit, mitochondrial precursor, putative	Respiratory chain NADH | mitochondrial	LmjF27.0740
Phosphofructokinase	Carbohydrate glycolysis	LmjF29.2510
Phosphoglucomutase	Ascorbate biosynthesis animals 1 | carbohydrate metabolism, glycosylation | glycolysis	LmjF21.0640

Table 7.1 Continued

Enzyme name	Pathway	GeneDB accession
Phosphoglucomutase	Carbohydrate metabolism gluconeogenesis glycosylation	LmjF21.0640
Phosphoglycerate kinase, cytosolic	Carbohydrate glycolysis Calvin gluconeogenesis	LmjF20.0110
Phosphoglycerate kinase, glycosomal	Carbohydrate glycolysis gluconeogenesis	LmjF30.3380
Phosphoglycerate kinase, glycosomal	Carbohydrate glycolysis Calvin gluconeogenesis	LmjF20.0100
Phosphoglycerate kinase, glycosomal PGKC	Carbohydrate glycolysis Calvin gluconeogenesis	LmjF20.0100
Phosphoglycerate mutase cofactor dependent-like	Carbohydrate | glycolysis | gluconeogenesis	LmjF08.0060
Phosphomannomutase	Carbohydrate metabolism, glycosylation | ascorbate biosynthesis plants 3	LmjF34.3780
Phosphomannomutase, putative	Carbohydrate metabolism, glycosylation | ascorbate biosynthesis plants 3	LmjF36.1960
Phosphomannose isomerase	Glycolysis | carbohydrate | glycolipid anchor | ascorbate biosynthesis plants 2	LmjF32.1580
Putative cytochrome c oxidase subunit vi	Respiratory chain| mitochondrial	LmjF21.1710
Putative ubiquinol – cytochrome-c reductase, probable	Respiratory chain complex III | mitochondrial	LmjF31.2580
Pyruvate carrier, mitochondrial possible	Glycolysis Krebs carbohydrate mitochondrion metabolite carrier	LmjF32.1110
Pyruvate kinase	Glycolysis	LmjF35.0020
Reiske iron–sulphur protein precursor	Respiratory chain mitochondrial | mitochondrial	LmjF35.1510
Similarity with triose p/pi translocator cptr_soltu, possible	Glycolysis Calvin	LmjF22.1010
Succinate dehydrogenase	Krebs| mitochondrion	LmjF24.1630
Succinyl-CoA ligase alpha-chain, mitochondrial precursor), putative	Krebs| mitochondrion	LmjF25.2140
Succinyl-CoA ligase beta subunit, putative	Krebs | mitochondrion	LmjF36.2950
Sucrase-like protein	Glycolysis	LmjF27.2340
Sugar transporter-like	Glycolysis	LmjF24.0680
Triosephosphate isomerase	Glycolysis Calvin gluconeogenesis	LmjF24.0850
Ubiquinol-cytochrome c reductase complex core protein i-like protein	Respiratory chain complex III | mitochondrial	LmjF35.1370
UTP-glucose-1-phosphate uridylyltransferase 2, putative	Ascorbate biosynthesis animals 1 | glycosylation |	LmjF18.0990

enzymes (hexokinase, phosphofructokinase) and glucose-6-phosphate dehydrogenase were somewhat lower in amastigotes, while that of pyruvate kinase was drastically reduced. In contrast, the activities of malate dehydrogenase, phosphoenolpyruvate carboxykinase and glucose-6-phosphate isomerase were greatly increased. The subcellular organization of carbohydrate metabolism in *Leishmania* parasites is similar to that in procyclic *T. brucei*, with the major part of the glycolytic pathway, malate dehydrogenase and phosphoenolpyruvate carboxykinase in glycosomes, and a dual distribution of enzymes of the pentosephosphate pathway (PPP) (Coombs *et al.*, 1982; Mottram and Coombs, 1985a, 1985b; Hart and Opperdoes, 1984; Maugeri *et al.*, 2003; Veitch *et al.*, 2004). All of the genes encoding proteins necessary for the uptake and degradation of glucose via the glycolytic pathway were already identified prior to the arrival of the *Leishmania* genome sequences (reviewed by Opperdoes and Michels, 2001 and Michels *et al.*, 2006). The C5-sugar kinases carry a peroxisome-targeting signal (PTS) and thus are most likely present inside glycosomes as well and they allow the C5 sugars to feed into the glycolytic pathway via the PPP enzymes.

Gluconeogenesis

While promastigotes feed mainly on sugars, the amastigote stage in the phagosome feeds probably mainly on fatty acids and amino acids and may have limited access to glucose and other sugars generated from glycosylated proteins and glycolipids being degraded in the phagosomal compartment. Sugar residues required by *Leishmania* for protein glycosylation, glycophospholipid-anchor formation and polysaccharide biosynthesis thus have to be formed *de novo* from oxaloacetate via the gluconeogenic pathway (Fig. 7.1). The enzymes essential for gluconeogenesis (i.e. besides many glycolytic enzymes also phosphoenolpyruvate carboxykinase and fructose-1,6-bisphosphatase), as well as the genes involved in glycosylation and polysaccharide formation have all been identified and some carry PTSs. Also genes for two enzymes involved in the antagonistic regulation of both glycolysis and gluconeogenesis, 6-phosphofructo-2-kinase and fructose-2,6-bisphosphatase, are present (Chevalier *et al.*, 2005). These two enzymes are responsible for the formation and degradation of fructose-2,6-bisphosphate, which in many organisms is an allosteric regulator of phosphofructokinase. However, in trypanosomatids this compound regulates the activity of the cytosolic pyruvate kinase (Van Schaftingen *et al.*, 1985). In agreement with the cytosolic function of this regulator these two enzymes lack a canonical PTS and have been shown to be cytosolic (Van Schaftingen *et al.*, 1985).

The capacity of glucose uptake of amastigotes is considerably less than that of promastigotes (Burchmore and Hart, 1995). Uptake is mediated via three glucose transporters encoded by homologous genes (LmGT1–3). In the case of *L. mexicana* LmGT1 and LmGT3, having different substrate affinities, are constitutively expressed in, respectively, the flagellar and the plasma membrane, while LmGT2 is a high affinity transporter that is only expressed in the plasma membrane of the promastigote. Thus each glucose transporter isoform has a distinct biological function in the parasite. (Burchmore and Landfear, 1998). These transporters are capable of importing a variety of hexoses (glucose, mannose, fructose, galactose) into the cells (Rodriguez-Contreras *et al.*, 2007). Thus also the amastigote is capable of catabolizing hexose sugars via the glycolytic and pentose-phosphate pathways although at a reduced rate. Because of the apparent low pyruvate kinase activity in the intracellular stage, glucose degradation does not necessarily result in the production of pyruvate, but rather succinate via the phosphoenolpyruvate carboxykinase/malate dehydrogenase branch in the glycosome, which also comprises fumarate hydratase and NADH-dependent fumarate reductase. This enzyme which is involved in an essential pathway of glycosomal NADH reoxidation, is not found in higher eukaryotes and thus could be an interesting drug target. Indeed a number of inhibitors of fumarate reductase have been shown to exert potent anti-leishmanial activity (Chen *et al.*, 2001). Succinate can also be produced by mitochondrial metabolism, very similar as has been demonstrated for procyclic *T. brucei* (Bringaud *et al.*, 2006; Michels *et al.*, 2006). Another end-product, alanine, is produced via the transamination of pyruvate by alanine aminotransferase, an enzyme which is expressed

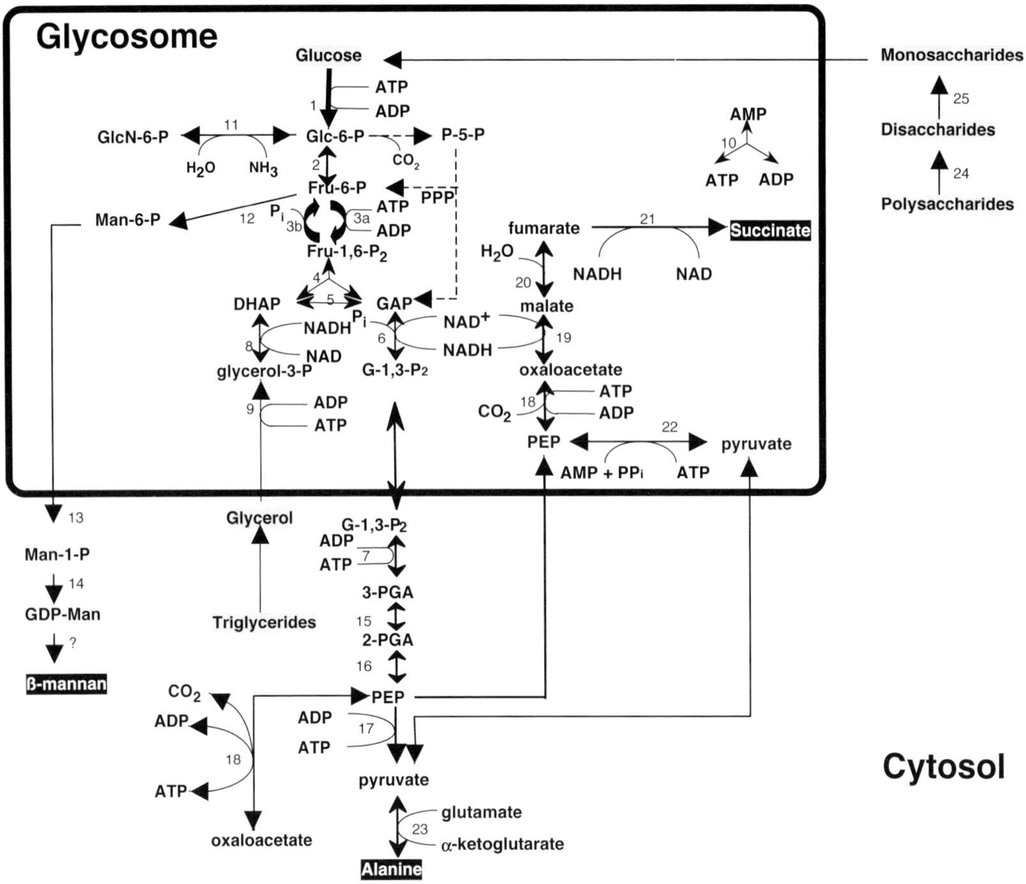

Figure 7.1 Gluconeogenesis and related pathways in *Leishmania*. Boxed metabolites are nutrients (in grey) or end-products (in black) of metabolism. PPP, pentose-phosphate pathway. Enzymes: 1, hexokinase; 2, phosphoglucose isomerase; 3a, phosphofructokinase; 3b, fructose-1,6-bisphosphatase ; 4, fructosebisphosphate aldolase; 5, triosephosphate isomerase; 6, glyceraldehyde-3-phosphate dehydrogenase; 7, phosphoglycerate kinase; 8, glycerol-3-phosphate dehydrogenase; 9 glycerol kinase; 10, adenylate kinase; 11, glucosamine-6-phosphate deaminase; 12, mannose-6-phosphate isomerase; 13, phosphomannomutase; 14, GDP-mannose pyrophosphorylase; 15, phosphoglycerate mutase; 16, enolase; 17, pyruvate kinase; 18, phosphoenolpyruvate carboxykinase; 19, malate dehydrogenase; 20, fumarate hydratase; 21, NADH-dependent fumarate reductase; 22, pyruvate phosphate dikinase; 23, alanine aminotransferase; 24, amylase-like protein; 25, sucrase-like protein.

in both life-cycle stages (Mottram and Coombs, 1985b). Although hexoses are probably less abundant in the phagosomes than other nutrients such as amino acids or lipids, monosaccharides too may be made available to amastigotes by the hydrolytic activity of phagosomal glucuronidase on glycosylated proteins delivered by the variety of vesicles fusing with the phagosome.

Recent research has provided another, possibly more important, role for the glucose transporters and various glycolytic enzymes in amastigotes. Both infectious metacyclic promastigotes and amastigotes synthesize large amounts of β-mannan, that was shown to be essential for amastigote replication and thus virulence (Ralton *et al.*, 2003). Glucose-transporter null mutants of *L. mexicana* are still viable as promastigotes because they can still grow on other nutrients than glucose, but amastigotes don't (Rodriguez-Contreras *et al.*, 2007; Burchmore *et al.*, 2003; Rodriguez-Contreras and Landfear, 2006); macrophage infectivity is not affected, but the intracellular parasites don't replicate. Mutant promastigotes synthesize carbohydrate-containing macromolecules from substrates such as alanine, glycerol, aspartate and acetate, but particularly the synthesis of β-mannan is significantly reduced (Rodriguez-Contreras and Landfear, 2006). This synthesis can be accounted for by gluconeogenesis, but the data suggest that this process cannot fully compensate for the reduced β-mannan synthesis resulting from the loss of hexose uptake. Indeed, a null mutant without the gluconeogenic enzyme fructose-1,6-bisphosphatase could grow as promastigote on hexose-containing medium and synthesize β-mannan, from the hexoses, but not produce this oligosaccharide from gluconeogenic substrates (Naderer *et al.*, 2006). The mutant promastigotes were internalized by macrophages but were unable to replicate in the phagosome and generate lesions in mice. These data validate enzymes responsible for mannan synthesis from either hexoses or gluconeogenic substrates as drug targets. These involve the hexose transporters, hexokinase, glucose-6-phoshate isomerase, fructose-1,6-bisphosphatase and enzymes involved in β-mannan synthesis from fructose 6-phosphate, such as mannose-6-phosphate isomerase, phosphomannomutase and GDP-mannose pyrophophorylase, but probably also a variety of other glycolytic enzymes acting in the gluconeogenic direction (aldolase, glyceraldehyde-3-phosphate dehydrogenase, phosphoglycerate kinase, phosphoglycerate mutase, enolase, phosphoenolpyruvate carboxykinase, malate dehydrogenase and glycerol-3-phosphate dehydrogenase). Which enzyme would be the best drug target will depend on the substrates available inside the phagosome and this has to be established first. For several of the enzymes of *Leishmania* carbohydrate metabolism, or their homologues from *T. brucei* or *T. cruzi*, crystal structures have already been determined and selective inhibitors developed. For some of them it has been shown that they kill cultured parasites (reviewed in Verlinde *et al.*, 2001; Opperdoes and Michels, 2001).

Methylglyoxal bypass

Leishmania has been reported to produce D-lactate (Darling and Blum, 1988), an end product of methylglyoxal metabolism (Fig. 7.2). In bacteria, under conditions of phosphate starvation, methylglyoxal is formed from triosephosphates by means of a methylglyoxal synthase. However, a gene encoding such an enzyme appears to be absent from *Leishmania*, as well as from the

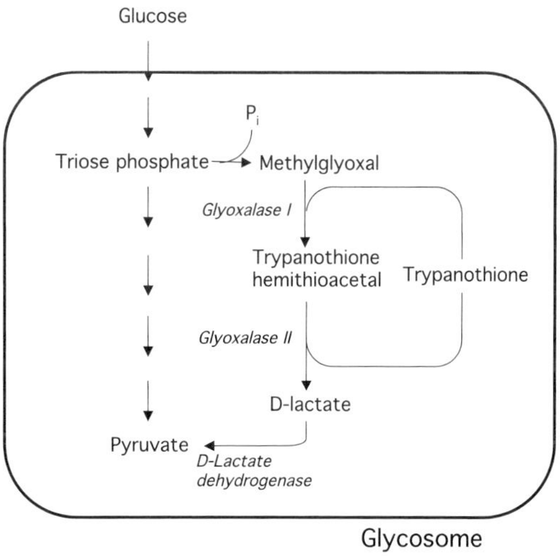

Figure 7.2 Glyoxalase pathway. Enzyme names are in italics. Glyoxalase I is absent from *T. brucei*. D-lactate dehydrogenase is only present in *Leishmania*.

other trypanosomatids (Table 7.2). Nevertheless, methylglyoxal can also be formed by fragmentation of triosephosphates through a spontaneous reaction or via a side reaction of the enzyme triosephosphate isomerase. Methylglyoxal is converted to D-lactate by a thiol-dependent glyoxalase system. Interestingly the trypanosomatid glyoxalases I and II use trypanothione, rather than glutathione, as the essential cofactor (Irsch and Krauth-Siegel, 2004). Both enzymes are present in *L. major*, in agreement with the production of D-lactate by this organism. Glyoxalase I seems to have prokaryotic ancestry (Vickers *et al.*, 2004). *L. brasiliensis*, however, seems to lack the glyoxalase II gene, which would suggest that also this organism would be unable to form D-lactate. *T. brucei*, which has a much more active glycolysis and therefore is thought to produce more methylglyoxal, lacks the glyoxylase I gene and does not form any D-lactate. Methylglyoxal is thought to be formed mainly inside glycosomes, the site of the first part of the glycolytic pathway, but the two enzymes of the glyoxalase system do not carry an identifiable PTS.

Interestingly, in contrast to *Trypanosoma* species, *Leishmania* possesses a D-lactate dehydrogenase. This enzyme may convert the D-lactate into pyruvate for further metabolism. The D-lactate dehydrogenases of three *Leismania*

Table 7.2 Trypanosomatid enzymes of the methylglyoxal bypass

	GeneDB accession		
Enzyme name	*T. brucei*	*T. cruzi*	*L. major*
Methylglyoxal synthase	Absent	Absent	Absent
GLO1||glyoxalase I, trypanothione-dependent	None	Tc00.1047053510743.70	LmjF35.3010
Glyoxalase II, trypanothione-dependent	Tb06.3A7.500	Tc00.1047053507603.230	LmjF12.0220
D-Lactate dehydrogenase-like protein	Absent	Absent	LmjF27.2020

species (*L. major*, *L. infantum* and *L. brasiliensis*) have a PTS, which supports the idea that the glyoxalase pathway is also glycosomal. The fact that methylglyoxal is toxic and mutagenic, that the methylglyoxalase system of *Leishmania* significantly differs from that of humans, and that D-lactate dehydrogenase is absent from humans, makes this pathway an interesting drug target (Ariza *et al.*, 2006).

Glycosome biogenesis

The unique sequestering of glycolysis and other pathways of carbohydrate metabolism in glycosomes in trypanosomatids and the vital importance of these processes to kinetoplastid parasites have led to the notion that this metabolic compartmentation itself might be essential as well. *T. brucei* homologues of 12 proteins, designated 'peroxins (acronym 'PEXs'), involved in biogenesis of peroxisomes of mammalian cells or yeasts, have been identified so far (reviewed in Moyersoen *et al.*, 2004). Indeed, depletion of each of them by RNAi affected the proper localization of glycosomal enzymes and led to death of bloodstream-form trypanosomes and, at growth conditions requiring glycolytic activity, of procyclic forms too (Moyersoen *et al.*, 2004; Furuya *et al.*, 2002). The importance of proper compartmentation has been attributed to the specific kinetic features and lack of activity regulation of the trypanosomatid glycolytic enzymes; as a result, the glycolytic flux can only be controlled under the conditions pertaining inside the organelle, not under those of the cytosol (Bakker *et al.*, 1997).

For *Leishmania*, the same 12 peroxin homologues have been identified in the genome, but only three of them (PEX2, PEX5, PEX14) have been experimentally analysed (Table 7.3). The essentiality of two peroxins (PEX2 and PEX5) has been confirmed for *L. donovani* promastigotes (Flashpohler *et al.*, 1997, 1999; Jardim, *et al.*, 2000). The fact that several glycosomally located enzymes of glycolysis and gluconeogenesis are involved in β-mannan synthesis, a process that is vital to amastigotes (see section on gluconeogenesis), strongly suggests that glycosome biogenesis will also be essential to this human disease-causing stage of the parasites and thus a candidate drug target.

The peroxins responsible for different aspects of glycosome biogenesis in trypanosomatids and the homologous peroxins involved in peroxisome biogenesis in humans are poorly conserved. The amino-acid sequences of those peroxins that have been identified have overall identities varying between 12 and 34% (Moyersoen *et al.*, 2004), offering great perspectives for the discovery of compounds that specifically affect the function of the trypanosomatid proteins but not the human ones. Indeed, current research attempts to solve three-dimensional structures of trypanosomatid peroxins that interact with each other in the cascade of reactions leading to the organelle's synthesis, and to develop compounds that prevent such interactions (Moyersoen *et al.*, 2004, Choe *et al.*, 2003).

Mitochondrial metabolism

The glycolytic end-product pyruvate is further metabolized in the mitochondrion to acetyl-CoA.

Table 7.3 Trypanosomatid peroxins identified in the TriTryp database

			GeneDB accession		
Peroxin	Function	Protein family	*T. bucei*	*T. cruzi*	*L. major*
PEX1	Import glycosomal matrix proteins	AAA+ family ATPase-like protein	Tb06.4F7.680	Tc00.1047053511745.20	LmjF30.2540
PEX2	Import glycosomal matrix proteins	Glycosomal membrane protein | RING-finger protein	Tb03.48O8.580	Tc00.1047053508479.230	LmjF25.2230
PEX4	Import glycosomal matrix proteins	Ubiquitin-conjugating protein family protein (E2)	Tb08.12O16.620	Tc00.1047053509607.10	LmjF07.0850
PEX5	Import glycosomal matrix proteins	Peroxisomal targeting signal-1 receptor | TPR protein	Tb05.28F8.110	Tc00.1047053508299.70	LmjF35.1420
PEX6	Import glycosomal matrix proteins	AAA+ family ATPase-like protein	Tb05.6E7.930	Tc00.1047053506947.40	LmjF16.0060
PEX7	Import glycosomal matrix proteins	Peroxisomal targeting signal type 2 receptor | WD-repeat protein	Tb03.28C22.1010	Tc00.1047053506989.160	LmjF29.0740
PEX10	Import glycosomal matrix proteins	Glycosomal membrane protein |RING-finger protein	Tb03.48O8.510	Tc00.1047053508479.190	LmjF25.2290
PEX11	Glycosome proliferation	Glycosomal membrane protein	Tb11.01.3370	Tc00.1047053511017.50	LmjF28.2260
PEX12	Import glycosomal matrix proteins	Glycosomal membrane protein | RING-finger protein	Tb10.61.0440	Tc00.1047053503809.20	LmjF19.1250
PEX14	Import glycosomal matrix proteins	Glycosomal membrane protein	Tb10.100.0130	Tc00.1047053504069.50	LmjF21.1840
PEX19	Insertion of glycosomal membrane proteins		Tb09.211.3300	Tc00.1047053510737.100	LmjF35.3260

A large part of the acetyl-CoA is converted to acetate by the TCA-cycle enzyme succinyl-CoA ligase and a unique acetate:succinate CoA transferase (ASCT) (Van Hellemond *et al.*, 1998; Rivière *et al.*, 2004). Together, the last two enzymes catalyse a cycle leading to the net production of one mole of ATP per mole of acetate produced (Van Hellemond *et al.*, 1998; Bochud-Allemann and Schneider, 2002). Outside of the Trypanosomatidae, ASCT activity has only ever been found in certain kinds of anaerobic mitochondria, including hydrogenosomes (Tielens *et al.*, 2002). The fact that this enzyme is absent from the mammalian host and is involved in ATP generation would render it an interesting drug target. The remainder of the acetyl-CoA can be metabolized via the enzymes of the TCA cycle, which are all present. An active respiratory chain comprising complex I (NADH dehydrogenase), complex II (succinate dehydrogenase), complex III (cytochrome b/c oxidoreductase), cytochrome c and complex IV (cytochrome oxidase), is present and takes care of the reoxidation of mitochondrially produced NADH and succinate and the generation of a proton gradient over the mitochondrial inner membrane. ATP synthase is also present. It utilizes the proton gradient for the synthesis of ATP. Complex I, although larger than in *E. coli* (14 subunits), seems to be rather small. Only 18 NADH dehydrogenase subunits could be identified in the genome database against some 46 for mammalian Complex I (Opperdoes and Michels unpublished). No evidence was found for the involvement of this complex in the generation of the proton gradient, because a number of the integral membrane subunits supposed to be involved in proton translocation could not be identified. Delta-1-pyrroline-5-carboxylate dehydrogenase, which receives its reducing equivalents directly from proline, and electron transfer flavoprotein (ETF)-ubiquinone oxidoreductase, which accepts reducing equivalents through ETF from long, medium and short/branched chain fatty acid-specific mitochondrial acyl-CoA dehydrogenases, and which were reported recently for the trypanosomatids (Allen *et al.*, 2004; van Weelden *et al.*, 2005; Ginger *et al.*, 2007) have also been identified in *L. major*.

Even though the presence of a complete set of TCA cycle enzymes can be predicted from the *Leishmania* genome sequence, they are apparently not used for the complete oxidation of acetyl-CoA to CO_2 as would occur if all enzymes are fully operational within the cycle. Indeed the fact that the mitochondrial isocitrate dehydrogenase utilizes NADP rather than NAD as cofactor suggests that this enzyme may have a function other than in the TCA cycle. This situation is reminiscent of what has been shown for the procyclic stage of *T. brucei* (Besteiro *et al.*, 2002; van Weelden *et al.*, 2003; van Weelden *et al.*, 2005), where the primary role of the TCA cycle in the midgut stage parasite is the provision of precursors for biosynthetic pathways, such as gluconeogenesis, fatty acid synthesis and the oxidation of amino acids such as threonine and proline. Also in *Leishmania* promastigotes this seems to be the most important function of the TCA-cycle enzymes, as suggested by the large amounts of succinate and acetate produced under aerobic conditions. Since the amastigote is also actively involved in gluconeogenesis it is not expected that mitochondrial metabolism here will be fundamentally different from what is encountered in procyclic *T. brucei*.

Lack of an alternative oxidase

An unusual and key feature of the energy metabolism of bloodstream-form *T. brucei* is the mitochondrially located alternative terminal oxidase which is involved in the reoxidation of NADH, produced in glycolysis, via a dihydroxyacetone-phosphate:glycerol-3-phosphate shuttle. A homologue of this oxidase was not found in *L. major*. Yet, *L. major* has in its mitochondrial membrane the FAD-dependent glycerol-3-phosphate dehydrogenase, which in trypanosomes is also part of the shuttle, but in *Leishmania* amastigotes most likely functions exclusively in the catabolism or use for gluconeogenesis of glycerol 3-phosphate derived from triglycerides being degraded in the phagosome (Guerra *et al.*, 2006).

Iron–sulphur synthesis

Iron–sulphur clusters form redox prosthetic groups for a wide range of proteins, many of which are localized to the mitochondrion. The *L. major* genome appears to encode enzymes required for the synthesis of these molecules.

There include a cysteine desulphurase, which may be targeted to the mitochondrion, and a mitochondrial iron–sulphur cluster NifU-like protein, suggesting that the organelle is indeed involved in the assembly of Fe–S cluster proteins using elemental sulphur.

Mitochondrial transporters

On the basis of genome analysis, *L. major* possesses a complete set of homologues of mammalian and yeast mitochondrial solute carriers. These are required for the transport of pyruvate, the end-product of glycolysis, and inorganic phosphate, and for the exchange of dicarboxylates and tricarboxylates. Without appropriate biochemical analysis it is not possible to determine the substrate specificities of individual carrier proteins on the basis of sequence homology alone, but the dicarboxylate and tricarboxylate carriers are probably involved in transport of TCA cycle intermediates and of aspartate and glutamate across the mitochondrial membrane. An ATP/ADP exchanger for the export of newly synthesized ATP from the mitochondrial matrix into the cytoplasm is also present.

Cofactor synthesis

Many enzymes use cofactors such as biotin, coenzyme A, FMN and FAD, NAD and NADP, pyridoxal phosphate and thiamine pyrophosphate. Humans obtain precursors of them as vitamins, required in small amounts in their diet, while many microorganisms synthesize them *de novo*. The trypanosomatid genome database did not reveal the presence of any of the pathways involved in the *de novo* synthesis of these cofactors. However, trypanosomatid genome searches suggested that these parasites contain many of the enzymes involved in the final stages of the cofactors synthesis as known from other organisms, although the situation for biotin and thiamine is unclear (Table 7.4). In contrast to the human situation, in which the last two enzymes of CoA biosynthesis have been fused to form a bifunctional enzyme, in *Leishmania* and the trypanosomes these are separate enzymes and, moreover, the last enzyme, dephospho-CoA kinase, is apparently of bacterial origin. Also the ATP-NAD kinase, responsible for the formation of NADP from NAD, seems of bacterial origin. This is an interesting aspect of *Leishmania* metabolism that deserves further study. Also specific transporters for precursors of cofactors should be interesting drug targets, but so far they have not yet been identified.

Folate metabolism

While in many other microorganisms such as bacteria and *Plasmodium* the folate pathway has proven to be a valuable target for drug intervention, no compounds targeting enzymes of folate metabolism have been found to be effective drugs against *Leishmania* infections (reviewed in Nare *et al.*, 1997; Ouellette *et al.*, 2002). There appear to be several reasons for this. Firstly, *Leishmania* are auxotrophes for folates (folate and biopterin) and must import these metabolites from an exogenous source (Nare *et al.*, 1997; Ouellette *et al.*, 2002). Consistent with this, no genes encoding enzymes of folate biosynthesis are present in the *L. major* genome, whereas there are as many as 12 genes encoding a novel class of transport membrane proteins responsible for folate transport (Richard *et al.*, 2002; Cunningham and Beverley, 2001). Such transporters were previously only known to occur in cyanobacteria and in plant plastids (Klaus *et al.*, 2005). Secondly, the enzymatic reduction of folate to become active as tetrahydrofolate, a coenzyme required for one carbon (C1) transfer reactions, can be catalysed by both the bifunctional dihydrofolate reductase-thymidylate synthase (DHFR-TS) (Beverley *et al.*, 1986) and pteridine reductase 1 (PTR1). The latter is readily overexpressed if DHFR-TS is inhibited, thus it may be necessary to block both DHFR-TS and PTR1 simultaneously for effective interference with folate metabolism. *Leishmania* contains a complete set of genes for the activation and interconversion of C1 units (Table 7.4). For a full account of this metabolism the reader is referred to Opperdoes and Coombs (2007).

Requirement of haem

Haem is required for the synthesis of haemoproteins, such as the cytochromes *a*, *b*, *c* and P450. Endosymbiont-containing trypanosomatids are able to synthesize their own haem, but all other trypanosomatids, including *Leishmania*, are auxotrophes for haem which either has to be added to the culture medium in the form of haemin or as

Table 7.4 Enzymes involved in the synthesis of biotin, folate, FAD, NAD, thiamine pyrophosphate and pyridoxal phosphate

Pathways	Enzyme name	EC number	Accession
biotin	Biotin/lipoate protein ligase-like protein		LmjF31.1070
Coenzyme A biosynthesis 1	Pantothenate kinase subunit, putative	2.7.1.33	LmjF28.0140
Coenzyme A biosynthesis 2	Hypothetical protein, conserved | phosphopantothenate – cysteine ligase (phosphopantothenoylcysteine synthetase) (PPC synthetase)	6.3.2.5	LmjF25.1900
Coenzyme A biosynthesis 3	Hypothetical protein, conserved | phosphopantothenoylcysteine decarboxylase (PPC-DC) (CoaC)	4.1.1.36	LmjF30.1540
Coenzyme A biosynthesis 4	Phosphopantetheine adenylyltransferase | hypothetical protein, conserved	2.7.7.3	LmjF32.2070
Coenzyme A biosynthesis 5	Dephospho-CoA kinase, putative	2.7.1.24	LmjF22.1530
Flavine metabolism 1 | FAD synthesis | cofactor | vitamin B2	Riboflavin kinase/FMN adenylyltransferase-like protein	2.7.1.26	LmjF35.3160
Flavine metabolism 2 | FAD synthesis | cofactor | vitamin B2	Phosphoadenosine phosphosulphate reductase-like protein | FAD synthetase (FMN adenylyltransferase) (FAD pyrophosphorylase) (flavin adenine dinucleotide synthetase)	2.7.7.2	LmjF28.0270
Folate metabolism | amino acid Threonine serine glycine pyruvate |cofactor	Glycine dehydrogenase	2.1.2.10	LmjF26.0030
Folate metabolism | amino acid Threonine serine glycine pyruvate | cofactor	Serine hydroxymethyltransferase, cytosolic	2.1.2.1	LmjF14.1320
Folate metabolism | amino acid Threonine serine glycine pyruvate | cofactor	Serine hydroxymethyltransferase, mitochondrial	2.1.2.1	LmjF28.2370
Folate metabolism | biopterin | pteridin | pterin | amino acid || cofactor	QDPR || quinonoid dihydropteridine reductase (7 copies)	1.5.1.34	LmjF34.4510
Folate metabolism	Biopterin transporter, putative (12 copies)		LmjF35.5150
Folate metabolism	DHFR-TS || dihydrofolate reductase-thymidylate synthase	1.5.1.3 | 2.1.1.45	LmjF06.0860
Folate metabolism	FTHS (c-1-tetrahydrofolate synthase, cytoplasmic, putative)	6.3.4.3|	LmjF30.2600
Folate metabolism	Methionine synthase, putative	2.1.1.13	LmjF07.0090
Folate metabolism	Methionyl-tRNA formyltransferase, putative	2.1.2.9	LmjF32.2240
Folate metabolism	Methylenetetrahydrofolate dehydrogenase-like protein	1.5.1.5	LmjF22.0340
Folate metabolism	Methylenetetrahydrofolate reductase, putative	1.5.1.20	LmjF36.6390
Folate metabolism	MTFC/MTFD (C-1-tetrahydrofolate synthase, cytoplasmic, putative)	3.5.4.9 | 1.5.1.5	LmjF26.0320

Table 7.4 Continued

Pathways	Enzyme name	EC number	Accession
Folate metabolism	Pteridine transporter ft5, putative		LmjF35.5150
Folate metabolism	Pteridine reductase PTR1	1.5.1.33	LmjF23.0270
Folate metabolism	FPGS || folylpolyglutamate synthetase	6.3.2.17	LmjF36.2610
Folate metabolism | glycine cleavage system | cofactor	Aminomethyltransferase, mitochondrial precursor, putative (subunit of the glycine cleavage system) (two copies)	2.1.2.10	LmjF36.3810
Folate metabolism glycine Formate | cofactor	Glycine cleavage system H protein, putative	2.1.2.10	LmjF35.4720
Folate metabolism | Pteridine | |biopterin	Pterin-4-alpha-carbinolamine dehydratase, putative	4.2.1.96	LmjF11.0220
NAD biosynthesis 1	Nicotinate phosphoribosyltransferase, putative	2.4.2.11	LmjF33.0960
NAD biosynthesis 2 | cofactor	Nicotinamide mononucleotide adenylyltransferase 2 (EC 2.7.7.1) | NAD(+) pyrophosphorylase	2.7.7.1	LmjF17.1340
NAD biosynthesis 3	NAD+ synthase	6.3.1.5	LmjF33.0960
NAD biosynthesis 4 | NADP	ATP-NAD kinase-like protein	2.7.1.23	LmjF06.0460
NAD biosynthesis | tryptophan	Kynureninase, possible	3.7.1.3	LmjF26.2240
Pyridoxal phosphate synthesis | vitamine B6	Pyridoxal kinase, putative	2.7.1.35	LmjF30.1250

a component of the serum required for growth. In eukaryotes haem synthesis starts from succinyl CoA produced by the mitochondrial TCA cycle and includes a total of seven enzymatic steps of which the first and the last three take place in mitochondria. The other steps occur in the cytosol. In agreement with the haemin requirement of trypanosomatids the genes encoding the first four enzymes in the haem biosynthetic pathway, aminolaevulinate synthase and aminolevulinate dehydratase (also known as porphobilinogen synthase) uroporphyrinogen-III synthase and uroporphyrinogen-III decarboxylase, have been detected neither in *Leishmania* (Sah *et al.*, 2002) nor in other trypanosomatids. Protoporphyrin IX may substitute for haemin to grow *Leishmania* in chemically defined medium, suggestive of ferrochelatase functionality in these cells (Chang and Chang, 1985). Indeed a ferrochelatase gene is present. However, the discovery, in three different species of *Leishmania*, of two genes homologous to coproporphyrinogen III oxidase and protoporphyrinogen oxidase, preceding ferrochelatase in the pathway of haem biosynthesis, has come as a surprise. Moreover, these typical mitochondrial enzymes have no detectable mitochondrial transit peptide. These two oxidase genes, together with a pteridine transporter gene, are juxtaposed as a separate transcription unit at one end of chromosome 6 and, together with the ferrochelatase gene, seem all to be of bacterial origin. Does this mean that *Leishmania* has acquired by horizontal transfer the ability to synthesize haem from precursors either in the sandfly midgut or in the phagosome? The macrophage has as one of its functions the removal from the circulation of erythrocytes that are at the end of their life time. Their subsequent degradation inside the phagolysosomes leads to the massive release of haem. Although haem degradation in mammalian cells does not include the production of coproporphyrinogen or protoporphyrinogen, maybe in the acidic environment of the lysosome some of the haem becomes chemically modified, escapes degradation to biliverdin and bilirubin and can be captured by the amastigote which then would require these three enzymes for its repair. It could be imagined that the acquisition of the three haem biosynthesis enzymes of bacterial origin has aided *Leishmania* to adapt to the extreme conditions prevailing in the macrophage phagolysosome (see also the section on oxidant stress protection below). Little is known about the structure of the haem incorporated into the *Leishmania* cytochromes other than that cytochrome c_1 has an atypical difference absorption spectrum and seems to be attached to the protein backbone via one, rather than two thioester linkages (Hill and White, 1968 ; Priest and Hajduk, 1992). Haem structure and synthesis in *Leishmania* certainly deserves further study.

Lipid metabolism

Beta oxidation

Genome analysis suggests that *L. major* should be capable of oxidizing fatty acids via *beta*-oxidation in two separate cellular compartments: the glycosomes and the mitochondrion (Table 7.5). Four homologues of the fatty acyl-CoA synthetases already identified for *T. brucei* (Jiang and Englund, 2001) are also present in *L. major*, each of them with a different chain-length specificity. Although no fatty acyl-CoA oxidase or dehydrogenase homologue with a PTS was found in the database, genes for the other enzymes involved in the *beta*-oxidation spiral, such as a bifunctional enzyme (with enoyl-CoA hydratase/enoyl-CoA isomerase/3-hydroxyacyl-CoA dehydrogenase activities, LmjF33.2600) and a 3-ketoacyl-CoA thiolase (LmjF23.0690), both having a PTS, were identified (Opperdoes and Szikora, 2006). Also four acyl-CoA dehydrogenases with specificity for, respectively, very long, long, intermediate and short/branched fatty acids were found in the genome. Three of them have a predicted mitochondrial transit peptide. Genes encoding a mitochondrial bifunctional enzyme and additional mitochondrial thiolases were also identified indicating that the mitochondrial compartment is also involved in beta-oxidation. The presence of genes coding for a 3,2-trans-enoyl-CoA isomerase and 2,4 dienoyl-CoA reductase suggests that *L. major* is also able to cope with unsaturated fatty acids. This resembles the situation found in mammals where the peroxisomes are involved in oxidation of long chain fatty acids which, after shortening, are further degraded in the mitochondrion to acetyl-CoA. Apparently *L. major* is able to oxidize fatty acids of a wide variety in length and composition, but how each of the two compartments contribute to the oxidation of these fatty acids is far from clear and requires further experimentation. An additional argument for the involvement of the glycosome in lipid metabolism is the fact that in each of the three trypanosomatid genomes genes for three so-called half ABC transporters (GAT1–3) have been identified homologous to the peroxisomal transporters involved in fatty-acid transport. Indeed, for *T. brucei* it was confirmed that these transporters are associated with the glycosomal membrane (Yernaux *et al.*, 2006). Preliminary data suggest that these transporters are involved in fatty acid transport through the glycosomal membrane.

Fatty acid synthesis

In addition to *beta*-oxidation of fatty acids, *L. major* is capable of fatty-acid biosynthesis. Mitochondrially produced citrate can be exported to the cytosol where it is cleaved by a citrate lyase into oxaloacetate and acetyl CoA. The latter can be converted by acetyl-CoA carboxylase to malonyl CoA for elongation in a novel type of fatty-acid biosynthesis pathway. A combination of comparative genomics and biochemical studies has revealed that the pathway used by trypanosomatids for fatty acid synthesis is most unusual (Lee *et al.*, 2006; 2007; Table 7.5). There is no evidence for the presence of a cytosolic type I fatty acid synthase, which is a multifunctional enzyme used for fatty-acid biosynthesis in many eukaryotes including mammals. Instead, three elongase genes seem to be responsible for the *de novo* synthesis of fatty acids in *T. brucei* (Lee *et al.*, 2006; 2007). In mammals and yeast, elongases function to extend, via CoA-linked intermediates in a stepwise fashion, the end product (usually palmitate or stearate) of cytosolic type I fatty

Table 7.5 Enzymes of lipid metabolism in trypanosomatids

<table>
<tr><th></th><th></th><th></th><th colspan="3">GeneDB accession</th></tr>
<tr><th>Pathways</th><th>Enzyme name</th><th>EC number</th><th>T. brucei</th><th>T. cruzi</th><th>L. major</th></tr>
<tr><td>Ether lipid</td><td>Alkyldihydroxyacetonephosphate synthase (EC 2.5.1.26) (alkyl-dhapsynthase) (alkylglycerone-phosphate synthase)</td><td></td><td>Tb06.28P18.170</td><td>Tb06.28P18.170</td><td>LmjF30.0120</td></tr>
<tr><td>Fatty acid lipid</td><td>3,2-Trans-enoyl-CoA isomerase</td><td></td><td>228.m01202 || Tb08.30P3. temporary.140 | RH73277p</td><td>7212.m00004 || Tc00.1047053507107.40</td><td>LmjF31.2330</td></tr>
<tr><td>Fatty acid | lipid metabolism
Beta oxidation</td><td>Acyl-CoA dehydrogenase (medium chain specific)</td><td>1.3.99.3</td><td>291.m00026 || Tb11.55.0026 ?</td><td>8359.m00029 || Tc00.1047053510303.290</td><td>LmjF06.0880</td></tr>
<tr><td>Fatty acid synthesis lipid</td><td>beta-Ketoacyl synthase family protein</td><td>2.3.1.41</td><td>199.m00383 | Tb927.2.3910 | 28H13.305</td><td>4937.m00002 || Tc00.1047053504157.20</td><td>LmjF33.2720</td></tr>
<tr><td>Lipid</td><td>2,4-Dienoyl-CoA reductase-like</td><td>4.2.1.17 | 1.1.1.35</td><td>227.m00864 || Tb07.10C21.490</td><td>8359.m00021 || Tc00.1047053510303.210</td><td>LmjF06.0930</td></tr>
<tr><td>Lipid</td><td>3,2-Trans-enoyl-CoA isomerase, mitochondrial precursor-like</td><td>4.2.1.17 | 1.1.1.35</td><td>169.m00967 || Tb04.3M17.150</td><td>7212.m00004 || Tc00.1047053507107.40</td><td>LmjF31.2330</td></tr>
<tr><td>Lipid</td><td>3-Hydroxyacyl-CoA dehydrogenase family-like protein (trifunctional enzyme)</td><td>4.2.1.17 | 1.1.1.35</td><td>Tb927.2.4130</td><td>7378.m00004</td><td>LmjF33.2600</td></tr>
<tr><td>Lipid</td><td>Enoyl-CoA hydratase/Enoyl-CoA isomerase/3-hydroxyacyl-CoA dehydrogenase, putative bifunctional or trifunctional enzyme</td><td>4.2.1.17 | 1.1.1.35</td><td>199.m00405 |Tb927.2.4130 | 28H13.415</td><td>7730.m00007 || Tc00.1047053508441.70</td><td>LmjF33.2600</td></tr>
<tr><td>Lipid</td><td>Peroxisomal enoyl-CoA hydratase</td><td>4.2.1.17</td><td>224.m00842 || Tb03.48K5.550</td><td>8643.m00021 || Tc00.1047053511277.210</td><td>LmjF18.0580</td></tr>
<tr><td>Lipid</td><td>Peroxisomal enoyl-CoA hydratase</td><td>4.2.1.17</td><td>224.m00842 || Tb03.48K5.550</td><td>8643.m00021 || Tc00.1047053511277.210</td><td>LmjF18.0580</td></tr>
<tr><td>Lipid</td><td>Peroxisomal enoyl-CoA hydratase</td><td>4.2.1.17</td><td>None</td><td>8643.m00021 || Tc00.1047053511277.210</td><td>LmjF18.0580</td></tr>
</table>

Table 7.5 Continued

<table>
<tr><th></th><th></th><th></th><th colspan="3">GeneDB accession</th></tr>
<tr><th>Pathways</th><th>Enzyme name</th><th>EC number</th><th>T. brucei</th><th>T. cruzi</th><th>L. major</th></tr>
<tr><td>Lipid</td><td>Peroxisomal enoyl-CoA hydratase, probable</td><td>4.2.1.17 | 1.1.1.35 | 5.3.3.-)</td><td>None</td><td>8643.m00021 || Tc00.1047053511277.210</td><td>LmjF18.0580</td></tr>
<tr><td>Lipid</td><td>Possible mitochondrial trifunctional enzyme alpha subunit precursor</td><td></td><td>199.m00405 |Tb927.2.4130 | 28H13.415</td><td>7378.m00004 ||Tc00.1047053507547.40</td><td>LmjF33.2600</td></tr>
<tr><td>Lipid isoprenoids sterol</td><td>Mevalonate kinase</td><td>2.7.1.36</td><td>Tb04.1D20.470</td><td>8046.m00001</td><td>LmjF31.0560</td></tr>
<tr><td>Lipid metabolism</td><td>3-Ketoacyl-CoA thiolase, possible</td><td>2.3.1.16</td><td>228.m00365 || Tb08.26A17.500</td><td>8107.m00003 || Tc00.1047053509463.30</td><td>LmjF23.0690</td></tr>
<tr><td>Lipid metabolism</td><td>3-Ketoacyl-CoA thiolase, possible</td><td></td><td>228.m00365 || Tb08.26A17.500</td><td>8107.m00003 || Tc00.1047053509463.30</td><td>LmjF23.0690</td></tr>
<tr><td>Lipid metabolism</td><td>Acetyl-coenzyme A carboxylase like carboxylase, possible</td><td></td><td>228.m01122 || Tb08.10K10.720</td><td>7702.m00005 || Tc00.1047053508369.50</td><td>LmjF31.2970</td></tr>
<tr><td>Lipid metabolism</td><td>Acyl-COA binding protein, probable</td><td></td><td>Tb04.29M18.660</td><td>Tc00.1047053510877.55</td><td>LmjF34.2600</td></tr>
<tr><td>Lipid metabolism</td><td>Acyl-CoA dehydrogenase, putative (very long chain)</td><td>1.3.99.3</td><td>237.m00364 || Tb11.01.3640</td><td>7887.m00004 || Tc00.1047053508827.40</td><td>LmjF28.2510</td></tr>
<tr><td>Lipid metabolism</td><td>Acyltransferase</td><td></td><td>Tb03.48K5.450</td><td>Tc00.1047053508153.100</td><td>LmjF29.2280</td></tr>
<tr><td>Lipid metabolism</td><td>Beta-ketoacyl (ACP) synthase family protein</td><td>2.3.1.41</td><td>199.m00383 | Tb927.2.3910</td><td>8600.m00012 || Tc00.1047053511109.120</td><td>LmjF33.2720</td></tr>
<tr><td>Lipid metabolism</td><td>Beta-ketoacyl synthase family protein, putative, probable</td><td>2.3.1.42</td><td>199.m00383 | Tb927.2.3910 | 28H13.305</td><td>8600.m00012 || Tc00.1047053511109.120</td><td>LmjF33.2720</td></tr>
<tr><td>Lipid metabolism</td><td>C8 sterol isomerase</td><td></td><td>Tb03.48K5.130</td><td>8364.m00009</td><td>LmjF29.2140</td></tr>
<tr><td>Lipid metabolism</td><td>C8 sterol isomerase</td><td></td><td>224.m00800 || Tb03.48K5.130</td><td>8364.m00009 || Tc00.1047053510329.90</td><td>LmjF29.2140</td></tr>
</table>

<table>
<tr><td>Lipid metabolism</td><td>Fatty acid elongase (very long chain)</td><td></td><td>131.t00014</td><td>Tc00.1047053509539.30</td><td>Tb05.26C7.100</td></tr>
<tr><td>Lipid metabolism</td><td>Thiolase protein-like (most likely trifunctional protein Beta subunit)</td><td></td><td>None</td><td>8680.m00015 || Tc00.1047053511389.150</td><td>LmjF31.1640</td></tr>
<tr><td>Lipid metabolism beta oxidation</td><td>Acyl-CoA dehydrogenase (long-chain specific)</td><td>1.3.99.3</td><td>291.m00026 || Tb11.55.0026</td><td>7029.m00022 || Tc00.1047053506629.220</td><td>LmjF27.0930</td></tr>
<tr><td>Lipid metabolism beta oxidation</td><td>Acyl-CoA dehydrogenase (short/branched chain specific)</td><td>1.3.99.3</td><td>291.m00026 || Tb11.55.0026 ?</td><td>7168.m00008 || Tc00.1047053507001.80</td><td>LmjF35.2730</td></tr>
<tr><td>Lipid metabolism Ether-lipid</td><td>Alkyl Dihydroxyacetonephosphate synthase</td><td>2.5.1.26</td><td>Tb06.28P18.170</td><td>4766.m00001</td><td>LmjF30.0120</td></tr>
<tr><td>Lipid metabolism ether-lipid</td><td>Alkyldihydroxyacetonephosphate synthase</td><td>2.5.1.26</td><td>226.m00335 || Tb06.28P18.170</td><td>6711.m00011 || Tc00.1047053505807.110</td><td>LmjF30.0120</td></tr>
<tr><td>Lipid metabolism ether-lipid</td><td>DHAP acyltransferase</td><td>2.3.1.42</td><td>169.m00598 || Tb04.2H8.550</td><td>4886.m00004 || Tc00.1047053504055.40</td><td>LmjF34.1090</td></tr>
<tr><td>Lipid metabolism ether-lipid</td><td>Dihydroxyacetone phosphate acyltransferase, possible</td><td>2.3.1.42</td><td>169.m00598 || Tb04.2H8.550</td><td>4886.m00004 || Tc00.1047053504055.40</td><td>LmjF34.1090</td></tr>
<tr><td>Lipid metabolism ether-lipid Phospholipid</td><td>1-Acyl-sn-glycerol-3-phosphate acyltransferase-like protein</td><td>2.3.1.23</td><td>237.m00680 || Tb11.01.6800</td><td>8017.m00006 || Tc00.1047053509157.60</td><td>LmjF32.1960</td></tr>
<tr><td>Lipid metabolism ether-lipid Phospholipid</td><td>Dihydroxyacetone phosphate acyltransferase, possible</td><td>2.3.1.42</td><td>Tb04.2H8.550</td><td>4886.m00004</td><td>LmjF34.1090</td></tr>
<tr><td>Lipid metabolism|folate metabolism|pteridine</td><td>Pteridine reductase?</td><td>1.5.1.33</td><td>228.m00311 || Tb08.26N11.790</td><td>none</td><td>LmjF23.0270 |PTR1</td></tr>
<tr><td>Lipid synthesis eukaryotic</td><td>3-Oxoacyl-(acyl-carrier protein) reductase, probable</td><td></td><td>none</td><td>none</td><td>LmjF24.2030</td></tr>
<tr><td>Oxidoreductase | lipid metabolism</td><td>Putative electron transfer flavoprotein beta-subunit
beta-etf-flavoprotein small subunit</td><td></td><td>Tb11.03.0950</td><td>Tc00.1047053504109.160</td><td>LmjF25.0120</td></tr>
</table>

acid synthesis, but in *T. brucei* these elongases, which have an overlapping specificity, function in extending butyryl-CoA stepwise to myristate or stearate (Paul *et al.*, 2001). A similar set of elongases is present in the *L. major* genome, except that *L. major* has 12 tandemly linked homologous genes, rather than four, which may explain why in *Leishmania* much longer fatty acids can be synthesized. Despite its uniqueness in nature it remains to be seen whether in *Leishmania* this pathway of fatty acid synthesis could function as a drug target. As will be discussed below, the oxidation of fatty acids is an import process in the amastigote and therefore fatty acid synthesis is expected to play a minor role in the metabolism of this pathogenic stage.

A type II fatty acid synthesis pathway with an acetyl-CoA carboxylase, acyl carrier protein (ACP), a keto-acyl (ACP) synthase, a keto-acyl (ACP) reductase and *trans*-2-enoyl (ACP) reductase (van Weelden *et al.*, 2005) is also present in *Leishmania*. The latter gene predicts a mitochondrial transit peptide, which suggests that the *Leishmania* mitochondrion contains a separate pathway for the synthesis of fatty acids, as has been shown in the case of *T. brucei*. Most likely this pathway is used for the synthesis of lipoic acid, an essential cofactor of the mitochondrial enzymes pyruvate dehydrogenase and 2-ketoglutarate dehydrogenase.

In addition, *L. major* synthesizes polyunsaturated fatty acids by using Δ6, Δ5 and Δ4 desaturases, and elongases specific for C_{18} Δ6 and C_{20} Δ5 polyunsaturated fatty acids. This suggests that a complete pathway for polyunsaturated fatty acid biosynthesis is also present (Livore *et al.*, 2007). Whether this pathway operates in the amastigote is not known.

Lipids as carbon sources for energy generation and biosynthetic pathways

L. mexicana amastigotes have been shown to contain higher activities than promastigotes of the enzymes that catalyse the β-oxidation of fatty acids. This suggests that β-oxidation of fatty acids is relatively more important to intracellular amastigotes living in a glucose-poor environment than promastigotes (Hart and Coombs, 1982; Coombs *et al.*, 1982). A comparison of several pathogens, including *Leishmania*, that occupy phagosomal compartments inside mammalian macrophages (Lorenz and Fink, 2002; Lorenz *et al.*, 2004; Munoz-Elias and McKinney, 2005) suggests that for organisms thriving in such a seemingly rather inhospitable niche, fatty acid oxidation is critical for viability and virulence. Genome analysis shows that, in comparison with that of the other trypanosomatids, the *Leishmania* genome is particularly rich in genes involved in lipid metabolism, probably allowing it to degrade a wide variety of fatty acids. Acetyl-CoA, the end-product of β-oxidation, is in general oxidized to carbon dioxide and water by the TCA cycle. There are some doubts as to whether the cycle functions as a cycle (see above). Acetyl-CoA can also be converted to acetate with the net synthesis of one molecule of ATP, via the succinyl-CoA synthase/ acetate-succinate CoA transferase cycle. The *L. major* genome has not revealed the presence of a functional glyoxylate cycle. This suggests that *Leishmania* would be unable to form sugars from fatty acids as sole carbon source and requires in addition an oxaloacetate-forming amino acid. After condensation of acetyl-CoA with oxaloacetate, the resulting citrate can be transported to the cytoplasm for the use in other reactions.

Synthesis of isoprenoids

Major components of cellular membranes in eukaryotic cells include sterols and phospholipids. Metabolic routes leading to the formation of sterols and other isoprenoids in *Leishmania* have been the subject of many biochemical studies that predated the availability of the trypanosomatid genome sequences and have been reviewed in detail elsewhere (Urbina, 1997). The availability of genome sequences now provides a fast track for cloning enzymes from an isoprenoid biosynthetic pathway that has considerable potential as a target for new trypanocidal drugs.

Synthesis of phospholipids

The first steps in the synthesis of phosphatidic acid, the intermediate common to triacylglycerol and glycerophospholipids biosynthesis, take place inside glycosomes. Glycerol 3-phosphate (G3P) is either formed by reduction of dihydroxyacetone-phosphate (DHAP), catalysed by a glycosomal glycerol-3-phosphate dehydrogenase (G3PDH), or by the phosphorylation of glycerol using a glycosomal glycerol kinase. Subsequently, either G3P or DHAP undergo an initial acylation reaction that uses a fatty acyl-CoA molecule as acyl group donor. Only one gene encoding an appropriate putative acyltransferase, glycosomal DHAP acyltransferase, was detected in the *L. major* genome (Table 7.6). This enzyme, which

Table 7.6 Trypanosomatid enzymes involved in phospholipd metabolism

Enzyme name	EC number	GeneDB accession		
		T. brucei	*T. cruzi*	*L. major*
1-Acyl-*sn*-glycerol-3-phosphate acyltransferase	2.3.1.23	Tb11.01.6800	Tc00.1047053509157.60	LmjF32.1960
Alkyldihydroxyacetonephosphate synthase	2.5.1.26	Tb06.28P18.170	Tc00.1047053505807.110	LmjF30.0120
cdp-Diacylglycerol synthetase	2.7.7.41	Tb07.8P12.300	Tc00.1047053511237.40	LmjF26.1620
Cholinephosphate cytidylyltransferase	2.7.7.15	Tb10.389.0730	Tc00.1047053509805.220	LmjF18.1330
DHAP acyltransferase	2.3.1.42	Tb04.2H8.550	Tc00.1047053504055.40	LmjF34.1090
Diacylglycerol cholinephosphotransferase	2.7.8.2	Tb10.6k15.1570	Tc00.1047053509791.150	LmjF36.5900
Dihydroxyacetone phosphate acyltransferase	2.3.1.42	Tb04.2H8.550	Tc00.1047053504055.40	LmjF34.1090
Ethanolamine-phosphate cytidylyltransferase	2.7.7.14	Tb11.01.5730	Tc00.1047053511727.120	LmjF32.0890
Glycerophosphoryl diester phosphodiesterase	3.1.4.46	Tb11.01.0580	Tc00.1047053511071.100	LmjF36.5960
Lysophospholipase	3.1.1.5	Tb08.11J15.160	Tc00.1047053511907.50	LmjF24.1840
Myo-inositol monophosphatase	3.1.3.25	Tb05.26K5.720	Tc00.1047053507047.120	LmjF17.1390
Phosphatidic acid phosphatase	3.1.3.4	Tb08.25E9.660	Tc00.1047053503809.110	LmjF19.1350
Phosphatidylethanolamine N-methyltransferase	2.1.1.17	Absent	Absent	LmjF31.3120
Phosphatidylinositol-specific phospholipase	3.1.4.11	Tb11.02.3780	Tc00.1047053504149.160	LmjF35.0040
Phosphatidylinositol synthase	2.7.8.11	Tb09.160.0530	Tc00.1047053510349.50	LmjF26.2480
Phosphatidylserine decarboxylase proenzyme	4.1.1.65	Tb09.211.1610	Absent	LmjF35.4590
Phosphatidylserine synthase	2.7.8.8	Tb07.28B13.50	Tc00.1047053509937.30	LmjF14.1200
Phosphoenolpyruvate mutase	5.4.2.9	Absent	Tc00.1047053511589.140	Absent
Phospholipase A1	3.1.1.32	Tb927.1.4830	Absent	Absent
Phospholipase A2	3.1.1.4	Tb09.211.3650	Tc00.1047053510743.60	LmjF35.3020
Phospholipase C	3.1.4.11	Tb11.02.3780	Tc00.1047053504149.160	LmjF30.2950
Phospholipase C	3.1.4.3	Absent	Absent	Absent
Phospholipase C glycosylphosphatidylinositol-specific (GPI-PLC)	3.1.4.3	Tb927.2.6000	Tc00.1047053506617.70	Absent
Phospholipase D	3.1.4.4	Absent	Absent	Absent

previously has been shown to be involved in the synthesis of ether lipids too (Heise and Opperdoes, 1997), resembles in many ways another, homologous G3P acyltransferase that is present in other organisms. However, no gene encoding an enzyme orthologous to the latter type of G3P acyltransferase was found in any trypanosomatid genome. Thus, it cannot be excluded that the glycosomal acyltransferase (LmjF34.1090) identified here has a wide substrate specificity accepting both DHAP and G3P as acyl acceptor. The second acyltransferase reaction is catalysed by lysophospholipid acyltransferase, or 1-acyl-sn-glycerol-3-phosphate acyltransferase. This enzyme, however, lacks a typical PTS and therefore one might assume that reactions of phospholipid assembly subsequent to that catalysed by DHAP acyltransferase probably do not occur in glycosomes. The formation of triacylglycerols and neutral phospholipids from phosphatidic acids begins with dephosphorylation, catalysed by a phosphatidate phosphatase. The resulting diacylglycerol can be directly acetylated to form a triacylglycerol by an acyltransferase or can react with CDP-choline or CDP-ethanolamine to form phosphatidylcholine or phosphatidylethanolamine, respectively. Ethanolamine-phosphate cytidylyltransferase and cholinephosphate cytidylyltransferase homologues were identified. In most organisms phosphatidylcholine can also be synthesized from phosphatidylethanolamine by a series of methylation reactions initiated by the enzyme phosphatidylethanolamine N-methyltransferase and this enzyme was detected in *L. major*, but not in *T. brucei* and *T. cruzi*. Finally, acidic phospholipids can be derived from phosphatidic acids after their reaction with CTP to form CDP-diacylglycerols. Both phosphatidylserine and phosphatidylinositol can be formed from the latter compounds since the corresponding phosphatidylserine and phosphatidylinositol synthases are present. Phosphatidylserine can be converted to phosphatidylethanolamine by a phosphatidylserine decarboxylase. The latter enzyme may be mitochondrial because it carries a predicted mitochondrial transit peptide. (Table 7.6).

Miltefosin (hexadecylphosphocholine, Impavido) is a lysophospholipid, which has been shown to interfere with the synthesis of phospholipids and sterols in both mammalian cells and *Leishmania* (Lux *et al.*, 2002; Rakotomanga *et al.*, 2007). The drug also inhibits cytochrome oxidase activity in promastigotes. (Luque-Ortega and Rivas, 2007). In *L. donovani* promastigotes miltefosin has been shown to significantly reduce the phosphatidylcholine content and to increase the phosphatidylethanolamine (PE) content in parasite membranes, suggesting that the drug may cause a partial inactivation of PE-N-methyltransferase (Rakotomanga *et al.*, 2007). Inhibition of N-methyltransferase activity by hexadecylphosphocholine has also been described in tumour cells (Jimenez-Lopez *et al.*, 2003). This enzyme was found in the *L. major* genome but interestingly it is absent in *T. cruzi* and *T. brucei* (see above). This may explain why miltefosin only finds application as a drug in leishmaniasis and not in trypanosomiasis. In hepatocytes the miltefosine molecule appeared to be slowly degraded by phospholipase-D-like cleavage resulting in the release of choline. Although a phospholipase D activity involved in signal transduction has been reported in the case of *L. donovani* (Blum *et al.*, 2001; Rakotomanga *et al.*, 2007), a corresponding gene was neither detected in the *L. major* genome, nor any of the other trypanosomatid genomes (see Table 7.6). This suggests that such an activity in *Leishmania* may be catalysed by a non-homologous enzyme. Also a phospholipase C gene was not detected. If indeed *Leishmania* possesses such an isofunctional protein with phospholipase activity, but without the ability to degrade this lysophospholipid analogue, then this could be an additional factor contributing to the efficacy of this drug against the parasite.

Amino acid metabolism

Catabolism

Amino acids (particularly proline, threonine and glutamic acid) may serve as important energy substrates for *Leishmania* in the midgut of the fly vector. L-Proline, the most abundant amino acid, is oxidized to glutamate via mitochondrial Δ-1-pyrroline-5-carboxylate dehydrogenase and pyrroline-5-carboxylate synthetase. The resulting glutamic acid is deaminated by a mitochondrial NAD-dependent glutamate dehydrogenase and then fed into the TCA cycle. All three trypanosomatids have a mitochondrial NAD-dependent glutamate dehydrogenase, but *T. cruzi* and *L. major* have an additional gene for a NADP-dependent isoenzyme which is probably present in the cytosol because it lacks a recognisable targeting peptide. This enzyme is closely related to the enzyme from γ-proteobacteria and has probably entered via lateral gene transfer. Aspartic acid is transaminated by a mitochondrial aspartate aminotransferase to oxaloacetate and fed into the cycle as well. Aspartate may also be converted to fumarate via a functional purine-nucleotide cycle, comprising the enzymes adenylosuccinate synthase, adenylosuccinate lyase and AMP-deaminase.

Glycine can be split by the glycine cleavage system into carbon dioxide and formic acid. Methionine is first converted in the cytosol to 2-ketobutyrate and the latter is converted in the mitochondrion to the TCA-cycle intermediate succinyl CoA. Isoleucine and valine, after transamination in the cytosol to their corresponding ketocarboxylic acids by a branched-chain aminotransferase, can be further oxidized in the mitochondrion via a short/branched-chain acyl-CoA dehydrogenase and a hydratase, to succinyl CoA and acetyl CoA. These enter the TCA cycle or the cycle catalysed by succinyl-CoA ligase and acetate: succinate CoA transferase, leading to the formation of acetate and ATP.

Branched-chain amino acid metabolism has been the subject of several studies in *Leishmania* (Blum, 1991; Ginger *et al.*, 1999; Ginger *et al.*, 2000; Ginger *et al.*, 2001). In animals and plants branched-chain amino acid metabolism occurs by a mitochondrial pathway and in trypanosomatids metabolism of these ketogenic amino acids provides substrates for the TCA and ASCT cycles. Some genes required for branched-chain amino acid metabolism (branched-chain 2-ketoamino-acid decarboxylase and 3-hydroxy-3-methylglutaryl (HMG) CoA lyase) encode potential mitochondrial transit peptides. In addition to a possible role as an energy-generating substrate, leucine, which is metabolized through the intermediacy of HMG-CoA, can also be incorporated directly into *Leishmania*'s sterol biosynthetic pathway (Ginger *et al.*, 2000; Ginger *et al.*, 2001). This direct route from leucine into the mevalonate pathway (which is the sole route used for isoprenoid biosynthesis in the trypanosomatids) has not been observed to operate in any other eukaryote, but is the major route for sterol biosynthesis for several *Leishmania* species, at least in cultured promastigotes. Previous immunolocalization of trypanosomatid HMG-CoA reductase, the key enzyme required for mevalonate formation, indicated that this enzyme is present in the mitochondrion (Pena-Diaz *et al.*, 2004). Thus this organelle seems to be a sub-cellular location for both HMGCoA reductase and leucine catabolism, which may be the explanation for the incorporation of leucine carbon into sterol by trypanosomatids.

Most of the genes of the classical pathways for aromatic amino acid oxidation are missing. Phenylalanine can be converted to tyrosine by a hydroxylase which is not present in the other trypanosomatids. Some genes from the anaerobic phenylalanine degradation pathway are present. Tyrosine is most likely converted to 4-hydroxyphenylpyruvate by two enzymes, a true tyrosine aminotransferase, probably also acting on phenylalanine and tryptophan, and an aspartate aminotransferase which in *T. cruzi* has been shown to have a broad substrate specificity and is capable of acting on tryptophan as well (Vernal *et al.*, 1998). The corresponding keto acids are most likely reduced by an aromatic hydroxyacid dehydrogenase, closely related to malate dehydrogenases. In *T. cruzi* this enzyme has been shown to have changed its substrate specificity and is called an aromatic α-hydroxyacid dehydrogenase (Vernal *et al.*, 2002). A homologue is present in the *L. major* genome as well. No enzymes of the classical pathway converting tyrosine to fumarate and acetoacetate were detected.

Leishmania contains several enzymes of the urea cycle, although a fully functional urea cycle is missing from *L. major*. The possible functions of these enzyme is discussed in the section oxidant stress protection below.

Leishmania lacks the enzymes of the major route of threonine degradation in mammals which includes the enzymes threonine dehydrogenase and aminoacetone synthase. Instead it has two alternative pathways. In the first one, it can convert threonine through a tetrahydrofolate (THF)-dependent pathway to glycine using the enzyme serine hydroxymethyltransferase

(SHMT) (also known as threonine aldolase). In addition to the demethylation of serine, SHMT is able to cleave the Cα–Cβ bond of threonine. The resulting glycine can be converted through serine to pyruvate by the THF-dependent glycine cleavage system (GCS) followed by a serine/threonine dehydratase (STD). Alternatively, the same dehydratase might convert threonine to 2-ketobutyrate, which is then oxidized to succinyl CoA. *T. brucei* lacks the dehydratase and SHMT, and seems to metabolize threonine using the alternative aminoacetone pathway, which involves a mitochondrial threonine dehydrogenase (TDH) and an aminoacetone synthase (2-amino-3-ketobutyrate CoA ligase). Thus, despite the fact that the parasite's threonine metabolism differs from that in the host, this would be less suitable as drug target because of the presence in *Leishmania* of alternative pathways.

Synthesis

The capacity for the synthesis of amino acids by *L. major* is limited to the non-essential amino acids *plus* threonine and methionine. The gene for D-3-phosphoglycerate dehydrogenase, the first enzyme committed to the synthesis of serine was found in *L. major* and a possible phosphoserine phosphatase homologue was detected with a BlastP e-value of 3e-5. Cysteine can be produced either from homocysteine by the trans-sulfuration pathway or by *de novo* synthesis from serine. Glycine can be formed from serine by a serine hydroxymethyl transferase. Alanine is formed from pyruvate by a cytosolic alanine aminotransferase and aspartate and asparagine can be formed from oxaloacetate by a mitochondrial aspartate aminotransferase and an asparagine synthase. Glutamate, glutamine and proline can be formed from 2-ketoglutarate, since the genes for all the necessary enzymes were found. Of the so-called essential amino acids, all genes coding for the enzymes for the synthesis of threonine were detected, but neither the aromatic amino acids (phenylalanine, tyrosine or tryptophan) nor any of the branched amino acids (leucine, isoleucine and valine) can be synthesized by *L. major*. Lysine and histidine cannot be formed as well. Methionine can both be synthesized *de novo* from homoserine or salvaged from homocysteine.

Purine salvage and pyrimidine synthesis

Several studies have shown that Trypanosomatidae are incapable of *de novo* purine synthesis, in common with other obligate parasitic protozoa (reviewed in: Marr and Berens, 1983). This has now been confirmed by the release of the trypanosomatid nuclear genome sequences. Only one of the 10 enzymes required to make inosine monophosphate (IMP) from phosphoribosyl pyrophosphate, adenylosuccinate lyase, is encoded in the *L. major* genome, and its presence is explained by the fact that this enzyme also plays a role in purine salvage by converting IMP to AMP (Van den Berghe *et al.*, 1997). Instead, genes encoding several enzymes involved in the interconversion of purine bases and nucleosides are present. Moreover, several of these purine salvage enzymes carry a PTS, confirming earlier claims that some of them are glycosomal (Hammond *et al.*, 1981). Trypanosomatids are able to synthesize pyrimidines *de novo* from glutamine, bicarbonate and aspartate by a pathway that involves six genes – carbamoylphosphate synthase, aspartate carbamoyltransferase, dihydroorotase, dihydroorotate dehydrogenase and orotidine-5′-phosphate decarboxylase/orotate phosphoribosyltransferase. In the *L. major* genome these genes are all present on chromosome 16 and tandemly linked. In mammals the first three enzyme activities are present as a multifunctional protein that arose as a consequence of gene fusions. The dihydroorotate product of these reactions is oxidized by a cytosolic dihydroorotate dehydrogenase, and the last two steps of pyrimidine synthesis are catalysed inside glycosomes by orotidine-5′-phosphate decarboxylase/orotate phosphoribosyltransferase, the genes for which are fused to encode a bifunctional protein. Interestingly, the orientation of enzyme activities on this fusion protein is opposite to the orientation present on the analogous multifunctional enzyme that is found in higher eukaryotes (Gao *et al.*, 1999). Thus far, this 'reverse fusion' is only found to be shared with the cyanobacteria, and may therefore be relevant to any discussion of the impact that lateral gene fusion has had upon the evolution of metabolism in the Trypanosomatidae (see below). In all eukaryotes, except the Kinetoplastida and yeasts, the enzyme dihydroorotae dehydrogenase is also intimately associated with the respiratory chain and utilizes ubiquinone as electron acceptor for the oxidation of dihydroorotate to orotic acid. *Leishmania* and all other Kinetoplastida (including the Bodonidae) have lost this enzyme in exchange for a cytosolic isofunctional enzyme of bacterial origin, which utilizes fumarate as electron acceptor. It has recently been shown that erythrocytic stages of the human malaria parasite *Plasmodium falciparum* maintain an active mitochondrial electron transport chain to serve just the regeneration of ubiquinone required for the proper functioning of this enzyme. Expression of the soluble yeast dihydroorotate dehydrogenase rendered *Plasmodium* completely resistant to inhibitors of mitochondrial electron transport (Painter *et al.*, 2007). This observation is of utmost importance, not only because it illustrates the vital role of this enzyme for the survival of a parasite and its potential as a drug target, but also because it helps us to understand a peculiarity of the biology of the trypanosomatids; it explains why dyskinetoplastic trypanosomatids, which have lost large parts of their mitochondrial genome and lack a functional mitochondrial respiratory chain, do occur in nature. The *L. major* enzyme has been overexpressed and characterized (Feliciano *et al.*, 2006) and the crystal structures of the corresponding *T. cruzi* (Inaoka *et al.*, 2005) and *T. brucei* enzyme (2B4G) with bound orotic acid and flavin mononucleotide have been solved. RNA interference against expression of the dihydroorotate dehydrogenase gene in bloodstream forms of *T. brucei* resulted in inhibition of parasite growth in culture medium depleted of pyrimidines (Fred Bruckner, personal communication). Thus, dihydroorotate dehydrogenase inhibitors can be useful, but only when combined with inhibitors of pyrimidine transport.

Protection against oxidant stress

Enzymes of urea cycle and putrescine formation

In all trypanosomatids amino acid catabolism leads to the formation of free ammonia which is excreted, but *Leishmania, Crithidia and Leptomonas* excrete urea, in addition to ammonia, which could be an indication for the presence of a functional urea cycle (Yoshida and Camargo, 1978). Indeed the *L. major* genome contains several genes for enzymes of the urea cycle which apparently have been acquired by horizontal gene transfer (e.g. an argininosuccinate synthase of clear bacterial signature and an arginase which clusters with 100% bootstrap support with the enzymes found in fungi). The latter enzyme is normally cytosolic, but in *Leishmania* it is present in the glycosomes (Roberts *et al.*, 2004). A fully functional urea cycle seems to be missing (Table 7.7 and Fig. 7.3). *L. major* lacks the genes coding for ornithine carbamoyltransferase and argininosuccinate lyase while *T. brucei and T. cruzi* lack all enzymes of the cycle. Thus in *Leishmania* arginine, ornithine and urea can be formed and this explains why *Leishmania* excretes urea when arginine is added to the growth medium, while arginine addition to members of the genus *Trypanosoma* does not lead to production of urea (Yoshida and Camargo,

Table 7.7 Enzymes of the urea cycle and putrescine synthesis

Enzyme name	EC number	GeneDB accession *T. brucei*	*T. cruzi*	*L. major*
Agmatinase	3.5.3.11	Tb08.26N11.490	Tc00.1047053510947.40	LmjF23.0070
Arginase	3.5.3.1	Absent	Absent	LmjF35.1480
Argininosuccinate lyase	4.3.2.1	Absent	Absent	Absent
Argininosuccinate synthase	6.3.4.5	Absent	Absent	LmjF23.0260
Carbamoyl-phosphate synthase (ammonia)	6.3.4.16	Tb05.6E7.720	Tc00.1047053508373.10	LmjF16.0590
Ornithine carbamoyltransferase	2.1.3.3	Absent	Absent	Absent
Ornithine decarboxylase	4.1.1.17	Tb11.01.5300	Absent	LmjF12.0280

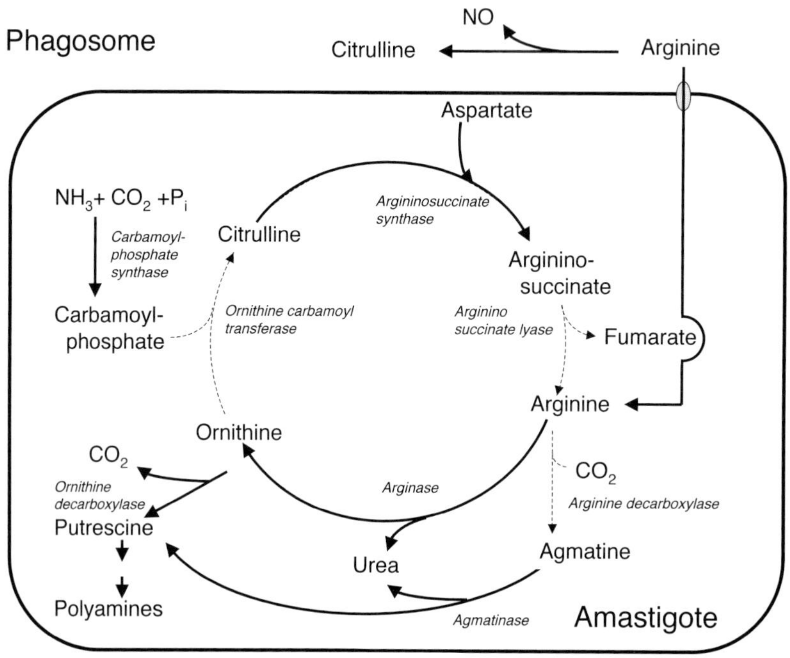

Figure 7.3 Enzymes of the urea cycle and polyamine biosynthesis in *Leishmania*. Broken arrows represent enzymes for which the corresponding genes were not detected. Enzyme names in italics.

1978). It has been suggested that the production of urea by *Leishmania* may be a special case of adaptation to life in an acidic phagolysosome because the release of ammonia in quantity could severely affect the vacuolar pH (Opperdoes and Coombs, 2007), but there may be another explanation for the acquisition of a few urea cycle enzymes by *Leishmania*.

Interferon-gamma, a Th1 cytokine normally associated with classical macrophage activation and killing of intracellular pathogens, induces a cationic amino acid transporter, which is responsible for cytokine-stimulated import of L-arginine by the macrophage. Macrophages require L-arginine for the production of nitric oxide (NO) by another inducible enzyme: NO synthase. Activated macrophages also produce the superoxide radical which combines with NO to form the even more toxic peroxynitrite ($OONO^-$). The latter rapidly decomposes into the highly reactive hydroxyl radical, OH· and NO_2. OH· radicals effectively kill most ingested pathogens. However, *Leishmania* survive this killing attack and the addition of arginine to infected macrophages even leads to a stimulation of parasite growth (Wanasen *et al.*, 2007). Little is known as to how *Leishmania* accomplishes this. The *Leishmania* genome gives some insight into this and even provides an explanation for how macrophage stimulation may support the growth of the intracellular parasite. The *L. major* genome has revealed the presence of a high affinity arginine transporter with acidic pH optimum which is most likely expressed in the intracellular stage (Shaked-Mishan *et al.*, 2006). This permease would allow the amastigote to scavenge from the macrophage the arginine destined for NO production and then, via its glycosomal arginase, neutralize the arginine by converting it into ornithine and urea. Ornithine is normally broken down via glutamate to ketoglutarate, but there is no evidence that this pathway operates in any of the trypanosomatids. One of the necessary enzymes, ornithine aminotransferase, was not detected. Yet, in view of the stimulation of parasite growth by arginine, it is tempting to assume that ornithine can be metabolized by an as yet unknown pathway. If not, ornithine is most likely excreted, or converted to polyamines, which either are utilized for trypanothione synthesis (see below) or excreted. Thus it is hypothesized that the function of the arginine transporter together with the glycosomal arginase are the lowering of the phagosomal arginine concentration by which the amastigote is able to protect itself against production of NO inside the phagosomes.

NO is a diatomic gaseous molecule similar to dioxygen and carbon monoxyde. These molecules can all be bound by haemoproteins such as haemoglobin and cytochromes. Indeed, in mammalian cells NO is known to function as a signal transducer by binding to a haem-containing high-affinity-binding site of a soluble guanylate cyclase (sGC) and so regulates cGMP formation. NO forms a tight 5-coordinate haem adduct with sGC (Lawson *et al.*, 2000). Thus, haem's strong affinity for NO may explain why *Leishmania*, in addition to two enzymes of the urea cycle, also has acquired by lateral gene transfer, several genes involved in haem biosynthesis. Maybe, in addition to the reduction of phagosomal arginine concentrations by the parasite's arginase, an as yet unidentified haemoprotein functions as an effective sink for any NO still being produced by the macrophage.

Thus, if confirmed, the arginase and the hypothetical haemoprotein could be interesting drug targets. Creation of arginase-deficient mutant parasites has indeed established that the enzyme is essential for *L. mexicana* promastigotes, but this is because they become auxotroph for ornithine (Roberts *et al.*, 2004). Unfortunately the effect of such deletion on the survival of amastigotes inside macrophages was not tested.

Polyamines

Putrescine can be synthesized by decarboxylation of ornithine (see above and section on oxidative stress), both in *Leishmania* and in *T. brucei*. However, an ornithine decarboxylase is absent from *T. cruzi*. There is some controversy as to whether in *T. cruzi* putrescine is synthesized by an alternative pathway comprising arginine decarboxylase plus agmatinase, instead of by the obviously absent ornithine decarboxylase (Hernandez *et al.*,1999). However, while a gene coding for an agmatinase is present, an arginine decarboxylase was not found in any of the trypanosomatids. This is in good agreement with the notion that *T. cruzi* is auxotrophic for polyamine and is able to salvage spermidine from the medium (Carrillo *et al.*, 2003).

Polyamines are essential for cell growth and differentiation. The amino acids arginine and ornithine are precursors for polyamine biosynthesis, and are necessary for the synthesis of trypanothione (Schmidt and Krauth-Siegel, 2002), a dithiol unique to trypanosomatids and which has a major role in protection against oxidative stress. The polyamine biosynthetic pathway is a potential target for drugs against various diseases caused by these protozoan parasites. The target has been validated using different enzyme inhibitors (Müller *et al.*, 2001), as well as null mutants of *L. donovani* (Roberts *et al.*, 2002, Jiang *et al.*, 1999, Roberts *et al.*, 2001). The first step in polyamine biosynthesis is carried out by ornithine decarboxylase and this enzyme is the target for the drug difluoromethylornithine (DFMO, eflornithine) in human sleeping sickness (Phillips *et al.*, 1987). The enzyme is also present in *L. major* (see also the section on the urea cycle). However, DFMO, which irreversibly inactivates its target enzyme, is not very effective against trypanosomatids other than *T. brucei*, indicating important differences. Indeed the *T. brucei* enzyme is the result of a relatively recent event of horizontal gene transfer from a vertebrate to an African trypanosome and is distantly related to the enzyme from the other trypanosomatids (Steglich and Schaeffer, 2005). Another difference may be in the half-life of ornithine decarboxylase. A difference in turnover of ornithine decarboxylases from various trypanosomatids has been demonstrated and the *C. fasciculata* enzyme seems especially sensitive to degradation. Whether this is also the case for the *Leishmania* enzyme is likely but remains to be demonstrated (Persson, 2007). S-Adenosine methionine decarboxylase catalyses the production of decarboxylated adenosyl methionine, which provides the aminopropyl groups in the synthesis of spermidine and spermine from putrescine and spermidine respectively. S-Adenosine methionine decarboxylase in trypanosomatids has been demonstrated to be a potential target for drugs, but the information about the enzyme is still limited. A candidate aminopropyltransferase and a trypanothione synthetase are present in *L. major* as well.

Cysteine can be produced from homocysteine by the trans-sulfuration pathway present in all three organisms. *De novo* synthesis from serine appears possible in *L. major*. The enzyme responsible for the conversion, cysteine synthase, is a typical plant-like enzyme, further only found in cyanobacteria, algae and plants. *L. major* can also interconvert glycine and serine via serine hydroxymethyl transferase. Methionine (as decarboxylated AdoMet), cysteine, glutamate and glycine are the precursor amino acids for glutathione. Although the latter is present in trypanosomatids, it is also used, together with spermidine, for the synthesis of trypanothione. This unique thiol has taken over many of the protective, metabolic and antioxidant functions of glutathione.

Trypanothione

Trypanothione is essential for the defence against chemical and oxidant stress in *Leishmania*. It is also the target for trivalent antimonial drugs which form conjugates with trypanothione and/or glutathione (Fairlamb, 2003). Methylglyoxal, a reactive by-product of the glycolytic pathway, can be converted via a trypanothione-dependent glyoxalase pathway (Vickers *et al.*, 2004) to D-lactate, an end product of *Leishmania* carbohydrate catabolism. Homologues of the *T. cruzi* glyoxalase I and II are present in *L. major*. See the section on the methylglyoxal bypass above.

Superoxide dismutase

Leishmania has four superoxide dismutases (SODA, two SODBs and an SODC). They are all of the iron type (Dufernez *et al.*, 2005). These Fe-SODs are typical prokaryotic enzymes and again could have entered the Trypanosomatidae by one or more events of horizontal transfer followed by gene duplications (Dufernez *et al.*, 2005). SODA and C are putative mitochondrial isoenzymes and the SODBs are predicted to be glycosomal and cytoplasmic, respectively. There is no evidence for the presence of either an Mn-dependent enzyme (Dufernez *et al.*, 2005) or Cu/Zn-superoxide dismutase. Several SOD deletion mutants have been generated. In general such mutants have shown to be viable, although they exhibit an increased sensitivity to oxidative stress. All pathogenic trypanosomatids lack catalase. For removal of peroxides they are dependent on trypanothione-linked peroxidases. Tryparedoxins are homologous to thioredoxin and trypanosomes contain multiple copies of both kinds of redox protein. Thioredoxin reductase is absent and the trypanosomatid peroxidases are coupled to trypanothione/trypanothione reductase (Opperdoes and Szikora, 2006).

NADPH production

Promastigotes of glucose-transporter null-mutants have a reduced capacity to generate the NADPH required for protection against reactive oxygen species and are significantly more susceptible to oxidative stress (Rodriguez-Contreras *et al.*, 2007), suggesting that enzymes of the oxidative branch of the pentose-phosphate pathway play a vital role in these cells. It is likely that this is even more the case for amastigotes which, in the phagosome, are exposed to even much higher levels of oxidative stress (Haas, 2007). However, because of the fact that in *Leishmania* multiple copies of the various enzymes involved in oxidative stress protection were identified this may not be an easy drug target (see above).

Ascorbate metabolism

Almost all animals synthesize ascorbate but primates (including humans) need vitamin C (ascorbic acid) as an external source. Ascorbate is essential for the production and maintenance of collagen, the hydroxylation of amino acids and the defence against oxidants. Trypanosomatids have no collagen. Since the only amino acid hydroxylation reaction identified in *Leishmania* is the conversion of phenylalanine into tyrosine, a reaction that does not occur in the trypanosomes and whose function in *L. major* is far from clear, ascorbate's only obvious function in trypanosomatids should be oxidative stress protection. This is supported by the presence of an ascorbate-dependent peroxidase in *L. major* and *T. cruzi* (Wilkinson *et al.*, 2002). *Leishmania* amastigotes survive inside the phagosome of the macrophage, which is a hostile environment known to be the site of generation of reactive oxygen species. Genome mining has identified a gulono-1,4-lactone oxidase homologue, the last enzyme in the formation of ascorbate. This enzyme has a PTS in both trypanosomes and

Leishmania, strongly suggesting that it is a glycosomal enzyme. Experimental evidence obtained for trypanosomes indicates that a fungal type of ascorbate biosynthesis may be operational in their glycosomes (Wilkinson *et al.*, 2005). Although many enzymes of both the animal-type and plant-type ascorbate synthetic pathways leading to the synthesis of ascorbic acid could be identified, the database lacked one or more essential enzymes of each pathway. Moreover, the trypanosome oxidase has been shown to prefer D-arabino-1,4-lactone rather than gulono-1,4-lactone as substrate (Wilkinson *et al.*, 2005). This is suggestive evidence for the operation of a yeast-type rather than an animal- or plant-type pathway in trypanosomatids, because the yeast pathway results in the formation of erythroascorbate from D-arabinose. Indeed, the trypanosomatid genomes contain a gene that encodes a homologue of NADP-dependent arabinose dehydrogenase which catalyses the formation of D-arabinono-1,4-lactone from D-arabinose (this enzyme is annotated as prostaglandin f2-alpha synthase, LmjF31.2150). The *Leishmania* enzyme also has a potential PTS, suggesting that erythroascorbate, rather than ascorbate, is generated within glycosomes as an antioxidant. D-arabinose is a rare sugar which is only known in polysaccharides of plants and so far there are no obvious candidate enzymes in trypanosomatids that are able to form D-arabinose from other C5 sugars. Interestingly, however, *L. major* has recently been reported to incorporate D-arabinose into lipophosphoglycan side chains (Dobson *et al.*, 2003). Since humans (and primates) have lost the capacity to synthesize ascorbic acid, due to a mutation in the gulonolactone oxidase gene, the trypanosomatid enzyme may be an interesting drug target. It also remains to be seen whether the only known amino-acid hydroxylation reaction to occur in *L. major*, the conversion of phenylalanine to tyrosine by phenylalanine hydroxylase, requires (erythro)ascorbate for catalysis. A thiol-dependent enzyme possibly involved in the reduction of dehydroascorbic acid has also been reported (Denton *et al.*, 2004), Interestingly, it is upregulated in amastigotes and could be involved in mediating sensitivity of this stage to antimonial drugs.

Acknowledgements

The authors like to thank the support and advice by the teams of the TIGR and Sanger centres for genome sequencing and Jean-Pierre Szikora for his bioinformatic support. The authors are supported by an IUAP grant from the Belgian government and received grants from the European Commission through its International Scientific Cooperation Projects (INCO).

References

Allen, J.W., Daltrop, O., Stevens, J.M., and Ferguson, S.J. (2003). C-type cytochromes: diverse structures and biogenesis systems pose evolutionary problems. Philos. Trans. R. Soc. Lond. B Biol. Sci. *358*, 255–266.

Ariza, A., Vickers, T.J., Greig, N., Armour, K.A., Dixon, M.J., Eggleston, I.M., Fairlamb, A.H., and Bond, C.S. (2006). Specificity of the trypanothione-dependent *Leishmania major* glyoxalase I: structure and biochemical comparison with the human enzyme. Mol. Microbiol. *59*, 1239–1248

Bakker, B.M., Michels, P.A.M., Opperdoes, F.R., and Westerhoff, H.V. (1997). Glycolysis in bloodstream form *Trypanosoma brucei* can be understood in terms of the kinetics of the glycolytic enzymes. J. Biol. Chem. *272*, 3207–3215.

Basselin, M., Denise, H., Coombs, G.H., and Barrett, M.P. (2002). Resistance to pentamidine in *Leishmania mexicana* involves exclusion of the drug from the mitochondrion. Antimicrob. Agents Chemother. *46*, 3731–3738.

Berriman, M., Ghedin, E., Hertz-Fowler, C., Blandin, G., Renauld, H., Bartholomeu, D.C., Lennard, N.J., Caler, E., Hamlin, N.E., Haas, B., Bohme, U., Hannick, L., Aslett, M.A., Shallom, J., Marcello, L., Hou, L., Wickstead, B., Alsmark, U.C., Arrowsmith, C., Atkin, R.J., Barron, A.J., Bringaud, F., Brooks, K., Carrington, M., Cherevach, I., Chillingworth, T.J., Churcher, C., Clark, L.N., Corton, C.H., Cronin, A., Davies, R.M., Doggett, J., Djikeng, A., Feldblyum, T., Field, M.C., Fraser, A., Goodhead, I., Hance, Z., Harper, D., Harris, B.R., Hauser, H., Hostetler, J., Ivens, A., Jagels, K., Johnson, D., Johnson, J., Jones, K., Kerhornou, A.X., Koo, H., Larke, N., Landfear, S., Larkin, C., Leech, V., Line, A., Lord, A., Macleod, A., Mooney, P.J., Moule, S., Martin, D.M., Morgan, G.W., Mungall, K., Norbertczak, H., Ormond, D., Pai, G., Peacock, C.S., Peterson, J., Quail, M.A., Rabbinowitsch, E., Rajandream, M.A., Reitter, C., Salzberg, S.L., Sanders, M., Schobel, S., Sharp, S., Simmonds, M., Simpson, A.J., Tallon, L., Turner, C.M., Tait, A., Tivey, A.R., Van Aken, S., Walker, D., Wanless, D., Wang, S., White, B., White, O., Whitehead, S., Woodward, J., Wortman, J., Adams, M.D., Embley, T.M., Gull, K., Ullu, E., Barry, J.D., Fairlamb, A.H., Opperdoes, F., Barrell, B.G., Donelson, J.E., Hall, N., Fraser, C.M., Melville, S.E., and El-Sayed, N.M. (2005). The genome of the African trypanosome *Trypanosoma brucei*. Science *309*, 416–422.

Besteiro, S., Biran, M., Biteau, N., Coustou, V., Baltz, T., Canioni, P., and Bringaud, F. (2002). Succinate secreted by *Trypanosoma brucei* is produced by a novel and unique glycosomal enzyme, NADH-dependent fumarate reductase. J. Biol. Chem. *277*, 38001–38012.

Beverley, S.M., Ellenberger, T.E., and Cordingley, J.S. (1986). Primary structure of the gene encoding the bifunctional dihydrofolate reductase-thymidylate synthase of *Leishmania major*. Proc. Natl. Acad. Sci. USA *83*, 2584–2588.

Blum, J.J. (1991). Oxidation of leucine by *Leishmania donovani*. J. Protozool. *38*, 527–531.

Blum, J.J., and Opperdoes, F.R. (1994). Secretion of sucrase by *Leishmania donovani*. J. Eukaryot. Microbiol. *41*, 228–231.

Blum, J.J., Lehman, J.A., Horn, J.M., and Gomez-Cambronero, J. (2001). Phospholipase D (PLD) is present in *Leishmania donovani* and its activity increases in response to acute osmotic stress. J. Eukaryot. Microbiol. *48*, 102–110.

Bochud-Allemann, N., and Schneider, A. (2002). Mitochondrial substrate level phosphorylation is essential for growth of procyclic *Trypanosoma brucei*. J. Biol. Chem. *277*, 32849–32854.

Bringaud, F., Riviere L. and Coustou, V. (2006). Energy metabolism of trypanosomatids: adaptation to available carbon sources. Mol. Biochem. Parasitol. *149*, 1–9.

Burchmore, R.J., and Barrett, M.P. (2001). Life in vacuoles – nutrient acquisition by Leishmania amastigotes. Int. J. Parasitol. *31*, 1311–1320.

Burchmore, R.J., and Hart, D.T. (1995). Glucose transport in amastigotes and promastigotes of *Leishmania mexicana mexicana*. Mol. Biochem. Parasitol. *74*, 77–86.

Burchmore, R.J., and Landfear, S.M. (1998). Differential regulation of multiple glucose transporter genes in *Leishmania mexicana*. J. Biol. Chem. *273*, 29118–29126.

Burchmore, R.J., Rodriguez-Contreras, D., McBride, K., Merkel, P., Barrett, M.P., Modi, G., Sacks, D., and Landfear, S.M. (2003). Genetic characterization of glucose transporter function in *Leishmania mexicana*. Proc. Natl. Acad. Sci. USA. 100 : 3901–3906.

Carrillo, C., Cejas, S., Huber, A., Gonzalez, N.S., and Algranati, I.D. (2003). Lack of arginine decarboxylase in Trypanosoma cruzi epimastigotes. J. Eukaryot. Microbiol. *50*, 312–316.

Chang, C.S., and Chang, K.P. (1985). Heme requirement and acquisition by extracellular and intracellular stages of *Leishmania mexicana amazonensis*. Mol. Biochem. Parasitol. *16*, 267–276.

Chen, M., Zhai, L., Christensen, S.B., Theander, T.G., and Kharazmi, A. (2001). Inhibition of fumarate reductase in *Leishmania major* and, *L. donovani* by chalcones. Antimicrob. Agents Chemother. *45*, 2023–2029.

Chevalier, N., Bertrand, L., Rider, M.H., Opperdoes, F.R., Rigden, D.J., and Michels, P.A.M. (2005). 6-Phosphofructo-2-kinase and fructose-2, 6-bisphosphatase in Trypanosomatidae. Molecular characterization, database searches, modelling studies and evolutionary analysis. FEBS J. *272*, 3542–3560.

Choe, J., Moyersoen, J., Roach, C., Carter, T.L., Fan, E., Michels, P.A.M., and Hol, W.G.J. (2003). Analysis of the sequence motifs responsible for the interactions of peroxins 14 and 5, which are involved in glycosome biogenesis in *Trypanosoma brucei*. Biochemistry *42*, 10915–10922.

Coombs, G.H., Craft, J.A., and Hart, D.T. (1982). A comparative study of *Leishmania mexicana* amastigotes and promastigotes. Enzyme activities and subcellular locations. Mol. Biochem. Parasitol. 5, 199–211.

Coombs, G.H., Tetley, L., Moss, V.A., and Vickerman, K. (1986). Three-dimensional structure of the *Leishmania* amastigote as revealed by computer-aided reconstruction from serial sections. Parasitology 92, 13–23.

Cunningham, M.L., and Beverley, S.M. (2001). Pteridine salvage throughout the *Leishmania* infectious cycle: implications for antifolate chemotherapy. Mol. Biochem. Parasitol. *113*, 199–213.

Darling, T.N., and Blum, J.J. (1988). D-lactate production by *Leishmania braziliensis* through the glyoxalase pathway. Mol. Biochem. Parasitol. *28*, 121–127.

Denton, H., McGregor, J.C., and Coombs, G.H. (2004). Reduction of anti-leishmanial pentavalent antimonial drugs by a parasite-specific thiol-dependent reductase, TDR1. Biochem. J. *381*, 405–412.

Dobson, D.E., Mengeling, B.J., Cilmi, S., Hickerson, S., Turco, S.J., and Beverley, S.M. (2003). Identification of genes encoding arabinosyltransferases (SCA) mediating developmental modifications of lipophosphoglycan required for sandfly transmission of *Leishmania major*. J. Biol. Chem. *278*, 28840–28848.

Dufernez, F., Yernaux, C., Gerbod, D., Noel, C., Chauvenet, M., Wintjens, R., Edgcomb, V.P., Capron, M., Opperdoes, F.R., and Viscogliosi, E. (2005). The presence of four iron-containing superoxide dismutase isozymes in Trypanosomatidae: Characterization, subcellular localization, and phylogenetic origin in *Trypanosoma brucei*. Free Radic. Biol. Med. *40*, 210–225.

Fairlamb, A.H. (2003). Chemotherapy of human African trypanosomiasis: current and future prospects. Trends Parasitol. *19*, 488–494.

Feliciano, P.R., Cordeiro, A.T., Costa-Filho, A.J., and Nonato, M.C. (2006). Cloning, expression, purification, and characterization of *Leishmania major* dihydroorotate dehydrogenase. Protein Expr. Purif. *48*, 98–103.

Flaspohler, J.A., Rickoll, W.L., Beverley, S.M., and Parsons, M. (1997). Functional identification of a *Leishmania* gene related to the peroxin 2 gene reveals common ancestry of glycosomes and peroxisomes. Mol. Cell. Biol. *17*, 1093–1101.

Flaspohler, J.A., Lemley, K., and Parsons, M. (1999). A dominant negative mutation in the *GIM1* gene of *Leishmania donovani* is responsible for defects in glycosomal protein localization. Mol. Biochem. Parasitol. *99*, 117–128.

Furuya, T., Kessler, P., Jardim, A., Schnaufer, A., Crudder, C., and Parsons, M. (2002). Glucose is toxic to glycosome-deficient trypanosomes. Proc. Natl. Acad. Sci. USA 99, 14177–14182.

Gao, G., Nara, T., Nakajima-Shimada, J., and Aoki, T. (1999). Novel organization and sequences of five genes encoding all six enzymes for *de novo* pyrimidine biosynthesis in *Trypanosoma cruzi*. J. Mol. Biol. *285*, 149–161.

Gardner, M.J., Hall, N., Fung, E., White, O., Berriman, M., Hyman, R.W., Carlton, J.M., Pain, A., Nelson, K.E., Bowman, S., Paulsen, I.T., James, K., Eisen, J.A., Rutherford, K., Salzberg, S.L., Craig, A., Kyes, S., Chan, M.S., Nene, V., Shallom, S.J., Suh, B., Peterson, J., Angiuoli, S., Pertea, M., Allen, J., Selengut, J., Haft, D., Mather, M.W., Vaidya, A.B., Martin, D.M., Fairlamb, A.H., Fraunholz, M.J., Roos, D.S., Ralph, S.A., McFadden, G.I., Cummings, L.M., Subramanian, G.M., Mungall, C., Venter, J.C., Carucci, D.J., Hoffman, S.L., Newbold, C., Davis, R.W., Fraser, C.M., and Barrell, B. (2002). Genome sequence of the human malaria parasite *Plasmodium falciparum*. Nature *419*, 498–511.

Ginger, M.L., Chance, M.L., and Goad, L.J. (1999). Elucidation of carbon sources used for the biosynthesis of fatty acids and sterols in the trypanosomatid *Leishmania mexicana*. Biochem. J. *342*, 397–405.

Ginger, M.L., Chance, M.L., Sadler, I.H., and Goad, L.J. (2001). The biosynthetic incorporation of the intact leucine skeleton into sterol by the trypanosomatid *Leishmania mexicana*. J. Biol. Chem. *276*, 11674–11682.

Ginger, M.L., Fairlamb, A.H., and Opperdoes, F.R. (2007). Comparative genomics of trypanosome metabolism. In: Trypanosomes after the Genome, Barry, D., McCulloch, R., Mottram, J., and Acosta-Serrano, A., eds. Horizon Bioscience, pp 373–416.

Ginger, M.L., Prescott, M.C., Reynolds, D.G., Chance, M.L., and Goad, L.J. (2000). Utilization of leucine and acetate as carbon sources for sterol and fatty acid biosynthesis by Old and New World *Leishmania* species, *Endotrypanum monterogeii* and *Trypanosoma cruzi*. Eur. J. Biochem. *267*, 2555–2566.

Guerra, D.G., Decottignies, A., Bakker, B.M., and Michels, P.A. (2006). The mitochondrial FAD-dependent glycerol-3-phosphate dehydrogenase of Trypanosomatidae and the glycosomal redox balance of insect stages of *Trypanosoma brucei* and *Leishmania* spp. Mol. Biochem. Parasitol. *149*, 155–169.

Haas, A. (2007). The Phagosome: Compartment with a License to Kill. Traffic *8*, 311–330.

Hammond, D.J., Gutteridge, W.E., and Opperdoes, F.R. (1981). A novel location for two enzymes of de novo pyrimidine biosynthesis in trypanosomes and *Leishmania*. FEBS Lett. *128*, 27–29.

Hart, D.T., and Coombs, G.H. (1982). *Leishmania mexicana*: energy metabolism of amastigotes and promastigotes. Exp. Parasitol. *54*, 397–409.

Hart, D.T., and Opperdoes, F.R. (1984). The occurrence of glycosomes (microbodies) in the promastigote stage of four major *Leishmania* species. Mol. Biochem. Parasitol. *13*, 159–172.

Heise, N., and Opperdoes, F.R. (1997). The dihydroxyacetonephosphate pathway for biosynthesis of ether lipids in *Leishmania mexicana* promastigotes. Mol. Biochem. Parasitol. *89*, 61–72.

Hernandez, S., and Schwarcz de Tarlovsky, S. (1999). Arginine decarboxylase in *Trypanosoma cruzi*, characteristics and kinetic properties. Cell. Mol. Biol. (Noisy-le-grand). 45, 383–391.

Hill, G.C., and White, D.C. (1968). Respiratory pigments of *Crithidia fasciculata*. J. Bacteriol. 95, 2151–2157.

Inaoka, D.K., Takashima, E., Osanai, A., Shimizu, H., Nara, T., Aoki, T., Harada, S., and Kita, K. (2005). Expression, purification and crystallization of *Trypanosoma cruzi* dihydroorotate dehydrogenase complexed with orotate. Acta Crystallograph Sect F Struct. Biol. Cryst. Commun. *61*, 875–878.

Irsch, T., and Krauth-Siegel, R.L. (2004). Glyoxalase II of African trypanosomes is trypanothione-dependent. J. Biol. Chem. *279*, 22209–22217.

Jardim, A., Liu, W., Zheleznova, E., and Ullman, B. (2000). Peroxisomal targeting signal-1 receptor protein PEX5 from *Leishmania donovani*. Molecular, biochemical, and immunocytochemical characterization. J. Biol. Chem. *275*, 13637–13644.

Jiang, Y., Roberts, S.C., Jardim, A., Carter, N.S., Shih, S., Ariyanayagam, M., Fairlamb, A.H., and Ullman, B. (1999). Ornithine decarboxylase gene deletion mutants of *Leishmania donovani*. J. Biol. Chem. *274*, 3781–3788.

Jiang, D.W., and Englund, P.T. (2001). Four *Trypanosoma brucei* fatty acyl-CoA synthetases: fatty acid specificity of the recombinant proteins. Biochem. J. *358*, 757–761.

Jimenez-Lopez, J.M., Carrasco, M.P., Segovia, J.L., and Marco, C. (2003). Hexadecylphosphocholine inhibits phosphatidylcholine synthesis via both the methylation of phosphatidylethanolamine and CDPcholine pathways in HePG2 cells. Int. J. Biochem. Cell Biol. *36*, 153–161.

Keegan, F.P., and Blum, J.J. (1992). Utilization of a carbohydrate reserve comprised primarily of mannose by *Leishmania donovani*. Mol. Biochem. Parasitol. 53, 193–200.

Keegan, F.P., and Blum, J.J. (1993). Incorporation of label from acetate and laurate into the mannan of *Leishmania donovani* via the glyoxylate cycle. J. Eukaryot. Microbiol. *40*, 730–732.

Klaus, S.M., Kunji, E.R., Bozzo, G.G., Noiriel, A., de la Garza, R.D., Basset, G.J., Ravanel, S., Rebeille, F., Gregory, J.F. 3rd and Hanson, A.D. (2005). Higher plant plastids and cyanobacteria have folate carriers related to those of trypanosomatids. J. Biol. Chem. *280*, 38457–38463.

Lawson, D.M., Stevenson, C.E., Andrew, C.R., and Eady, R.R. (2000). Unprecedented proximal binding of nitric oxide to heme: implications for guanylate cyclase. EMBO J. 19, 5661–5671.

Lee, S.H., Stephens, J.L., and Englund, P.T. (2007). A fatty-acid synthesis mechanism specialized for parasitism. Nat .Rev. Microbiol. 5, 287–297.

Lee, S.H., Stephens, J.L., Paul, K.S., and Englund, P.T. (2006). Fatty acid synthesis by elongases in trypanosomes. Cell *126*, 691–699.

Livore, V.I., Tripodi, K.E., and Uttaro, A.D. (2007). Elongation of polyunsaturated fatty acids in trypanosomatids. FEBS J. *274*, 264–274.

Lorenz, M.C., Bender, J.A., and Fink, G.R. (2004). Transcriptional response of *Candida albicans* upon internalization by macrophages. Eukaryot. Cell 3, 1076–1087.

Lorenz, M.C., and Fink, G.R. (2002). Life and death in a macrophage: role of the glyoxylate cycle in virulence. Eukaryot. Cell *1*, 657–662.

Luque-Ortega, J.R., and Rivas, L. (2007). Miltefosine (hexadecylphosphocholine) inhibits cytochrome c oxidase in *Leishmania donovani* promastigotes. Antimicrob. Agents Chemother. *51*, 1327–1332

Lux, H., Heise, N., Klenner, T., Hart, D.T., and Opperdoes, F.R. (2000). Ether-lipid (alkyl-phospholipid) metabolism and the mechanism of action of ether-lipid analogues in *Leishmania*. Mol. Biochem. Parasitol. *111*, 1–14.

Marr, J.J., and Berens, R.L. (1983). Pyrazolopyrimidine metabolism in the pathogenic trypanosomatidae. Mol. Biochem. Parasitol. *7*, 339–356.

Maugeri, D.A., Cazzulo, J.J., Burchmore, R.J., Barrett, M.P., and Ogbunude, P.O. (2003). Pentose phosphate metabolism in *Leishmania mexicana*. Mol. Biochem. Parasitol. *31*, 117–125.

McNicoll, F., Drummelsmith, J., Muller, M., Madore, E., Boilard, N., Ouellette, M., and Papadopoulou, B. (2006). A combined proteomic and transcriptomic approach to the study of stage differentiation in *Leishmania infantum*. Proteomics 6, 3567–3581.

Michels, P.A., Bringaud, F., Herman, M., and Hannaert, V. (2006). Metabolic functions of glycosomes in trypanosomatids. Biochim. Biophys. Acta *91763*, 1463–1477.

Mottram, J.C., and Coombs, G.H. (1985a). *Leishmania mexicana*: enzyme activities of amastigotes and promastigotes and their inhibition by antimonials and arsenicals. Exp. Parasitol. *59*, 151–160.

Mottram, J.C., and Coombs, G.H. (1985b). *Leishmania mexicana*: subcellular distribution of enzymes in amastigotes and promastigotes. Exp. Parasitol. *59*, 265–274.

Moyersoen, J., Choe, J., Fan, E., Hol, W.G.J., and Michels, P.A.M. (2004). Biogenesis of peroxisomes and glycosomes: trypanosomatid glycosome assembly is a promising new drug target. FEMS Microbiol. Rev. *28*, 603–643.

Mukherjee, A., Padmanabhan, P.K., Sahani, M.H., Barrett, M.P., and Madhubala, R. (2006). Roles for mitochondria in pentamidine susceptibility and resistance in *Leishmania donovani*. Mol. Biochem. Parasitol. *145*, 1–10.

Müller, S., Coombs, G.H., and Walter, R.D. (2001). Targeting polyamines of parasitic protozoa in chemotherapy. Trends Parasitol. *17*, 242–249.

Munoz-Elias, E.J., and McKinney, J.D. (2005). *Mycobacterium tuberculosis* isocitrate lyases 1 and 2 are jointly required for in vivo growth and virulence. Nat. Med. *11*, 638–644.

Naderer, T., Ellis, M.A., Sernee, M.F., De Souza, D.P., Curtis, J., Handman, E., and McConville, M.J. (2006). Virulence of *Leishmania major* in macrophages and mice requires the gluconeogenic enzyme fructose-*1*, 6-bisphosphatase. Proc. Natl. Acad. Sci. USA *103*, 5502–5507.

Nare, B., Hardy, L.W., and Beverley, S.M. (1997). The roles of pteridine reductase 1 and dihydrofolate reductase-thymidylate synthase in pteridine metabolism in the protozoan parasite *Leishmania major*. J. Biol. Chem. *272*, 13883–13891.

Nolan, D.P., and Voorheis, H.P. (1992). The mitochondrion in bloodstream forms of *Trypanosoma brucei* is energized by the electrogenic pumping of protons catalysed by the F1F0-ATPase. Eur. J. Biochem. *209*, 207–216.

Opperdoes, F.R., and Coombs, G.H. (2007). Metabolism of *Leishmania*: proven and predicted. Trends Parasitol. *23*, 149–158.

Opperdoes, F.R., and Michels, P.A. (2001). Enzymes of carbohydrate metabolism as potential drug targets. Int. J. Parasitol. *31*, 482–490.

Opperdoes, F.R., and Szikora, J.P. (2006). In silico prediction of the glycosomal enzymes of *Leishmania major* and trypanosomes. Mol. Biochem. Parasitol. *147*, 193–206.

Ouellette, M., Drummelsmith, J., El-Fadili, A., Kundig C, Richard, D., and Roy, G. (2002). Pterin transport and metabolism in *Leishmania* and related trypanosomatid parasites. Int. J. Parasitol. *32*, 385–398.

Painter, H.J., Morrisey, J.M., Mather, M.W., and Vaidya, A.B. (2007). Specific role of mitochondrial electron transport in blood-stage *Plasmodium falciparum*. Nature. (2007). *446*, 88–91.

Paul, K.S., Jiang, D., Morita, Y.S., and Englund, P.T. (2001). Fatty acid synthesis in African trypanosomes: a solution to the myristate mystery. Trends Parasitol. *17*, 381–387.

Pena-Diaz, J., Montalvetti, A., Flores, C.L., Constan, A., Hurtado-Guerrero, R., De Souza, W., Gancedo, C., Ruiz-Perez, L.M., and Gonzalez-Pacanowska, D. (2004). Mitochondrial localization of the mevalonate pathway enzyme 3-Hydroxy-3-methyl-glutaryl-CoA reductase in the Trypanosomatidae. Mol. Biol. Cell *15*, 1356–1363.

Persson L. (2007). Ornithine decarboxylase and S-adenosylmethionine decarboxylase in trypanosomatids. Biochem. Soc. Trans. *35*, 314–317.

Phillips, M.A., Coffino, P., and Wang, C.C. (1987). Cloning and sequencing of the ornithine decarboxylase gene from *Trypanosoma brucei*. Implications for enzyme turnover and selective difluoromethylornithine inhibition. J. Biol. Chem. *262*, 8721–8727.

Priest, J.W., and Hajduk, S.L. (1992). Cytochrome c reductase purified from *Crithidia fasciculata* contains an atypical cytochrome c_1. J. Biol. Chem. *267*, 20188–20195.

Rakotomanga, M., Blanc, S., Gaudin, K., Chaminade, P., and Loiseau, P.M. (2007). Miltefosine affects the lipid metabolism in *Leishmania donovani* promastigotes. Antimicrob. Agents Chemother. *51*, 1425–1430

Ralton, J.E., Naderer, T., Piraino, H.L., Bashtannyk, T.A., Callaghan, J.M., and McConville, M.J. (2003). Evidence that intracellular beta1–2 mannan is a virulence factor in *Leishmania* parasites. J. Biol. Chem. *278*, 40757–40763.

Richard, D., Kundig, C., and Ouellette, M. (2002). A new type of high affinity folic acid transporter in the protozoan parasite *Leishmania* and deletion of its gene in methotrexate-resistant cells. J. Biol. Chem. *277*, 29460–29467.

Riviere, L., van Weelden, S.W., Glass, P., Vegh, P., Coustou, V., Biran, M., van Hellemond, J.J., Bringaud, F., Tielens, A.G., and Boshart, M. (2004). Acetyl: succinate CoA-transferase in procyclic *Trypanosoma brucei*. Gene identification and role in carbohydrate metabolism. J. Biol. Chem. *279*, 45337–45346.

Roberts, C.W., McLeod, R., Rice, D.W., Ginger, M., Chance, M.L., and Goad, L.J. (2003). Fatty acid and sterol metabolism: potential antimicrobial targets in apicomplexan and trypanosomatid parasitic protozoa. Mol. Biochem. Parasitol. *126*, 129–142.

Roberts, S.C., Scott, J., Gasteier, J.E., Jiang, Y., Brooks Bv, Jardim Av, Carter, N.S., Heby, O., and Ullman, B. (2002). S-adenosylmethionine decarboxylase from *Leishmania donovani*. Molecular, genetic, and biochemical characterization of null mutants and overproducers. J. Biol. Chem. *277*, 5902–5909.

Roberts, S.C., Jiang, Y., Jardim, A., Carter, N.S., Heby, O., and Ullman, B. (2001). Genetic analysis of spermidine synthase from *Leishmania donovani*. Mol. Biochem. Parasitol. *115*, 217–226.

Roberts, S.C., Tancer, M.J., Polinsky, M.R., Gibson, K.M., Heby, O., and Ullman, B. (2004). Arginase plays a pivotal role in polyamine precursor metabolism in *Leishmania*. Characterization of gene deletion mutants. J. Biol. Chem. *279*, 23668–23678.

Rodriguez-Contreras, D., Feng, X., Keeney, K.M., Bouwer, H.G., and Landfear, S.M. (2007). Phenotypic characterization of a glucose transporter null mutant in *Leishmania mexicana*. Mol. Biochem. Parasitol. 153, 9–18.

Rodriguez-Contreras, D., and Landfear, S.M. (2006). Metabolic changes in glucose transporter-deficient *Leishmania mexicana* and parasite virulence. J. Biol. Chem. *281*, 20068–20076.

Sah, J.F., Ito, H., Kolli, B.K., Peterson, D.A., Sassa, S., and Chang, K.P. (2002). Genetic rescue of *Leishmania* deficiency in porphyrin biosynthesis creates mutants suitable for analysis of cellular events in uroporphyria and for photodynamic therapy. J. Biol. Chem. *277*, 14902–14909.

Schmidt, A., and Krauth-Siegel, R.L. (2002). Enzymes of the trypanothione metabolism as targets for antitrypanosomal drug development. Curr. Top. Med. Chem. 2, 1239–1259.

Shaked-Mishan, P., Suter-Grotemeyer, M., Yoel-Almagor, T., Holland, N., Zilberstein, D., and Rentsch, D. (2006). A novel high-affinity arginine transporter from the human parasitic protozoan *Leishmania donovani*. Mol. Microbiol. *60*, 30–38.

Steglich, C., and Schaeffer, S.W. (2006). The ornithine decarboxylase gene of *Trypanosoma brucei*: Evidence for horizontal gene transfer from a vertebrate source. Infect. Genet. Evol. 6, 205–219.

Tetley, L., and Vickerman, K. (1991). The glycosomes of trypanosomes: number and distribution as revealed by electron spectroscopic imaging and 3-D reconstruction. J. Microsc. *162*, 83–90.

Tielens, A.G., Rotte, C., van Hellemond, J.J., and Martin, W. (2002). Mitochondria as we don't know them. Trends Biochem. Sci. *27*, 564–572.

Urbina, J.A. (1997). Lipid biosynthesis pathways as chemotherapeutic targets in kinetoplastid parasites. Parasitology. *114*, S91-S99.

Van Den Berghe, G., Vincent, M.F., and Jaeken, J. (1997). Inborn errors of the purine nucleotide cycle: adenylosuccinase deficiency. J. Inherit. Metab. Dis. *20*, 193–202.

Van Hellemond, J.J., Opperdoes, F.R., and Tielens, A.G. (1998). Trypanosomatidae produce acetate via a mitochondrial acetate:succinate CoA transferase. Proc. Natl. Acad. Sci. USA *95*, 3036–3041.

Van Schaftingen, E., Opperdoes, F.R., and Hers, H.G. (1985). Stimulation of *Trypanosoma brucei* pyruvate kinase by fructose 2, 6-bisphosphate. Eur. J. Biochem. *153*, 403–406.

Veitch, N.J., Maugeri, D.A., Cazzulo, J.J., Lindqvist, Y., and Barrett, M.P. (2004). Transketolase from *Leishmania mexicana* has a dual subcellular localization. Biochem. J. *382*, 759–767.

Verlinde, C.L., Hannaert, V., Blonski, C., Willson, M., Perie, J.J., Fothergill-Gilmore, L.A., Opperdoes, F.R., Gelb, M.H., Hol, W.G., and Michels, P.A. (2001). Glycolysis as a target for the design of new anti-trypanosome drugs. Drug. Resist. Updat. *4*, 50–65.

Vernal, J., Cazzulo, J.J., and Nowicki, C. (1998). Isolation and partial characterization of a broad specificity aminotransferase from *Leishmania mexicana* promastigotes. Mol. Biochem. Parasitol. *96*, 83–92.

Vernal, J., Fiser, A., Sali, A., Muller, M., Cazzulo, J.J., and Nowicki, C. (2002). Probing the specificity of a trypanosomal aromatic alpha-hydroxy acid dehydrogenase by site-directed mutagenesis. Biochem. Biophys. Res. Commun. *293*, 633–639.

Vickers, T.J., Greig, N., and Fairlamb, A.H. (2004). A trypanothione-dependent glyoxalase I with a prokaryotic ancestry in *Leishmania major*. Proc. Natl. Acad. Sci. USA *101*, 13186–13191.

Wanasen, N., MacLeod, C.L., Ellies, L.G., and Soong, L. (2007) L-arginine and cationic amino acid transporter 2B regulate growth and survival of *Leishmania amazonensis* amastigotes in macrophages. Infect. Immun. *75*, 2802–2810.

van Weelden, S.W., Fast, B., Vogt, A., van der Meer, P., Saas, J., van Hellemond, J.J., Tielens, A.G., and Boshart, M. (2003). Procyclic *Trypanosoma brucei* do not use Krebs cycle activity for energy generation. J. Biol. Chem. *278*, 12854–12863.

van Weelden, S.W., van Hellemond, J.J., Opperdoes, F.R., and Tielens, A.G. (2005). New functions for parts of the Krebs cycle in procyclic *Trypanosoma brucei*, a cycle not operating as a cycle. J. Biol. Chem. *280*, 12451–12460.

Wilkinson, S.R., Obado, S.O., Mauricio, I.L., and Kelly, J.M. (2002). *Trypanosoma cruzi* expresses a plant-like ascorbate-dependent hemoperoxidase localized to the endoplasmic reticulum. Proc. Natl. Acad. Sci. USA *99*, 13453–13458.

Wilkinson, S.R., Prathalingam, S.R., Taylor, M.C., Horn, D., and Kelly, J.M. (2005). Vitamin C biosynthesis in trypanosomes: a role for the glycosome. Proc. Natl. Acad. Sci. USA *102*, 11645–11650.

Yernaux, C., Fransen, M., Brees, C., Lorenzen, S., and Michels, P.A. (2006). *Trypanosoma brucei* glycosomal ABC transporters: identification and membrane targeting. Mol. Membr. Biol. *23*, 157–172.

Yoshida, N., and Camargo, E.P. (1978). Ureotelism and ammonotelism in trypanosomatids. J. Bacteriol. *136*, 1184–1186.

Drug Resistance in *Leishmania*

8

Marc Ouellette, Jolyne Drummelsmith, Philippe Leprohon, Karima El Fadili, Aude Foucher, Baptiste Vergnes and Danielle Légaré

Abstract

Leishmaniasis is controlled mostly through the use of chemotherapy. Pentavalent antimonials have long been the mainstay treatment but resistance to this class of drugs is increasing throughout the endemic regions. Other useful drugs against leishmaniasis include amphotericin B, miltefosine, and to a lesser extent, pentamidine. Few novel drugs are currently in development. Whole genome approaches such as detection of gene amplification events, functional cloning, DNA microarrays and proteomics have already led to an increased understanding of resistance mechanisms induced under laboratory conditions. The current challenge is to understand resistance mechanisms occurring in the field, and whole genome approaches will be useful in this endeavour.

General introduction

The phenomenon of antimicrobial drug resistance has snowballed into a serious public health concern with economic and social implications that are global in scale. Some of the major challenges include penicillin-resistant streptococci, nosocomial infections, tuberculosis, malaria, leishmaniasis, and sexually transmitted viral infections (www.who-int/infectious-diseases-report/2000/). Unfortunately, with the limited drugs under investigation, few solutions are on the horizon. Studies of resistance mechanisms are important because (1) they permit the development of diagnostic tools for the early recognition of resistance, thereby preventing useless and often toxic chemotherapy; (2) they suggest more rational uses of drugs and drug combinations, hence minimizing the development of resistance; and (3) they pinpoint intracellular drug targets and defence mechanisms, allowing the development of drug analogues to circumvent the most common defences.

The protozoan parasite *Leishmania* is distributed worldwide and has a number of clinical manifestations including cutaneous, mucocutaneous and visceral leishmaniasis (see Chapter 2). These parasites have a digenetic life cycle and are found as motile flagellated promastigotes in the alimentary tract of the sandfly vector, where they can differentiate into the highly infectious metacyclic form. During a blood meal, the sandfly transmits the parasites to the animal host, where they are engulfed by macrophages, transform into amastigotes, and divide within the acidified host phagolysosomes (see Chapters 6, 10 & 11). There are as yet no satisfactory vaccines for protective immunity, although significant effort is being invested to try to improve this situation (see Chapter 13). Control of leishmaniasis therefore relies largely on chemotherapy, which for the most part hinges on antiquated drugs. To be effective, these drugs must target the intracellular amastigote parasite.

Available drugs against leishmaniasis are limited in number and each has various shortcomings. Pentavalent antimonials (Sb^V) have long been the pillar of anti-*Leishmania* chemotherapy and still play a leading role in treatment, but resistance can be found in Southern Europe (Faraut-Gambarelli *et al.*, 1997), in Iran (Hadighi *et al.*, 2006), in South America (Jackson *et al.*, 1990;

Rojas *et al.*, 2006), and it has reached epidemic proportions in North East India (Sundar *et al.*, 2000). Second line drugs include pentamidine and amphotericin B (AmB), which have not experienced widespread use due to factors such as toxicity and cost. Various lipid formulations of AmB were found to be associated with both high activity and low toxicity (Sundar *et al.*, 2004) and are now used extensively in regions where resistance to Sb^V is high. However, these formulations carry a prohibitive price tag for developing and emerging countries. Recently, the oral drug miltefosine was approved for the treatment of human visceral *Leishmania* infections (Sundar *et al.*, 2002), and fluconazole was shown to be effective against cutaneous leishmaniasis (Alrajhi *et al.*, 2002). It remains to be seen, however, how these oral agents will stand up in face of the pandemic of resistance. Finally, there are a few drugs in development, some of which exploiting biochemical pathways unique to the parasite that were revealed following the completion of the genome sequence (see Chapter 7). These new drugs could potentially be useful for the treatment of *Leishmania* infections.

Defining resistance

Treatment failure in the field of infectious diseases is not uncommon and can be due to several factors. *Leishmania* is no exception and treatment failure has been reported for many years. Despite their toxicity, antimonials have remained effective and available in endemic countries for over 60 years. To remain equally effective, the drug concentration and duration of therapy have increased over the years (Olliaro and Bryceson, 1993), suggesting that the parasites have been slowly acquiring mutations leading to decreased susceptibility to the drug. Other factors such as drug lots with various efficacies, decrease ability of the immune system of the human host, pharmacological deficiencies, or under-treatment could also explain failures (Croft, 2001; Sundar, 2001).

The analysis of resistance mechanisms in field isolates is complicated by the lack of a recognized *in vitro* drug testing assay able to link a resistance phenotype to drug response in patients. Indeed, susceptibility testing has been a challenging issue. Initially, susceptibility measurements were made in the promastigote stage but it is now clear that this form of the parasite is not relevant to studying susceptibility to antimonial drugs (Sereno *et al.*, 1998; Ephros *et al.*, 1999) and possibly to other drugs. However, recent studies have clearly demonstrated that *in vitro* intracellular testing of Sb^V susceptibility in infected macrophages is in good agreement with clinical response (Faraut-Gambarelli *et al.*, 1997; Lira *et al.*, 1999; Hadighi *et al.*, 2006; Mukherjee *et al.*, 2007; Rojas *et al.*, 2006). Similar relationships between *in vitro* testing and clinical response to AmB were also observed (C. Riera, personal communication; P. Bastien, personal communication). These *in vitro* susceptibility assays have been greatly facilitated by the use of parasites expressing reporter genes such as the firefly luciferase (Sereno *et al.*, 2001; Hadighi *et al.*, 2006). Analysis of luciferase-expressing parasites does not involve the labour intensive parasite staining and counting method and allows high throughput (Roy *et al.*, 2000; Sereno *et al.*, 2001). This assay requires transfection of the parasites, a relatively straightforward technique that does not modify the properties of the parasite related to macrophage infection and drug susceptibility (Gourbal *et al.*, 2004; Ashutosh *et al.*, 2005; Dube *et al.*, 2005;; Hadighi *et al.*, 2006).

The emergence of resistant parasites must have been facilitated by a number of phenomena including the high endemicity of the parasite, the high proportion of treated individuals and the almost exclusive anthroponotic (human to human) transmission cycle of *Leishmania* spp. in several regions of the world (Sundar *et al.*, 2000; Hadighi *et al.*, 2006; Rojas *et al.*, 2006). Antimonial drugs have long been the mainstay of anti-*Leishmania* therapy and not surprisingly it is resistance to this class of drug that is the most widespread in the field. Despite this, most studies on resistance to antimonials have been carried out on *Leishmania* promastigotes in which resistance was induced under laboratory conditions. Few studies have dealt with clinical isolates, although this is likely to change in the near future. Similarly, most of our understanding of resistance mechanisms to other anti-*Leishmania* drugs is also derived from parasites in which resistance was induced *in vitro* and mostly in the promastigote stage. In the following sections, we shall describe our current understanding of resistance mechanisms stemming mostly from *in vitro* work and see how this could compare with the situation in the field. We shall first highlight how whole genome approaches can contribute to our understanding of resistance mechanisms.

Whole genome approaches to study resistance

Detection of gene amplification

The *Leishmania* genome is small enough (33 Mb) to detect gene amplification (as small as a five- to tenfold increase in copy number) by simple examination of digested genomic DNAs of wild-type cells and drug-resistant mutants run in parallel on an agarose gel (Coderre *et al.*, 1983; Ouellette *et al.*, 1998). Given the high GC content of the *Leishmania* genome, digestion with enzymes recognizing A-T rich sequences such as *Hin*dIII and *Eco*RI facilitate the visualisation of the amplified band(s). This strategy was first used to detect gene amplification events in *Leishmania* promastigote cells resistant to methotrexate (MTX) (Coderre *et al.*, 1983) and afterwards was found useful to detect several loci amplified upon selection with a number of drugs, including those useful for the treatment of *Leishmania* infections (Table 8.1). Although MTX is not used against *Leishmania*, considerable work has dealt with resistance mechanisms to this model drug and has led to the development, or served as a proof of principle, for several genomic technologies that are used for the study of resistance. Amplified DNAs can be found either as circular DNAs or as linear amplicons, linear amplicons being sometimes precursors of circles (Grondin *et al.*, 1998). It is possible to isolate circular amplicons of *Leishmania* by using a standard alkaline lysis technique used to prepare bacterial plasmids, whereas linear amplicons can be visualized and isolated from pulse field gels (Ouellette *et al.*, 1998).

DNA amplification is seen frequently in field isolates. One of the first loci found to be amplified is the LD1/CD1 region (reviewed in Segovia and Ortiz, 1997). The driving force leading to the amplification of this region is likely be the presence of the biopterin transporter BT1 (Kündig *et al.*, 1999; Lemley *et al.*, 1999), as biopterin is a strong *Leishmania* growth promoter (Roy *et al.*, 2001). Another example includes a region equivalent to chromosome 9 of *Leishmania major* that was found amplified in field isolates resistant to Sb^V (Singh *et al.*, 2003). However, the mechanism by which genes present on this region leads to resistance still needs to be established. The *MRPA* gene of *Leishmania*, a gene well established as a contributor to antimonial resistance (Légaré *et al.*, 2001; El Fadili *et al.*, 2005) was also found amplified in several *L. donovani* kala-azar isolates coming from patients unresponsive to Sb^V therapy (Mukherjee *et al.*, 2007).

The presence of an amplified fragment of DNA associated with a resistance phenotype is suggestive of its implication in resistance, although its direct involvement in resistance needs to be proven either by gene overexpression or gene inactivation. Before the completion of the *Leishmania* genome project, it was necessary to isolate the amplicon, subclone relatively large DNA fragments in a *Leishmania* expression vector, transfect these plasmids and see whether recombinant parasites became resistant to the drug. Further subcloning would lead to the identification of the resistance gene (Callahan and Beverley, 1992; Papadopoulou *et al.*, 1992). With the availability of whole genome sequences of several *Leishmania* species (www.genedb.org), it is now possible to determine which portion of the genome is amplified by sequencing a small region of the amplified DNA and then matching it on the *Leishmania* genome. As amplified DNA regions in *Leishmania* are always relatively small, it is possible to find candidate genes near (± 50 kb) the region sequenced. Transfection of candidate gene(s) and susceptibility testing should then reveal its role in resistance.

The ability of DNA microarrays to detect gene amplification events is less exploited than for expression profiling, but it was used, for example, in the study of human breast tumours (Pollack *et al.*, 1999). The suitability of DNA microarrays to detect gene amplification events has been demonstrated for *Leishmania* (Guimond *et al.*, 2003). While simple techniques such as comparison of

Table 8.1 Gene amplification detected in drug-resistant *Leishmania*

Drugs	Genes	Function	References
Clinically useful drugs			
Antimonials	*MRPA*	ABC protein involved in the sequestration of thiol-Sb conjugates	(Anacleto *et al.*, 2003; El Fadili *et al.*, 2005; Ferreira-Pinto *et al.*, 1996; Haimeur *et al.*, 2000; Mukherjee *et al.*, 2007)
	ORFSbV	Unknown	(Haimeur and Ouellette, 1998)
	G circle	Unknown	(Arana *et al.*, 1998)
	GSH1	Rate limiting step of glutathione biosynthesis	(Haimeur *et al.*, 2000; Mukherjee *et al.*, 2007)
	ORF with no name	Unknown	(Singh *et al.*, 2003)
Amphotericin B	circles	Unknown	(Singh *et al.*, 2001)
Experimental model drugs			
Methotrexate	*DHFR-TS*	Dihydro folate reductase-thymidylate synthase (main target of MTX)	(Coderre *et al.*, 1983) (Garvey and Santi, 1986) (Kündig *et al.*, 1999)
	PTR1	Pteridine reductase (by-passes DHFR)	(Beverley *et al.*, 1984) (Papadopoulou *et al.*, 1992) (Callahan and Beverley, 1992) (Chiquero *et al.*, 1994)
Difluoro methyl ornithine	*ODC*	Ornithine decarboxylase	(Hanson *et al.*, 1992) (Sanchez *et al.*, 1997)
Tunicamycin	*NAGT*	N-acetylglucosamine-1-phosphate transferase	(Detke *et al.*, 1988)
Hydrophobic MDR drugs	*MDR1*	P-glycoprotein	(Henderson *et al.*, 1992) (Chow *et al.*, 1993) (Chiquero *et al.*, 1998) (Katakura *et al.*, 1999)
Mycophenolic acid	*IMPDH*	IMP dehydrogenase	(Wilson *et al.*, 1991)
Tubercidin	*TOR*	Toxic nucleoside Resistance gene	(Detke, 1997)
Hydroxyurea	*M2*	M2 subunit protein of ribonucleotide reductase	(Lye *et al.*, 1999)
Terbinafine	*HTBF*	Function unknown	(Marchini *et al.*, 2003)
Arsenite	*MRPA*	ABC protein	(Katakura and Chang, 1989) (Ouellette *et al.*, 1991) (Grondin *et al.*, 1993)
	GSH1	γ-Glutamyl cysteine synthetase	(Grondin *et al.*, 1997)

digested DNA on ethidium bromide-stained gels can lead to the detection of gene amplification events, in contrast to DNA arrays, this method cannot provide the number of unique amplicons nor on their identity. Moreover, the detection of gene amplification by classical methods can only serve in the case of an increase in copy number whereas DNA microarrays could also serve in the case of gene deletion, a phenomenon that occurs in resistant *Leishmania* cells (Richard *et al.*, 2004).

Use of functional cloning for isolating drug resistance genes

Functional cloning was pioneered for work on lipophosphoglycan biosynthesis (Ryan *et al.*, 1993). Functional cloning is a whole genome approach ideally suited for the isolation of drug resistance genes. The strategy is shown in Fig. 8.1. A genomic bank is made in one of the two versions of a cosmid vector (Ryan *et al.*, 1993; Kelly *et al.*, 1994). This genomic bank is then transfected in *Leishmania* cells and these transfectants are then tested for a change in drug susceptibility. Resistance can be due to a gain of function (e.g. over-expression of the drug target) or to a loss of function (e.g. down-regulation of a drug uptake system). If a gain of function is suspected, it would be more appropriate to transfect the cosmid bank into a wild-type background, select with the marker of the cosmid (e.g. hygromycin B) and then plate cells with varying concentrations of the drug being studied. This dominant positive strategy was first shown to work while using the model drug MTX (Cotrim *et al.*, 1999; Kündig *et al.*, 1999) and has since worked well for several drugs (Table 8.2). To investigate the potential of loss of function, it is usually possible to transfect a cosmid bank into a resistant mutant, select individual colonies on hygromycin B plates, pick individual colonies into duplicate 96 well plates. One plate is then grown in the presence of the marker drug used for the selection of the cosmid (e.g. hygromycin B) and the other plate with both hygromycin B and the drug being studied. Most parasites (since they are resistant) will grow in the presence of the selective drug but if a gene present on the cosmid restores (partly or in whole) sensitivity to the drug in the resistant mutant (Fig. 8.1, right panel) growth of that clone will be inhibited. The corresponding clone in the plate with hygromycin alone would then be further studied. This dominant negative approach has worked well for several drugs (Table 8.2). In both dominant positive and dominant negative strategies, it is necessary to isolate the cosmid thought to be responsible for the phenotype studied and retransfect it in the tested strain to ascertain that it is indeed implicated in the phenotype being investigated. This approach only works when a single gene is responsible for the phenotype being studied. As mentioned above, sequencing of one end of a cosmid and comparing it to the annotated genome can rapidly lead to putative candidates involved in resistance which can be formally tested by transfection as described above.

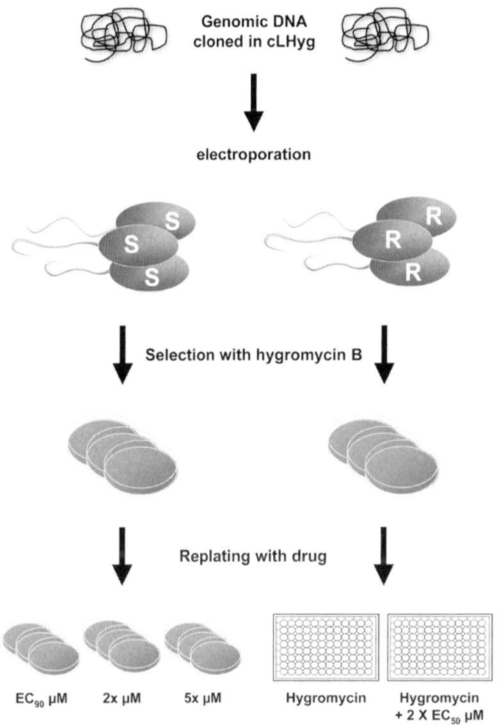

Figure 8.1 Functional cloning in *Leishmania*. Genomic banks of *Leishmania* cells (derived from either resistant or sensitive parasites) are made in an expression-based cosmid vector. These cosmids can be transfected into either sensitive or resistant isolates. Parasites are plated in the presence of the selective drug cosmid marker (hygromycin B on this figure). In the case of a dominant positive selection approach (left panel) the transfectants are collected and replated with the investigated drugs at the lowest concentration that would not allow the growth of the untransfected parasites. In the case of a dominant negative selection approach (right panel), transfected colonies are picked individually into duplicate 96-well plates and clones are selected for regained drug susceptibility.

Use of DNA microarrays

Resistance is often due to gene amplification (see Table 8.1) but there are several examples in drug-resistant parasites where specific RNAs are modulated (increased or decreased) without a change in the copy number of their respective gene (Haimeur *et al.*, 1999; Guimond *et al.*,

Table 8.2 Genes isolated by functional cloning

Functional cloning	Drugs	Genes	Function	References
Dominant positive				
	Methotrexate	BT1, PTR1, DHFR	Biopterin transporter, pteridine reductase and dihyrofolate reductase	(Kündig *et al.*, 1999; Cotrim *et al.*, 1999)
	Antimonials	HSP70	Heat shock protein 70	(Brochu *et al.*, 2004)
	terbinafine	SQS1	squalene synthase	(Cotrim *et al.*, 1999)
	tubercidin	TOR		(Cotrim *et al.*, 1999)
	miltefosine			(Clos and Choudhury, 2006)
	pentamidine	PRP1	ABC protein	(Coelho *et al.*, 2003)
Dominant negative				
	tubercidin	LdNT1.1 LdNT1.2	Nucleoside transporters	(Vasudevan *et al.*, 1998)
	formycin B	LdNT2	Nucleoside transporter	(Carter *et al.*, 2000)
	miltefosine	LdMT	P-type phospholipid translocase	(Perez-Victoria *et al.*, 2003)
	Antimonials	AQP1	Aquaglyceroporin 1	(Marquis *et al.*, 2005)

2003; Marquis *et al.*, 2005). Several techniques have been developed for detecting genes differentially expressed in *Leishmania* (Charest *et al.*, 1996; Lamontagne and Papadopoulou, 1999; Wu *et al.*, 2000; Quijada *et al.*, 2005) and one promising technique is DNA microarray analysis. DNA microarrays are powerful tools now being employed in the field of parasitology. Targeted cDNAs, random 1 kb genomic DNAs, PCR fragments, and 70-mer oligonucleotides have been spotted on microarrays and have provided useful results to study differential gene expression between different life stages of various *Leishmania* species (Saxena *et al.*, 2003; Akopyants *et al.*, 2004; Almeida *et al.*, 2004; Holzer *et al.*, 2006; McNicoll *et al.*, 2006; Salotra *et al.*, 2006;). While DNA microarrays have been used extensively for looking at gene expression changes triggered by exposure to antibiotics or antimicrobials (Wilson *et al.*, 1999; Gmuender *et al.*, 2001; Ng *et al.*, 2003;) their application to the study of drug resistance is emerging. Some examples of drug resistance study using DNA microarrays include the monitoring of gene expression in *Candida albicans* selected *in vitro* for resistance to fluconazole or amphotericin B (Cowen *et al.*, 2002; Barker *et al.*, 2004); in *Leishmania* cells made resistant to different class of drugs *in vitro* (Guimond *et al.*, 2003; El Fadili *et al.*, 2005; Leprohon *et al.*, 2006) and in a number of bacterial species (e.g. (Haas *et al.*, 2005; Mongodin *et al.*, 2003)). Microarrays have shown their usefulness in the study of resistance and consequently we recently generated 70-mer oligonucleotide *Leishmania* full genome arrays that should be useful for analysing resistance at the global RNA expression level. This will have the advantage of detecting several changes of gene expression at once (up- or down-regulation) at the whole genome level in a fairly comprehensive fashion. Obviously, changes in RNA expression are not always correlated to changes in protein levels (McNicoll *et al.*, 2006) but in the case of resistance this may differ as several differentially expressed RNAs have been linked to resistance. For example, in a study of MTX resistance in *Leishmania*, we found a good correlation between proteins and RNA overexpression using a small targeted DNA microarray (Fig. 8.2).

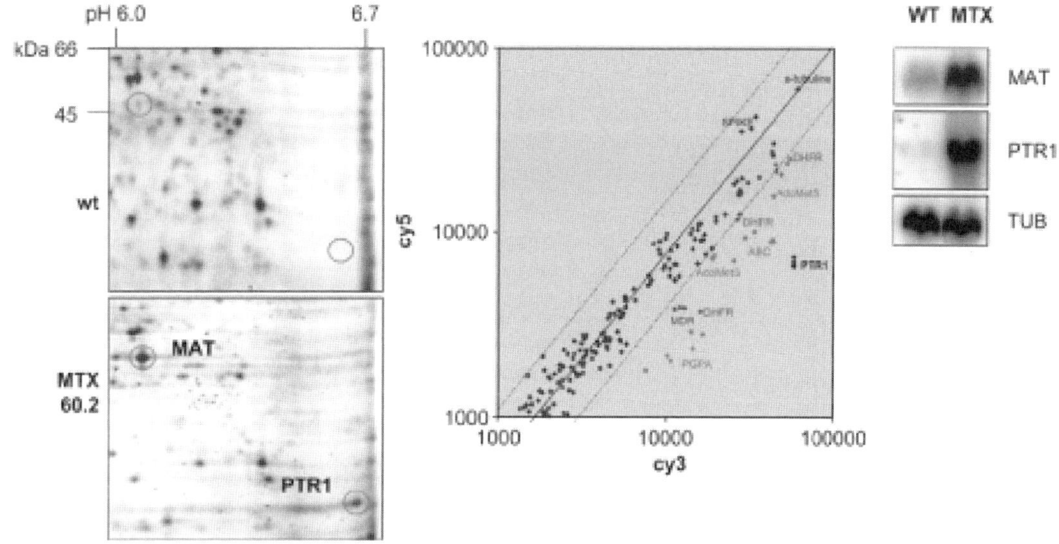

Figure 8.2 Combined proteomic and transcriptomic approaches for the study of drug resistance in *Leishmania*. *L. major* selected for MTX resistance (MTX60.2) was subjected to proteomic (Drummelsmith *et al.*, 2004) and transcriptomic (Guimond *et al.*, 2003) screens. In this case, PTR1 and MAT were detected by both techniques. The left panel shows a region of a 2D gel where both PTR1 and MAT can be observed as overexpressed proteins in *L. major* MTX 60.2. The middle panel shows a scatter plot of a targeted DNA microarray hybridization with the wild-type RNA labelled with Cy5 and the MTX60.2 RNA labelled with Cy3 (Guimond *et al.*, 2003). Genes overexpressed included *PTR1*, *PGPA* (linked to *PTR1*), *MAT* (AdoMetS) and *DHFR* (the main target of MTX). The *PTR1* and *MAT* microarray results were confirmed by Northern blot analysis (right panel).

Proteomic approaches for the study of resistance

Similarly to DNA microarrays, the use of proteomics for studying antimicrobial resistance is emerging (Papadopoulou *et al.*, 2004). One advantage of proteomics over DNA microarrays is that proteins are closer to function than RNA. Proteomics also allow the detection of post-translational modifications (PTM) (Drummelsmith *et al.*, 2003) and one recent report in *Trypanosoma brucei* suggests that resistance to an arsenical drug may be linked to a PTM (Foucher *et al.*, 2006). Another advantage of proteomic techniques includes the possibility of obtaining clues to function by, for example, cellular localisation (e.g. see Foucher *et al.*, 2006). One disadvantage of proteomic experiments is that they are usually more involved than using DNA microarrays due to both increased structural variability and dynamic range of cellular concentration displayed by proteins compared to RNA molecules. Specialized techniques are required to observe particular classes of proteins, such as lower abundance or hydrophobic proteins. Therefore, the combination of several techniques and fractionation approaches would be required for a good proteome representation (Righetti *et al.*, 2005).

Proteomic techniques have proved useful to find an alkyl hydroxyperoxide reductase possibly involved in metronidazole resistance in *Helicobacter pylori* (McAtee *et al.*, 2001); a protein involved in ergosterol biosynthesis in azole-resistant C. *albicans* (Hooshdaran *et al.*, 2004); and of the over-expression of a phosphate ABC transporter protein in penicillin-resistant *Streptococcus pneumoniae* (Soualhine *et al.*, 2005). Another example concerns our work with the parasite *Leishmania* where we have demonstrated the overproduction of the pterin reductase PTR1 (Drummelsmith *et al.*, 2003) and of the methionine adenosyl transferase (MAT; (Drummelsmith *et al.*, 2004)) in MTX-resistant parasites (see Fig. 8.2). PTR1 is directly involved in resistance where it bypasses the need for the main MTX target dihydrofolate reductase (DHFR), while

MAT facilitates the selection of resistant parasites, rather than conferring resistance directly. The latter type of mutation may have been missed by traditional methods.

Detection of point mutations

Point mutations are often linked to drug resistance and this can also be the case in drug-resistant *Leishmania* (Vasudevan *et al.*, 2001; Perez-Victoria *et al.*, 2003; Arrebola *et al.*, 1994). Currently, identifying point mutations by genomic strategies is challenging, although a loss or gain of function due to mutation leading to resistance may be inferred by functional complementation. Traditional expression profiling DNA microarrays are not well designed for the detection of point mutations. Interestingly, NimbleGen (www.nimblegen.com) has developed a technique using overlapping small oligonucleotides for the detection of point mutations and other polymorphisms in bacterial genomes and these have been used successfully in the context of resistant organisms (Albert *et al.*, 2005). This method consists of resequencing bacterial genomes using hybridization based-microarrays and seems like an attractive method for the detection of point mutations. It remains to be seen, however, how cost effective this technique would be for larger genomes such as that of *Leishmania*. The fact that *Leishmania* is a diploid organism may further complicate the use of this comparative genome sequencing technique.

The comparative proteomic approach by differential migration in 2D gels, may also be useful for the detection of some point mutations. Indeed, in ongoing work with *S. pneumoniae*, we were able to detect several point mutations including changes of as little as 0.01 pI unit (a D to a E change!). This is nonetheless labour-intensive, does not easily extrapolate to larger studies and will certainly not be comprehensive. Novel mass spectrometry techniques such as iTRAQ where multiple peptides per protein are identified and quantitated, could also serve to increase the odds of seeing point mutations at the protein level. Thus, at present, proteomics is possibly the technique that allows most easily the detection of point mutations, but the resequencing of whole genomes is an attractive alternative that may soon come available. The sequencing of a 33 Mb genome with few repetitive sequences, as it is the case with *Leishmania*, may become possible in a cost-effective fashion using massive parallel sequencing such as the 454-system (Margulies *et al.*, 2005) or polony sequencing (Shendure *et al.*, 2005).

Mode of action and resistance mechanisms to drugs used clinically against *Leishmania* infections

Sb^V

The efficacies of the current pentavalent antimonial formulations sodium stibogluconate (Pentostam) and meglumine antimoniate (Glucantime) were proven over 50 years ago (reviewed in Guerin *et al.*, 2002). Despite their extensive use, we remain uncertain of their mechanism of action, structures, and even the identities of the biologically active components of the optimal formulations (Ouellette *et al.*, 2004). Resistance mechanisms are well established *in vitro* (Ouellette *et al.*, 2004; Croft *et al.*, 2006), but less so in clinical isolates, although some of the general resistance strategies and proteins appear to be the same (Hadighi *et al.*, 2006; Mukherjee *et al.*, 2007) with some new twists (Decuypere *et al.*, 2005).

To be effective, Sb^V has to enter the host cell, cross the phagolysosomal membrane and be active against intracellular amastigotes. It is highly likely that to be active, Sb^V must also be converted to a trivalent form (Sb^{III}), thus leading to the definition of Sb^V as a pro-drug. The details of the activation mechanism are not known, and the exact site of this conversion (macrophages, phagolysosomes and/or parasites) is also unclear. Molecules possibly implicated in reduction of Sb^V to Sb^{III} include host and parasite thiols (Santos Ferreira *et al.*, 2003) and two newly discovered parasite enzymes called TDR1 (Denton *et al.*, 2004) and ACR2 (Zhou *et al.*, 2004). A loss of drug activation could lead to resistance, and indeed this was shown for *L. donovani* axenic amastigotes resistant to Sb^V which exhibit a diminished ability to reduce Sb^V to Sb^{III} (Shaked-Mishan *et al.*, 2001).

Recently, Sb^V (but not Sb^{III}) was found to be a potent inhibitor of host protein tyrosine phosphatases, leading to an increase in cytokine responses (Pathak and Yi, 2001). This result suggests that Sb^{V} may kill the parasites by both direct and indirect methods, and that the activation of the host response is implicated in the antileishmanial activity of Sb^{V}. One possible genomic strategy that could be exploited to understand the mode of action of Sb^{V} would be to use human DNA microarrays to monitor changes in gene expression of *Leishmania*-infected macrophages or without Sb^{V} treatment.

The route of entry of Sb^{V} into *Leishmania* cells is still unknown, but Sb^{III} enters through the aquaglyceroporin AQP1 (Gourbal *et al.*, 2004). AQPs are channels for small neutral molecules and are part of the large family of aquaporins (King *et al.*, 2004). Since intramacrophagic *Leishmania* transfected with *AQP1* are hypersensitive to Sb^{V} and Sb^{V} is not transported by AQP1, it suggests that some reduction of Sb^{V} to Sb^{III} occurs in the macrophages (Gourbal *et al.*, 2004). Recent evidence has suggested that part of the mode of action of Sb^{V} could be in depleting the cells of its reduced thiols (Wyllie *et al.*, 2004). Consistent with this, trypanothione, a glutathione-spermidine conjugate, the main reduced thiol of *Leishmania* (Fairlamb and Cerami, 1992) and other thiols are increased in *Leishmania* cells selected for resistance *in vitro* (Mukhopadhyay *et al.*, 1996; Haimeur *et al.*, 2000; El Fadili *et al.*, 2005). This increase is mediated by the overexpression or amplification of a number of different genes involved in the synthesis of glutathione or spermidine (Grondin *et al.*, 1997; Haimeur *et al.*, 1999; Guimond *et al.*, 2003), the two building blocks of trypanothione. Modulation in the expression levels of these genes has also been observed in clinical isolates (Decuypere *et al.*, 2005; Mukherjee *et al.*, 2007).

Recent work suggests that Sb^{V}, and also other anti-leishmania drugs, exerts some of their action by inducing an apoptotic-like process (Sereno *et al.*, 2001; Lee *et al.*, 2002). Further studies of these pathways should lead to a better understanding of the molecular mechanisms of drug action and possibly of resistance. The apoptotic pathway in *Leishmania* is not well understood but will probably differ from mechanisms found in metazoans (Zangger *et al.*, 2002; Debrabant *et al.*, 2003). We have recently obtained evidence for a difference in drug-mediated apoptosis between sensitive and resistant clinical isolates in which classical resistance mechanisms were not observed. Through a proteomic comparison of these isolates we have identified proteins potentially implicated in this apoptotic pathway (Vergnes, B. *et al.*, manuscript in preparation).

Reduced accumulation of metals, mediated either by reduced influx or increased efflux was observed in resistant *Leishmania* (Dey *et al.*, 1994; Dey *et al.*, 1996; Brochu *et al.*, 2003). In several strains, reduced accumulation was linked to the decreased expression of the aquaglyceroporin AQP1 (Marquis *et al.*, 2005). The first gene linked to antimonial resistance was the ABC protein MRPA (Ouellette *et al.*, 1991). ABC proteins are ubiquitous proteins with a conserved ATP-binding cassette domain of ~200 amino acids, and several are involved in drug resistance (Borst and Elferink, 2002). MRPA is located in the membrane of an intracellular organelle and confers resistance by sequestration of the Sb-thiol conjugate (Légaré *et al.*, 2001). Biochemical evidence for a distinct plasma membrane thiol metal efflux pump has also been obtained (Dey *et al.*, 1996), although the protein responsible for this activity has never been isolated and it is not clear whether this transport activity is modulated upon resistance.

The above-mentioned resistance mechanisms which were derived mostly from studies on cells selected for resistance in the laboratory have yielded a coherent model for antimony resistance in *Leishmania* (Ouellette *et al.*, 2004). Evidence further supporting this model has been obtained in kala-azar clinical isolates (Mukherjee *et al.*, 2007), although through the characterization of field isolates we have gathered considerable indirect evidence for additional resistance mechanisms as well.

The potential of whole genome approaches to pinpoint genes/proteins in resistance is high. Indeed, the study of gene amplification has already led to the detection of *MRPA* (Ouellette *et al.*, 1990; Ouellette *et al.*, 1991) and *GSH1* (Grondin *et al.*, 1997; Haimeur *et al.*, 2000) (Table 8.1). Functional cloning was also useful for isolating the Sb^{III} importer AQP1 (Marquis *et al.*, 2005) and has allowed us to find a novel role for HSP70 in Sb^{III} tolerance (Brochu *et al.*, 2004) (Table 8.2). A novel antimony resistance gene was also isolated by functional cloning in the amastigote stage of the parasite (manuscript in preparation). Targeted DNA microarrays have been useful to study RNA overexpression in antimony-resistant mutants and have allowed us to detect several genes already known to be implicated in resistance (Guimond *et al.*, 2003; El Fadili *et al.*, 2005; Leprohon *et al.*, 2006). Similarly, proteomic screens of resistant organisms have allowed the detection of proteins implicated in thiol synthesis (Foucher, A. manuscript in preparation) and found novel proteins differentially expressed in field isolates that are part of the apoptotic pathway and potentially linked to resistance. Thus, the availability of the genome sequence and the development of associated genomic and proteomic techniques have dramatically increased the potential to assess resistance mechanisms. Now that clinical isolates with proven resistance are available, it should be possible in the near future to understand the main resistance mechanism(s) to Sb^V occurring in the field.

Pentamidine

The cationic aromatic diamide pentamidine has been used as a second-line drug in the treatment of visceral leishmaniasis refractory to Sb^V treatment. However, the efficacy (and consequently the use) of pentamidine is rapidly declining in India, suggesting that parasites are becoming resistant (Sundar, 2001). As is the case with Sb^V, the cellular target of pentamidine is unknown. It has been suggested that pentamidine could enter cells through an arginine or polyamine transporter (Kandpal and Tekwani, 1997; Basselin *et al.*, 2000;) although this has recently been challenged (Basselin *et al.*, 2002).

Recent evidence would suggest that pentamidine targets a component of the mitochondria (Basselin *et al.*, 2002; Mukherjee *et al.*, 2005), which was further supported by the fact that it could greatly enhance the efficacy of mitochondrial respiratory chain complex II inhibitors (Mehta and Shaha, 2004). The exact route of pentamidine entry into mitochondria and its specific mitochondrial targets are, however, still unknown. In *in vitro* pentamidine-resistant *L. mexicana* mutants, the drug does not accumulate in the mitochondria and the cytosolic fraction of the drug is extruded outside the cell (Basselin *et al.*, 2002) although this does not seem to be the case for *L. donovani* (Mukherjee *et al.*, 2005).

Resistance mechanisms to pentamidine in *Leishmania* field isolates are unknown and our understanding stems from *in vitro* resistant isolates. Clearly, whole genome approaches could be useful for a better understanding of both the mode of action and of resistance to pentamidine. Gene amplification events have never been observed in pentamidine-resistant isolates, but the ABC transporter PRP1 was isolated by functional cloning in cells being selected for pentamidine resistance (Coelho *et al.*, 2003). This putative transporter protein could potentially confer resistance to pentamidine by extruding it outside the cell. DNA microarray work is in progress on pentamidine-resistant isolates. A large proteomic screen has revealed that several mitochondrial proteins were found to be differentially expressed in these pentamidine-resistant cells (R. Madhubala, personal communication), consistent with the hypothesized mode of action of this drug. Nonetheless, further work is required to isolate the main gene/protein implicated in pentamidine resistance.

Amphotericin B and other antifungal drugs

The polyene amphotericin B (AmB) interacts with fungal membrane sterols and preferentially with ergosterol. Like fungi, *Leishmania* have ergostane-based sterols as their major membrane sterol (Goad *et al.*, 1984) which likely explains the efficacy of AmB against *Leishmania*. While highly effective, AmB is toxic and is associated with severe side effects. However, in areas where there are high levels of Sb^V resistance, AmB is clearly the drug of choice, and lipid formulations of AmB, despite costs, are highly effective and less toxic (Sundar *et al.*, 2004). Resistance to AmB in field isolates does not yet seem to be a problem, although relapses are sometimes found after treatment, especially in HIV-positive patients. Successive relapses could contribute to an increase in AmB-resistant *Leishmania* isolates and *L. infantum* clinical isolates resistant to AmB are available (unpublished observations).

The specialized sterols of *Leishmania* and their close similarities to sterol composition and biosynthesis in yeast have led to the testing of a number of antifungal sterol biosynthesis inhibitors against *Leishmania* (Roberts *et al.*, 2003). The allylamine terbinafine and the cytochrome P450-dependent 14α-methylsterol 14-demethylase inhibitor ketoconazole have shown some efficacy against *Leishmania*. In line with these findings, the efficacy of a six-week oral course of fluconazole for treating cutaneous leishmaniasis caused by *L. major* was demonstrated (Alrajhi *et al.*, 2002).

A study correlating the emergence of AmB resistance with a change in membrane fluidity found that the major sterol of the resistant strain was an ergosterol precursor rather than ergosterol itself as found in the sensitive parent (Mbongo *et al.*, 1998). Resistance to fluconazole (Roberts *et al.*, 2003) also led to major changes in sterol composition, but *Leishmania* can survive with greatly altered sterol profiles. This suggests that a single drug targeting the sterol pathway may not be sufficient for treating *Leishmania* infections. Resistance to AmB was related to a defect in the expression of *ERG6*, an ergosterol biosynthetic gene (Pourshafie *et al.*, 2004).

Whole genomic approaches have been useful in the study of resistance to antifungal drugs useful against leishmaniasis. Gene amplification events were observed in *Leishmania* AmB mutants (Singh *et al.*, 2001), although the link between these amplicons and resistance has yet to be confirmed. The H locus of *Leishmania* was also amplified after selection with terbinafine (Ellenberger and Beverley, 1989) and subsequent work has shown that HTBF, an homologue of the yeast YIP1 protein, is involved in resistance by a mechanisms that still needs to be determined (Marchini *et al.*, 2003). Functional cloning was used to isolate terbinafine and itraconazole resistance genes. One resistance gene corresponded to squalene synthase (an enzyme upstream of the target of the drug) further demonstrating that resistance can occur by building up the level of ergosterol intermediates (Cotrim *et al.*, 1999). DNA microarrays and proteomic approaches were useful to monitor gene expression in *C. albicans* selected *in vitro* for resistance to fluconazole or AmB (Cowen *et al.*, 2002; Barker *et al.*, 2004; Hooshdaran *et al.*, 2004) and similar studies could be used for *Leishmania*.

Miltefosine

Miltefosine is an alkyl phospholipid initially developed as an antitumour agent. Miltefosine was first shown to be active against *L. donovani in vitro* (Croft *et al.*, 1987) and was rapidly shown in small studies to have a favourable therapeutic index for Indian visceral leishmaniasis. A phase III trial in adults has demonstrated that oral miltefosine is highly active against visceral leishmaniasis (Sundar *et al.*, 2002). The mode of action of miltefosine is not known precisely but is associated with changes in alkyl-lipid metabolism, and phospholipid biosynthesis (Lux *et al.*, 2000). In the promastigote stage it was observed that miltefosine could induce a cell death process reminiscent of apoptosis (Paris *et al.*, 2004; Verma and Dey, 2004). Due to the still investigational nature of miltefosine in the field, clinically resistant parasites have not yet been reported although reports of increasing relapse rates in phase IV trials suggest that resistance is progressing quickly (Croft *et al.*, 2006). This mirrors the situation in the laboratory where resistance can easily be induced *in vitro*.

In vitro resistant miltefosine-resistant mutants have been characterized (Perez-Victoria *et al.*, 2003; Seifert *et al.*, 2003) and possible resistance mechanisms include reduced drug uptake, differential plasma membrane permeability, faster drug metabolism and efflux of the drug. In one study, *Leishmania* cells selected for resistance to the anti-cancer agent daunomycin were overexpressing the ABC P-glycoprotein gene MDR1 and were cross-resistant to miltefosine (Perez-Victoria *et al.*, 2001). Compounds known to reverse multidrug resistance by interacting with the P-glycoprotein decreased miltefosine cross-resistance (Perez-Victoria *et al.*, 2001), suggesting that MDR1 may be implicated in miltefosine resistance. Furthermore, studies of the resistance mechanisms in *L. donovani* cells selected for resistance to miltefosine *in vitro* have revealed a correlation between resistance and reduced accumulation of the drug (Perez-Victoria *et al.*, 2003). This decreased accumulation was caused by a defect in the inward translocation of the drug; that is, movement from the outer to the inner leaflet of the plasma membrane.

Whole genome approaches have been useful in revealing some of the mechanisms of miltefosine resistance. The miltefosine translocase (LdMT) responsible for defective translocation in the mutants was isolated by functional rescue. The LdMT protein corresponds to a novel P-type ATPase belonging to the aminophospholipid translocase subfamily (Perez-Victoria *et al.*, 2003). Point mutations at two different positions in the protein were implicated in the loss of translocation of both the drug and glycerophospholipids. Miltefosine is a major breakthrough for the control of *Leishmania* infections, but the ease with which resistant mutants are obtained is worrying. It remains to be seen whether mutations in LdMT can also occur in the field, as parasites with crippled phospholipid translocators may also be less virulent. Proteomic screens have been done through collaborative studies with the Granada group, but these initial screens, using total proteins, were not useful in finding markers of resistance (unpublished observations).

Conclusions and perspectives

This chapter focussed on the few drugs currently in use in clinical settings, and clearly the number of alternatives is small. Drug combination is a useful strategy to slow down the potential emergence of resistant parasites (Ouellette and Ward, 2002), but this luxury almost does not exist for *Leishmania* although some drug combinations studies with miltefosine have recently been tested (Seifert and Croft, 2006). Few novel drugs are in the pipeline, but only two are somewhat advanced in the development process: paromomycin and sitamaquine. Paromomycin is an aminoglycoside aminocyclitol antibiotic and is thought to have as a primary target the *Leishmania* ribosomes (Maarouf *et al.*, 1995) but can also alter membrane fluidity and lipid metabolism, and may also target key mitochondrial activities (Maarouf *et al.*, 1997). The lack of funding from various organisations has complicated key final studies for paromomycin approval (reviewed in Guerin *et al.*, 2002) although this is changing (Croft *et al.*, 2006). Since paromomycin has not yet been used extensively, resistance is not a problem in the field, but some work has been done with parasites where resistance was induced *in vitro*. In *E. coli*, mutations in ribosomal RNA can be associated with resistance, but this was not the case in one *Leishmania* resistant cell line studied (Fong *et al.*, 1994). In another resistant *Leishmania* strain, a correlation was found between decreased drug accumulation and resistance (Maarouf *et al.*, 1998) although the protein implicated, if any, is still unknown. Sitamaquine is an oral 8-aminoquinoline that has been used in isolated patients or for compassionate use in HIV-related *Leishmania* infections, but little is known regarding its mode of action or resistance mechanisms. Future development and investigation will tell how useful this novel drug may be for treating *Leishmania* infections.

The availability of the sequences of the genome of several *Leishmania* species (Ivens *et al.*, 2005) (www.genedb.org) is a milestone in *Leishmania* research. This should expedite the discovery of novel drug targets as functional genomic technologies have a great potential for the discovery of the mode of action and resistance mechanisms to anti-*Leishmania* drugs. The understanding of resistance mechanisms in field isolates should now be a priority which will allow the generation of diagnostic tools. These tools would serve in rapid diagnosis of resistant infections but could also serve in assessing the prevalence of infection and in determining whether similar resistance mechanisms are operating in different *Leishmania* species.

Acknowledgements

The experimental work presented in this chapter is funded in part by a CIHR group grant GR 14500 and CIHR operating grants to M.O. J.D. was a CIHR post-doctoral fellow and P.L. is the recipient of a CIHR studentship. K.E. and A.F. are Strategic Training Fellows of the Strategic Training Program in Microbial Resistance (IRSC STP-53924), a partnership of the CIHR Institute of Infection and Immunity and the Fonds de Recherche en Santé du Québec', B.V. is a Fondation pour la Recherche Médicale post-doctoral fellow. M.O is a Burroughs Wellcome Fund Scholar in Molecular Parasitology and holds a Tier 1 Canada Research Chair in Antimicrobial resistance.

References

Akopyants, N.S., Matlib, R.S., Bukanova, E.N., Smeds, M.R., Brownstein, B.H., Stormo, G.D. and Beverley, S.M. (2004). Expression profiling using random genomic DNA microarrays identifies differentially expressed genes associated with three major developmental stages of the protozoan parasite *Leishmania major*. Mol. Biochem. Parasitol. *136*, 71–86.

Albert, T.J., Dailidiene, D., Dailide, G., Norton, J.E., Kalia, A., Richmond, T.A., Molla, M., Singh, J., Green, R.D. and Berg, D.E. (2005). Mutation discovery in bacterial genomes, metronidazole resistance in Helicobacter pylori. Nat. Methods 2, 951–953.

Almeida, R., Gilmartin, B.J., McCann, S.H., Norrish, A., Ivens, A.C., Lawson, D., Levick, M.P., Smith, D.F., Dyall, S.D., Vetrie, D., Freeman, T.C., Coulson, R.M., Sampaio, I., Schneider, H. and Blackwell, J.M. (2004). Expression profiling of the *Leishmania* life cycle, cDNA arrays identify developmentally regulated genes present but not annotated in the genome. Mol. Biochem. Parasitol. *136*, 87–100.

Alrajhi, A.A., Ibrahim, E.A., De Vol, E.B., Khairat, M., Faris, R.M. and Maguire, J.H. (2002). Fluconazole for the treatment of cutaneous leishmaniasis caused by *Leishmania major*. N. Engl. J. Med. *346*, 891–895.

Anacleto, C., Abdo, M.C., Ferreira, A.V., Murta, S.M., Romanha, A.J., Fernandes, A.P. and Moreira, E.S. (2003). Structural and functional analysis of an amplification containing a PGPA gene in a glucantime-resistant *Leishmania (Viannia) guyanensis* cell line. Parasitol. Res. *90*, 110–118.

Arana, F.E., Perez-Victoria, J.M., Repetto, Y., Morello, A., Castanys, S. and Gamarro, F. (1998). Involvement of thiol metabolism in resistance to glucantime in *Leishmania tropica* [In Process Citation]. Biochem. Pharmacol 56, 1201–1208.

Arrebola, R., Olmo, A., Reche, P., Garvey, E.P., Santi, D.V., Ruiz-Perez, L.M. and Gonzalez-Pacanowska, D. (1994). Isolation and characterization of a mutant dihydrofolate reductase-thymidylate synthase from methotrexate-resistant *Leishmania* cells. J. Biol. Chem. *269*, 10590–10596.

Ashutosh, Gupta, S., Ramesh, Sundar, S. and Goyal, N. (2005). Use of *Leishmania donovani* field isolates expressing the luciferase reporter gene in in vitro drug screening. Antimicrob. Agents Chemother. *49*, 3776–3783.

Barker, K.S., Crisp, S., Wiederhold, N., Lewis, R.E., Bareither, B., Eckstein, J., Barbuch, R., Bard, M. and Rogers, P.D. (2004). Genome-wide expression profiling reveals genes associated with amphotericin B and fluconazole resistance in experimentally induced antifungal resistant isolates of *Candida albicans*. J. Antimicrob. Chemother. *54*, 376–385.

Basselin, M., Coombs, G.H. and Barrett, M.P. (2000). Putrescine and spermidine transport in *Leishmania*. Mol. Biochem. Parasitol. *109*, 37–46.

Basselin, M., Denise, H., Coombs, G.H. and Barrett, M.P. (2002). Resistance to pentamidine in *Leishmania mexicana* involves exclusion of the drug from the mitochondrion. Antimicrob. Agents Chemother. *46*, 3731–3738.

Beverley, S.M., Coderre, J.A., Santi, D.V. and Schimke, R.T. (1984). Unstable DNA amplifications in methotrexate-resistant *Leishmania* consist of extrachromosomal circles which relocalize during stabilization. Cell 38, 431–439.

Borst, P. and Elferink, R.O. (2002). Mammalian ABC transporters in health and disease. Annu. Rev. Biochem. *71*, 537–592. Epub (2001). Nov 2009.

Brochu, C., Wang, J., Roy, G., Messier, N., Wang, X.Y., Saravia, N.G. and Ouellette, M. (2003). Antimony uptake systems in the protozoan parasite *Leishmania* and accumulation differences in antimony-resistant parasites. Antimicrob. Agents Chemother. *47*, 3073–3079.

Brochu, C., Haimeur, A. and Ouellette, M. (2004). The heat shock protein HSP70 and heat shock cognate protein HSC70 contribute to antimony tolerance in the protozoan parasite *Leishmania*. Cell Stress Chaperones 9, 294–303.

Callahan, H.L. and Beverley, S.M. (1992). A member of the aldoketo reductase family confers methotrexate resistance in *Leishmania*. J. Biol. Chem. *267*, 24165–24168.

Carter, N.S., Drew, M.E., Sanchez, M., Vasudevan, G., Landfear, S.M. and Ullman, B. (2000). Cloning of a novel inosine-guanosine transporter gene from *Leishmania donovani* by functional rescue of a transport-deficient mutant. J. Biol. Chem. *275*, 20935–20941.

Charest, H., Zhang, W.W. and Matlashewski, G. (1996). The developmental expression of *Leishmania donovani* A2 amastigote-specific genes is post-transcriptionally mediated and involves elements located in the 3′-untranslated region. J. Biol. Chem. *271*, 17081–17090.

Chiquero, M.J., Olmo, A., Navarro, P., Ruiz-Perez, L.M., Castanys, S., Gonzalez-Pacanowska, D. and Gamarro, F. (1994). Amplification of the H locus in *Leishmania infantum*. Biochim Biophys Acta *1227*, 188–194.

Chiquero, M.J., Perez-Victoria, J.M., F, O.V., Gonzalez-Ros, J.M., del Moral, R.G., Ferragut, J.A., Castanys, S. and Gamarro, F. (1998). Altered drug membrane permeability in a multidrug-resistant *Leishmania tropica* line [In Process Citation]. Biochem. Pharmacol 55, 131–139.

Chow, L.M., Wong, A.K., Ullman, B. and Wirth, D.F. (1993). Cloning and functional analysis of an extrachromosomally amplified multidrug resistance-like gene in *Leishmania enriettii*. Mol. Biochem. Parasitol. *60*, 195–208.

Clos, J. and Choudhury, K. (2006). Functional cloning as a means to identify *Leishmania* genes involved in drug resistance. Mini Rev. Med. Chem. *6*, 123–129.

Coderre, J.A., Beverley, S.M., Schimke, R.T. and Santi, D.V. (1983). Overproduction of a bifunctional thymidylate synthetase-dihydrofolate reductase and DNA amplification in methotrexate-resistant *Leishmania tropica*. Proc. Natl. Acad. Sci. USA *80*, 2132–2136.

Coelho, A.C., Beverley, S.M. and Cotrim, P.C. (2003). Functional genetic identification of PRP1, an ABC transporter superfamily member conferring pentamidine resistance in *Leishmania major*. Mol. Biochem. Parasitol. *130*, 83–90.

Cotrim, P.C., Garrity, L.K. and Beverley, S.M. (1999). Isolation of genes mediating resistance to inhibitors of nucleoside and ergosterol metabolism in *Leishmania* by overexpression/selection. J. Biol. Chem. *274*, 37723–37730.

Cowen, L.E., Nantel, A., Whiteway, M.S., Thomas, D.Y., Tessier, D.C., Kohn, L.M. and Anderson, J.B. (2002). Population genomics of drug resistance in *Candida albicans*. Proc. Natl. Acad. Sci. USA. *99*, 9284–9289.

Croft, S.L., Neal, R.A., Pendergast, W. and Chan, J.H. (1987). The activity of alkyl phosphorylcholines and related derivatives against *Leishmania donovani*. Biochem. Pharmacol 36, 2633–2636.

Croft, S.L. (2001). Monitoring drug resistance in leishmaniasis. Trop Med. Int. Health 6, 899–905.

Croft, S.L., Sundar, S. and Fairlamb, A.H. (2006). Drug resistance in leishmaniasis. Clin. Microbiol. Rev. *19*, 111–126.

Debrabant, A., Lee, N., Bertholet, S., Duncan, R. and Nakhasi, H.L. (2003). Programmed cell death in trypanosomatids and other unicellular organisms. Int. J. Parasitol. *33*, 257–267.

Decuypere, S., Rijal, S., Yardley, V., De Doncker, S., Laurent, T., Khanal, B., Chappuis, F. and Dujardin, J.C. (2005). Gene Expression analysis of the mechanism of natural Sb(V) resistance in *Leishmania donovani* Isolates from Nepal. Antimicrob. Agents Chemother. *49*, 4616–4621.

Denton, H., McGregor, J.C. and Coombs, G.H. (2004). Reduction of anti-leishmanial pentavalent antimonial drugs by a parasite-specific thiol dependent reductase TDR1. Biochem. J. 1.

Detke, S., Chaudhuri, G., Kink, J.A. and Chang, K.P. (1988). DNA amplification in tunicamycin-resistant *Leishmania mexicana*. Multicopies of a single 63-kilobase supercoiled molecule and their expression. J. Biol. Chem. *263*, 3418–3424.

Detke, S. (1997). Identification of a transcription factor like protein at the TOR locus in *Leishmania mexicana* amazonensis. Mol. Biochem. Parasitol. 90, 505–511.

Dey, S., Papadopoulou, B., Haimeur, A., Roy, G., Grondin, K., Dou, D., Rosen, B.P. and Ouellette, M. (1994). High level arsenite resistance in *Leishmania tarentolae* is mediated by an active extrusion system. Mol. Biochem. Parasitol. *67*, 49–57.

Dey, S., Ouellette, M., Lightbody, J., Papadopoulou, B. and Rosen, B.P. (1996). An ATP-dependent As(III)-glutathione transport system in membrane vesicles of *Leishmania tarentolae*. Proc. Natl. Acad. Sci. USA. 93, 2192–2197.

Drummelsmith, J., Brochu, V., Girard, I., Messier, N. and Ouellette, M. (2003). Proteome mapping of the protozoan parasite leishmania and application to the study of drug targets and resistance mechanisms. Mol. Cell Proteomics 2, 146–155.

Drummelsmith, J., Girard, I., Trudel, N. and Ouellette, M. (2004). Differential protein expression analysis of *Leishmania major* reveals novel roles for methionine adenosyltransferase and S-adenosylmethionine in methotrexate resistance. J. Biol. Chem. *279*, 33273–33280.

Dube, A., Singh, N., Sundar, S. and Singh, N. (2005). Refractoriness to the treatment of sodium stibogluconate in Indian kala-azar field isolates persist in in vitro and in vivo experimental models. Parasitol. Res. 96, 216–223.

El Fadili, K., Messier, N., Leprohon, P., Roy, G., Guimond, C., Trudel, N., Saravia, N.G., Papadopoulou, B., Legare, D. and Ouellette, M. (2005). Role of the ABC transporter MRPA (PGPA) in antimony resistance in *Leishmania infantum* axenic and intracellular amastigotes. Antimicrob. Agents Chemother. *49*, 1988-(1993).

Ellenberger, T.E. and Beverley, S.M. (1989). Multiple drug resistance and conservative amplification of the H region in *Leishmania major*. J. Biol. Chem. *264*, 15094–15103.

Ephros, M., Bitnun, A., Shaked, P., Waldman, E. and Zilberstein, D. (1999). Stage-Specific Activity of Pentavalent Antimony against *Leishmania donovani* Axenic Amastigotes. Antimicrob. Agents Chemother. *43*, 278–282.

Fairlamb, A.H. and Cerami, A. (1992). Metabolism and functions of trypanothione in the Kinetoplastida. Annu. Rev. Microbiol. *46*, 695–729.

Faraut-Gambarelli, F., Piarroux, R., Deniau, M., Giusiano, B., Marty, P., Michel, G., Faugere, B. and Dumon, H. (1997). In vitro and in vivo resistance of *Leishmania infantum* to meglumine antimoniate, a study of 37 strains collected from patients with visceral leishmaniasis. Antimicrob. Agents Chemother. *41*, 827–830.

Ferreira-Pinto, K.C., Miranda-Vilela, A.L., Anacleto, C., Fernandes, A.P., Abdo, M.C., Petrillo-Peixoto, M.L. and Moreira, E.S. (1996). *Leishmania (V.) guyanensis*, isolation and characterization of glucantime-resistant cell lines. Can J. Microbiol. 42, 944–949.

Fong, D., Chan, M.M., Rodriguez, R., Gately, L.J., Berman, J.D. and Grogl, M. (1994). Paromomycin resistance in *Leishmania tropica*, lack of correlation with mutation in the small subunit ribosomal RNA gene. Am. J. Trop Med. Hyg 51, 758–766.

Foucher, A., Papadopoulou, B. and Ouellette, M. (2006). Prefractionation by digitonin extraction increased representation of the cytosolic and intracellular proteome of *Leishmania infantum*. J. Proteome Res. 5, 1741–1750.

Foucher, A.L., McIntosh, A., Douce, G., Wastling, J., Tait, A. and Turner, C.M. (2006). A proteomic analysis of arsenical drug resistance in *Trypanosoma brucei*. Proteomics 6, 2726–2732.

Garvey, E.P. and Santi, D.V. (1986). Stable amplified DNA in drug-resistant *Leishmania* exists as extrachromosomal circles. Science *233*, 535–540.

Gmuender, H., Kuratli, K., Di Padova, K., Gray, C.P., Keck, W. and Evers, S. (2001). Gene expression changes triggered by exposure of *Haemophilus influenzae* to novobiocin or ciprofloxacin, combined transcription and translation analysis. Genome Res. *11*, 28–42.

Goad, L.J., Holz, G.G. and Beach, D.H. (1984). Sterols of *Leishmania* species. Implications for biosynthesis. Mol. Biochem. Parasitol. *10*, 161–170.

Gourbal, B., Sonuc, N., Bhattacharjee, H., Legare, D., Sundar, S., Ouellette, M., Rosen, B.P. and Mukhopadhyay, R. (2004). Drug uptake and modulation of drug resistance in *Leishmania* by an aquaglyceroporin. J. Biol. Chem. *279*, 31010–31017.

Grondin, K., Papadopoulou, B. and Ouellette, M. (1993). Homologous recombination between direct repeat sequences yields P-glycoprotein containing amplicons in arsenite resistant *Leishmania*. Nucleic Acids Res. *21*, 1895–1901.

Grondin, K., Haimeur, A., Mukhopadhyay, R., Rosen, B.P. and Ouellette, M. (1997). Co-amplification of the gamma-glutamylcysteine synthetase gene gsh1 and of the ABC transporter gene pgpA in arsenite-resistant *Leishmania tarentolae*. Embo J. *16*, 3057–3065.

Grondin, K., Kundig, C., Roy, G. and Ouellette, M. (1998). Linear amplicons as precursors of amplified circles in methotrexate-resistant *Leishmania tarentolae*. Nucleic Acids Res. *26*, 3372–3378.

Guerin, P.J., Olliaro, P., Sundar, S., Boelaert, M., Croft, S.L., Desjeux, P., Wasunna, M.K. and Bryceson, A.D. (2002). Visceral leishmaniasis, current status of control, diagnosis, and treatment, and a proposed research and development agenda. Lancet Infect. Dis. 2, 494–501.

Guimond, C., Trudel, N., Brochu, C., Marquis, N., Fadili, A.E., Peytavi, R., Briand, G., Richard, D., Messier, N., Papadopoulou, B., Corbeil, J., Bergeron, M.G., Legare, D. and Ouellette, M. (2003). Modulation of gene expression in *Leishmania* drug resistant mutants as determined by targeted DNA microarrays. Nucleic Acids Res. *31*, 5886–5896.

Haas, W., Kaushal, D., Sublett, J., Obert, C. and Tuomanen, E.I. (2005). Vancomycin stress response in a sensitive and a tolerant strain of Streptococcus pneumoniae. J. Bacteriol. *187*, 8205–8210.

Hadighi, R., Mohebali, M., Boucher, P., Hajjaran, H., Khamesipour, A. and Ouellette, M. (2006). Unresponsiveness to Glucantime treatment in Iranian cutaneous leishmaniasis due to drug resistant *Leishmania tropica* and *L. major* parasites. PLOS Med. 3, e162.

Haimeur, A. and Ouellette, M. (1998). Gene amplification in *Leishmania tarentolae* selected for resistance to sodium stibogluconate [In Process Citation]. Antimicrob. Agents Chemother. *42*, 1689–1694.

Haimeur, A., Guimond, C., Pilote, S., Mukhopadhyay, R., Rosen, B.P., Poulin, R. and Ouellette, M. (1999). Elevated levels of polyamines and trypanothione resulting from overexpression of the ornithine decarboxylase gene in arsenite-resistant *Leishmania*. Mol. Microbiol. *34*, 726–735.

Haimeur, A., Brochu, C., Genest, P., Papadopoulou, B. and Ouellette, M. (2000). Amplification of the ABC transporter gene PGPA and increased trypanothione levels in potassium antimonyl tartrate (SbIII) resistant *Leishmania tarentolae*. Mol. Biochem. Parasitol. *108*, 131–135.

Hanson, S., Beverley, S.M., Wagner, W. and Ullman, B. (1992). Unstable amplification of two extrachromosomal elements in alpha-difluoromethylornithine-resistant *Leishmania donovani*. Mol. Cell Biol. *12*, 5499–5507.

Henderson, D.M., Sifri, C.D., Rodgers, M., Wirth, D.F., Hendrickson, N. and Ullman, B. (1992). Multidrug resistance in *Leishmania donovani* is conferred by amplification of a gene homologous to the mammalian mdr1 gene. Mol. Cell Biol. *12*, 2855–2865.

Holzer, T.R., McMaster, W.R. and Forney, J.D. (2006). Expression profiling by whole-genome interspecies microarray hybridization reveals differential gene expression in procyclic promastigotes, lesion-derived amastigotes, and axenic amastigotes in *Leishmania mexicana*. Mol. Biochem. Parasitol. *146*, 198–218.

Hooshdaran, M.Z., Barker, K.S., Hilliard, G.M., Kusch, H., Morschhauser, J. and Rogers, P.D. (2004). Proteomic analysis of azole resistance in Candida albicans clinical isolates. Antimicrob. Agents Chemother. *48*, 2733–2735.

Ivens, A.C., Peacock, C.S., Worthey, E.A., Murphy, L., Aggarwal, G., Berriman, M., Sisk, E., Rajandream, M.A., Adlem, E., Aert, R., Anupama, A., Apostolou, Z., Attipoe, P., Bason, N., Bauser, C., Beck, A., Beverley, S.M., Bianchettin, G., Borzym, K., Bothe, G., Bruschi, C.V., Collins, M., Cadag, E., Ciarloni, L., Clayton, C., Coulson, R.M., Cronin, A., Cruz, A.K., Davies, R.M., De Gaudenzi, J., Dobson, D.E., Duesterhoeft, A., Fazelina, G., Fosker, N., Frasch, A.C., Fraser, A., Fuchs, M., Gabel, C., Goble, A., Goffeau, A., Harris, D., Hertz-Fowler, C., Hilbert, H., Horn, D., Huang, Y., Klages, S., Knights, A., Kube, M., Larke, N., Litvin, L., Lord, A., Louie, T., Marra, M., Masuy, D., Matthews, K., Michaeli, S., Mottram, J.C., Muller-Auer, S., Munden, H., Nelson, S., Norbertczak, H., Oliver, K., O'Neil, S., Pentony, M., Pohl, T.M., Price, C., Purnelle, B., Quail, M.A., Rabbinowitsch, E., Reinhardt, R., Rieger, M., Rinta, J., Robben, J., Robertson, L., Ruiz, J.C., Rutter, S., Saunders, D., Schafer, M., Schein, J., Schwartz, D.C., Seeger, K., Seyler, A., Sharp, S., Shin, H., Sivam, D., Squares, R., Squares, S., Tosato, V., Vogt, C., Volckaert, G., Wambutt, R., Warren, T., Wedler, H., Woodward, J., Zhou, S., Zimmermann, W., Smith, D.F., Blackwell, J.M., Stuart, K.D., Barrell, B. and Myler, P.J. (2005). The genome of the kinetoplastid parasite, *Leishmania major*. Science *309*, 436–442.

Jackson, J.E., Tally, J.D., Ellis, W.Y., Mebrahtu, Y.B., Lawyer, P.G., Were, J.B., Reed, S.G., Panisko, D.M. and Limmer, B.L. (1990). Quantitative in vitro drug potency and drug susceptibility evaluation of *Leishmania* ssp. from patients unresponsive to pentavalent antimony therapy. Am. J. Trop Med. Hyg *43*, 464–480.

Kandpal, M. and Tekwani, B.L. (1997). Polyamine transport systems of *Leishmania donovani* promastigotes. Life Sci. *60*, 1793–1801.

Katakura, K. and Chang, K.P. (1989). H DNA amplification in *Leishmania* resistant to both arsenite and methotrexate. Mol. Biochem. Parasitol. *34*, 189–191.

Katakura, K., Iwanami, M., Ohtomo, H., Fujise, H. and Hashiguchi, Y. (1999). Structural and functional analysis of the LaMDR1 multidrug resistance gene in *Leishmania amazonensis*. Biochem. Biophys Res. Commun. *255*, 289–294.

Kelly, J.M., Das, P. and Tomas, A.M. (1994). An approach to functional complementation by introduction of large DNA fragments into *Trypanosoma cruzi* and *Leishmania donovani* using a cosmid shuttle vector. Mol. Biochem. Parasitol. *65*, 51–62.

King, L.S., Kozono, D. and Agre, P. (2004). From structure to disease, the evolving tale of aquaporin biology. Nat. Rev. Mol. Cell Biol. *5*, 687–698.

Kündig, C., Haimeur, A., Légaré, D., Papadopoulou, B. and Ouellette, M. (1999). Increased transport of pteridines compensates for mutations in the high affinity folate transporter and contributes to methotrexate resistance in the protozoan parasite *Leishmania tarentolae*. Embo J. *18*, 2342–2351.

Kündig, C., Leblanc, E., Papadopoulou, B. and Ouellette, M. (1999). Role of the locus and of the resistance gene on gene amplification frequency in methotrexate resistant *Leishmania tarentolae*. Nucleic Acids Res. *27*, 3653–3659.

Lamontagne, J. and Papadopoulou, B. (1999). Developmental Regulation of Spliced Leader RNA Gene in *Leishmania donovani* Amastigotes Is Mediated by Specific Polyadenylation. J. Biol. Chem. *274*, 6602–6609.

Lee, N., Bertholet, S., Debrabant, A., Muller, J., Duncan, R. and Nakhasi, H.L. (2002). Programmed cell death in the unicellular protozoan parasite *Leishmania*. Cell Death Differ *9*, 53–64.

Légaré, D., Richard, D., Mukhopadhyay, R., Stierhof, Y.D., Rosen, B.P., Haimeur, A., Papadopoulou, B. and Ouellette, M. (2001). The *Leishmania* ABC protein PGPA is an intracellular metal-thiol transporter ATPase. J. Biol. Chem. *276*, 26301–26307.

Lemley, C., Yan, S., Dole, V.S., Madhubala, R., Cunningham, M.L., Beverley, S.M., Myler, P.J. and Stuart, K.D. (1999). The *Leishmania donovani* LD1 locus gene ORFG encodes a biopterin transporter (BT1) [In Process Citation]. Mol. Biochem. Parasitol. *104*, 93–105.

Leprohon, P., Légaré, D., Girard, I., Papadopoulou, B. and Ouellette, M. (2006). Modulation of *Leishmania* ABC protein gene expression through its life cycle and among drug resistant parasites. Euk. Cell 5, 1713–1725.

Lira, R., Sundar, S., Makharia, A., Kenney, R., Gam, A., Saraiva, E. and Sacks, D. (1999). Evidence that the High Incidence of Treatment Failures in Indian Kala-Azar Is Due to the Emergence of Antimony-Resistant Strains of *Leishmania donovani*. J. Infect. Dis. *180*, 564–567.

Lux, H., Heise, N., Klenner, T., Hart, D. and Opperdoes, F.R. (2000). Ether – lipid (alkyl-phospholipid) metabolism and the mechanism of action of ether – lipid analogues in *Leishmania*. Mol. Biochem. Parasitol. *111*, 1–14.

Lye, L.F., Hsu, J.Y., Singh, A.K., Su, K.E. and Lee, S.T. (1999). Characterization by pulse-field electrophoresis of a new region of DNA amplification containing the M2 subunit of ribonucleotide reductase in hydroxyurea-resistant *Leishmania*. Parasitol. Res. *85*, 188–193.

Maarouf, M., Lawrence, F., Croft, S.L. and Robert-Gero, M. (1995). Ribosomes of *Leishmania* are a target for the aminoglycosides. Parasitol. Res. *81*, 421–425.

Maarouf, M., Lawrence, F., Brown, S. and Robert-Gero, M. (1997). Biochemical alterations in paromomycin-treated *Leishmania donovani* promastigotes. Parasitol. Res. *83*, 198–202.

Maarouf, M., Adeline, M.T., Solignac, M., Vautrin, D. and Robert-Gero, M. (1998). Development and characterization of paromomycin-resistant *Leishmania donovani* promastigotes. Parasite 5, 167–173.

Marchini, J.F., Cruz, A.K., Beverley, S.M. and Tosi, L.R. (2003). The H region HTBF gene mediates terbinafine resistance in *Leishmania major*. Mol. Biochem. Parasitol. *131*, 77–81.

Margulies, M., Egholm, M., Altman, W.E., Attiya, S., Bader, J.S., Bemben, L.A., Berka, J., Braverman, M.S., Chen, Y.J., Chen, Z., Dewell, S.B., Du, L., Fierro, J.M., Gomes, X.V., Godwin, B.C., He, W., Helgesen, S., Ho, C.H., Irzyk, G.P., Jando, S.C., Alenquer, M.L., Jarvie, T.P., Jirage, K.B., Kim, J.B., Knight, J.R., Lanza, J.R., Leamon, J.H., Lefkowitz, S.M., Lei, M., Li, J., Lohman, K.L., Lu, H., Makhijani, V.B., McDade, K.E., McKenna, M.P., Myers, E.W., Nickerson, E., Nobile, J.R., Plant, R., Puc, B.P., Ronan, M.T., Roth, G.T., Sarkis, G.J., Simons, J.F., Simpson, J.W., Srinivasan, M., Tartaro, K.R., Tomasz, A., Vogt, K.A., Volkmer, G.A., Wang, S.H., Wang, Y., Weiner, M.P., Yu, P., Begley, R.F. and Rothberg, J.M. (2005). Genome sequencing in microfabricated high-density picolitre reactors. Nature *437*, 376–380.

Marquis, N., Gourbal, B., Rosen, B.P., Mukhopadhyay, R. and Ouellette, M. (2005). Modulation in aquaglyceroporin AQP1 gene transcript levels in drug-resistant *Leishmania*. Mol. Microbiol. *57*, 1690–1699.

Mbongo, N., Loiseau, P.M., Billion, M.A. and Robert-Gero, M. (1998). Mechanism of amphotericin B resistance in *Leishmania donovani* promastigotes. Antimicrob. Agents Chemother. *42*, 352–357.

McAtee, C.P., Hoffman, P.S. and Berg, D.E. (2001). Identification of differentially regulated proteins in metronidozole resistant Helicobacter pylori by proteome techniques. Proteomics *1*, 516–521.

McNicoll, F., Drummelsmith, J., Muller, M., Madore, E., Ouellette, M. and Papadopoulou, B. (2006). A combined proteomic and transcriptomic approach to the study of stage differentiation in *Leishmania infantum*. Proteomics 6, 3567–3581.

Mehta, A. and Shaha, C. (2004). Apoptotic death in *Leishmania donovani* promastigotes in response to respiratory chain inhibition, complex II inhibition results in increased pentamidine cytotoxicity. J. Biol. Chem. *279*, 11798–11813. Epub 1(2003). Dec 11716.

Mongodin, E., Finan, J., Climo, M.W., Rosato, A., Gill, S. and Archer, G.L. (2003). Microarray transcription analysis of clinical Staphylococcus aureus isolates resistant to vancomycin. J. Bacteriol. *185*, 4638–4643.

Mukherjee, A., Padmanabhan, P.K., Sahani, M.H., Barrett, M.P. and Madhubala, R. (2005). Roles for mitochondria in pentamidine susceptibility and resistance in *Leishmania donovani*. Mol. Biochem. Parasitol.

Mukherjee, A., Padmanabhan, P.K., Singh, S., Roy, G., Girard, I., Chatterjee, M., Ouellette, M. and Madhubala, R. (2007). Molecular analysis of antimony-resistant Indian field isolates of *Leishmania donovani*. J. Antimicrob. Chemother. 59, 204–211.

Mukhopadhyay, R., Dey, S., Xu, N., Gage, D., Lightbody, J., Ouellette, M. and Rosen, B.P. (1996). Trypanothione overproduction and resistance to antimonials and arsenicals in *Leishmania*. Proc. Natl. Acad. Sci. USA. 93, 10383–10387.

Ng, W.L., Kazmierczak, K.M., Robertson, G.T., Gilmour, R. and Winkler, M.E. (2003). Transcriptional regulation and signature patterns revealed by microarray analyses of *Streptococcus pneumoniae* R6 challenged with sublethal concentrations of translation inhibitors. J. Bacteriol. *185*, 359–370.

Olliaro, P.L. and Bryceson, A.D.M. (1993). Practical progress and new drugs for changing patterns of leishmaniasis. Parasitol. Today 9, 323–328.

Ouellette, M., Fase-Fowler, F. and Borst, P. (1990). The amplified H circle of methotrexate-resistant leishmania tarentolae contains a novel P-glycoprotein gene. Embo J. *9*, 1027–1033.

Ouellette, M., Hettema, E., Wust, D., Fase-Fowler, F. and Borst, P. (1991). Direct and inverted DNA repeats associated with P-glycoprotein gene amplification in drug resistant *Leishmania*. Embo J. *10*, 1009–1016.

Ouellette, M., Haimeur, A., Grondin, K., Legare, D. and Papadopoulou, B. (1998). Amplification of ABC transporter gene pgpA and of other heavy metal resistance genes in *Leishmania tarentolae* and their study by gene transfection and gene disruption [In Process Citation]. Methods Enzymol *292*, 182–193.

Ouellette, M. and Ward, S. (2002). Drug resistance in parasites. In, Molecular Medical Parasitology. J. Marr, T. Nielsen and R. Komuniecki ed. Academic Press. p. 395–430.

Ouellette, M., Drummelsmith, J. and Papadopoulou, B. (2004). Leishmaniasis, drugs in the clinic, resistance and new developments. Drug Resist Updat *7*, 257–266.

Papadopoulou, B., Roy, G. and Ouellette, M. (1992). A novel antifolate resistance gene on the amplified H circle of Leishmania. Embo J. *11*, 3601–3608.

Papadopoulou, B., Drummelsmith, J. and Ouellette, M. (2004). Proteomics to Explore Pathogenesis and Drug Resistance Mechanisms in Protozoan Parasites. In, Biomediacal and Pharmaceutical applications of proteomics. H. Hondermarck ed. Kluwer Academic Publisher.

Paris, C., Loiseau, P.M., Bories, C. and Breard, J. (2004). Miltefosine induces apoptosis-like death in *Leishmania donovani* promastigotes. Antimicrob. Agents Chemother. *48*, 852–859.

Pathak, M.K. and Yi, T. (2001). Sodium stibogluconate is a potent inhibitor of protein tyrosine phosphatases and augments cytokine responses in hemopoietic cell lines. J. Immunol. *167*, 3391–3397.

Perez-Victoria, F.J., Castanys, S. and Gamarro, F. (2003). *Leishmania donovani* resistance to miltefosine involves a defective inward translocation of the drug. Antimicrob. Agents Chemother. *47*, 2397–2403.

Perez-Victoria, F.J., Gamarro, F., Ouellette, M. and Castanys, S. (2003). Functional cloning of the miltefosine transporter. A novel P-type phospholipid translocase from *Leishmania* involved in drug resistance. J. Biol. Chem. *278*, 49965–49971. Epub 4(2003). Sep 49927.

Perez-Victoria, J.M., Perez-Victoria, F.J., Parodi-Talice, A., Jimenez, I.A., Ravelo, A.G., Castanys, S. and Gamarro, F. (2001). Alkyl-Lysophospholipid Resistance in Multidrug-Resistant *Leishmania tropica* and Chemosensitization by a Novel P-Glycoprotein-Like Transporter Modulator. Antimicrob. Agents Chemother. *45*, 2468–2474.

Pollack, J.R., Perou, C.M., Alizadeh, A.A., Eisen, M.B., Pergamenschikov, A., Williams, C.F., Jeffrey, S.S., Botstein, D. and Brown, P.O. (1999). Genome-wide analysis of DNA copy-number changes using cDNA microarrays. Nat. Genet. *23*, 41–46.

Pourshafie, M., Morand, S., Virion, A., Rakotomanga, M., Dupuy, C. and Loiseau, P.M. (2004). Cloning of S-adenosyl-L-methionine,C-24-Delta-sterol-methyltransferase (ERG6) from *Leishmania donovani* and characterization of mRNAs in wild-type and amphotericin B-resistant promastigotes. Antimicrob. Agents Chemother. *48*, 2409–2414.

Quijada, L., Soto, M. and Requena, J.M. (2005). Genomic DNA macroarrays as a tool for analysis of gene expression in *Leishmania*. Exp. Parasitol. *111*, 64–70.

Richard, D., Leprohon, P., Drummelsmith, J. and Ouellette, M. (2004). Growth phase regulation of the main folate transporter of *Leishmania infantum* and its role in methotrexate resistance. J. Biol. Chem. *279*, 54494–54501.

Righetti, P.G., Castagna, A., Antonioli, P. and Boschetti, E. (2005). Prefractionation techniques in proteome analysis, the mining tools of the third millennium. Electrophoresis *26*, 297–319.

Roberts, C.W., McLeod, R., Rice, D.W., Ginger, M., Chance, M.L. and Goad, L.J. (2003). Fatty acid and sterol metabolism, potential antimicrobial targets in apicomplexan and trypanosomatid parasitic protozoa. Mol. Biochem. Parasitol. *126*, 129–142.

Rojas, R., Valderrama, L., Valderrama, M., Varona, M.X., Ouellette, M. and Saravia, N.G. (2006). Resistance to antimony and treatment failure in human L. viannia infection. J. Infect. Dis. in press.

Roy, G., Dumas, C., Sereno, D., Wu, Y., Singh, A.K., Tremblay, M.J., Ouellette, M., Olivier, M. and Papadopoulou, B. (2000). Episomal and stable expression of the luciferase reporter gene for quantifying *Leishmania* spp. infections in macrophages and in animal models [In Process Citation]. Mol. Biochem. Parasitol. *110*, 195–206.

Roy, G., Kundig, C., Olivier, M., Papadopoulou, B. and Ouellette, M. (2001). Adaptation of leishmania cells to in vitro culture results in a more efficient reduction and transport of biopterin. Exp. Parasitol. *97*, 161–168.

Ryan, K.A., Dasgupta, S. and Beverley, S.M. (1993). Shuttle cosmid vectors for the trypanosomatid parasite *Leishmania*. Gene *131*, 145–150.

Ryan, K.A., Garraway, L.A., Descoteaux, A., Turco, S.J. and Beverley, S.M. (1993). Isolation of virulence genes directing surface glycosyl-phosphatidylinositol synthesis by functional complementation of *Leishmania*. Proc. Natl. Acad. Sci. USA. *90*, 8609–8613.

Salotra, P., Duncan, R.C., Singh, R., Subba Raju, B.V., Sreenivas, G. and Nakhasi, H.L. (2006). Upregulation of surface proteins in *Leishmania donovani* isolated from patients of post kala-azar dermal leishmaniasis. Microbes Infect.

Sanchez, C.P., Mucci, J., Gonzalez, N.S., Ochoa, A., Zakin, M.M. and Algranati, I.D. (1997). Alpha-difluoromethylornithine-resistant cell lines obtained after one-step selection of *Leishmania mexicana* promastigote cultures. Biochem. J. *324*, 847–853.

Santos Ferreira, C., Martins, P.S., Demicheli, C., Brochu, C., Ouellette, M. and Frezard, F. (2003). Thiol-induced reduction of antimony(V) into antimony(III), a comparative study with trypanothione, cysteinyl-glycine, cysteine and glutathione. Biometals *16*, 441–446.

Saxena, A., Worthey, E.A., Yan, S., Leland, A., Stuart, K.D. and Myler, P.J. (2003). Evaluation of differential gene expression in *Leishmania major* Friedlin procyclics and metacyclics using DNA microarray analysis. Mol. Biochem. Parasitol. *129*, 103–114.

Segovia, M. and Ortiz, G. (1997). LD1 amplifications in *Leishmania*. Parasitol. Today *13*, 342–348.

Seifert, K., Matu, S., Javier Perez-Victoria, F., Castanys, S., Gamarro, F. and Croft, S.L. (2003). Characterisation of *Leishmania donovani* promastigotes resistant to hexadecylphosphocholine (miltefosine). Int. J. Antimicrob. Agents *22*, 380–387.

Seifert, K. and Croft, S.L. (2006). In vitro and in vivo interactions between miltefosine and other antileishmanial drugs. Antimicrob. Agents Chemother. *50*, 73–79.

Sereno, D., Cavaleyra, M., Zemzoumi, K., Maquaire, S., Ouaissi, A. and Lemesre, J.L. (1998). Axenically Grown Amastigotes of *Leishmania infantum* Used as an In Vitro Model To Investigate the Pentavalent Antimony Mode of Action. Antimicrob. Agents Chemother. *42*, 3097–3102.

Sereno, D., Holzmuller, P., Mangot, I., Cuny, G., Ouaissi, A. and Lemesre, J.L. (2001). Antimonial-mediated DNA fragmentation in *Leishmania infantum* amastigotes. Antimicrob. Agents Chemother. *45*, 2064–2069.

Sereno, D., Roy, G., Lemesre, J.L., Papadopoulou, B. and Ouellette, M. (2001). DNA Transformation of *Leishmania infantum* axenic amastigotes and their use in drug screening. Antimicrob. Agents Chemother. *45*, 1168–1173.

Shaked-Mishan, P., Ulrich, N., Ephros, M. and Zilberstein, D. (2001). Novel intracellular SbV reducing activity correlates with antimony susceptibility in *Leishmania donovani*. J. Biol. Chem. *276*, 3971–3976.

Shendure, J., Porreca, G.J., Reppas, N.B., Lin, X., McCutcheon, J.P., Rosenbaum, A.M., Wang, M.D., Zhang, K., Mitra, R.D. and Church, G.M. (2005). Accurate multiplex polony sequencing of an evolved bacterial genome. Science *309*, 1728–1732.

Singh, A.K., Papadopoulou, B. and Ouellette, M. (2001). Gene amplification in amphotericin B-resistant *Leishmania tarentolae*. Exp. Parasitol. *99*, 141–147.

Singh, N., Singh, R.T. and Sundar, S. (2003). Novel mechanism of drug resistance in kala-azar field isolates. J. Infect. Dis. *188*, 600–607.

Soualhine, H., Brochu, V., Menard, F., Papadopoulou, B., Weiss, K., Bergeron, M.G., Legare, D., Drummelsmith, J. and Ouellette, M. (2005). A proteomic analysis of penicillin resistance in *Streptococcus pneumoniae* reveals a novel role for PstS, a subunit of the phosphate ABC transporter. Mol. Microbiol. *58*, 1430–1440.

Sundar, S., More, D.K., Singh, M.K., Singh, V.P., Sharma, S., Makharia, A., Kumar, P.C. and Murray, H.W. (2000). Failure of Pentavalent Antimony in Visceral Leishmaniasis in India, Report from the Center of the Indian Epidemic. Clin. Infect. Dis. *31*, 1104–1107.

Sundar, S. (2001). Drug resistance in Indian visceral leishmaniasis. Trop Med. Int. Health *6*, 849–854.

Sundar, S., Jha, T.K., Thakur, C.P., Engel, J., Sindermann, H., Fischer, C., Junge, K., Bryceson, A. and Berman, J. (2002). Oral miltefosine for Indian visceral leishmaniasis. N. Engl. J. Med. *347*, 1739–1746.

Sundar, S., Mehta, H., Suresh, A.V., Singh, S.P., Rai, M. and Murray, H.W. (2004). Amphotericin B treatment for Indian visceral leishmaniasis, conventional versus lipid formulations. Clin. Infect. Dis. *38*, 377–383. Epub (2004). Jan 2013.

Vasudevan, G., Carter, N.S., Drew, M.E., Beverley, S.M., Sanchez, M.A., Seyfang, A., Ullman, B. and Landfear, S.M. (1998). Cloning of leishmania nucleoside transporter genes by rescue of a transport-deficient mutant [In Process Citation]. Proc. Natl. Acad. Sci. USA. *95*, 9873–9878.

Vasudevan, G., Ullman, B. and Landfear, S.M. (2001). Point mutations in a nucleoside transporter gene from *Leishmania donovani* confer drug resistance and alter substrate selectivity. Proc. Natl. Acad. Sci. USA. *98*, 6092–6097.

Verma, N.K. and Dey, C.S. (2004). Possible mechanism of miltefosine-mediated death of *Leishmania donovani*. Antimicrob. Agents Chemother. *48*, 3010–3015.

Wilson, K., Collart, F.R., Huberman, E., Stringer, J.R. and Ullman, B. (1991). Amplification and molecular cloning of the IMP dehydrogenase gene of *Leishmania donovani*. J. Biol. Chem. *266*, 1665–1671.

Wilson, M., DeRisi, J., Kristensen, H.H., Imboden, P., Rane, S., Brown, P.O. and Schoolnik, G.K. (1999). Exploring drug-induced alterations in gene expression in *Mycobacterium tuberculosis* by microarray hybridization. Proc. Natl. Acad. Sci. USA. *96*, 12833–12838.

Wu, Y., El Fakhry, Y., Sereno, D., Tamar, S. and Papadopoulou, B. (2000). A new developmentally regulated gene family in *Leishmania* amastigotes encoding a homolog of amastin surface proteins. Mol. Biochem. Parasitol. *110*, 345–357.

Wyllie, S., Cunningham, M.L. and Fairlamb, A.H. (2004). Dual action of antimonial drugs on thiol redox metabolism in the human pathogen *Leishmania donovani*. J. Biol. Chem. in press.

Zangger, H., Mottram, J.C. and Fasel, N. (2002). Cell death in *Leishmania* induced by stress and differentiation, programmed cell death or necrosis? Cell Death Differ *9*, 1126–1139.

Zhou, Y., Messier, N., Ouellette, M., Rosen, B.P. and Mukhopadhyay, R. (2004). *Leishmania major* LmACR2 is a pentavalent antimony reductase that confers sensitivity to the drug pentostam. J. Biol. Chem. *279*, 37445–37451.

Leishmania Surface Proteins

9

Emanuela Handman, Anthony T. Papenfuss, Terence P. Speed and James W. Goding

Abstract

Leishmania are polarized flagellated cells with surface membrane domains of distinct composition and function. Specific surface proteins are distributed on the flagellum, the flagellar pocket and around the cell body, and they contribute to the functional specialization of these regions. The *Leishmania* genome database suggests that a few hundred different proteins are expected to be displayed on the parasite surface, but the number of well-characterized proteins has been disappointingly small.

The best characterized proteins are anchored into the membrane with GPI anchors. Much less is known about those that are held in the membrane with stretches of hydrophobic amino acids. The well-characterized surface proteins include ligands for the insect or the mammalian host. Remarkably, some of these proteins are heavily decorated with complex glycans that are similar or identical to those found on lipophosphoglycans that contain no protein.

Knowledge of the genomes of several *Leishmania* species, as well as those of *T. brucei* and *T. cruzi*, combined with the rapid development of algorithms for use in bioinformatics is providing tools of extraordinary power for the identification of the genes encoding all membrane proteins. However, as has been the case with most other eukaryotic genomes, a large percentage of predicted proteins have not yet been assigned a function. This remains the major challenge of the post-genomic era. New developments in gene deletion, modulation and overexpression will greatly facilitate the task. However, enthusiasm for a large-scale 'global' or 'systems' approach to membrane proteins should be moderated by knowledge that the remarkably diverse post-translational modifications seen in *Leishmania* and other species cannot be predicted from gene sequences. Biochemistry, cell biology and physiology are not obsolete. We will still need to study proteins of interest in detail, one by one.

Leishmania and their hosts: sandflies and vertebrates

Leishmania are members of the *Trypanosomatidae*, a large group of parasitic protozoa characterized by their digenetic life style. *Leishmania* are parasitic in blood-sucking sandflies which transmit them to vertebrate hosts. In the flies, they reside as flagellated promastigotes in the gut. In the vertebrate host they become obligatory intracellular amastigotes residing in macrophages and dendritic cells. As the *Leishmania* shuttle between the sandfly vectors and mammals, they undergo a series of developmental changes within each host. These biochemical and structural modifications have major physiological implications. They lead to the ability of the organisms to move from one host to the other and to establish infection in two very different environments. The structure and composition of the parasite membrane have been shown to play an important role in the adaptation to these environments.

Numerous adaptations have evolved to allow the specific colonization of the *Phlebotomus* or *Lutzomyia* sandflies by *Leishmania*, which are, to date, the only known vectors. Some *Leishmania* species and strains show strict preferences for particular *Phlebotomus* subspecies and do not survive in other related flies (Kamhawi *et al.*, 2000). This specific relationship has been shown to depend critically on the presence of a structurally polymorphic lipophosphoglycan (LPG) on the promastigote surface. LPG is thought to anchor the dividing procyclic promastigotes onto the gut epithelium in the insect vector and prevent their expulsion with the gut content.

Differentiation of the rapidly dividing procyclic promastigotes into the mature non-dividing metacyclic stage is accompanied by modifications to the structure of LPG, leading to masking of the oligosaccharides involved in binding to the gut and allowing the release of the parasites from their anchorage and migration to the mouth parts of the insect prior to inoculation into the vertebrate host at the next blood meal.

It is quite likely that parasite surface molecules other than LPG also play roles in the physiological adaptations to the fly environment. One of these is proteophosphoglycan (PPG), a mucin which is decorated with glycans similar to those of LPG (Ilg, 2000). PPG may bind to receptors similar to those for LPG on the gut. Another important contributor to the colonization of the gut may be the surface protease leishmanolysin/gp63 (Ramamoorthy *et al.*, 1992; Roberts *et al.*, 1995). These molecules will be discussed in more detail later in this review.

Turning to the mammalian host, *Leishmania* promastigotes, as well as amastigotes, show specificity for the macrophage, their main host cell. However, in contrast to the insect stage, there is a remarkably broad mammalian host range, including dogs, gerbils and hyraxes. Parasite surface molecules including LPG, leishmanolysin/gp63, PPG, and the Parasite Surface Antigen-2 family (PSA-2/gp46) have been shown to play a role in macrophage binding and establishment of infection (Handman *et al.*, 2002). Most of the studies that established a role for these molecules have involved infecting mice with promastigotes, although mice are not the natural host for *Leishmania*. It remains to be determined if and how these molecules function in the amastigote and in the natural hosts. It is tempting to speculate that the plasticity of *Leishmania*-mammalian host range is a reflection of the variety of surface ligands for macrophage receptors in different hosts.

Host lipids may also contribute to the ability of the amastigotes to colonize the diverse range of mammalian hosts. Unlike mammals, *Leishmania* synthesize inositol phosphorylceramide instead of sphingomyelin or glycosphingolipids. However, the parasites scavenge host sphingolipids, remodel them and use them in their surface membrane and organelles such as the acidocalcisomes (Zhang *et al.*, 2005). These remodelled sphingolipids seem to be particularly important for the survival of the amastigote in the phagolysosome (Zhang *et al.*, 2005). They may compensate for the lack of a thick glycocalyx and the loss of many of the surface glycoproteins and glycolipids during the transformation of the promastigotes into amastigotes (Winter *et al.*, 1994).

Most reviews on parasite proteins have focussed on the molecules and mechanisms used by the parasite to evade the host immune response (Schaible *et al.*, 2005). What is often ignored is the fact that some molecules and mechanisms have evolved to allow the parasite to exploit nutrients available only in a particular niche in the host (Schaible *et al.*, 2005). In the case of *Leishmania*, this niche may be the phagolysosome. For example, *Leishmania* cannot synthesize purines, but they can access the macrophage autophagosomes, where degradation and recycling of cellular material takes place, and from where they can scavenge purines (Schaible *et al.*, 2005). Uptake of purines requires specific receptors, which will be discussed later in this review.

Leishmania are polarized cells with functionally distinct membrane domains

The *Leishmania* is a polarized cell defined by an intact flagellum emerging from an open flagellar pocket at one end of the promastigote, and a rudimentary flagellum budding from a closed pocket in the amastigote (Fig. 9.1). As for all polarized cells, the surface membrane of the parasite can be divided into distinct functional domains. In *Leishmania*, the domains are morphologically distinct, and comprise the flagellar membrane, the flagellar pocket and the pellicular plasma membrane (Landfear *et al.*, 2001). Although

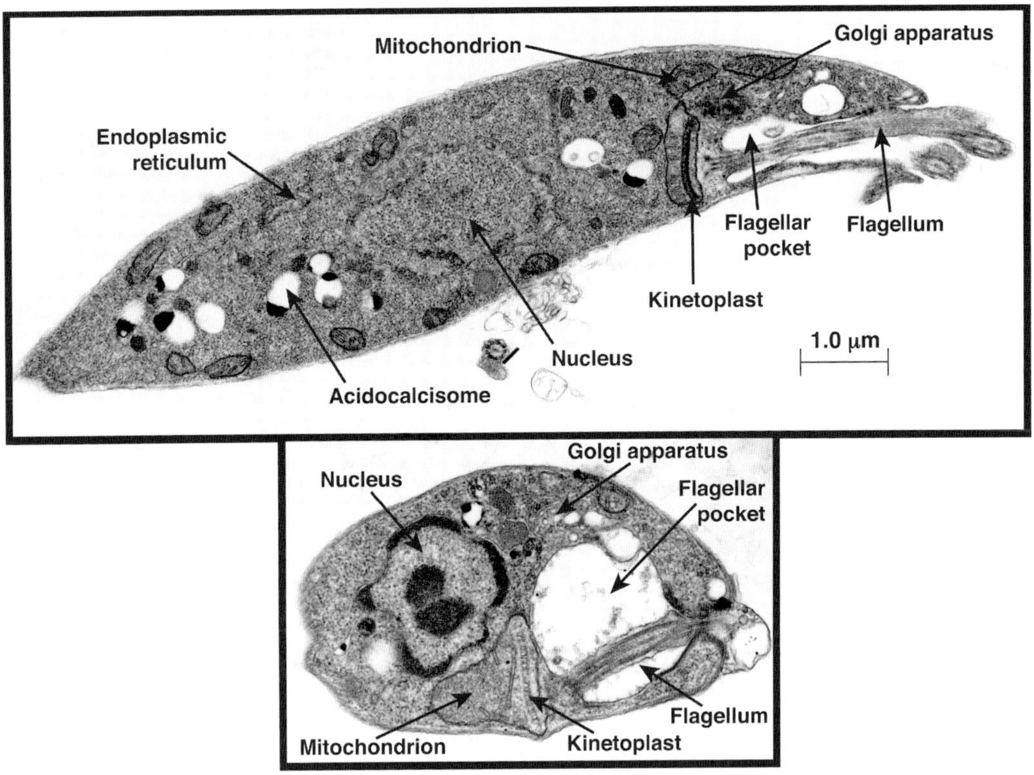

Figure 9.1 Ultrastructure of longitudinal sections through a *Leishmania* promastigote and amastigote showing the polarized cell structure. Also shown are the flagellum and flagellar pocket, the kinetoplast and mitochondrion, the nucleus, endoplasmic reticulum and the Golgi apparatus.

contiguous, these domains have unique proteins and lipids and perform specialized functions.

Flagellar membrane proteins

The flagellum and associated regions have a surprisingly diverse range of functions. The flagellum pulls the parasite forward, but it is also involved in anchoring the parasite to the gut of the fly. In trypanosomes, the flagellar basal body, which is connected to the mitochondrion, has been shown to play an unexpected role in cell division (Liu *et al.*, 2005; Ogbadoyi *et al.*, 2003). It is likely that it fulfils a similar function in *Leishmania* promastigotes, but it is not known what functions it may fulfil in the amastigote.

The flagellar membrane has some distinct proteins, which presumably fulfil specialized functions. One of these is isoform ISO1 or LeGT1 of the glucose transporters encoded by the PRO1 gene, originally identified in *L. enriettii* (Landfear, 2000) and the *L. mexicana LmGT1* homologue (Burchmore *et al.*, 2003). A closely related isoform (ISO2 or LmGT2) which differs from ISO1 in the amino terminal cytoplasmic domain is present in *L. mexicana* but is displayed around the entire promastigote body (see below). In ISO1, five amino acids in the cytoplasmic domain are necessary, but not sufficient, to target the protein to the flagellum. The mechanism responsible for flagellar membrane targeting has not been identified, but it has been suggested that specific motifs play a role together with a specialized intraflagellar transport system, as described in other organisms (Snapp *et al.*, 1999). Both isoforms of the glucose transporters seem to be preferentially expressed in promastigotes, and it is not clear if amastigotes use different pathways or different transporters (see below). The unusual location of the flagellar glucose transporter may indicate that its function is not transport, but rather glucose sensing, as has been shown for the mammalian GLUT1 (Bandyopadhyay *et al.*, 2000).

Another family of proteins present on the flagellum are receptor adenylate cyclases (Sanchez *et al.*, 1995) (see below). To date, neither the ligands nor their signalling pathways have been established.

Flagellar pocket proteins

The flagellar pocket is a deep invagination at the base of the flagellum, and it is responsible for both receptor-mediated endocytosis and export of proteins destined for secretion or incorporation into the cell membrane. Unlike the cytoplasmic face of the membrane surrounding the cell body, which is attached to a corset of subpellicular tubules, the flagellar pocket is free of microtubules and it has been suggested that endocytosis and secretion are restricted to this area (Gull, 1999). Nonetheless, there is some evidence that the microtubules may be actively involved in directing the vesicular traffic to the flagellar pocket (McConville *et al.*, 2002).

While all secreted and membrane proteins seem to transit through the flagellar pocket on their way to their final destination, very few membrane proteins have been described that are specifically retained in the flagellar pocket membrane. One protein of *T. brucei* has been localized to the flagellar pocket and shown to be targeted there by a combination of protein motifs and interaction with the cytoskeleton (Hill *et al.*, 1999).

Leishmania have been shown to contain a number of ATP-binding cassette (ABC) transporters (Ivens *et al.*, 2005), the most intensely studied are the P-glycoproteins which are involved in drug resistance, although this is obviously not their physiological function (Ouellette *et al.*, 2004). Most of these are expressed in the Golgi apparatus and endoplasmic reticulum, but one has been localized to the plasma membrane in the flagellar pocket (Dodge *et al.*, 2004). This is a member of the ABCA family, which is postulated to regulate the distribution of phospholipids in the membrane. Its precise function is not clear, but overexpression led to loss of virulence in macrophages (Dodge *et al.*, 2004).

Membrane proteins on the cell body

As mentioned above, the plasma membrane around the cell body is attached to a network of subpellicular microtubules which confer shape to the parasite. This array comprises more than 100 microtubules that are aligned along the anterior-posterior axis of the cell and are attached to the plasma membrane via acylated microtubule-binding proteins (Schneider *et al.*, 1988) or multi-span integral membrane proteins (Snapp *et al.*, 1997).

In the promastigote, the plasma membrane is decorated with a glycocalyx of at least 20–40 nm thickness (Pimenta *et al.*, 1991) or even thicker if the approximately 300 nm long PPG is considered (Ilg, 2000). Its major components are LPG and PPG, although many other glycolipids (for example, glycoinositol phospholipids or GIPLs) and glycoproteins contribute. The glycocalyx contains many nutrient and ion transporters, as well as ligands for host cell receptors, in both the sandfly vector and the mammalian macrophage. The glycocalyx is also thought to protect the promastigote in the hydrolytic environment of the fly gut, although no mechanism has been established. It is not clear if it fulfils a protective role in amastigotes, which have a much reduced glycocalyx thickness due to the downregulation of many glycoconjugates (Ilgoutz *et al.*, 2001). The amastigote may compensate for the loss of surface molecules by acquisition and remodelling of host-derived glycolipids (see below) (Denny *et al.*, 2004; Zhang *et al.*, 2005).

In addition to LPG, the *Leishmania* glycocalyx has been shown to contain at least three GPI-anchored proteins, the zinc metalloprotease gp63/leishmanolysin, the membrane proteophosphoglycan (mPPG), and the parasite surface antigen-2 complex (PSA-2/gp46).

GPI-anchored proteins and glycolipids

All the ligands currently known to bind to host macrophages are GPI-anchored glycoconjugates (Fig. 9.2). Some are glycolipids such as LPG, while others are glycoproteins.

The major surface proteins of *Leishmania* that are GPI-anchored are initially synthesized with a cleavable amino terminal signal sequence and a carboxyl terminal hydrophobic sequence which is rapidly removed in the endoplasmic reticulum and replaced with a GPI anchor, followed by rapid movement to the cell surface by an unknown mechanism (Ralton *et al.*, 2002).

GPI anchor precursors are synthesized in the endoplamic reticulum and transferred

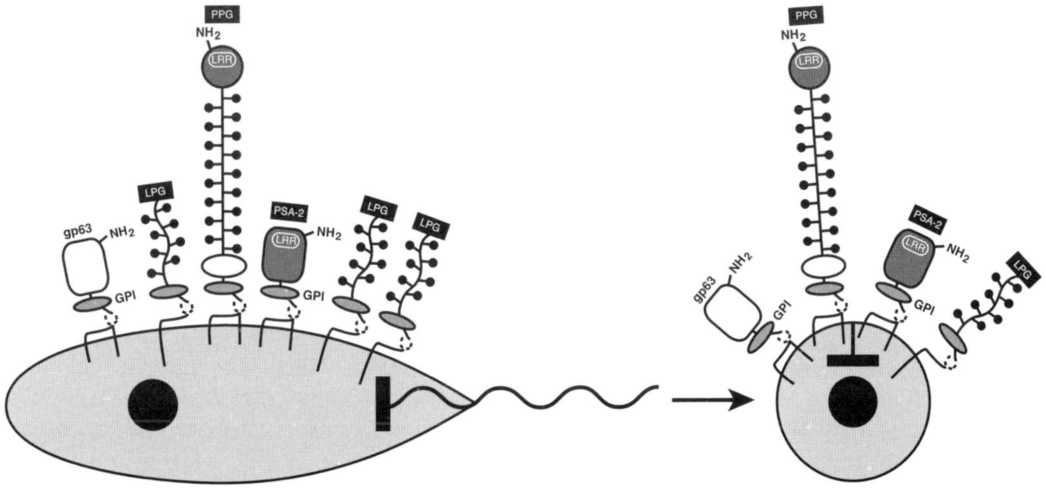

Figure 9.2 Schematic representation of the promastigote and amastigote surface glycocalyx containing the GPI-anchored glycoconjugates known to act as ligands for host macrophages, lipophosphoglycan (LPG), gp63, parasite surface antigen-2 (PSA-2), and proteophosphoglycan (PPG).

to newly synthesized proteins by a luminally oriented transamidase complex (Ferguson, 1999; McConville *et al.*, 2000). The anchor biosynthesis has been studied extensively in *T. brucei*. It is considered that a similar, but not identical, pathway is used by *Leishmania* (McConville *et al.*, 2000). Synthesis of the GPI anchor involves the addition of N-acetylglucosamine (GlcNAc) to phosphatidylinositol (PI) and subsequent acetylation of GlcNAc to form GlcN-PI. This is followed by the transfer of three mannose residues from dolichol phosphate mannose to form Man_3Glc-PI, and transfer of ethanolamine phosphate to form EtN-P-Man_3-GlcN-acyl-PI. In yeast and mammalian cells, the attachment of the anchor to the protein is catalysed by a heterodimeric complex of membrane proteins GPI8p and gaa1p. This enzyme complex removes the C-terminal hydrophobic sequence of the nascent protein and attaches the ethanolamine of the GPI precursor (Ferguson, 1999; McConville *et al.*, 2000). In contrast to the yeast and mammalian proteins, the *L. mexicana* GPI8 gene encodes a hydrophilic protein which lacks an endoplasmic reticulum retention sequence, suggesting that it may interact with other proteins of the GPI transamidase complex for its activity (Hilley *et al.*, 2000; Sharma *et al.*, 2000).

The *L. major* promastigote gp63 isoforms are N-glycosylated with three high mannose glycans (McGwire *et al.*, 1996; Olafson *et al.*, 1990), but little is known about the glycosylation of most other GPI-anchored proteins. Some *Leishmania* membrane proteins such as PPG have highly unusual glycosylation termed phosphoglycosylation in which LPG-like glycans are attached to serine residues (Ilg, 2000).

Membrane lipids have been reviewed elsewhere (Ilgoutz *et al.*, 2001; Naderer *et al.*, 2004), but require at least a mention here because the remarkable sharing of glycan moieties among certain membrane lipids and glycoproteins suggests a strong functional linkage between them. The most likely basis for such a linkage is a role of the shared glycans as ligands for receptors on the surface of cells lining the gut of the insect vector and membrane receptors on cells of the vertebrate host.

LPG is a GPI-anchored polysaccharide that is the most abundant component of the promastigote glycocalyx and plays a critical role in anchoring the parasite into the sandfly gut. LPG has also been shown to be critical for survival and transmission (Stierhof *et al.*, 1999). It is also considered an important ligand for macrophages and, at least in *L. major*, it is a virulence factor in the vertebrate host (Kelleher *et al.*, 1992; Kelleher *et al.*, 1995).

The structure and organization of LPG glycans is developmentally regulated during the progression of the promastigote from the rapidly dividing procyclic stage to the mature metacyclic stage. LPG is present, albeit in low abundance, in *L. major* amastigotes, but it seems to be absent from the *L. donovani* amastigote surface (Bahr *et al.*, 1993; McConville *et al.*, 1991; Moody *et al.*, 1991; Moody *et al.*, 1993). In addition to quantitative changes, the structure of the *L. major* amastigote LPG is distinct from that of the promastigote (Moody *et al.*, 1991; Moody *et al.*, 1993).

The structure of *Leishmania* LPG shows remarkable genetic polymorphism. There are major differences in its structure and composition in different species and even strains of the same species isolated from different geographical areas (reviewed in Naderer *et al.*, 2004). The role of LPG in the parasite life cycle and virulence shows considerable variation between *Leishmania* species. LPG seems essential for virulence in *L. major*, but not in *L. mexicana* (reviewed in Naderer *et al.*, 2004). It is likely, but unproven, that the difference in virulence between the two species is related to the different structure of LPG.

Gp63/leishmanolysin

Gp63 was the first surface GPI-anchored membrane protein of *Leishmania* to be investigated in detail (Bouvier *et al.*, 1989; Bouvier *et al.*, 1987; Bouvier *et al.*, 1985; Bouvier *et al.*, 1990; Etges *et al.*, 1986; Etges *et al.*, 1986; Yao *et al.*, 2003).

Gp63 is encoded by multiple genes. In *L. major*, there is a tandem array of gp63 genes on chromosome 10, a single gene on chromosome 28, and a related gene on chromosome 31. gp63 orthologues have been identified in *T. cruzi* and *T. brucei* (El-Sayed *et al.*, 2005; Ivens *et al.*, 2005). It is of interest that in the *T. cruzi* genome there are 420 gp63-like genes and pseudogenes, some dispersed throughout the genome and some clustered. The reason for this massive expansion of gp63-like genes is unknown but will probably become apparent once their function is determined.

Gp63 is a metalloprotease. It has been shown to be a promastigote virulence factor by virtue of its binding to several macrophage receptors including CR3, the mannose-fucose receptor and the fibronectin receptor, but the relationship, if any, between the protease activity and macrophage-binding capability has not been clarified. It has been suggested that gp63 uses a Ser-Arg-Tyr-Asp sequence for host cell binding. This sequence mimics the mammalian cell adhesion motif Arg-Gly-Asp-Ser (RGDS) used by fibronectin to bind to its receptor on macrophages (Brittingham *et al.*, 1999). Gp63 can also bind to complement receptors CR1 and CR3 after opsonization with the complement components C3b and C3bi.

Gp63 is thought to protect the parasite from complement lysis by cleaving the membrane attack complex from the surface of the parasite. (Brittingham *et al.*, 1995; Yao *et al.*, 2003). This role of gp63 is supported by the observation that gp63 knockout parasites were significantly more sensitive to complement than wild type parasites (Hilley *et al.*, 2000; Yao *et al.*, 2003).

These gp63 activities may allow the initial infection by the invading promastigote to take hold, but it seems that gp63 is not essential for amastigote survival in the mammalian host, because although deletion of all gp63 in *L. major* led to a significant reduction in parasite fitness and virulence, the amastigotes which managed to survive were virulent (Joshi *et al.*, 2002). It is therefore surprising that deletion of all GPI-anchored proteins from *L. mexicana* by gene targeting of GPI8, the enzyme transamidase which is essential for their biosynthesis, did not affect parasite virulence (Hilley *et al.*, 2000). This observation may reflect major differences in the biology of *L. major* and *L. mexicana* (McMahon-Pratt *et al.*, 2004). Although these species cause similar cutaneous diseases in humans, they interact with the host in distinct ways (Antoine *et al.*, 2004; McMahon-Pratt *et al.*, 2004).

Gp63 is expressed on the surface of promastigotes of all *Leishmania* species, but it can also be secreted, either as a result of proteolysis or by direct release of an unanchored protein (Naderer *et al.*, 2004). The secreted form of gp63 may be important in hydrolysis of proteins in the sandfly gut, or the mammalian host. It may also degrade complement components in the initial pool of blood into which the parasites are delivered by the sandfly bite.

As is the case with most *Leishmania* proteins, much less is known about the expression of gp63 in amastigotes (Frommel *et al.*, 1989; Voth *et al.*, 1998). Stage-specific expression of gp63 is regulated by changes in mRNA as well as protein stability and processing (Yao *et al.*, 2003; Yao *et al.*, 2005). The expression of gp63 on the surface of amastigotes is still debated, but transfection and over-expression of the *L. major* amastigote gp63 in *L. donovani* has shown that it is capable of being displayed on the amastigote surface (Brittingham *et al.*, 1995). A biological role for gp63 in amastigotes has not been established, but it may help in nutrient acquisition in the phagolysosome.

Proteophosphoglycans

PPGs are a heterogeneous family of highly glycosylated mucins that exist in both secreted and GPI-anchored forms (Ilg, 2000; Ivens *et al.*, 2005). The genome database predicts the presence of at least four PPG genes in a tandem array on chromosome 35. The structure of the PPGs varies significantly in different *Leishmania* species and developmental stages, but it is not known if the different forms are functionally distinct. The PPG secreted by promastigotes consists of long filaments which encase promastigotes in the sandfly gut in a tangled mass (Stierhof *et al.*, 1999). This mass blocks the stomodeal valve, obstructing feeding. It is thought that this promotes multiple bites and increases the probability of parasite transmission into the mammalian host.

PPG may also improve survival of the parasites in the mammalian host by an unknown mechanism affecting the host macrophages (Piani *et al.*, 1999; Rogers *et al.*, 2004). *L. mexicana*, but not *L. major*, amastigotes secrete large amounts of PPG into the communal parasitophorous vacuole (Peters *et al.*, 1997), where it is thought to protect them from the host microbicidal mechanisms (Peters *et al.*, 1997).

A gene encoding a GPI-anchored form of PPG has been isolated from *L. major* (Ilg *et al.*, 1999). It encodes a polypeptide of 2300 amino acids. When analysed by SDS-polyacrylamide gel electrophoresis, it displays a larger than expected size and a polydisperse pattern due to its abundant glycosylation. Membrane-bound PPG (mPPG) is expressed by both promastigotes and amastigotes, and is present on the parasite surface as well as in the flagellar pocket, the Golgi and the lysosome (Foth *et al.*, 2002). The gene encoding GPI-linked PPG has homologues in *L. mexicana* and *L. donovani*, but so far their protein products have not been characterized. Studies using surface biotinylation and immunoprecipitation have detected several other GPI-anchored PPGs on *L. major* promastigotes, but these have also not been characterized in detail (Montgomery *et al.*, 2002).

mPPG contains two striking clusters of internal repeats, including a core of 107 [APSSSSA(P/S)SSSSS] motifs, most of which are serine-phosphoglycosylated with glycans shared with LPG, a series of leucine-rich repeats (LRR) and a di-cysteine motif in the amino terminal domain (Kedzierski *et al.*, 2004; Montgomery *et al.*, 2000; Montgomery *et al.*, 2002).

There is a remarkable similarity between the LRRs and di-cysteine motifs in PPG and those in the unrelated GPI-anchored PSA-2 (Kedzierski *et al.*, 2004; Montgomery *et al.*, 2000). Since most LRRs have been shown to mediate protein–protein interactions, this similarity suggested that they may bind to the same ligand and have a similar function (Kajava, 1998). Early studies indicated that PPG binds to macrophages (Piani *et al.*, 1999). It had been assumed that this binding was mediated by the glycans shared by PPG and LPG because these were known to be ligands for macrophages (Piani *et al.*, 1999). However, subsequent studies have demonstrated that the LRRs contribute significantly to this binding, and that the parasite LRRs bind to the complement receptor 3 on the host macrophage (Kedzierski *et al.*, 2004). It would therefore appear that PPG facilitates binding of the parasite to macrphophages by at least two distinct mechanisms.

Bacterial LRRs have been shown to play similar roles in adhesion to host cells and establishment of infection (Inohara *et al.*, 2005; Kedzierski *et al.*, 2004). It may be speculated that the parasite interaction with macrophages is similar to the two-phase interaction of white blood cells with capillary endothelial cells, where leukocyte glycans (LPG in the case of *Leishmania*) bind to endothelial cells with rapid kinetics but low affinity, followed by high affinity binding (by LRRs in the case of *Leishmania*), firm adhesion and diapedesis (Fig. 9.3) (Springer, 1990; Springer, 1994).

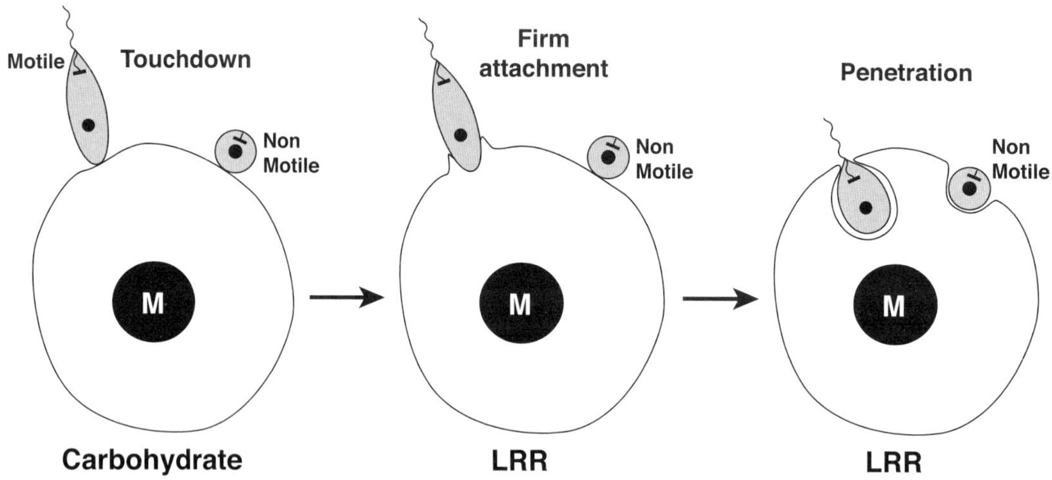

Figure 9.3 Schematic representation of the potential two-step mechanism of parasite invasion of host macrophages, from glycan binding with rapid kinetics but low affinity (LPG and/or PPG) to LRR binding with high affinity (PPG and/or PSA-2).

There is strong evidence that PPGs are important for parasite virulence in the mammalian host, at least in some species, but as is the case with many other *Leishmania* proteins, the role and importance of PPG may be different in different species. For example, PPG null *L. mexicana* are as virulent as wild type organisms, at least in mice (Ilg *et al.*, 2001), but *L. major* mutants lacking PPG fail to survive in macrophages (Späth *et al.*, 2003). Surprisingly, some of these mutant parasites could be recovered from infected, but asymptomatic, mice (Späth *et al.*, 2003).

It was postulated that the mutant *L. major* parasites lacking PPG had colonized cells other than macrophages, but there was no indication of their identity. It has been shown that wild type parasites persist in the otherwise immune host, possibly in cells which are not macrophages (Aebischer *et al.*, 1993; Bogdan *et al.*, 2000; Moll *et al.*, 1995; Rittig *et al.*, 2000). The initial inflammatory host cells recruited to the injection site in the skin are neutrophils and macrophages (Baldwin *et al.*, 2004), suggesting the possibility of an obligatory passage through these cell types before other cells may become infected. These cells are also the most abundant in the natural infection where a blood pool is produced by the sandfly bite. The journey of the parasite from its injection into the skin to its final destination in the cell in which it can persist indefinitely represents the major gap in our knowledge of the parasite life cycle. The identification of the long sought-after safe haven where the virulent parasites persist is a worthy but elusive goal, in particular because most of the data describing the infection process have been obtained by infecting mice with parasites via a syringe, and the relevance to human disease is not certain.

Parasite surface antigen-2 (gp46)

The PSA-2 protein family includes both secreted and membrane proteins. The salient features of the membrane PSA-2 members are an LRR domain followed, on almost all predicted PSA-2 sequences, by a threonine-rich repeat region and a GPI anchor to the surface membrane. The structure of PSA-2 with its threonine-rich region (including T_7KP_2 repeats) has significant similarity to some members of the very large family of mucins expressed in the mammalian stages of *T. cruzi* (El-Sayed *et al.*, 2005). It will be of great interest to compare the sequences and the function of these mucins in the two organisms.

The PSA-2 family is found in all species of *Leishmania* examined, but not in species of the smaller subgenus *Viannia*, which comprises only the *L. braziliensis* and *L. guyanensis* complexes (McMahon-Pratt *et al.*, 1992). PSA-2 genes are multicopy and highly variable within and among species. There are variations in the number of LRRs that make up their LRR domain, as well as in the threonine-rich region and in the presence or absence of a GPI anchor (Kedzierski *et al.*, 2004; McMahon-Pratt *et al.*, 1992; Montgomery *et al.*, 2000).

The recently released *L. major* genome sequence has identified 32 PSA-2-related genes (Ivens *et al.*, 2005), but their expression pattern and function are currently unknown. There is an urgent need to determine the relationship between individual genes and their products. At present, this information is almost totally lacking.

In *L. major*, the PSA-2 family of surface GPI-anchored glycoproteins was originally identified on the promastigote, but one member of the family has been shown to be present on the amastigote (Handman *et al.*, 1995; Symons *et al.*, 1994). Our early studies indicated that in the *L. major* promastigote, three PSA-2 isoforms are expressed on the surface. Several other isoforms are secreted but these have not been characterized. The three PSA-2 surface isoforms differ significantly in molecular weight (95,000, 80,000 and 50,000 kDa) and in their N-terminal sequences as determined by Edman degradation (Handman *et al.*, 1995). The *L. major* amastigote expresses a single major isoform of M_r 50,000 which is distinct from the promastigote molecule. This work needs to be revisited and extended to include efforts to match the identity of each protein product with the relevant gene. In view of the large number of genes and the small number of characterized proteins this will be no small task.

L. chagasi promastigotes express two major PSA-2 polypeptides of M_r 44,000 and 66,000 (Beetham *et al.*, 2003; Beetham *et al.*, 1997). At present, there is no information on PSA-2 expression in *L. chagasi* amastigotes. PSA-2 protein expression is upregulated in the infectious-stage promastigotes of *L. chagasi*, where it appears to provide protection from complement lysis (Beetham *et al.*, 2003; Lincoln *et al.*, 2004). The precise domain which confers resistance to complement lysis has not been identified, but it is hypothesized that the LRRs may be involved as well as the threonine-rich stalk because of its similarity to a domain in the mammalian decay accelerating factor which fulfils a similar role (Coyne *et al.*, 1992). The mechanism by which resistance acts is not understood, but it may be speculated that extensive glycosylation of the threonines projects the protein away from the plasma membrane where the LRRs can interact with complement components.

Genes and proteins belonging to the PSA-2 family have also been isolated and described for *L. amazonensis* (Lohman *et al.*, 1990), *L. mexicana*, *L. infantum* (Jimenez-Ruiz *et al.*, 1998), and *L. tropica* (Gamarro, unpublished).

Integral membrane proteins

The *Leishmania* plasma membrane also contains type 1 and multi-spanning membrane proteins which are anchored in the membrane via hydrophobic transmembrane sequences, but few have been characterized in detail. The export of these proteins to the plasma membrane occurs through the flagellar pocket, and the targeting to the plasma membrane appears to be determined by sorting motifs. As discussed previously, such motifs have been described for the glucose transporters and for the 3′ nucleotidase, but more work is necessary to determine the general applicability of this mechanism.

Proteins involved in nutrient and ion transport

Leishmania express a number of membrane transporters involved in nutrient acquisition and in the maintenance of ion balance in the cell. Some of these proteins are expressed constitutively, but others are expressed in a stage-specific manner, reflecting the specific requirements of the promastigote in the sandfly gut and the amastigote in the phagolysosome. In addition to their role in nutrient transport, some transporters facilitate the uptake of drugs and therefore could be useful as potential drug-delivery vehicles or as drug targets in their own right.

In some cases. transport activities have been characterized in detail, but the relevant proteins and genes have not been isolated. For example, active transport of putrescine and spermidine has been described in *L. donovani* and *L. mexicana*. Pentamidine is the drug of choice against antimony-resistant leishmaniasis and although its precise mechanism of action is not fully understood, it inhibited both transporters (Basselin *et al.*, 2000). Interestingly, there was a striking difference between the promastigote and the amastigote transporters in their susceptibility to external pH. The amastigote transporters were active at acidic pH, similar to the environment in the parasitophorous vacuole, while the optimal pH for the promastigote transporters was pH 7.4 (Basselin *et al.*, 2000). The data are compatible with the hypothesis that distinct polyamine transporters are expressed in promastigotes and amastigotes, and that these have evolved to function optimally in the relevant environments.

Hexose transporters

Hexose transporters belong to a major superfamily of proteins that mediate active and facilitated transport of nutrients in both prokaryotic and eukaryotic cells. They are polytopic proteins with numerous membrane-spanning regions. Hexose transporters of *Leishmania* are able to transport glucose, mannose and galactose (Landfear, 2000). Rates of transport in amastigotes are much reduced compared to promastigotes, perhaps reflecting an overall metabolic downshift in this life cycle stage (see Chapter 5).

As mentioned earlier, two glucose transporters have been described in *L. enrietti* promastigotes, LeGT1 (ISO1) and LeGT2 (ISO2). These have been shown to mediate uptake of mannose and fructose in addition to glucose (Landfear, 2000). Three related transporters have been described in *L. mexicana*. LmGT1, like LeGT1, is expressed on the flagellum, while LmGT2 and LmGT3 are localized to the cell body (Burchmore *et al.*, 2003). Deletion of all three genes in *L. mexicana* suggested that glucose uptake is not essential for growth in culture, but may well be critical for the completion of the promastigote developmental programme and acquisition of virulence in the sandfly. There is a substantial body of evidence indicating that the ingestion of a sugar meal is necessary for parasite transmission by sandflies (Schlein, 1993).

Gene deletion experiments suggest that glucose transport is essential for amastigote survival in macrophages and for the transformation of promastigotes to amastigotes in axenic cultures (Burchmore *et al.*, 2003). Gene 'add-back' experiments using LmGT2 could restore, to some extent, the ability of the gene deletion mutants lacking all three transporters to survive as amastigotes in macrophages. However, only LmGT3 could restore survival of amastigotes to wild type level, indicating that this is the main transporter in this life cycle form of the parasite.

The pattern of expression and function of the glucose transporters indicates that, like most *Leishmania* genes, the regulation of expression must occur at the level of RNA stability and/or processing level because the mRNA level of LmGT2 is higher in promastigotes than in amastigotes, while that of LmGT1 and LmGT3 is similar in the two developmental stages (Burchmore *et al.*, 1998).

Another member of the glucose transporter superfamily is the myo-inositol/H^+ symporter, which uses the proton-electrochemical gradient across the plasma membrane to transport the organic substrate (Drew *et al.*, 1995). This protein is expressed constitutively in both promastigotes and amastigotes.

Biopterin/folate transporters

Leishmania are pterin and folate auxotrophs and have to scavenge them from the host. *Leishmania* have a family of 13 pteridine transporter genes (Ivens *et al.*, 2005). Some are *L. major*-specific while others are conserved in all trypanosomatids. Among those already characterized, distinct entities specifically transport pterins and folate (Richard *et al.*, 2002).

Pterin uptake is mediated by a high-affinity transporter, biopterin transporter 1 (BT1), but also, at least to some extent, by diffusion (Kundig *et al.*, 1999). The differentiation from procyclic to metacyclic promastigotes is accompanied by a down-regulation of BT1 expression (Cunningham *et al.*, 2001a; Cunningham *et al.*, 2001b). Parasites in which the BT1 gene had been deleted transformed into amastigotes more efficiently than wild type parasites. A similar phenotype was observed in parasites lacking pteridine reductase (*PTR1*), the enzyme responsible for reducing biopterin to the active tetrahydrobiopterin (Cunningham *et al.*, 2001a; Cunningham *et al.*, 2001b). As expected, in view of the central role of pterins, the *L. donovani BT1* null mutants were attenuated and gave rise to reduced infections in mice (Papadopoulou *et al.*, 2002). Surprisingly, an *L. major* pteridine reductase (*PTR1*) null mutant was more infective in the mouse model (Cunningham *et al.*, 2001). The increased virulence was ascribed in part to the larger numbers of metacyclic promastigotes, but also to an interplay between transport of biopterin and its metabolism (Cunningham *et al.*, 2001).

Folate uptake is mediated by other members of the pteridine transporter family, one of which has been shown to be a high affinity receptor for folate in *L. tarentolae* (FT5) (Richard *et al.*, 2002).

Nucleoside transporters

Nucleoside transporters are of particular importance because *Leishmania* are not able to synthesize purines and have to acquire them from their host.

Two linked genes with a similar sequence encoding adenosine/pyrimidine nucleoside transporters from *L. donovani* were cloned by their ability to complement a mutant parasite line deficient in adenosine/pyrimidine transport (LdNT1.1 and LdNT1.2) (Vasudevan *et al.*, 1998; Vasudevan *et al.*, 2001). The LdNT1.1 is expressed around the promastigote body, but not the flagellum. Both transporters have 11 predicted transmembrane domains, and are similar to the mammalian equilibrative nucleoside transporters.

LdBT2, a gene encoding an inosine/guanosine transporter with similarity to LdNT1 was cloned by genetic rescue of a mutant parasite deficient in inosine/guanosine transport (Stein *et al.*, 2003). LdNT1 and LdNT2 appear to be the active transporters in the promastigote. The number and type of nucleoside transporters active in the amastigote is not known. Related transporters have been identified in *T. brucei* and have been shown to mediate uptake of adenosine/adenine as well as the anti-trypanosome drugs pentamidine and melarsoprol (Maser *et al.*, 1999). Interestingly, pentamidine and melarsoprol are not nucleoside drugs.

Transferrin receptor

Iron is required for many cellular processes. In the mammalian host, iron is transported bound to the carrier protein transferrin, which has an extremely high affinity for iron, such that there may be less than one molecule of iron per litre of plasma in free form. The extreme sequestration of iron by transferrin may constitute a defence mechanism against microbes and creates a problem for parasitic organisms. In trypanosomatids, the problem has been solved by the evolution of specific parasite receptors for transferrin, which appear to mediate its uptake by receptor-mediated endocytosis and represent the first stage of a mechanism to transport iron into the parasite cytosol. While this process is known in considerable detail for the mammalian transferrin receptor, detailed knowledge of post-receptor binding events in *Leishmania* is almost totally lacking. The sequence of the *Leishmania* transferrin receptor shows no homology to the mammalian receptor, and presumably arose independently and not as a result of lateral gene transfer.

In mammalian cells, iron is released from internalized transferrin in the acidic environment of the endosome, from which it is transported into the cytosol by the divalent cation transporter Nramp2 (natural resistance associated macrophage protein 2) (Fleming *et al.*, 1998). Nramp2 is closely related to Nramp1, which was first identified from linkage analysis of resistance or susceptibility of inbred strains of mice to *L. donovani*, *Salmonella* and BCG (Blackwell, 1989; Govoni *et al.*, 1998; Picard *et al.*, 2000). Like Nramp2, Nramp1 (also known as Slc11a1 or solute carrier 11a1) is also expressed in the endosome, but the mechanism by which it controls resistance or susceptibility to *L. donovani* is still not fully understood. It is thought to regulate the concentration of iron taken up by phagocytosis or degradation of cells in the phagolysosome, but it is still unclear whether this is the mechanism by which it influences susceptibility to infection (Mulero *et al.*, 2002; Mulero *et al.*, 2002).

Both promastigotes and amastigotes bind diferric transferrin through receptors on their surface. A 70 kDa type 1 integral membrane protein which binds transferrin has been isolated from promastigotes of *L. major* and *L. infantum*, and shown to be distinct from either the *T. brucei* or the mammalian receptors (Voyiatzaki *et al.*, 1992). In the case of the amastigote, integration of the *Leishmania*-containing phagosome into the host endosomal system allows the amastigote to subvert the host endosomal process, and to bind the iron which is released in that compartment. Amastigotes have been shown to bind and internalize transferrin, so presumably they have a functional receptor (Borges *et al.*, 1998). However, in contrast to promastigotes, where transferrin binding is limited to the flagellar pocket, in amastigotes transferrin binding sites are distributed over the entire surface. Whether the amastigote receptor is the same as that described for the promastigote is not known. Also not known is whether the receptor is recycled to the cell surface, as is the case in mammalian cells. In mammalian cells, apotransferrin is returned to the cell surface intact, and released in a pH-dependent cycle (Morgan, 1983; van Driel *et al.*, 1984). Data from Borges and colleagues indicated that in the parasite, the internalized transferrin is directed to the lysosomal compartment, possibly for degradation (Borges *et al.*, 1998).

An additional and novel mechanism of iron acquisition has been recently discovered. It has been known for some time that *Leishmania* have an NADPH-dependent iron reductase activity, which converts Fe^{3+} into the more soluble Fe^{2+} (Wilson *et al.*, 2002). This suggested that a form of iron able to be taken up directly by membrane transporters would be available to the parasite. Indeed, one such ferrous iron transporter designated LIT1 has been identified recently in *L. amazonensis* (Huynh *et al.*, 2006). LIT1 is the product of one of two identical genes on chromosome 31. LIT1 is a 432 amino acid protein with eight predicted transmembrane domains, and belongs to the ZIP family of iron transporters. Although the mRNA is present in both promastigotes and amastigotes, the expression of the protein is regulated post-translationally by iron. LIT1 is detected exclusively on the intracellular amastigotes, which live in an iron poor environment (Huynh *et al.*, 2006). LIT1 is not necessary for promastigote growth in culture, or for the initial infection of macrophages, but it is required for the intracellular replication of amastigotes. As predicted from their inability to grow in macrophages, LIT1 knockout parasites did not cause lesions in mice, but, surprisingly they persisted for long periods in infected animals (Huynh *et al.*, 2006).

Ectoenzymes

3′/5′ Nucleotidase

3′/5′ Nucleotidase activities were first described in the early 1980s and shown to be displayed on the exterior aspect of the promastigote surface and in the flagellar pocket (Dwyer *et al.*, 1985). The most likely function of these enzymes is the salvage of host-derived purines that *Leishmania* are not able to synthesize by converting them into forms that can be taken up by nucleoside transporters. The *L. donovani* 3′ nucleotidase is a 43 kDa glycoprotein conserved in all *Leishmania* (Debrabant *et al.*, 2001). In the case of *L. mexicana*, and probably in all other species, the enzyme has been shown to be expressed in promastigotes, but not in amastigotes, suggesting that the intracellular form of the parasite must use alternative means for the acquisition of purines (Sopwith *et al.*, 2002). To date the 5′ nucleotidase has not been characterized.

ATPase

ATPase genes have been identified in the *L. major* genome, but the characterization of the proteins lags behind. An ouabain-sensitive $Na^+,K(^+)$-ATPase in the plasma membrane of *L. mexicana* has been described and is involved in the intracellular regulation of these cations (Felibertt *et al.*, 1995). While it is expected that ATPases fulfil an important role in the polarization of the membrane which is critical for the uptake of sugars and amino acid, knowledge of their precise structure and localization is still lacking (reviewed in Naderer *et al.*, 2004).

Acid phosphatase

L. donovani and *L. mexicana* promastigotes and amastigotes display acid phosphatase activity on the external surface of their membrane (Shakarian *et al.*, 2003). This histidine-acid phosphatase is a type 1 membrane protein of 53 kDa in promastigotes with a 64 kDa form in amastigotes. It is not clear if the difference in molecular weight between the two forms is due to post translational modifications or reflects distinct gene products. The gene for acid phosphatase as well as the enzyme activity is present in all *Leishmania* species examined as well as in *Crithidia*, but not in *T.*

brucei or *T. cruzi*. Although its precise function has not been determined, it has been postulated that it may be important for growth and survival in the host (Shakarian *et al.*, 2003).

Proteins involved in signalling pathways

Several classes of important signalling molecules that are widely expressed in most eukaryotic cells are absent in trypanosomatids, including heterotrimeric G proteins, most classes of catalytic receptors, and SH2 and SH3 interaction domains (El-Sayed *et al.*, 2005). Genomic analysis showed that the trypanosomatids have a large set of eukaryotic type protein kinases but they lack receptor-linked tyrosine and tyrosine kinase-like kinases (Parsons *et al.*, 2005). They also possess few membrane-bound protein kinases. Eleven genes encoding catalytic receptors identified in the *L. major* genome database are all adenylate cyclases (El-Sayed *et al.*, 2005). Their predicted proteins contain an extracellular N-terminal domain, a single transmembrane domain and a cytoplasmic catalytic domain. The expression pattern of these genes is not known, but there is evidence that two adenylate cyclases from *L. donovani* are only expressed in the promastigote (Sanchez *et al.*, 1995). To date, no activating molecules have been described.

Developmentally regulated proteins of unknown function

Metacyclic promastigotes are the infectious form of the parasite transmitted by the sandfly. They are thought to have evolved adaptations necessary for the establishment of infection in the mammalian host. Similarly, amastigotes are thought to have developed adaptations to the microenvironment of the macrophage phagolysosome. In both cases, the expression of metacyclic-specific or amastigote-specific surface proteins has been considered a major contributor to the establishment of infection. Several such developmentally regulated proteins have been described, but in most cases their precise function is not known.

Amastin

As its name suggests amastin was first described as an abundant surface protein in amastigotes of *T. cruzi* (Teixeira *et al.*, 1994). The *L. major* genome contains 57 amastin genes, most of which are located in tandem clusters on several different chromosomes (Ivens *et al.*, 2005). As is the case in *T. cruzi*, the amastin genes are interspersed with genes of unknown function called tuzins (Ivens *et al.*, 2005; Rochette *et al.*, 2005). The genomic association of these two families of genes is intriguing and may suggest functional interactions between the proteins. Amastins vary in size between 160 and 540 amino acids. All *L. major* amastins have predicted four or five transmembrane domains, but some also have canonical GPI anchor sequences (Ivens *et al.*, 2005). Expression analysis in *L. major* and *L. infantum* showed that some are amastigote-specific, but others are expressed constitutively in both life cycle stages (Rochette *et al.*, 2005). Since the function of the amastins is not known, the relevance of the differential expression of the genes and its control remains fascinating but obscure.

KMP-11

KMP-11 was discovered in the 1990s as an unusual protein of 11 kDa tightly associated with LPG on the surface of *L donovani* promastigotes (Jardim *et al.*, 1995; Jardim *et al.*, 1995). However, KMP-11 is also present on amastigotes, despite the lack of detectable LPG on this life cycle stage of *L. donovani*. This highly acidic alpha helical protein has been shown to form an amphipathic structure with a high affinity for lipid bilayers. It has been suggested that it may function to stabilize membrane molecules, of which LPG may be only one example. It is possible that in the amastigote, KMP-11 associates with the most abundant type of glycolipids, the glycoinositol phospholipids (GIPLs) (McConville *et al.*, 1990).

Hydrophilic acylated surface proteins

Hydrophilic Acylated Surface Proteins (HASPs) are a family of three small hydrophilic proteins, which find their way onto the surface of metacyclic promastigotes of all *Leishmania* species through an unusual mechanism involving covalent attachment of lipid (McKean *et al.*, 1997; McKean *et al.*, 2001). The function of these proteins is not known. Surprisingly, deletion of the genes had no effect on the parasite biology, but over-expression led to loss of virulence and loss of cell structure, suggesting that the concentration of the protein is important (McKean *et al.*, 1997; McKean *et al.*, 2001).

Many highways, but no royal road to function

Prior to the advent of gene cloning, functions were assigned to discrete proteins by a diversity of approaches, which generally worked from biological observations towards identification of the relevant proteins. This process was often facilitated by the use of antibodies, and the advent of monoclonal antibodies made it far more powerful. In a few cases, enzyme activities were identified and the classical approaches of protein purification could be used. Ultimately, many of these approaches culminated in the identification of the relevant genes.

Genetic complementation

Biological function is now commonly revealed by genetic manipulation, including analysis of the phenotype of random or directed mutations. Rescue (complementation) of the resultant deleterious phenotypes by 'throwing in a gene library' and fishing out the gene that is retained in surviving cells has often provided a shortcut to the relevant gene, and thereby greatly illuminated function. This approach has been particularly successful in the isolation and characterization of several genes encoding enzymes in the LPG biosynthetic pathway (Beverley *et al.*, 1995; Descoteaux *et al.*, 1994). Starting with mutants defective in the production of LPG, genes encoding a number of sugar transporters, galactosyl, mannosyl and arabinosyltransferases have been isolated and their precise function mapped (Descoteaux *et al.*, 1998; Dobson *et al.*, 2003; Dobson *et al.*, 2003; Hong *et al.*, 2000; Ma *et al.*, 1997). Similarly, complementation of an adenosine/pyrimidine transporter-deficient mutant led to the identification and isolation of the two nucleoside transporter genes described above (Vasudevan *et al.*, 1998).

Gene deletion and gene silencing, knockout and knockdown

Although gene deletion by homologous recombination has also proven a very powerful tool in the analysis of gene function in *Leishmania*, this approach seems to work much more easily in some species than in others. For example, although it works efficiently in *L. mexicana*, its efficiency seems to be lower in *L. major* for reasons that are currently unknown (Handman unpublished data and Debbie Smith, personal communication).

Similarly, the technique of complete or partial gene silencing or 'knockdown' via RNA interference (RNAi) has proven to be very effective in *T. brucei* (Motyka *et al.*, 2004), but it does not seem to work in *T. cruzi* and *Leishmania*, possibly due to the absence of the Dicer gene which encodes an enzyme that cuts mRNA into short oligonucleotides (Robinson *et al.*, 2003). Whether or not a way can be found to exploit this powerful approach in *Leishmania* remains to be seen. In contrast, the use of antisense RNA has had some success in reducing both the expression of gp63 as well as parasite virulence (Chen *et al.*, 2000). Antisense RNA was also able to transiently reduce expression of PPG in transfected *L. major* promastigotes, but not for long enough to assess the biological effect on the parasites (Handman and Chang, unpublished data).

Mining the *Leishmania* genome

Perhaps the most radical challenge facing the post-genomic era is to exploit the riches that lie buried within the genomes. However, deciphering complex eukaryotic genomes will be a slow process and a diversity of approaches will be required. Many years have passed since the first eukaryotic genomes were sequenced, but the function of almost half of the genes that were identified by sequence remains unknown. Another point worth making is that one of the most important post-transcriptional modifications of proteins, glycosylation, is not under the direct control of the genome (Merry *et al.*, 2005). Although the glycosyltransferases are defined by the genome, their activity and therefore their final product at any given time cannot be predicted. The variability of the glyco-repertoire is the result of the dynamics of sugar metabolism where a small number of enzymes and monosaccharides are able to generate multiple sugar structures depending on factors such as their distribution within the Golgi apparatus and their relative abundance (de Graffenried *et al.*, 2004). The functional importance in the biology of *Leishmania* of different glycoforms decorating the same molecule has been demonstrated most clearly for LPG, but may also be critical for glycoproteins such as PPG and PSA-2 (Sacks *et al.*, 2000; Soares *et al.*, 2004).

One of the most powerful and informative approaches in the search for protein function has been the search for sequence homology with proteins of known function in other species. Even short stretches of sequence similarity between species may provide sufficient clues concerning function, particularly if the motif in question already has a function assigned to it, such as the active site of an enzyme. Because the functions of proteins are constantly being discovered, it is essential to conduct homology searches on a frequent and ongoing basis. It would be a shame to miss a long-sought function because of failure to look.

Computer algorithms are getting better and better at detecting distant relationships, but this approach has its limitations. The vast evolutionary times since divergence of species means that for some proteins, the homology is so low that evolutionary and functional relationships have become obscured by genetic drift, and homology searches may not be fruitful (Najmanovich *et al.*, 2005). In addition, some proteins may be multi-functional, or may have diverged to related functions or even to totally unrelated functions (Jeffrey, 1999; Tompa *et al.*, 2005).

Sequence analysis

Sequence analysis can be a useful tool for the identification of new proteins belonging to a particular class by using the amino acid signature that characterizes the proteins in that particular class, and searching for these signatures among proteins of unknown function. For example, the discovery of novel targeting motifs permits the prediction of protein localization in a particular cellular compartment or membrane of interest. Other examples include identification of potential secreted proteins, prediction of integral membrane proteins, GPI-anchored proteins, metal-binding motifs such as EF hands and putative active sites of enzymes.

These predictions are usually based on assignation of scores to the proteins. The score measures the presence and quality of the characteristic motif or signal. More precisely, proteins are scored against a model, or a set of rules that describe the motif or signal. Signals can comprise quite short regions or they can be large and complex.

A variety of tools to score proteins for their characteristic signals have been developed, such as hidden Markov models (HMM), neural networks models, window-based methods and rule-based methods. Models typically have some free parameters, which are estimated using a set of known and experimentally validated proteins from the class of interest. Such sets are referred to as 'training' or 'learning' sets. It is often the case that training sets consist of proteins from many species, with relatively few, or perhaps no examples from the species of interest. An alternative approach presented below, is to develop methods that use species-specific knowledge and training data from the species of interest, for example *Leishmania*.

The validity of such 'training' is, of course, dependent on the reliability of the material used to teach it. The only truly rigorous form of training would be sequences which have been independently established by direct experimentation. Anything else would be subject to the criticism of 'the blind leading the blind'.

The scores from prediction methods need to be calibrated by applying the model to a negative control data set, i.e. proteins known not to belong to the class of interest. The calibration can provide a hard threshold to declare whether or not the protein is predicted to be in the class of interest, or it can be used to calculate p-values indicating the significance of the protein's score in relation to a chance model.

Resources for the prediction of GPI-anchored proteins

Since many of the known biologically important *Leishmania* proteins have been shown to be GPI-anchored, we set out to explore the potential of the bioinformatics tools described above to identify novel GPI-anchored proteins in the database.

Existing online resources are already available. For example, GeneDB (Hertz-Fowler *et al.*, 2004) is a genome database for prokaryotic and eukaryotic organisms, including *L. major*, *L. infantum* and several trypanosome species. In addition to sequence data, genomic position and Gene Ontology (GO) annotation, it also provides predictions of signal peptides using a program called SIGNALP, transmembrane domains using TMHMM and GPI-anchor signals using DGPI.

SIGNALP (Bendtsen *et al.*, 2004; Nielsen *et al.*, 1999; Nielsen *et al.*, 1997) identifies classically secreted proteins and predicts the location of the signal peptidase I cleavage site in the protein. It combines a neural network and a hidden Markov model of both a signal peptide and a signal anchor model. Enthusiasm for identification of signal peptides by computation should be tempered by the experimental demonstration that many random sequences can act as functional signal sequences (Kaiser *et al.*, 1987). SIGNALP has the advantage that it was trained independently on large databases of experimentally verified secreted proteins from eukaryotes, Gram-negative and Gram-positive bacteria. The eukaryotic dataset includes 1192 proteins from multiple species including several species of *Leishmania*. In the October 2005 version of GeneDB, there were 1491 *L. major* proteins with a predicted signal peptide.

TMHMM (Krogh *et al.*, 2001; Sonnhammer *et al.*, 1998) is a predictor of transmembrane domains based on a hidden Markov model (HMM) that closely models the composition and grammar of membrane spanning proteins. The HMM was trained on a set of 160 known membrane spanning proteins. The latest available version of GeneDB (October 2005) has 1571 *L. major* proteins with between 1 and 20 predicted transmembrane domains. When using the predictions of this program, one needs to be aware that it sometimes misidentifies the hydrophobic regions of signal peptides and GPI-anchor signals as transmembrane domains, and that in many cases the exact topography of a membrane protein is not known. According to Krogh *et al.*, 2001, about 20% of signal peptides from eukaryotic and Gram-negative bacteria are identified as transmembrane helices, while the error rate is as high as 60% for signal peptides in Gram-positive bacteria.

DGPI is a commonly used prediction program for GPI-anchored protein, which uses a rule-based approach to predict the cleavage site for the hydrophobic sequence in the protein which is replaced by the GPI anchor (Kronegg *et al.*, 1999). The protein must have a predicted signal peptide (von Heijne, 1986) and a hydrophobic region of at least 13 amino acids which is identified using a sliding window. The cleavage site is positioned about eight amino acids from the end of the hydrophobic region and fine-tuned using a small position-specific weight matrix estimated from 172 GPI-anchored proteins with validated cleavage sites obtained from multiple species. GeneDB contains 165 GPI-anchored proteins which have been predicted by DGPI, although only 82 proteins have both a predicted signal peptide (predicted with SIGNALP v2.0) and a predicted GPI cleavage site.

SIGNALP, TMHMM and DGPI are all trained on proteins from multiple species, which makes it substantially easier to collect a large data set of appropriate proteins for training. This is especially useful when searching new genomes, where there may be few well-characterized membrane proteins with signal peptides, transmembrane domains or GPI-anchor signals. On the other hand, by using a training set from a single species, one might hope that subtle species-specific differences may yield a more sensitive and accurate method for detecting relevant proteins in the species of interest. Moreover, a species-specific approach can be iteratively improved as more proteins are validated by direct experimental determination of the protein sequence.

Leishmania-specific GPI-anchor prediction

A new predictor for GPI-anchored proteins, named GPIHMM based on a hidden Markov model was developed for the species-specific prediction of GPI-anchored proteins in *Plasmodium falciparum* (Vukcevic, 2004; Gilson *et al.*, 2006). Its design allows it to be easily retrained on new species. It has been tested on protozoan, metazoan and fungal proteins and its performance was compared to DGPI and BIG-π (Eisenhaber *et al.*, 1998; Eisenhaber *et al.*, 1999; Eisenhaber *et al.*, 2000; Sunyaev *et al.*, 1999). When the program was trained on the same datasets as BIG-π, it was found to perform at least as well as BIG-π and substantially better than DGPI (Vukcevic, 2004).) On *P. falciparum* it performed substantially better than any other predictor of GPI-anchored proteins (Gilson *et al.*, 2006).

The GPIHMM model consists of three parts: a signal peptide at the start, a middle section of variable length with a background distribution of amino acids that is intended to model the bulk of the protein, and the GPI anchor signal at the end. The model used for the signal peptide is the same as that used by SignalP-HMM. The GPI-anchor signal is broken down into four parts with the physical properties stemming purely from the training set:

(a) a flexible, polar region before the ω-site (10 amino acids);
(b) a region of small amino acids around the ω-site (four amino acids);
(c) a moderately polar region between the ω-site and the hydrophobic tail (seven amino acids); and
(d) a hydrophobic tail of variable length (≤ 35 amino acids) whose length distribution is modelled explicitly.

We have re-trained GPIHMM on representatives from each of the known *L. major* GPI-anchored protein families: gp63 (LmjF10.0460), PPG1 (LmjF35.0550), and PSA-2 (LmjF12.0810). Since the number of known GPI-anchored proteins from *L. major* is very small, the dataset was augmented with two VSG proteins from the *Trypanosomes*, ILTAT 1.3 from *T. brucei* and YNAT 1.1 from *T. congolense*. We used the same negative control data set used by (Vukcevic, 2004) to calibrate our model. This consisted of 1800 randomly chosen proteins from the Swiss-Prot database that do not have an annotated GPI anchor signal, and are long enough to span the HMM. GPIHMM was run on all *L. major* proteins available in GeneDB.

Table 9.1 lists the *L. major* proteins predicted by GPIHMM to be GPI-anchored with p-values less than 0.01. The predictions of TMHMM (v2.0) and DGPI are also shown for each protein. In order to provide stronger and more specific support for the presence of a signal peptide, SIGNALP-HMM (v3.0) was used to independently predict the presence or absence of a signal peptide, despite the fact that GPIHMM includes a signal peptide model. Most of the predicted GPI-anchored proteins show the presence of 1 or 2 predicted transmembrane domains, which typically coincide with the hydrophobic regions of the signal peptide and the GPI anchor signal. Since the GPI anchor addition occurs with cleavage from a transmembrane sequence, this finding is reassuring.

The list of predicted GPI-anchored proteins generated by GPIHMM overlaps with, but is also substantially different from the results of DGPI. Of the 61 predicted GPI-anchored proteins with $p<0.01$ obtained using GPIHMM, 36 were also identified by DGPI. The signal peptide HMM does not perform well on proteins containing long serine-rich low complexity regions, so PPG1 (a known GPI-anchored protein), PPG4, PPG5 and LmjF33.2000 were filtered from the list. The cut-off of $p<0.01$ used here is a soft one and chosen largely for reasons of space. GPIHMM predicted a further 415 proteins with p-values in the range of 0.01 to 0.05 (data not shown), and of these only 27 were also identified by DGPI. As expected, the remaining predicted GPI-anchored proteins in GeneDB were assigned p-values greater than 0.05 by GPIHMM. It is usually the case that generic methods will do as well as species-specific methods on the easiest targets. The harder to identify cases will usually benefit the most from a species-specific approach.

The approach described above is very encouraging, but it is only the first stage of analysis. We are a long way from the ability to use such computing tools to make accurate predictions, and the 'wet lab' approach is essential to generate definitive data. Further work is required to validate the results obtained using GPIHMM and its more species-specific approach to *Leishmania* GPI-anchor prediction. Validated GPI-anchored proteins can then be used to improve the accuracy of parameter estimates by re-training and improving the model predictions. This illustrates the additional advantage of a careful, species-specific approach over the use of a generic tool like DGPI. The species-specific approach enables an

Table 9.1 Putative *L. Major* GPI-anchored proteins predicted using GPIHMM with a cut-off of $P < 0.01$

Accession number	Name	Score	*P*-value	ω-site	Signalp	#TM	DGPI
LmjF12.0810	Promastigote surface antigen protein 2 PSA2	66.682	0.001	360	Yes	2	Yes
LmjF12.1000	Promastigote surface antigen protein 2 PSA2	66.682	0.001	360	Yes	2	Yes
LmjF12.0980	Promastigote surface antigen protein 2 PSA2	66.682	0.001	360	Yes	2	Yes
LmjF12.0940	Promastigote surface antigen protein 2 PSA2	66.682	0.001	360	Yes	2	Yes
LmjF12.0780	Surface antigen protein 2 precursor, CHR12_tmp.1150	66.682	0.001	350	Yes	2	Yes
LmjF12.0730	Surface antigen protein 2, putative, CHR12_tmp.1010	66.217	0.001	522	Yes	2	Yes
LmjF12.0890	Surface antigen protein 2, putative, CHR12_tmp.1220	61.97	0.001	370	Yes	2	Yes
LmjF12.0990	Surface antigen protein, putative, CHR12_tmp.1250	61.967	0.001	585	Yes	2	Yes
LmjF12.0760	Surface antigen protein 2, putative, CHR12_tmp.1170	61.507	0.001	562	Yes	2	Yes
LmjF10.0480	GP63-4, GP63, leishmanolysin	59.684	0.001	577	Yes	1	Yes
LmjF10.0460	GP63-1, GP63, leishmanolysin	59.684	0.001	577	Yes	1	Yes
LmjF10.0465	GP63–2, GP63, leishmanolysin	56.828	0.001	577	Yes	1	Yes
LmjF12.0850	Surface antigen protein 2, putative, CHR12_tmp.1080	55.925	0.001	604	Yes	2	Yes
LmjF12.0830	Surface antigen protein 2, putative, CHR12_tmp.1100	55.922	0.001	623	Yes	2	Yes
LmjF12.0960	Surface antigen protein 2, putative, CHR12_tmp.0970	55.922	0.001	643	Yes	2	Yes
LmjF12.0740	Surface antigen protein, putative, CHR12_tmp.1180	53.93	0.001	735	Yes	2	Yes
LmjF12.1070	Surface antigen protein 2, putative, CHR12_tmp.1070	53.535	0.001	679	Yes	2	Yes
LmjF12.0755	Surface antigen protein 2, putative, CHR12_tmp.1140	53.464	0.001	545	Yes	2	Yes
LmjF12.0765	Surface antigen protein 2, putative, CHR12_tmp.1140	53.464	0.001	547	Yes	2	Yes
LmjF12.0870	Surface antigen protein 2, putative, CHR12_tmp.1230	52.842	0.001	522	Yes	2	Yes
LmjF35.0290	Hypothetical protein, conserved	49.734	0.001	1502	Yes	0	No
LmjF12.1005	Surface antigen protein 2, putative	38.096	0.001	278	No	1	Yes
LmjF05.0900	Surface antigen-like protein, LM5.90, L5856.05	28.691	0.001	263	Yes	1	No

LmjF04.0140	Hypothetical protein, conserved in *Leishmania*, P131.05	28.341	0.001	511	Yes	0	Yes
LmjF04.0160	Hypothetical protein, conserved in *Leishmania*, P131.07	28.341	0.001	511	Yes	0	Yes
LmjF04.0150	Hypothetical protein, conserved in *Leishmania*, P131.06	28.341	0.001	511	Yes	0	Yes
LmjF04.0170	Hypothetical protein, conserved in *Leishmania*, P131.08	28.341	0.001	511	Yes	0	Yes
LmjF04.0130	Hypothetical protein, conserved in *Leishmania*, P131.04	28.341	0.001	511	Yes	0	Yes
LmjF04.0210	Surface antigen-like protein, L302.04	26.153	0.001	253	Yes	2	Yes
LmjF28.0570	Major surface protease gp63, putative, CHR28_tmp.1700	24.313	0.001	545	Yes	2	Yes
LmjF04.0180	Surface antigen-like protein, L302.01	24.233	0.001	483	Yes	1	Yes
LmjF31.1440	Hypothetical protein, unknown function	23.469	0.001	506	Yes	2	No
LmjF15.0570	hypothetical protein, unknown function	23.26	0.001	312	Yes	0	Yes
LmjF04.0190	Surface antigen-like protein, L302.02	21.922	0.001	688	Yes	1	Yes
LmjF09.0580	Surface antigen-like protein	19.558	0.001	260	Yes	2	Yes
LmjF31.2000	GP63-like protein, leishmanolysin-like protein	18.548	0.001	612	Yes	2	No
LmjF35.5240	Hypothetical protein, unknown function	18.185	0.001	248	Yes	0	Yes
LmjF19.0170	Hypothetical protein, conserved	17.899	0.001	1479	Yes	1	No
LmjF04.0200	Surface antigen-like protein, L302.03	15.359	0.001	163	Yes	0	Yes
LmjF23.1170	MBAP2, membrane-bound acid phosphatase 2	13.301	0.001	528	Yes	2	No
LmjF09.0890	Phospholipid transporting ATPase-like protein, putative	13.246	0.001	2508	Yes	7	No
LmjF34.2200	Hypothetical protein, conserved, CHR34_tmp.1790	11.763	0.002	420	Yes	1	No
LmjF34.2650	Hypothetical protein, conserved, CHR34_tmp.2250	11.058	0.002	595	Yes	2	No
LmjF34.3630	Hypothetical protein, conserved, CHR34_tmp.3210	11.051	0.002	503	Yes	2	No
LmjF14.0770	Hypothetical protein, unknown function	10.684	0.003	371	Yes	0	Yes
LmjF32.3720	Hypothetical protein, unknown function	10.651	0.003	459	Yes	12	No
LmjF08.1080	Cathepsin L-like protease	9.727	0.006	430	Yes	1	No
LmjF08.1030	Cathepsin L-like protease	9.727	0.006	430	Yes	1	No

Table 9.1 Continued

Accession number	Name	Score	*P*-value	ω-site	Signalp	#TM	DGPI
LmjF08.1060	Cathepsin L-like protease	9.727	0.006	430	Yes	1	No
LmjF13.0480	Hypothetical protein, conserved	9.334	0.008	497	Yes	0	Yes
LmjF06.1120	Hypothetical protein, conserved, CHR6_tmp.102c	9.321	0.008	253	Yes	2	No
LmjF20.0360	Hypothetical protein, conserved, CHR20_tmp.0350	8.665	0.008	303	Yes	2	No
LmjF27.0070	Hypothetical protein, conserved	8.296	0.009	301	Yes	1	No
LmjF12.0400	3′-Nucleotidase/nuclease, putative, CHR12_tmp.0400	8.282	0.009	364	Yes	2	No
LmjF35.3220	Hypothetical protein, conserved	8.243	0.009	152	No	0	No
LmjF01.0640	HSP70-like protein, LmjF1.0640	8.212	0.009	1088	No	1	No
LmjF30.0805	Hypothetical protein, conserved, 30.t00398	7.989	0.009	186	Yes	0	No
LmjF03.0470	Hypothetical protein, conserved, LmjF3.0470	7.783	0.009	992	No	6	No
LmjF30.2990	Hypothetical protein, conserved	7.415	0.009	422	Yes	1	No
LmjF10.0540	Hypothetical protein, unknown function	7.355	0.009	1076	Yes	1	No
LmjF10.0170	Nucleoside phosphatase, putative	7.266	0.009	661	Yes	1	No

The score generated by GPIHMM, the *P*-value of this score and the location of the predicted cleavage site (ω-site) are shown in the table. While GPIHMM includes a signal peptide model, the presence of a putative signal peptide was independently predicted using SignalP-HMM (v3.0). The number of predicted transmembrane domains (#TM) was determined using TMHMM (v2.0). The DGPI prediction for each of these proteins is also shown. The signal peptide model performs poorly on proteins containing long Serine-rich low complexity regions, substantially over-estimating their scores, thus PPG1 (a known GPI-anchored protein), PPG4 PPG5 and LmjF33.2000 were removed from the list.

iterative process of laboratory work, model improvement and new rounds of prediction. These computational approaches can only provide a set of hypotheses, and thus far their impact on experimental determination has been low. The hope is that this situation will change in the near future.

'To understand function, study structure' (Maizels, 1990)

A complementary approach to sequence analysis is the comparison of the three-dimensional structure of proteins in different species. The advent of high throughput crystallisation and structure determination holds the promise of revealing common folding patterns which may be obvious even when genetic drift has been so great that sequence homology is no longer recognizable. In a few cases, this promise has been realized. The combination of structural comparisons with newer computer algorithms for detecting distant sequence relationships is particularly powerful (Yakunin *et al.*, 2004; Bradley *et al.*, 2005; Najmanovich *et al.*, 2005; Socolich *et al.*, 2005; see also Kuznetsova *et al.*, 2005). The main limitation is that many proteins are difficult or impossible to crystallize in forms that diffract to high resolution. Nonetheless, solving the structure of large numbers of 'low hanging fruit' that crystallize easily is likely to provide the key to identifying many structural and functional relationships between proteins in different species.

Protein–protein Interactions

Yet another approach that can be used on a large scale involves the use of solid phase microarrays of proteins in their native conformation. These can be used to identify interacting proteins, from which function might be inferred (Delehanty *et al.*, 2003; Haab, 2001). Similarly, chips for high throughput screening of specific carbohydrate-binding proteins are becoming available (Park *et al.*, 2004). Identification of interactions between proteins has been facilitated by the use of the yeast two-hybrid system (Gietz, 2005; Ito *et al.*, 2001), and has recently culminated in comprehensive 'interactome' maps (Ito *et al.*, 2002; LaCount *et al.*, 2005; Rual *et al.*, 2005; Suthram *et al.*, 2005). When interactions are found in yeast, there is a reasonable chance that homologues in other species will also interact (Huang *et al.*, 2004). Surprisingly, a recent study using a high throughput version of the yeast two-hybrid system and bioinformatics found little conservation of interacting protein clusters between *P. falciparum* and other eukaryotic organisms (LaCount *et al.*, 2005; Suthram *et al.*, 2005). This may be due to the fact that little experimental validation was done on this presumed *P. falciparum* interactome (LaCount *et al.*, 2005; Suthram *et al.*, 2005). Databases of protein–protein interactions should be searched frequently, because newly discovered interactions in one species may provide the key to function in other species.

Enthusiasm for the yeast two-hybrid system should be tempered however, by an understanding of its limitations. Detailed comparison of the interactions that were identified in putative 'interactome' maps has shown a remarkable lack of agreement between different studies. There is a high risk of false positive identifications, and any potential interactions must be verified by independent means. The yeast two-hybrid system is also limited by the fact that many protein–protein interactions in *Leishmania* may require post-translational modifications which may not occur in yeast.

Conclusions

The genomes of the three trypanosomatids, *T. brucei*, *T. cruzi* and *L. major* are predicted to encode approximately 9100, 12,000 and 8300 proteins, respectively. These encode a core of about 6200 proteins common to all three species, which show an average of 44% identity between *L. major* and the two trypanosome species (El-Sayed *et al.*, 2005). The remainder constitute species-specific genes of which *T. brucei* and *T. cruzi* have a greater proportion than *L. major* (El-Sayed *et al.*, 2005). Interestingly, the majority of the species-specific genes appear to encode members of surface protein families, suggesting that they reflect adaptations to the different invasion and survival strategies used by each organism.

Assuming that surface membrane proteins represent 1–10% of the total proteome, one would expect about 80–800 such proteins in *L. major*. As described in this chapter, the number of well-characterized proteins is still small. The current map of the *Leishmania* surface is low resolution, derived from the identification, isolation and characterization of individual proteins and genes. The availability of the genomes of several *Leishmania* species, as well as those of *T. brucei* and *T. cruzi*, combined with the rapid development of better algorithms for use in bioinformatics is providing tools of extraordinary power for the identification of novel membrane proteins and ultimately holds the promise of a global view of the parasite membrane. The challenge of the post-genome era is to assign function. In the final analysis, we will still need to study our proteins of interest, as we always have had to do. And let's not forget the power of some good biochemistry (Zick *et al.*, 2005).

References

Aebischer, T., Moody, S.F., and Handman, E. (1993). Persistence of virulent *Leishmania major* in murine cutaneous leishmaniasis, a possible hazard for the host. Infect. Immun. *61*, 220–226.

Antoine, J.C., Prina, E., Courret, N., and Lang, T. (2004). *Leishmania* spp., on the interactions they establish with antigen-presenting cells of their mammalian hosts. Adv. Parasitol. 58, 1–68.

Bahr, V., Stierhof, Y.-K., Ilg, T., Demar, M., Quinten, M., and Overath, P. (1993). Expression of lipophosphoglycan, high-molecular weight phosphoglycan and glycoprotein 63 in promastigotes and amastigotes of *Leishmania mexicana*. Mol. Biochem. Parasitol. *58*, 107–122.

Baldwin, T., Curtis, J., O'Keeffe, M., Vremec, D., Shortman, K., and Handman, E. (2004). Dendritic cell populations in the *Leishmania major* infected skin and draining lymph nodes. Infect. Immun. *72*, 1991–2001.

Bandyopadhyay, G., Sajan, M.P., Kanoh, Y., Standaert, M.L., Burke, T.R., Jr., Quon, M.J., Reed, B.C., Dikic, I., Noel, L.E., Newgard, C.B., and Farese, R. (2000). Glucose activates mitogen-activated protein kinase (extracellular signal-regulated kinase) through proline-rich tyrosine kinase-2 and the Glut1 glucose transporter. J. Biol. Chem. *275*, 40817–40826.

Basselin, M., Coombs, G.H., and Barrett, M.P. (2000). Putrescine and spermidine transport in *Leishmania*. Mol. Biochem. Parasitol. *109*, 37–46.

Beetham, J.K., Donelson, J.E., and Dahlin, R.R. (2003). Surface glycoprotein PSA (GP46) expression during short- and long-term culture of *Leishmania chagasi*. Mol. Biochem. Parasitol. *131*, 109–117.

Beetham, J.K., Myung, K.S., McCoy, J.J., Wilson, M.E., and Donelson, J.E. (1997). Glycoprotein 46 mRNA abundance is post-transcriptionally regulated during development of *Leishmania chagasi* promastigotes to an infectious form. J. Biol. Chem. *272*, 17360–17366.

Bendtsen, J.D., Nielsen, H., von Heijne, G., and Brunak, S. (2004). Improved prediction of signal peptides, SignalP 3.0. J. Mol. Biol. *340*, 783–795.

Beverley, S.M., and Turco, S.J. (1995). Identification of genes mediating lipophosphoglycan (LPG) biosynthesis by functional complementation of *Leishmania donovani* mutants. Ann. Trop. Med. Parasitol. *89*, 11–17.

Blackwell, J.M. (1989). The macrophage resistance gene Lsh/Ity/Bcg. Res. Immunol. *140*, 767–769.

Bogdan, C., Donhauser, N., Döring, R., Röllinghoff, M., Diefenbach, A., and Rittig, M.G. (2000). Fibroblasts as host cells in latent Leishmaniosis. J. Exp. Med. *191*, 2121–2129.

Borges, V.M., Vannier-Santos, M.A., and de Souza, W. (1998). Subverted transferrin trafficking in *Leishmania*-infected macrophages. Parasitol. Res. *84*, 811–822.

Bouvier, J., Bordier, C., Vogel, H., Reichelt, R., and Etges, R. (1989). Characterization of the promastigote surface protease of *Leishmania* as a membrane-bound zinc endopeptidase. Mol. Biochem. Parasitol. *37*, 235–245.

Bouvier, J., Etges, R., and Bordier, C. (1987). Identification of the promastigote surface protease in seven species of *Leishmania*. Mol. Biochem. Parasitol. *24*, 73–79.

Bouvier, J., Etges, R.J., and Bordier, C. (1985). Identification and purification of membrane and soluble forms of the major surface protein of *Leishmania* promastigotes. J. Biol. Chem. *260*, 15504–15509.

Bouvier, J., Schneider, P., Etges, R., and Bordier, C. (1990). Peptide substrate specificity of the membrane-bound metalloprotease of *Leishmania*. Biochemistry *29*, 10113–10119.

Bradley, P., Misura, K.M., and Baker, D. (2005). Toward high-resolution de novo structure prediction for small proteins. Science *309*, 1868–1871.

Brittingham, A., Chen, G., McGwire, B.S., Chang, K.P., and Mosser, D.M. (1999). Interaction of *Leishmania* gp63 with cellular receptors for fibronectin. Infect. Immun. *67*, 4477–4484.

Brittingham, A., Morrison, C.J., McMaster, W.R., McGwire, B.S., Chang, K.P., and Mosser, D.M. (1995). Role of the *Leishmania* surface protease gp63 in complement fixation, cell adhesion, and resistance to complement-mediated lysis. J. Immunol. *155*, 3102–3111.

Burchmore, R.J., and Landfear, S.M. (1998). Differential regulation of multiple glucose transporter genes in *Leishmania mexicana*. J. Biol. Chem. *273*, 29118–29126.

Burchmore, R.J., Rodriguez-Contreras, D., McBride, K., Merkel, P., Barrett, M.P., Modi, G., Sacks, D., and Landfear, S.M. (2003). Genetic characterization of glucose transporter function in *Leishmania mexicana*. Proc. Natl. Acad. Sci. USA. *100*, 3901–3906.

Chen, D.Q., Kolli, B.K., Yadava, N., Lu, H.G., Gilman-Sachs, A., Peterson, D.A., and Chang, K.P. (2000). Episomal expression of specific sense and antisense mRNAs in *Leishmania amazonensis*, modulation of gp63 level in promastigotes and their infection of macrophages in vitro. Infect. Immun. *68*, 80–86.

Coyne, K.E., Hall, S.E., Thompson, S., Arce, M.A., Kinoshita, T., Fujita, T., Anstee, D.J., Rosse, W., and Lublin, D.M. (1992). Mapping of epitopes, glycosylation sites, and complement regulatory domains in human decay accelerating factor. J. Immunol. *149*, 2906–2913.

Cunningham, M.L., and Beverley, S.M. (2001a). Pteridine salvage throughout the *Leishmania* infectious cycle, implications for antifolate chemotherapy. Mol. Biochem. Parasitol. *113*, 199–213.

Cunningham, M.L., Titus, R.G., Turco, S.J., and Beverley, S.M. (2001b). Regulation of differentiation to the infective stage of the protozoan parasite *Leishmania major* by tetrahydrobiopterin. Science *292*, 285–287.

de Graffenried, C.L., and Bertozzi, C.R. (2004). The roles of enzyme localisation and complex formation in glycan assembly within the Golgi apparatus. Curr. Opin. Cell Biol. *16*, 356–363.

Debrabant, A., Bastien, P., and Dwyer, D.M. (2001). A unique surface membrane anchored purine-salvage enzyme is conserved among a group of primitive eukaryotic human pathogens. Mol. Cell. Biochem. *220*, 109–116.

Delehanty, J.B., and Ligler, F.S. (2003). Method for printing functional protein microarrays. Biotechniques *34*, 380–385.

Denny, P.W., Goulding, D., Ferguson, M.A., and Smith, D.F. (2004). Sphingolipid-free *Leishmania* are defective in membrane trafficking, differentiation and infectivity. Mol. Microbiol. *52*, 313–327.

Descoteaux, A., Garraway, L.A., Ryan, K.A., Garrity, L.K., Turco, S.J., and Beverley, S.M. (1994). Identification of genes by functional complementation in protozoan parasite *Leishmania*. Meths. Mol. Genetics *3*, 22–48.

Descoteaux, A., Mengeling, B.J., Beverley, S.M., and Turco, S.J. (1998). *Leishmania donovani* has distinct mannosylphosphoryltransferases for the initiation and elongation phases of lipophosphoglycan repeating unit biosynthesis. Mol. Biochem. Parasitol. *94*, 27–40.

Dobson, D.E., Mengeling, B.J., Cilmi, S., Hickerson, S., Turco, S.J., and Beverley, S.M. (2003). Identification of genes encoding arabinosyltransferases (SCA) mediating developmental modifications of lipophosphoglycan required for sandfly transmission of *Leishmania major*. J. Biol. Chem. *278*, 28840–28848.

Dobson, D.E., Scholtes, L.D., Valdez, K.E., Sullivan, D.R., Mengeling, B.J., Cilmi, S., Turco, S.J., and Beverley, S.M. (2003). Functional identification of galactosyltransferases (SCGs) required for species-specific modifications of the lipophosphoglycan adhesin controlling *Leishmania major*-sandfly interactions. J. Biol. Chem. *278*, 15523–15531.

Dodge, M.A., Waller, R.F., Chow, L.M., Zaman, M.M., Cotton, L.M., McConville, M.J., and Wirth, D.F. (2004). Localization and activity of multidrug resistance protein 1 in the secretory pathway of *Leishmania* parasites. Mol. Microbiol. *51*, 1563–1575.

Drew, M.E., Langford, C.K., Klamo, E.M., Russell, D.G., Kavanaugh, M.P., and Landfear, S.M. (1995). Functional expression of a myo-inositol/H+ symporter from *Leishmania donovani*. Mol. Cell. Biol. *15*, 5508–5515.

Dwyer, D.M., and Gottlieb, M. (1985). The biochemistry of *Leishmania* surface membranes. In Chang, K.-P. and Bray, R. S. (eds.), Leishmaniasis. Elsevier, Amsterdam, Vol. 1, p. 31–48.

Eisenhaber, B., Bork, P., and Eisenhaber, F. (1998). Sequence properties of GPI-anchored proteins near the omega-site, constraints for the polypeptide binding site of the putative transamidase. Protein Eng. *11*, 1155–1161.

Eisenhaber, B., Bork, P., and Eisenhaber, F. (1999). Prediction of potential GPI-modification sites in proprotein sequences. J. Mol. Biol. *292*, 741–758.

Eisenhaber, B., Bork, P., Yuan, Y., Loffler, G., and Eisenhaber, F. (2000). Automated annotation of GPI anchor sites, case study C. elegans. Trends Biochem. Sci. *25*, 340–341.

El-Sayed, N.M., Myler, P.J., Bartholomeu, D.C., Nilsson, D., Aggarwal, G., Tran, A.N., Ghedin, E., Worthey, E.A., Delcher, A.L., Blandin, G., Westenberger, S.J., Caler, E., Cerqueira, G.C., Branche, C., Haas, B., Anupama, A., Arner, E., Aslund, L., Attipoe, P., Bontempi, E., Bringaud, F., Burton, P., Cadag, E., Campbell, D.A., Carrington, M., Crabtree, J., Darban, H., da Silveira, J.F., de Jong, P., Edwards, K., Englund, P.T., Fazelina, G., Feldblyum, T., Ferella, M., Frasch, A.C., Gull, K., Horn, D., Hou, L., Huang, Y., Kindlund, E., Klingbeil, M., Kluge, S., Koo, H., Lacerda, D., Levin, M.J., Lorenzi, H., Louie, T., Machado, C.R., McCulloch, R., McKenna, A., Mizuno, Y., Mottram, J.C., Nelson, S., Ochaya, S., Osoegawa, K., Pai, G., Parsons, M., Pentony, M., Pettersson, U., Pop, M., Ramirez, J.L., Rinta, J., Robertson, L., Salzberg, S.L., Sanchez, D.O., Seyler, A., Sharma, R., Shetty, J., Simpson, A.J., Sisk, E., Tammi, M.T., Tarleton, R., Teixeira, S., Van Aken, S., Vogt, C., Ward, P.N., Wickstead, B., Wortman, J., White, O., Fraser, C.M., Stuart, K.D., and Andersson, B. (2005). The genome sequence of *Trypanosoma cruzi*, etiologic agent of Chagas disease. Science *309*, 409–415.

El-Sayed, N.M., Myler, P.J., Blandin, G., Berriman, M., Crabtree, J., Aggarwal, G., Caler, E., Renauld, H., Worthey, E.A., Hertz-Fowler, C., Ghedin, E., Peacock, C., Bartholomeu, D.C., Haas, B.J., Tran, A.N., Wortman, J.R., Alsmark, U.C., Angiuoli, S., Anupama, A., Badger, J., Bringaud, F., Cadag, E., Carlton, J.M., Cerqueira, G.C., Creasy, T., Delcher, A.L., Djikeng, A., Embley, T.M., Hauser, C., Ivens, A.C., Kummerfeld, S.K., Pereira-Leal, J.B., Nilsson, D., Peterson, J., Salzberg, S.L., Shallom, J., Silva, J.C., Sundaram, J., Westenberger, S., White, O., Melville, S.E., Donelson, J.E., Andersson, B., Stuart, K.D., and Hall, N. (2005). Comparative genomics of trypanosomatid parasitic protozoa. Science *309*, 404–409.

Etges, R., Bouvier, J., and Bordier, C. (1986). The major surface protein of *Leishmania* promastigotes is a protease. J. Biol. Chem. *261*, 9098–9101.

Etges, R., Bouvier, J., and Bordier, C. (1986). The major surface protein of *Leishmania* promastigotes is anchored in the membrane by a myristic acid-labelled phospholipid. EMBO J. 5, 597–601.

Felibertt, P., Bermudez, R., Cervino, V., Dawidowicz, K., Dagger, F., Proverbio, T., Marin, R., and Benaim, G. (1995). Ouabain-sensitive Na^{+},$K(^{+})$-ATPase in the plasma membrane of *Leishmania mexicana*. Mol. Biochem. Parasitol. *74*, 179–187.

Ferguson, M.A.J. (1999). The structure, biosynthesis and functions of glycosylphosphatidylinositol anchors, and the contributions of typanosome research. J. Cell Sci. *112*, 2799–2809.

Fleming, M.D., Romano, M.A., Su, M.A., Garrick, L.M., Garrick, M.D., and Andrews, N.C. (1998). Nramp2 is mutated in the anemic Belgrade (b) rat, evidence of a role for Nramp2 in endosomal iron transport. Proc. Natl. Acad. Sci. USA. *95*, 1148–1153.

Foth, B., Piani, A., Curtis, J.M., Ilg, T., McConville, M., and Handman, E. (2002). *Leishmania major* proteophosphoglycans exist as membrane-bound and soluble forms and localise to the cell membrane, the flagellar pocket and the lysosome. Int. J. Parasitol. *32*, 1701–1708.

Frommel, T.O., Button, L.L., Fujikura, Y., and McMaster, W.R. (1989). The major surface glycoprotein (GP63) is present in both life stages of *Leishmania*. Mol. Biochem. Parasitol. *38*, 25–32.

Gietz, I.D. (2005). Yeast two-hybrid system screening. Methods Mol. Biol. *313*, 345–372.

Gilson, P.R., Nebl, T., Vikcevic, D., Moritz, R.L., Sargeant, T., Speed, T.P., Schofield, L., and Crabb, B.S. (2006). Identification and stoichiometry of glycosylphosphatidylinositol-anchored membrane proteins of the human malaria parasite *Plasmodium falciparum*. Mol. Cell Proteomics 5, 1286–1299.

Govoni, G., and Gros, P. (1998). Macrophage NRAMP1 and its role in resistance to microbial infections. Inflamm. Res. *47*, 277–284.

Gull, K. (1999). The cytoskeleton of trypanosomatid parasites. Annu. Rev. Microbiol. 53, 629–655.

Haab, B.B. (2001). Advances in protein microarray technology for protein expression and interaction profiling. Curr. Opin. Drug Discov. Dev. *4*, 116–123.

Handman, E., and Bullen, D.V. (2002). Interaction of *Leishmania* with the host macrophages. Trends Parasitol. *18*, 332–334.

Handman, E., Osborn, A.H., Symons, F., van, D.R., and Cappai, R. (1995). The *Leishmania* promastigote surface antigen 2 complex is differentially expressed during the parasite life cycle. Mol. Biochem. Parasitol. *74*, 189–200.

Hertz-Fowler, C., Peacock, C.S., Wood, V., Aslett, M., Kerhornou, A., Mooney, P., Tivey, A., Berriman, M., Hall, N., Rutherford, K., Parkhill, J., Ivens, A.C., Rajandream, M.-A., and Barrell, B. (2004). GeneDB, a resource for prokaryotic and eukaryotic organisms. Nucleic Acids Res. *32*, D339–343.

Hill, K.L., Hutchings, N.R., Russell, D.G., and Donelson, J.E. (1999). A novel protein targeting domain directs proteins to the anterior cytoplasmic face of the flagellar pocket in African trypanosomes. J. Cell Sci. 112 Pt *18*, 3091–3101.

Hilley, J.D., Zawadzki, J.L., McConville, M.J., Coombs, G.H., and Mottram, J.C. (2000). *Leishmania mexicana* mutants lacking glycosylphosphatidylinositol (GPI), protein transamidase provide insights into the biosynthesis and functions of GPI-anchored proteins. Mol. Biol. Cell *11*, 1183–1195.

Hong, K., Ma, D., Beverley, S.M., and Turco, S.J. (2000). The *Leishmania* GDP-mannose transporter is an autonomous, multi-specific, hexameric complex of LPG2 subunits. Biochemistry 39, 2013–2022.

Huang, T.W., Tien, A.C., Huang, W.S., Lee, Y.C., Peng, C.L., Tseng, H.H., Kao, C.Y., and Huang, C.Y. (2004). POINT, a database for the prediction of protein–protein interactions based on the orthologous interactome. Bioinformatics *20*, 3273–3276.

Huynh, C., Sacks, D.L., and Andrews, N.W. (2006). A *Leishmania amazonensis* ZIP family iron transporter is essential for parasite replication within macrophage phagolysosomes. J. Exp. Med. *203*, 2363–2375.

Ilg, T. (2000). Proteophosphoglycans of *Leishmania*. Parasitol. Today *16*, 489–497.

Ilg, T., Demar, M., and Harbecke, D. (2001). Phosphoglycan repeat-deficient *Leishmania mexicana* parasites remain infectious to macrophages and mice. J. Biol. Chem. *276*, 4988–4997.

Ilg, T., Montgomery, J., Stierhof, Y.D., and Handman, E. (1999). Molecular cloning and characterization of a novel repeat-containing *Leishmania major* gene, *ppg*1, that encodes a membrane-associated form of proteophosphoglycan with a putative glycosylphosphatidylinositol anchor. J. Biol. Chem. *274*, 31410–31420.

Ilgoutz, S.C., and McConville, M.J. (2001). Function and assembly of the *Leishmania* surface coat. Int. J. Parasitol. *21*, 899–908.

Inohara, Chamaillard, McDonald, C., and Nunez, G. (2005). NOD-LRR proteins, role in host-microbial interactions and inflammatory disease. Annu. Rev. Biochem. *74*, 355–383.

Ito, T., Chiba, T., Ozawa, R., Yoshida, M., Hattori, M., and Sakaki, Y. (2001). A comprehensive two-hybrid analysis to explore the yeast protein interactome. Proc. Natl. Acad. Sci. USA. *98*, 4569–4574.

Ito, T., Ota, K., Kubota, H., Yamaguchi, Y., Chiba, T., Sakuraba, K., and Yoshida, M. (2002). Roles for the two-hybrid system in exploration of the yeast protein interactome. Mol. Cell Proteomics *1*, 561–566.

Ivens, A.C., Peacock, C.S., Worthey, E.A., Murphy, L., Aggarwal, G., Berriman, M., Sisk, E., Rajandream, M.A., Adlem, E., Aert, R., Anupama, A., Apostolou, Z., Attipoe, P., Bason, N., Bauser, C., Beck, A., Beverley, S.M., Bianchettin, G., Borzym, K., Bothe, G., Bruschi, C.V., Collins, M., Cadag, E., Ciarloni, L., Clayton, C., Coulson, R.M., Cronin, A., Cruz, A.K., Davies, R.M., De Gaudenzi, J., Dobson, D.E., Duesterhoeft, A., Fazelina, G., Fosker, N., Frasch, A.C., Fraser, A., Fuchs, M., Gabel, C., Goble, A., Goffeau, A., Harris, D., Hertz-Fowler, C., Hilbert, H., Horn, D., Huang, Y., Klages, S., Knights, A., Kube, M., Larke, N., Litvin, L., Lord, A., Louie, T., Marra, M., Masuy, D., Matthews, K., Michaeli, S., Mottram, J.C., Muller-Auer, S., Munden, H., Nelson, S., Norbertczak, H., Oliver, K., O'Neil, S., Pentony, M., Pohl, T.M., Price, C., Purnelle, B., Quail, M.A., Rabbinowitsch, E., Reinhardt, R., Rieger, M., Rinta, J., Robben, J., Robertson, L., Ruiz, J.C., Rutter, S., Saunders, D., Schafer, M., Schein, J., Schwartz, D.C., Seeger, K., Seyler, A., Sharp, S., Shin, H., Sivam, D., Squares, R., Squares, S., Tosato, V., Vogt, C., Volckaert, G., Wambutt, R., Warren, T., Wedler, H., Woodward, J., Zhou, S., Zimmermann, W., Smith, D.F., Blackwell, J.M., Stuart, K.D., Barrell, B. *et al.* (2005). The genome of the kinetoplastid parasite, *Leishmania major*. Science 309, 436–442.

Jardim, A., Funk, V., Caprioli, R.M., and Olafson, R.W. (1995). Isolation and structural characterization of the *Leishmania donovani* kinetoplastid membrane protein-11, a major immunoreactive membrane glycoprotein. Biochem. J. *305*, 307–313.

Jardim, A., Hanson, S., Ullman, B., McCubbin, W.D., Kay, C.M., and Ofafson, R.W. (1995). Cloning and structure-function analysis of the *Leishmania donovani* kinetoplastid membrane protein-11. Biochem. J. *305*, 315–320.

Jeffrey, C.J. (1999). Moonlighting proteins. TIBS *24*, 8–11.

Jimenez-Ruiz, A., Boceta, C., Bonay, P., Requena, J.M., and Alonso, C. (1998). Cloning, sequencing, and expression of the PSA genes from *Leishmania infantum*. Eur. J. Biochem. *251*, 389–397.

Joshi, P.B., Kelly, B.L., Kamhawi, S., Sacks, D.L., and McMaster, W.R. (2002). Targeted gene deletion in *Leishmania major* identifies leishmanolysin (GP63) as a virulence factor. Mol. Biochem. Parasitol. *120*, 33–40.

Kaiser, C.A., Preuss, D., Grisafi, P., and Botstein, D. (1987). Many random sequences functionally replace the secretion signal sequence of yeast invertase. Science *235*, 312–317.

Kajava, A.V. (1998). Structural diversity of leucine-rich repeat proteins. J. Mol. Biol. *277*, 519–527.

Kamhawi, S., Modi, G.B., Pimenta, P.F.P., Rowton, E., and Sacks, D.L. (2000). The vectorial competence of *Phlebotomus sergenti* is specific for *Leishmania tropica* and is controlled by species-specific, lipophosphoglycan-mediated midgut atttachment. Parasitology *121*, 25–33.

Kedzierski, L., Montgomery, J., Bullen, D., Curtis, J., Gardiner, E., Jimenez-Ruiz, A., and Handman, E. (2004). A leucine-rich repeat motif of *Leishmania* parasite surface antigen 2 binds to macrophages through the complement receptor 3. J. Immunol. *172*, 4902–4906.

Kedzierski, L., Montgomery, J., Curtis, J., and Handman, E. (2004). Leucine-rich repeats in host-pathogen interactions. Arch. Immunol. Ther. Exp. *52*, 104–112.

Kelleher, M., Bacic, A., and Handman, E. (1992). Identification of a macrophage-binding determinant on lipophosphoglycan from *Leishmania major* promastigotes. Proc. Natl. Acad. Sci. USA. *89*, 6–10.

Kelleher, M., Moody, S.M., Mirabile, P., Osborn, A.H., Bacic, A., and Handman, E. (1995). Lipophosphoglycan blocks attachment of *Leishmania major* amastigotes to macrophages. Infect. Immun. *63*, 43–50.

Krogh, A., Larsson, B., von Heijne, G., and Sonnhammer, E.L. (2001). Predicting transmembrane protein topology with a hidden Markov model, application to complete genomes. J. Mol. Biol. *305*, 567–580.

Kronegg, J., and Buloz, D. (1999). http://129.194.185.165/dgpi/

Kundig, C., Haimeur, A., Legare, D., Papadopoulou, B., and Ouellette, M. (1999). Increased transport of pteridines compensates for mutations in the high affinity folate transporter and contributes to methotrexate resistance in the protozoan parasite *Leishmania tarentolae*. EMBO J. *18*, 2342–2351.

Kuznetsova, E., Proudfoot, M., Sanders, S.A., Reinking, J., Savchenko, A., Arrowsmith, C.H., Edwards, A.M., and Yakunin, A.F. (2005). Enzyme genomics; application of general enzymatic screens to discover new enzymes. FEMS Microbiol. Rev. *29*, 263–279.

LaCount, D.J., Vignali, M., Chettier, R., Phansalkar, A., Bell, R., Hesselberth, J.R., Schoenfeld, L.W., Ota, I., Sahasrabudhe, S., Kurschner, C., Fields, S., and Hughes, R.E. (2005). A protein interaction network of the malaria parasite *Plasmodium falciparum*. Nature *438*, 103–107.

Landfear, S.M. (2000). Genetics and biochemistry of *Leishmania* membrane transporters. Curr. Opin. Microbiol. *3*, 417–421.

Landfear, S.M., and Ignatushchenko, M. (2001). The flagellum and flagellar pocket of trypanosomatids. Mol. Biochem. Parasitol. *115*, 1–17.

Lincoln, L.M., Ozaki, M., Donelson, J.E., and Beetham, J.K. (2004). Genetic complementation of *Leishmania* deficient in PSA (GP46) restores their resistance to lysis by complement. Mol. Biochem. Parasitol. *137*, 185–189.

Liu, B., Liu, Y., Motyka, S.A., Agbo, E.E., and Englund, P.T. (2005). Fellowship of the rings, the replication of kinetoplast DNA. Trends Parasitol. *21*, 363–369.

Lohman, K.L., Langer, P.J., and McMahon-Pratt, D. (1990). Molecular cloning and characterization of the immunologically protective surface glycoprotein GP46/M-2 of *Leishmania amazonensis*. Proc. Natl. Acad. Sci. USA. *87*, 8393–8397.

Ma, D., Russell, D.G., Beverley, S.M., and Turco, S.J. (1997). Golgi GDP-mannose uptake requires *Leishmania* LPG2, A member of a eukaryotic family of putative nucleotide-sugar transporters. J. Biol. Chem. *272*, 3799–3805.

McConville, M.J., and Bacic, A. (1990). Characterization and surface localization of galactofuranose-containing glycoinositolphospholipids of *Leishmania major*. Mol. Biochem. Parasitol. *38*, 57–68.

McConville, M.J., and Blackwell, J.M. (1991). Developmental changes in the glycosylated phosphatidylinositols of *Leishmania donovani*. Characterization of the promastigote and amastigote glycolipids. J. Biol. Chem. *266*, 15170–15179.

McConville, M.J., and Menon, A.K. (2000). Recent developments in the cell biology and biochemistry of glycosylphosphatidylinositol lipids (review). Mol. Membr. Biol. *17*, 1–16.

McConville, M.J., Mullin, K.A., Ilgoutz, S.C., and Teasdale, R.D. (2002). Secretory pathway of trypanosomatid parasites. Microbiol. Mol. Biol. Rev. *66*, 122–154.

McGwire, B.S., and Chang, K.P. (1996). Post-translational regulation of a *Leishmania* HEXXH metalloprotease (gp63). The effects of site-specific mutagenesis of catalytic, zinc binding, N-glycosylation, and glycosyl phosphatidylinositol addition sites on N-terminal end cleavage, intracellular stability, and extracellular exit. J. Biol. Chem. *271*, 7903–7909.

McKean, P.G., Delahay, R., Pimenta, P.F.P., and Smith, D.F. (1997). Characterisation of a second protein encoded by the differentially regulated LmcDNA16 gene family of *Leishmania major*. Mol. Biochem. Parasitol. *85*, 221–231.

McKean, P.G., Denny, P.W., Knuepfer, E., Keen, J.K., and Smith, D.F. (2001). Phenotypic changes associated with deletion and overexpression of a stage-regulated gene family in *Leishmania*. Cell. Microbiol. *3*, 511–523.

McMahon-Pratt, D., and Alexander, J. (2004). Does the *Leishmania major* paradigm of pathogenesis and protection hold for New World cutaneous leishmaniases or the visceral disease? Immunol. Rev. *201*, 206–224.

McMahon-Pratt, D., Traub-Cseko, Y., Lohman, K.L., Rogers, D.D., and Beverley, S.M. (1992). Loss of the GP46/M2 surface membrane glycoprotein gene family in the *Leishmania braziliensis* complex. Mol. Biochem. Parasitol. *50*, 151–160.

Maizels, N. (1990). To understand function, study structure. Cell *60*, 887–889.

Maser, P., Sutterlin, C., Kralli, A., and Kaminsky, R. (1999). A nucleoside transporter from *Trypanosoma brucei* involved in drug resistance. Science *285*, 242–244.

Merry, A.H., and Merry, C.L. (2005). Glycoscience finally comes of age. EMBO Rep. *6*, 900–903.

Moll, H., Flohe, S., and Rollinghoff, M. (1995). Dendritic cells in *Leishmania major*-immune mice harbor persistent parasites and mediate an antigen-specific T cell immune response. Eur. J. Immunol. *25*, 693–699.

Montgomery, J., Curtis, C., and Handman, E. (2002). Genetic and structural heterogeneity of proteophosphoglycan in *Leishmania*. Mol. Biochem. Parasitol. *121*, 75–85.

Montgomery, J., Ilg, T., Thompson, J.K., Kobe, B., and Handman, E. (2000). Identification and predicted structure of a leucine-rich repeat motif shared by *Leishmania major* proteophosphoglycan and parasite surface antigen 2. Mol. Biochem. Parasitol. *107*, 289–295.

Moody, S.F., Handman, E., and Bacic, A. (1991). Structure and antigenicity of the lipophosphoglycan from *Leishmania major* amastigotes. Glycobiology *1*, 419–424.

Moody, S.F., Handman, E., McConville, M.J., and Bacic, A. (1993). The structure of *Leishmania major* amastigote lipophosphoglycan. J. Biol. Chem. *268*, 18457–18466.

Morgan, E.H. (1983). Effect of pH and iron content of transferrin on its binding to reticulocyte receptors. Biochim. Biophys. Acta *762*, 498–502.

Motyka, S.A., and Englund, P.T. (2004). RNA interference for analysis of gene function in trypanosomatids. Curr. Opin. Microbiol. *7*, 362–368.

Mulero, V., Searle, S., Blackwell, J.M., and Brock, J.H. (2002). Solute carrier 11a1 (Slc11a1; formerly Nramp1) regulates metabolism and release of iron acquired by phagocytic, but not transferrin-receptor-mediated, iron uptake. Biochem. J. *363*, 89–94.

Mulero, V., Wei, X.Q., Liew, F.Y., and Brock, J.H. (2002). Regulation of phagosomal iron release from murine macrophages by nitric oxide. Biochem. J. *365*, 127–132.

Naderer, T., Vince, J.E., and McConville, M.J. (2004). Surface determinants of *Leishmania* parasites and their role in infectivity in the mammalian host. Curr. Mol. Med. *4*, 649–665.

Najmanovich, R.J., Torrance, J.W., and Thornton, J.M. (2005). Prediction of protein function from structure, insights from methods for the detection of local structural similarities. Biotechniques 38, 847, 849, 851.

Nielsen, H., Brunak, S., and von Heijne, G. (1999). Machine learning approaches for the prediction of signal peptides and other protein sorting signals. Protein Eng. *12*, 3–9.

Nielsen, H., Engelbrecht, J., Brunak, S., and von Heijne, G. (1997). Identification of prokaryotic and eukaryotic signal peptides and prediction of their cleavage sites. Protein Eng. *10*, 1–6.

Ogbadoyi, E.O., Robinson, D.R., and Gull, K. (2003). A high-order trans-membrane structural linkage is responsible for mitochondrial genome positioning and segregation by flagellar basal bodies in trypanosomes. Mol. Biol. Cell *14*, 1769–1779.

Olafson, R.W., Thomas, J.R., Ferguson, M.A., Dwek, R.A., Chaudhuri, M., Chang, K.P., and Rademacher, T.W. (1990). Structures of the N-linked oligosaccharides of Gp63, the major surface glycoprotein, from *Leishmania mexicana amazonensis*. J. Biol. Chem. *265*, 12240–12247.

Ouellette, M., Drummelsmith, J., and Papadopoulou, B. (2004). Leishmaniasis, drugs in the clinic, resistance and new developments. Drug Resist. Updat. *7*, 257–266.

Papadopoulou, B., Roy, G., Breton, M., Kundig, C., Dumas, C., Fillion, I., Singh, A.K., Olivier, M., and Ouellette, M. (2002). Reduced infectivity of a *Leishmania donovani* biopterin transporter genetic mutant and its use as an attenuated strain for vaccination. Infect. Immun. *70*, 62–68.

Parsons, M., Worthey, E., Ward, P.N., and Mottram, J.C. (2005). Comparative analysis of the kinomes of three pathogenic trypanosomatids; *Leishmania major*, *Trypanosoma brucei*, and *Trypanosoma cruzi*. BMC Genomics 6, 126–137.

Park, S., Lee, M.R., Pyo, S.J., and Shin, I. (2004). Carbohydrate chips for studying high-throughput carbohydrate-protein interactions. J. Am. Chem. Soc. *126*, 4812–4819.

Peters, C., Kawakami, M., Kaul, M., Ilg, T., Overath, P., and Aebischer, T. (1997). Secreted proteophosphoglycan of *Leishmania mexicana* amastigotes activates complement by triggering the mannan binding lectin pathway. Eur. J. Immunol. *27*, 2666–2672.

Peters, C., Stierhof, Y.-D., and Ilg, T. (1997). Proteophosphoglycan secreted by *Leishmania mexicana* amastigotes causes vacuole formation in macrophages. Infect. Immun. *65*, 783–786.

Piani, A., Ilg, T., Elefanty, A.G., Curtis, J., and Handman, E. (1999). *Leishmania major* proteophosphoglycan is expressed by amastigotes and has an immunomodulatory effect on macrophage function. Microbes and Infection *1*, 589–599.

Picard, V., Govoni, G., Jabado, N., and Gros, P. (2000). Nramp 2 (DCT1/DMT1) expressed at the plasma membrane transports iron and other divalent cations into a calcein-accessible cytoplasmic pool. J. Biol. Chem. *275*, 35738–35745.

Pimenta, P.F., Saraiva, E.B.B., and Sacks, D.L. (1991). The comparative fine structure and surface glycoconjugate expression of three life stages of *Leishmania major*. Exp. Parasitol. *72*, 191–204.

Ralton, J.E., Mullin, K.A., and McConville, M.J. (2002). Intracellular trafficking of glycosylphosphatidylinositol (GPI)-anchored proteins and free GPIs in *Leishmania mexicana*. Biochem. J. *363*, 365–375.

Ramamoorthy, R., Donelson, J.E., Paetz, K.E., Maybodi, M., Roberts, S.C., and Wilson, M.E. (1992). Three distinct RNAs for the surface protease gp63 are differentially expressed during development of *Leishmania donovani chagasi* promastigotes to an infectious form. J. Biol. Chem. *267*, 1888–1895.

Richard, D., Kundig, C., and Ouellette, M. (2002). A new type of high affinity folic acid transporter in the protozoan parasite *Leishmania* and deletion of its gene in methotrexate-resistant cells. J. Biol. Chem. *277*, 29460–29467.

Rittig, M.H., and Bogdan, C. (2000). *Leishmania*-host-cell interaction, complexities and alternative views. Parasitol. Today *16*, 292–297.

Roberts, S.C., Wilson, M.E., and Donelson, J.E. (1995). Developmentally regulated expression of a novel 59-kDa product of the major surface protease (Msp or gp63) gene family of *Leishmania chagasi*. J. Biol. Chem. *270*, 8884–8892.

Robinson, K.A., and Beverley, S.M. (2003). Improvements in transfection efficiency and tests of RNA interference (RNAi) approaches in the protozoan parasite *Leishmania*. Mol. Biochem. Parasitol. *128*, 217–228.

Rochette, A., McNicoll, F., Girard, J., Breton, M., Leblanc, E., Bergeron, M.G., and Papadopoulou, B. (2005). Characterization and developmental gene regulation of a large gene family encoding amastin surface proteins in *Leishmania* spp. Mol. Biochem. Parasitol. *140*, 205–220.

Rogers, M.E., Ilg, T., Nikolaev, A.V., Ferguson, M.A., and Bates, P.A. (2004). Transmission of cutaneous leishmaniasis by sandflies is enhanced by regurgitation of fPPG. Nature *430*, 463–467.

Rual, J.F., Venkatesan, K., Hao, T., Hirozane-Kishikawa, T., Dricot, A., Li, N., Berriz, G.F., Gibbons, F.D., Dreze, M., Ayivi-Guedehoussou, N., Klitgord, N., Simon, C., Boxem, M., Milstein, S., Rosenberg, J., Goldberg, D.S., Zhang, L.V., Wong, S.L., Franklin, G., Li, S., Albala, J.S., Lim, J., Fraughton, C., Llamosas, E., Cevik, S., Bex, C., Lamesch, P., Sikorski, R.S., Vandenhaute, J., Zoghbi, H.Y., Smolyar, A., Bosak, S., Sequerra, R., Doucette-Stamm, L., Cusick, M.E., Hill, D.E., Roth, F.P., and Vidal, M. (2005). Towards a proteome-scale map of the human protein–protein interaction network. Nature *437*, 1173–1178.

Sacks, D.L., Modi, G., Rowton, E., Spath, G., Epstein, L., Turco, S.J., and Beverley, S.M. (2000). The role of phosphoglycans in *Leishmania*-sandfly interactions. Proc. Natl. Acad. Sci. USA. *97*, 406–411.

Sanchez, M.A., Zeoli, D., Klamo, E.M., Kavanaugh, M.P., and Landfear, S.M. (1995). A family of putative receptor-adenylate cyclases from *Leishmania donovani*. J. Biol. Chem. *270*, 17551–17558.

Schaible, U.E., and Kaufmann, S.H. (2005). A nutritive view on the host-pathogen interplay. Trends Microbiol. *13*, 373–380.

Schlein, Y. (1993). *Leishmania* and sandflies, interactions in the life cycle and transmission. Parasitol. Today *9*, 255–258.

Schneider, A., Eichenberger, W., and Seebeck, T. (1988). A microtubule-binding protein of *Trypanosoma brucei* which contains covalently bound fatty acid. J. Biol. Chem. *263*, 6472–6475.

Shakarian, A.M., Joshi, M.B., Yamage, M., Ellis, S.L., Debrabant, A., and Dwyer, D.M. (2003). Members of a unique histidine acid phosphatase family are conserved amongst a group of primitive eukaryotic human pathogens. Mol. Cell. Biochem. *245*, 31–41.

Sharma, D.K., Hilley, J.D., Bangs, J.D., Coombs, G.H., Mottram, J.C., and Menon, A.K. (2000). Soluble GPI8 restores glycosylphosphatidylinositol anchoring in a trypanosome cell-free system depleted of lumenal endoplasmic reticulum proteins. Biochem. J. 351 Pt 3, 717–722.

Snapp, E.L., and Landfear, S.M. (1997). Cytoskeletal association is important for differential trageting of glucose transporter isoforms in *Leishmania*. J. Cell Biol. *139*, 1775–1783.

Snapp, E.L., and Landfear, S.M. (1999). Characterization of a targeting motif for a flagellar membrane protein in *Leishmania enriettii*. J. Biol. Chem. *274*, 29543–29548.

Soares, R.P., Barron, T., McCoy-Simandle, K., Svobodova, M., Warburg, A., and Turco, S.J. (2004). *Leishmania tropica*, intraspecific polymorphisms in lipophosphoglycan correlate with transmission by different Phlebotomus species. Exp. Parasitol. *107*, 105–114.

Socolich, M., Lockless, S.W., Russ, W.P., Lee, H., Gardner, K.H., and Ranganathan, R. (2005). Evolutionary information for specifying a protein fold. Nature *437*, 512–518.

Sonnhammer, E.L., von Heijne, G., and Krogh, A. (1998). A hidden Markov model for predicting transmembrane helices in protein sequences. Proc. Int. Conf. Intell. Syst. Mol. Biol. *6*, 175–182.

Sopwith, W.F., Debrabant, A., Yamage, M., Dwyer, D.M., and Bates, P.A. (2002). Developmentally regulated expression of a cell surface class I nuclease in *Leishmania mexicana*. Int. J. Parasitol. *32*, 449–459.

Späth, G.F., Lye, L.-F., Segawa, H., Sacks, D., Turco, S.J., and Beverley, S.M. (2003). Persistence without pathology in phosphoglycan-deficient *Leishmania major*. Science *301*, 1241–1243.

Springer, T.A. (1990). The sensation and regulation of interactions with the extracullular environment, the cell biology of lymphocyte adhesion receptors. Annu. Rev. Cell Biol. *6*, 359–402.

Springer, T.A. (1994). Traffic signals for lymphocyte recirculation and leukocyte emigration, a multistep paradigm. Cell *76*, 301–314.

Stein, A., Vaseduvan, G., Carter, N.S., Ullman, B., Landfear, S.M., and Kavanaugh, M.P. (2003). Equilibrative nucleoside transporter family members from *Leishmania donovani* are electrogenic proton symporters. J. Biol. Chem. *278*, 35127–35134.

Stierhof, Y.D., Bates, P.A., Jacobson, R.L., Rogers, M.E., Schlein, Y., Handman, E., and Ilg, T. (1999). Filamentous proteophosphoglycan secreted by *Leishmania* promastigotes forms gel-like three-dimensional networks that obstruct the digestive tract of infected sandfly vectors. Eur. J. Cell Biol. *78*, 675–689.

Sunyaev, S.R., Eisenhaber, F., Rodchenkov, I.V., Eisenhaber, B., Tumanyan, V.G., and Kuznetsov, E.N. (1999). PSIC, profile extraction from sequence alignments with position-specific counts of independent observations. Protein Eng. *12*, 387–394.

Suthram, S., Sittler, T., and Ideker, T. (2005). The Plasmodium protein network diverges from those of other eukaryotes. Nature *438*, 108–112.

Symons, F.M., Murray, P.J., Osborn, A.H., Cappai, R., and Handman, E. (1994). Characterisation of a polymorphic family of integral membrane proteins in promastigotes of different *Leishmania* species. Mol. Biochem. Parasitol. *67*, 103–113.

Teixeira, S.M.R., Russell, D.G., Kirchhoff, L.V., and Donelson, J.E. (1994). A differentially expressed gene family encoding 'Amastin' a surface protein of *Trypanosoma cruzi* amastigotes. J. Biol. Chem. *269*, 20509–20516.

Tompa, P., Szasz, C., and Buday, L. (2005). Structural disorder throws new light on moonlighting. Trends Biochem. Sci. *30*, 484–489.

van Driel, I.R., Stearne, P.A., Grego, B., Simpson, R.J., and Goding, J.W. (1984). The receptor for transferrin on murine myeloma cells, one-step purification based on its physiology, and partial amino acid sequence. J. Immunol. *133*, 3220–3224.

Vasudevan, G., Carter, N.S., Drew, M.E., Beverley, S.M., Sanchez, M.A., Seyfang, A., Ullman, B., and Landfear, S.M. (1998). Cloning of *Leishmania* nucleoside transporter genes by rescue of a transport-deficient mutant. Proc. Natl. Acad. Sci. USA. 95, 9873–9878.

Vasudevan, G., Ullman, B., and Landfear, S.M. (2001). Point mutations in a nucleoside transporter gene from *Leishmania donovani* confer drug resistance and alter substrate selectivity. Proc. Natl. Acad. Sci. USA. *98*, 6092–6097.

von Heijne, G. (1986). A new method for predicting signal sequence cleavage sites. Nucleic Acids Res. *14*, 4683–4690.

Voth, B.R., Kelly, B.L., Joshi, P.B., Ivens, A.C., and McMaster, W.R. (1998). Differentially expressed *Leishmania major* gp63 genes encode cell surface leishmanolysin with distinct signals for glycosylphosphatidylinositol attachment. Mol. Biochem. Parasitol. *93*, 31–41.

Voyiatzaki, C.S., and Soteriadou, K.P. (1992). Identification and isolation of the *Leishmania* transferrin receptor. J. Biol. Chem. *267*, 9112–9117.

Vukcevic, D. (2004). Predicting GPI-anchored proteins using a hidden Markov model. PhD Thesis University of Melbourne. http,//bioinf.wehi.edu.au/gpihmm/.

Wilson, M.E., Lewis, T.S., Miller, M.A., McCormick, M.L., and Britigan, B.E. (2002). *Leishmania chagasi*, uptake of iron bound to lactoferrin or transferrin requires an iron reductase. Exp. Parasitol. *100*, 196–207.

Winter, G., Fuchs, M., McConville, M.J., Stierhof, Y.-D., and Overath, P. (1994). Surface antigens of *Leishmania mexicana* amastigotes, characterization of glycoinositol phospholipids and a macrophage-derived glycosphingolipid. J. Cell Sci. *107*, 2471–2482.

Yakunin, A.F., Yee, A.A., Savchenko, A., Edwards, A.M., and Arrowsmith, C.H. (2004). Structural proteomics; a tool for genome annotation. Curr. Opin. Chem. Biol. *8*, 42–48.

Yao, C., Donelson, J.E., and Wilson, M.E. (2003). The major surface protease (MSP or GP63) of *Leishmania* sp. Biosynthesis, regulation of expression, and function. Mol. Biochem. Parasitol. *132*, 1–16.

Yao, C., Luo, J., Hsiao, C., Donelson, J.E., and Wilson, M.E. (2005). Internal and surface subpopulations of the major surface protease (MSP) of *Leishmania chagasi*. Mol. Biochem. Parasitol. *139*, 173–183.

Zhang, K., Hsu, F.F., Scott, D.A., Docampo, R., Turk, J., and Beverley, S.M. (2005). *Leishmania* salvage and remodelling of host sphingolipids in amastigote survival and acidocalcisome biogenesis. Mol. Microbiol. *55*, 1566–1578.

Zick, A., Onn, I., Bezalel, R., Margalit, H., and Shlomai, J. (2005). Assigning functions to genes, identification of S-phase expressed genes in *Leishmania major* based on post-transcriptional control elements. Nucleic Acids Res. 33, 4235–4242.

The Biology of *Leishmania*–Sandfly Interactions

10

David Sacks, Phillip Lawyer and Shaden Kamhawi

Abstract

Phlebotomine sandflies are the only known natural vectors of *Leishmania*, and of the nearly 800 phlebotomine species described, fewer than 50 are known to be involved in the transmission cycle of these parasites. Furthermore, some vectors species are highly restricted to the species of *Leishmania* that they transmit in nature. This chapter reviews the natural habitat of *Leishmania* amastigotes and promastigotes during their transformation, growth, differentiation, and migration in the alimentary tract of their competent vectors, and the barriers to survival that are encountered in refractory flies. Most importantly, the chapter details the *Leishmania* species-, strain- and stage-specific molecules that are known or postulated to permit the development of transmissible infections to proceed. The chapter will hopefully serve to better inform the search of the *Leishmania* genome so that a fuller accounting of the molecules controlling parasite-vector interactions can be achieved, with a longer view towards the development of a successful transmission blocking vaccine(s).

Diversity and distribution of natural vectors

Phlebotomine sandflies are the only known natural vectors of *Leishmania*, meaning that they are essential to support the life cycle of this parasite. It follows then that, with few exceptions, wherever there is leishmaniasis there are sandflies. These strictly terrestrial dipterans are found worldwide mostly in the tropics and subtropics, with a few hardy species ranging into temperate zones in both northern (to 50°N) and southern (40°S) hemispheres. Of the roughly 800 recognized species of sandflies, approximately 425 are found in the New World and 375 in the Old World (Seccombe *et al.*, 1992; Young and Duncan, 1994). Over a hundred of these have been implicated in the transmission of *Leishmania* but fewer than fifty have actually been proven with reasonable certainty to be vectors. Proven vector species belonging to the genera *Phlebotomus* in the Old World and *Lutzomyia* in the New World are given in Table 10.1.

Leishmania transmission occurs in a wide variety of habitats in both the Old and New Worlds. Old World sandflies are primarily savannah and desert species associated with warm areas of low rainfall. Most New World species inhabit forests in warm areas of higher rainfall. However, several in North and South America inhabit drier, savannah-like regions. Sandfly habitats are distributed in elevation from well below sea level in the Jordan Valley to nearly 3800 m above sea level in the Peruvian Andes. As a rule, sandflies are limited to areas where daytime temperatures are above 15.6°C for at least 3 months of the year (Lane, 1993). Marked seasonal changes in sandfly population densities occur in areas where there are significant seasonal fluctuations in temperature and precipitation. In arid or semi-arid regions, sandfly populations usually reach their peak toward the end of the rainy season and ebb at the end of the dry season. In hot, dry deserts or in dry temperate zones with hot summers and

Table 10.1 Old and New World *Leishmania* species and their vectors

(a) Old World *Leishmania* species and their vectors

Leishmania species (associated clinical forms)	Proven or suspected vector sandfly species	Places of known or suspected transmission	References supporting suspicion or incrimination
L. aethiopica (CL, DCL)	*P. (La.) longipes*	Kenya, Ethiopia	(Foster, 1972; Ashford, 1977)
	P. (La.) pedifer	Kenya, Ethiopia	(Ashford *et al.*, 1973)
L. donovani (VL, PKDL)	*P. (P.) alexandri*	North Africa to Western China	(Guan *et al.*, 1986);
	P. (E.) argentipes	Bangladesh, Nepal, India	(Schorscher and Goris, 1992); (Shortt, 1931)
	P. (S.) celiae	Kenya, Southern Ethiopia	(Gebre-Michael and Lane, 1996)
	P. (S.) martini	Kenya, Ethiopia	(Perkins *et al.*, 1988)
	P.(La.) orientalis	Sudan, Ethiopia	(Hailu *et al.*, 1995)
L. infantum (VL, CL)	*P. (P.) alexandri*	Iran, Middle East	(Azizi *et al.*, 2006)
	P. (La.) ariasi	Western Mediterranean	(Killick-Kendrick and Rioux, 1981)
	P. (A.) brevis	Northern Iran to Caucasus	(Lewis and Ward, 1987)
	P. (La.) langeroni	Egypt-Tunisia littoral	(el Sawaf *et al.*, 1985)
	P. (La.) neglectus	Eastern Mediterranean	(Adler, 1964); (Leger *et al.*, 1979); (Lewis and Ward, 1987)
	P. (La.) perfiliewi	Mediterranean Basin, Algeria	(Lewis and Ward, 1987); (Maroli *et al.*, 1987)
	P. (La.) perniciosus	Western Mediterranean	(Lewis and Ward, 1987)
L. major (CL, DCL)	*P. (Ph.) duboscqi*	Sahelian Africa, Kenyan Rift Valley	(Dedet, 1979); (Lawyer *et al.*, 1990)
	P. (Ph.) papatasi	North Africa, Middle East	(Killick-Kendrick, 1990a); (Lane and Fritz, 1986); (Perfil'ev, 1966)
Leishmania sp. (Namibia) (CL)	*P. near (S.) rossi*	Namibia	(Davidson, 1987); (Killick-Kendrick, 1990a)
L. tropica (CL, LR)	*P. (La.) aculeatus*	Kenya	(Johnson *et al.*, 1993)
	P. (La.) guggisbergi	Kenya	(Lawyer *et al.*, 1991)
	P. (P.) sergenti	South Central Asia, Middle East, North Africa	(Adler, 1957); (Lewis and Ward, 1987); (Saf'janova, 1979)

(b) New World *Leishmania* species and their vectors

Leishmania species (associated clinical forms)	Proven or suspected vector sandfly species	Places of known or suspected transmission	References supporting suspicion or incrimination
L. amazonensis (CL, DCL, MCL)	*Lu. (N.) flaviscutellata*	Northern South America, Brazil	(Lainson and Shaw, 1968); (Pifano *et al.*, 1973)

Table 10.1 (b) Continued

Leishmania species (associated clinical forms)	Proven or suspected vector sandfly species	Places of known or suspected transmission	References supporting suspicion or incrimination
L. braziliensis (CL, MCL)	*Lu. (Ps.) carrerrai*	Western Amazon Basin	(Le Pont and Desjeux, 1986)
	Lu. (Ps.) complexa	Para, Brazil	(de Souza *et al.*, 1996)
	Lu. (N.) intermedia	Southern Brazil	(Forattini *et al.*, 1972); (Rangel *et al.*, 1984); (Deane and Grimaldi, 1985)
	Lu. (?) ovallesi	Guatemala, Venezuela	(Rowton *et al.*, 1992); (Feliciangeli *et al.*, 1988); (Feliciangeli and Rabinovich, 1998)
	Lu. (Ps.) wellcomei	Para, Brazil	(Lainson *et al.*, 1973); (Lainson and Shaw, 1987); (Ryan *et al.*, 1987)
	Lu. (N.) whitmani	Eastern Brazil	(de Queiroz *et al.*, 1994)
L. chagasi (VL, CL)	*Lu. (Lu.) longipalpis*	Central America, Brazil, Venezuela	(Deane and Deane, 1964); (Lainson *et al.*, 1977)
L. colombiensis (CL)	*Lu. (H.) hartmanni*	Colombia	(Lainson *et al.*, 1979; Kreutzer *et al.*, 1991)
L. guyanensis (CL, MCL)	*Lu. (N.) anduzei*	Northern South America, Brazil	(Arias and Freitas, 1978); (Lainson *et al.*, 1976); (Killick-Kendrick, 1999)
	Lu. (N.) umbratilis	Amazon Basin, Brazil, French Guiana	(Lainson *et al.*, 1981)
L. lainsoni (CL)	*Lu. (T.) ubiquitalis*	Brazil, Peru	(Lainson *et al.*, 1992)
L. mexicana (CL, DCL)	*Lu. (D.) anthophora*	Southern Texas, USA	(Young and Lawyer, 1987); (McHugh *et al.*, 1993)
	Lu. (H.) ayacuchensis	Ecuador, Peru	(Gomez and Hashiguchi, 1991)
	Lu. (Lu.) diabolica	Southern Texas, USA	(Young and Lawyer, 1987); (Lawyer *et al.*, 1987)
	Lu. (N.) olmeca olmeca	Central America	(Fairchild and Theodor, 1971)
L. mexicana (?Dominica?) (DCL)	*Lu. (C.) christophei*	Dominican Republic	(Young and Lawyer, 1987)
L. panamensis (CL, MCL)	*Lu. (V.) ovallesi*	Panama, Venezuela	(istensen and Herrer, 1973)
	Lu. (N.) trapidoi	Central America, Colombia	(Johnson *et al.*, 1963); (istensen *et al.*, 1983); (Morales *et al.*, 1981)
	Lu. (N.) ylephiletor	Central America	(istensen *et al.*, 1983)
L. peruviana (CL)	*Lu. (H.) peruensis*	Northern Andes, Peru	(Herrer, 1982); (Lainson and Shaw, 1987); (Villaseca *et al.*, 1993)
	Lu. (V.) verrucarum	Northern Andes, Peru	(Lainson and Shaw, 1987); (Villaseca *et al.*, 1993); (Davies *et al.*, 1993)
L. shawi (CL)	*Lu. (N.) whitmani*	Brazil	(Lainson *et al.*, 1989)
L. venezuelensis (CL, DCL)	*Lu. (N.) olmeca*	Northern South America	(Lainson and Shaw, 1987)

Based on earlier tables by (Young and Lawyer, 1987; Killick-Kendrick, 1990a; Lane, 1993; Lawyer and Perkins, 2000), with modifications and updates by the authors.
Proven vectors listed meet the following criteria: Multiple isolations and typing of parasites from the sandfly, human and reservoir contact with the vector have been demonstrated, the parasite has been shown to develop to its metacylic stage in the sandfly midgut; experimental transmission by bite has been demonstrated in some cases. Places of known or suspected transmission are given in broad terms and a particular vector may occur in only part of the country or region mentioned.

cold winters, adults of some species may disappear entirely during the driest and/or coldest seasons of the year.

Sandfly biology and physiology

Biology

Relative to most flies, sandflies are small, usually no more than 3 mm long, and fragile. Their body colour ranges from light brown to black (depending on the species), the effect of numerous hairs and cuticular colour patterns. Sandflies characteristically rest with wings held upward and outward above the body so that the costal (front) margins form an angle of approximately 60° (Fig. 10.1). They are usually dormant during the day, resting in cool, humid sites that provide protection from the elements. Resting sites vary according to species and include animal burrows, bird nests, termitaria, animal stables, human dwellings, cellars, latrines, caves, crevices in rocks, walls and soil, dense vegetation, leaf litter, fissures in tree bark, in tree holes and buttresses. The tropical forest offers the greatest variety of resting sites, accounting for the greater diversity of New World sandfly species (Killick-Kendrick, 1999; Lawyer and Perkins, 2000).

Sandflies are weak fliers and travel in short hops rather than in sustained flight as they seek sugar or blood meals, mates, or resting or oviposition sites. Most sandfly species fly horizontally very near ground level, where they feed on ground-dwelling rodents and other mammals. Some forest inhabiting sandflies also move vertically between forest floor and canopy where they may occasionally encounter a human host (Chaniotis *et al.*, 1974; Ready *et al.*, 1986). Forest-inhabiting sandflies, which include mostly Neotropical species, rarely fly more than 200 m from their resting site (Chaniotis *et al.*, 1974; Chippaix *et al.*, 1984). In savannah and desert-type habitats typical of the Old World, flight range may be much longer, depending on host availability. *P. ariasi* was observed to fly up to 2.2 km in the Cevennes Mountain region of southern France (Killick-Kendrick *et al.*, 1984). In the flat, open *Acacia* woodlands in sub-Saharan Sudan, Quate (Quate, 1964) recovered *P. orientalis* 730 m from its release site. In the *Acacia*-thornbush savannah of Kenya's Rift Valley Province a female *P. martini* was recaptured 750 m from a release point after 12 hours, and another female 1 km from the same release point after 19 days (Lawyer and Perkins, 2000).

Figure 10.1 *Phlebotomus papatasi*, the principal vector of the Old World parasite *Leishmania major.*

Both male and female sandflies require carbohydrates for energy, the source of which is poorly known in most species. It has been reported that sandflies feed on aphid or Homoptera honeydew (Killick-Kendrick and Killick-Kendrick, 1987), flower nectar, fruits and other plant juices (Schlein and Warburg, 1986). In addition to a sugar meal, female sandflies require a blood meal before they will lay eggs, although a few species are autogenous (produce eggs without a blood meal) in nature in the first gonotrophic cycle (the period between blood meals during which the ova develop and are subsequently deposited). Autogenous egg batches tend to be smaller than anautogenous batches (Johnson, 1961). In subsequent cycles a blood meal is required. Host blood meal preference differs among sandfly species, with some being limited to only a few host animals while others show broadly opportunistic feeding behaviour. This preference is the result of intrinsic and extrinsic factors that give rise to enzootic, zoonotic and anthroponotic leishmaniases. Clearly, opportunistic feeders are better vectors because they may be exposed to a wider range of potentially infected hosts. The majority of sandfly species are gonotrophically concordant, taking only one blood meal per egg batch. Species that take more than one blood meal per egg batch are said to be gonotrophically discordant. This behaviour results in increased vector-host contact, improving the chances of a fly becoming infected with, and then transmitting, *Leishmania*. Depending on the sandfly species and ambient conditions, a varying proportion of females survives oviposition and undergoes multiple gonotrophic cycles. Each additional cycle requires at least one more blood meal, thus increasing the capacity of the vector to transmit *Leishmania*. *P. papatasi* and *P. duboscqi*, the two proven vectors of *L. major*, are often gonotrophically discordant in nature and have been shown in laboratory experiments to be able to survive up to four and eight gonotrophic cycles, respectively (Mukhopadhyay and Ghosh, 1999).

Blood-feeding activity of sandflies usually begins at or after dusk, in the early evening, corresponding with a drop in temperature and a rise in relative humidity, and continues at varying levels through part or all of the night. Some forest species (e.g. *Lu. wellcomei*, *Lu. carrerai*, *Lu. pesoana*) also will attack during the daytime if their resting habitats are disturbed. Sandflies may be endophilic, biting humans inside their houses (e.g. *P. argentipes*) or exophilic, biting primarily out of doors (e.g. *Lu. trapidoi*) (Young and Duncan, 1994). They usually probe the surface of the skin several times to locate a suitable site before inserting their mouthparts. The bites can be quite painful, with insertion of the mouthparts disrupting cells and damaging capillaries. A potent vasodilatory component of the sandfly saliva induces blood and lymph to flow into the wound, forming a pool from which the blood is sucked (Ribeiro *et al.*, 1989).

Depending on the species, mating may occur before, during or after a blood meal is taken. The search for a receptive female may be aided by pheromones and by locating a resting site or vertebrate host where females are present (Ward *et al.*, 1991). In some important vector species such as *Lu. longipalpis* and *P. argentipes*, males congregate on the vertebrate host in large numbers ('lekking'), attracting the females and thereby enhancing vector–host contact (Jarvis and Rutledge, 1992; Lane, 1993).

Sandflies undergo complete metamorphosis with an egg stage, a larval stage consisting of four feeding instars, a pupal stage and an adult stage. Females usually lay between 30 and 70 eggs during a single gonotrophic cycle, and depending on species, size and nature of the previous blood meal, larval diet and other factors such as temperature, humidity, photoperiod and organic substrate, the complete life cycle may take three to eight weeks. Diapause or quiescence triggered by environmental extremes such as heat, cold, drought or reduced photoperiod may prolong development for months, resulting in overlapping generations that may become synchronized to produce a major peak of adults (Lawyer and Perkins, 2000). As this peak population feeds on potentially infected hosts and undergoes one or more gonotrophic cycles, the proportion of infected females in the population increases, thus impacting the transmission dynamics of the disease.

Physiology

The relationship between *Leishmania* parasites and phlebotomine sandflies appears to be commensal, i.e. of little consequence to the fly but absolutely essential for the parasite. To better understand the ecology of the *Leishmania* parasite in the sandfly, it may be helpful to consider the sequence of physiological events that delineate a normal gonotrophic cycle. In sandflies that ingest infected blood, this corresponds to the period when the parasite transforms from its mammalian form, the amastigote, to the infective metacyclic promastigote.

When a sandfly takes a blood meal, the ingested blood passes through the food canal and oesophagus into the posterior midgut via the stomodeal valve (SV), which regulates the flow of fluids into the gut (Fig. 10.2). At this time a type I peritrophic matrix (PM) is secreted by midgut cells and completely envelops the fresh blood meal in a delicate sac of proteins and glycoproteins held together by chitinous microfibrils (Blackburn *et al.*, 1988; Pimenta *et al.*, 1997). The PM protects the midgut epithelial cells from damage by blood meal contents and acts as a barrier to microflora that could cause infection. For digestion to occur, the PM must be permeable to the digestive proteases induced by the presence of blood (Dillon and Lane, 1993b; Ramalho-Ortigao *et al.*, 2003). An unfed midgut is acidic, having an estimated pH of 6 (Gontijo *et al.*, 1998). The influx of a relatively large volume of blood raises the pH, promoting the activity of digestive proteases

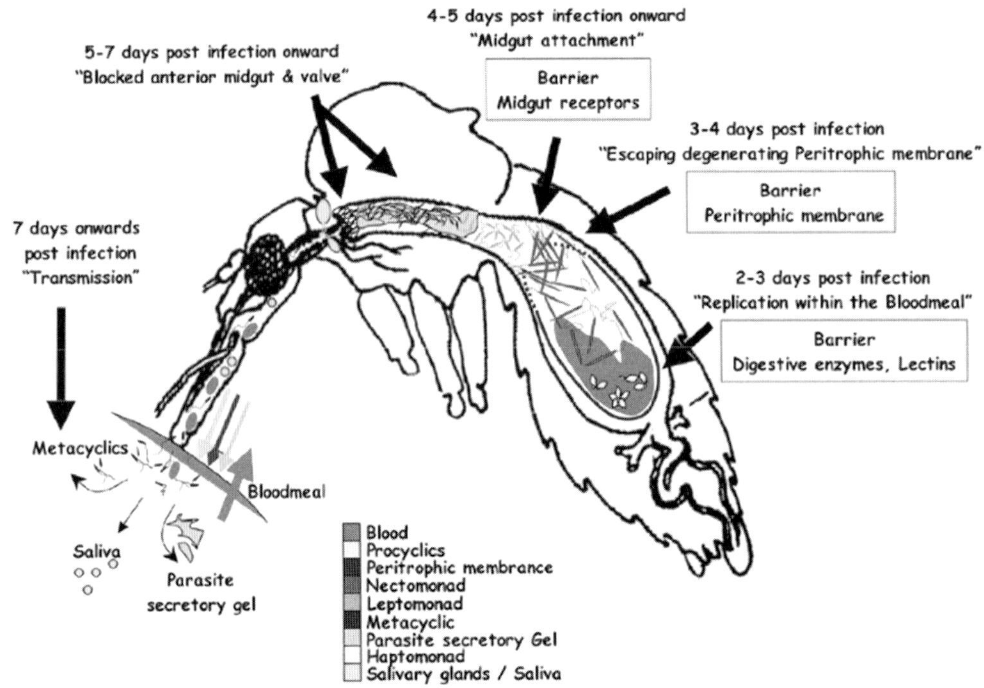

Figure 10.2 The life cycle of *Leishmania* from amastigotes to infective metacyclics in the digestive tract of a competent vector sandfly. A colour version of this figure is located in the plate section at the back of the book.

(Dillon and Lane, 1993b). Midgut protease activity peaks around 24–48 hours post blood meal, depending on the sandfly species and the blood source and returns to baseline levels at about 72 hours (Borovsky and Schlein, 1987; Dillon and Lane, 1993b; Ramalho-Ortigao *et al.*, 2003). The intake of blood also induces diuresis to re-establish homeostasis. Simultaneously, the resulting distension of the midgut induces peristalsis. As digestion proceeds, chitinase activity is detected as early as 48 hours post blood meal (Ramalho-Ortigao *et al.*, 2005). This activity peaks at 72 hours in both *Lu. longipalpis* and *P. papatasi* coincident with the degradation of the PM. The digested gut contents pass into the hindgut via the pyloric valve (PV) and into the ileum, and eventually out through the rectum. Concomitantly, sugars stored in the crop (diverticulum) are gradually dispensed into the midgut, producing an osmotic gradient that decreases in concentration from SV to PV (Tang and Ward, 1998). Nutrients derived from the blood meal are used in egg development and usually by the time the blood meal is completely digested, the eggs are mature and ready to be deposited. At this point the sandfly is ready to take another blood meal.

The life-cycle of *Leishmania* in natural vectors

When feeding on an infected host, the female sandfly may ingest amastigotes with the blood meal. The number of parasites taken in with the blood has not been quantified, and would in any case vary depending on the volume of blood and the degree of parasitaemia. Suffice it to say that the number is probably quite small. The leishmanias are broadly grouped in three sections based on recognized patterns of development in the gut of the sandfly (Lainson, 1982). Parasites in section Hypopylaria are primitive flagellates that develop in the hindgut and migrate posteriorly through the ileum and rectum. Formerly classified as *Leishmania*, these are now generally recognized as a separate genus, the *Sauroleishmania*, none of which are human pathogens. Their only known hosts are lizards of the Eastern Hemisphere (Old World). Because infection is limited to the hindgut of the sandfly, transmission probably occurs when the lizard eats infected sandflies.

In the Section Peripylaria the leishmania establish an initial infection in the hindgut (pylorus), where rounded promastigotes attach to the cuticular lining, migrating anteriorly during the course of development to the midgut and foregut. These include the *L. braziliensis* complex parasites, which are important pathogens of mammals, including man, in neotropical regions (Lainson, 1982). These parasites are now classified as a separate sub-genus, Vianna. Except for establishment of infections in the hindgut of the vector, these parasites develop much the same as other mammalian leishmanias.

The majority of human leishmanias are in the Section Suprapylaria. Their development is restricted to the midgut and foregut of the sandfly host, and as with the Peripylaria, transmission is by bite of an infected sandfly. Vertebrate hosts are limited to mammals, including man, in which the parasites are found only as amastigotes in macrophages of the skin, viscera and blood (Lainson, 1982). In this section are parasites of the *L. mexicana* and *L. hertigi* complexes of the New World, *L. tropica* and *L. major* species of the Old World, and the *L. donovani* complex of the Old and New Worlds.

Because most human leishmanias undergo suprapylarian development in the vector, we provide here a detailed account of the consistent features of the life cycle of suprapylarian parasites in their natural sandfly vectors. This life cycle can be divided into three stages relative to the digestion of the blood meal and progression of the vector's gonotrophic cycle.

First stage: 0–2 days

In the first stage, 0–2 days post feeding, all parasite development occurs in the blood meal within the endoperitrophic space of the abdominal midgut. Transformation of ingested amastigotes to promastigotes begins between 6 and 12 hours post feeding, while the blood meal, which occupies the entire abdominal midgut, is still red with intact cells and enveloped in the peritrophic sac (Lawyer *et al.*, 1987; Killick-Kendrick, 1990b; Lawyer *et al.*, 1990). During this time procyclic promastigotes develop from amastigotes and undergo multiple divisions. Procyclics are short, ovoid, sluggish forms with large, diffuse nuclei and short flagellae often seen dividing by binary fission. Within 36–48 hours, these early promastigotes begin to transform into long, slender, fast-moving forms with long flagellae, termed nectomonads.

Second stage: 2–5 days

In the second stage, 2–5 days post feeding, the peritrophic membrane degrades and ruptures and the partially digested gut contents begin their passage through the pylorus and into the ileum. This event is accompanied by intense multiplication of procyclic promastigotes and their complete transformation to nectomaonads, which fill the anterior abdominal midgut. Many become attached by interdigitating their flagellae with the microvillar lining of the gut wall and, in this way, avoid being passed out with the digested blood meal. As the blood meal diminishes, parasites migrate forward into the thoracic midgut and transform into actively dividing leptomonads that are embedded in and immobilized by a gel-like matrix produced by the parasites called the promastigote secretory gel (PSG) (Rogers *et al.*, 2002; Gossage *et al.*, 2003). Haptomonads, presumably differentiated from leptomonads, begin attaching to the cuticular intima of the cardia and stomodeal valve by foot-like enlargements of their flagella called hemidesmosomes (Killick-Kendrick *et al.*, 1974a), or bind to each other to form a concentric physical plug (Lawyer *et al.*, 1987; Lawyer *et al.*, 1990). At this point the sandfly's eggs are nearly mature.

Third stage: 5± days

During the third and final stage, 5 ± days post feeding, the gut is cleared of the blood meal, egg development is completed, and oviposition begins. Concomitantly, at the stomodeal valve, a dense lens- or ball-shaped mass of haptomonads impedes the flow of blood into the midgut. Meanwhile, the proliferating leptomonads remain imbedded in the PSG matrix, forming a sausage-shaped plug that extends throughout the thoracic midgut, augmenting the blockage formed by the amassed haptomonads at the stomodeal valve. Finally, non-dividing free metacyclics differentiate from leptomonads (Rogers *et al.*, 2002) and appear in large numbers behind the stomodeal valve. These unencumbered forms are free to swim forward into the pharynx, cibarium, and proboscis, as well as rearward into the midgut. By this time, the fly is ready to seek another blood meal and begins the next gonotrophic cycle (Lawyer *et al.*, 1987; Lawyer *et al.*, 1990).

Metacyclogenesis

The differentiation of promastigotes into the metacyclic stage coincident with the vector's search for another blood meal is a key factor in *Leishmania* transmission (Sacks and Perkins, 1984; Lawyer *et al.*, 1987; Sacks, 1989; Lawyer *et al.*, 1990). Metacyclogenesis has been described as a 'preadaptation to survival' within the vertebrate (Sacks, 1989). The infectivity of promastigotes taken from stationary cultures was found to greatly exceed that of promastigotes from cultures in the logarithmic phase (Sacks *et al.*, 1985), and similar developmental changes occur during the growth of promastigotes in the fly (Sacks and Perkins, 1984, 1985). Lawyer *et al.*, (1990) viewed metacyclogenesis not only as a preadaptation to 'survival' in the vertebrate, but a preadaptation to 'arrival' as well. From the standpoint of *Leishmania*, infectivity would seem irrelevant without an effective delivery mechanism. Unlike other developmental forms, which are immobilized because of attachment to the stomodeal valve or entrapment in the PSG behind the valve, metacyclic promastigotes are highly motile, unattached, and unencumbered, and thus are free to migrate anteriorly into the foregut, making this the ideal form for transmission. In fact this is the only form that has been observed in sandfly mouthparts or on the skin at the site of the bite (Killick-Kendrick *et al.*, 1979; Lawyer *et al.*, 1987; Lawyer *et al.*, 1990).

Mechanisms of transmission by bite.

Sandflies are telmophagic, lacerating superficial capillaries in the dermis to create a pool of blood upon which they feed (Ribeiro, 1987, 1995). (Killick-Kendrick and Molyneux, 1981) suggested a mechanism for transmission in which metacyclics interfere directly with sensilla on the mouthparts that control probing and engorgement, thus influencing the rate and direction of flow of material through the food canal, resulting in egestion of parasites into the skin. Others, however, described blocks, plugs or balls formed by stomodeal valve infections of adherent haptomonads and masses of leptomonds and metacyclics embedded in a gel-like matrix (Shortt and Swaminath, 1928; Beach *et al.*, 1984; Warburg *et al.*, 1986; Lawyer *et al.*, 1987; Walters *et al.*, 1987; Lawyer *et al.*, 1990; Stierhof *et al.*, 1999; Rogers *et al.*, 2002). Support for a blockage theory can be found in reports of experimental transmission by infected flies, which appear to have difficulty in engorging (Molyneux, 1977; Beach *et al.*, 1985; Stierhof *et al.*, 1999; Rogers *et al.*, 2002). Schlein *et al.* (1992) also described a dysfunction of the stomodeal valve in infected flies, where the authors observed deterioration of the cuticular lining of the stomodeal valve of *P. papatasi*. Recently, (Volf *et al.*, 2004) corroborated these observations in several vector sandfly species. In infected flies, the damaged stomodeal valve remains open; therefore the suction of the pump is in both directions and the midgut contents, including parasites, are drawn into the pumps and mixed with the ingested blood. This regurgitation engulfs parasites from the midgut and deposits them in the host tissue. Since metacyclics are the only parasites swimming freely at this point, they are inoculated into the host (Lawyer *et al.*, 1990). Therefore, efficient transmission is promoted by the probing behaviour intrinsic to the sandfly in concert with the blockage and damage to the valve caused by the parasites.

Species-restricted vector competence

The identification of *Leishmania* genes that may be directly relevant to the successful establishment, growth, differentiation, migration and ultimate transmission of *Leishmania* by sandflies would clearly be facilitated by an understanding of how these processes are compromised in non-permissive parasite-vector associations. While it might even be informative to investigate the barriers to *Leishmania* development that are inherent to other blood-sucking diptera (e.g. mosquitos), better advantage can be taken from the well described differences in vector competency that are inherent to different species of sandflies. Evidence suggesting a close evolutionary fit between *Leishmania* species and their vectors has been reviewed by Killick-Kendrick (1985). The most striking example is *P. papatasi*, which despite its wide distribution has not been incriminated in the natural transmission of any species other than *L. major* (Figs 10.3 and 10.4). Similarly, *P. sergenti* is a proven vector of *L. tropica* only, and only sandflies belonging to the *Larroussius* subgenus have been found to transmit *L. infantum*. In some instances, these restrictions might be explained by unique co-associations of parasite and vector, as for instance the arid habitats of *P. papatasi* and the giant gerbil reservoirs of *L. major* (Ready, 2000). But there are clear examples of sandfly vectors that are sympatric with *Leishmania* species that they appear unable to transmit, as for instance the abundant *P. papatasi* engorged on human blood found in households of active cases of visceral leishmaniasis due to *L. donovani* in India (Dhanda *et al.*, 1983). Still, the most conclusive evidence for species-restricted vector competence comes from laboratory-based studies, which have generally reproduced the field observations and reinforced a co-evolutionary perspective. Thus, *P. papatasi,* fed on either experimental lesions or through a membrane, will support the full growth and development of *L. major*, but not of other *Leishmania* species (Adler, 1927; Adler, 1938; Killick-Kendrick *et al.*, 1994; Pimenta *et al.*, 1994). An overview of the studies that have compared the development of particular *Leishmania* species within appropriate and inappropriate vectors is included below, with special emphasis on those findings that provide clues as to the nature of the selection pressures encountered by the parasite in the invertebrate host.

When and where do *Leishmania* infections fail in refractory sandflies?

It has been observed that *Leishmania* infections can fail at an early stage in unnatural vector sandfly species. In one of the earliest studies comparing the development of different species of *Leishmania* in *P. papatasi* , and the first to adapt a membrane feeding technique to blood containing cultured promastigotes, Adler and Theodor (Adler, 1927) observed dead or

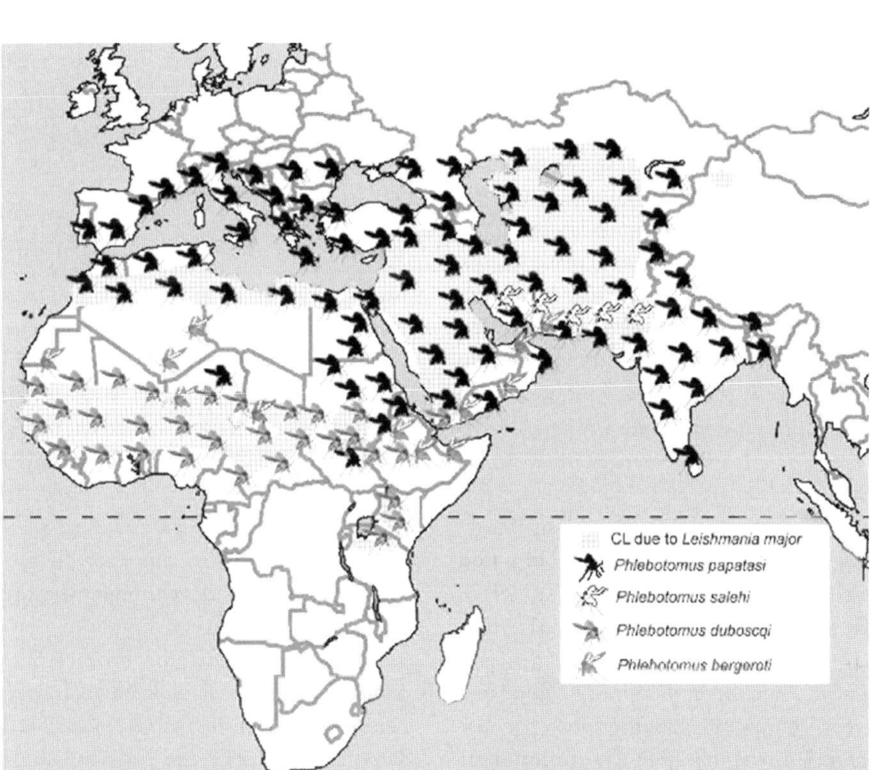

Figure 10.3 Geographic distribution of *Leishmania major* compared with the distribution of its proven and suspected vectors (Lewis, 1982; Magill, 1995).

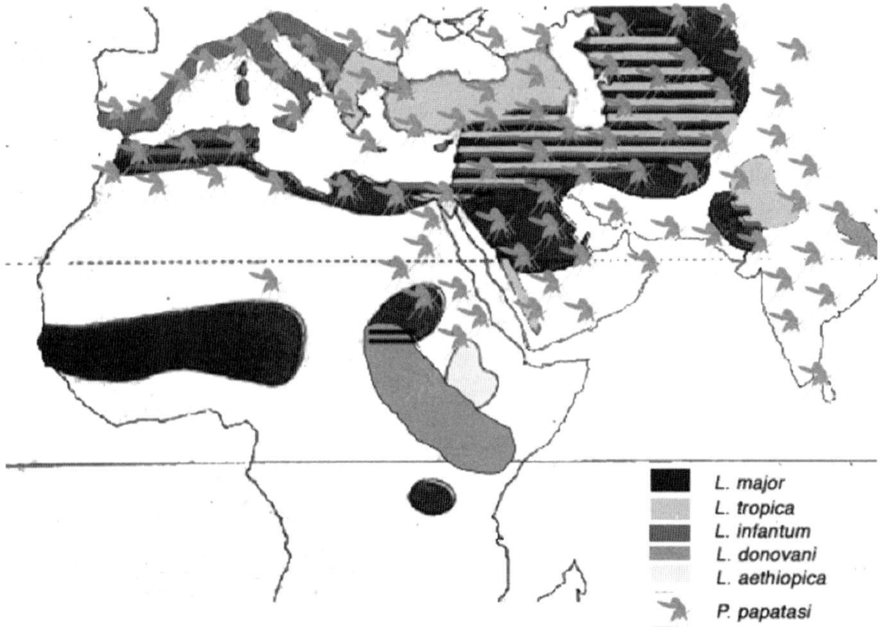

Figure 10.4 Geographic distribution of *Phlebotomus papatasi* compared with the distribution of Old World leishmanias other than *L. major*. A colour version of this figure is located in the plate section at the back of the book.

motionless flagellates confined to the midgut following infection with *L. infantum*. In contrast, infection with a presumed strain of *L. major* produced swarming infections in the midgut and cardia. Similarly, (Shatova, 1984) observed degenerating parasites by electron microscopy in *P. papatasi* midguts following transformation of *L. tropica* amastigotes, whereas *L. major* amastigotes transformed, divided and developed normally. (Schlein and Jacobson, 1998) observed a high proportion of promastigotes containing disintegrating organelles in *P. papatasi* guts three days after feeding on defibrinated blood seeded with *L. donovani* amastigotes. The only study to implicate an early barrier to parasite survival posed by a non-vector species other than *P. papatasi* was reported by Lawyer *et al.* (1990), who noted the slow multiplication of a Kenyan strain of *L. major* in *Sergentomyia schwetzi* compared to its natural vector, *P. duboscqi*.

The conclusion that the early blood-fed midgut presents an important barrier to parasite survival is reinforced by the observation that significant parasite mortality was also observed during the development of *L. major* in a permissive vector, i.e. *P. papatasi*. When viable midgut parasites were enumerated within the first few hours after engorgement, of the viable amastigotes present in the flies just following engorgement, approximately half were killed during the next 4 –8 hours, during the early transition of amastigotes to extracellular promastigotes (Pimenta *et al.*, 1997). The numbers of surviving parasites were sufficient, however, to establish fully transmissible infections. Similarly, the development of *L. (Viannia) braziliensis* and *L. amazonensis* in a New World vector *Lu. migoni* was associated with a high mortality of parasites inside the blood meal during digestion, but the residual parasites in each case were able to establish mature infections containing metacyclic promastigotes (Nieves and Pimenta, 2000). These data imply that in the case of inappropriate parasite strains, their early mortality in refractory flies exceeds the threshold needed for recovery of the high intensity of infection associated with transmission.

It should be noted that for each of the studies documenting the early killing of inappropriate *Leishmania* strains within *P. papatasi* midguts, the demonstration that the parasites could develop normally in their respective natural vectors was in no case provided. Thus the interpretation that *L. major* parasites have uniquely adapted to an especially hostile *P. papatasi* midgut environment is undermined by the possibility that the other strains were generally compromised for midgut survival, perhaps as a consequence of extensive *in vitro* or animal passage. This possibility becomes especially relevant in light of the numerous studies in which the early transformation and growth of inappropriate strains in *P. papatasi* appeared to be normal. Returning to the early studies by Adler and Theodor (Adler, 1927), infections with *L. braziliensis* were heavy up until days 3–4, after which time the flagellates became progressively less in number and remained confined to the stomach. Even in experiments involving *L. infantum* for which some flies clearly contained damaged cells, other flies harboured heavy infections, again confined to the stomach. When (Heyneman, 1963) carried out experimental *L. donovani* infections in a *P. papatasi* colony established from wild-caught flies sympatric with the parasite isolate used for study, dissections showed high intensity infection in all flies examined at day 2, followed by rapid loss of promastigotes through the hindgut after day 4. By contrast, no drop in parasite numbers was observed in older infections during monitoring of wild-caught, naturally-infected *P. orientalis*, the presumed vector of *L. donovani* in Sudan. In studies by (Walters *et al.*, 1989a; Walters *et al.*, 1992), it was observed that when *P. papatasi* were membrane fed on a suspension of washed human RBC and a macrophage cell line infected with *L. panamensis*, the survival and early growth and differentiation were generally similar to that observed in the natural vector, *Lu. gomezi*. The infection rate in *P. papatasi* was markedly reduced after 5 days, and the authors were the first to note that the loss coincided with the excretion of viable parasites along with the blood meal remnants. In the most comprehensive survey of *P. papatasi* vector competence for diverse species and strains of *Leishmania*, (Pimenta *et al.*, 1994) noted that in every instance the loss of inappropriate species coincided with blood meal excretion. Thus two days after membrane feeding on mouse blood containing tissue amastigotes or cultured promastigotes of *L. major*, *L. tropica*, *L. donovani*, or *L. amazonensis*, 90–100% of the flies were infected, and there was no significant difference in the number of midgut promastigotes. On day 4, shortly after the blood meals had been digested and passed, the infection rate remained 100% in *P. papatasi* infected with *L. major*, but was reduced to between 0–16% in flies infected with each of the other species.

In consideration of other species-restricted vectors, the early studies by (Hindle, 1931) documented a remarkably similar timing to the loss of a Chinese strain of *L. donovani* in a sympatric strain of *P. sergenti* as was observed for unnatural *Leishmania* species in *P. papatasi*. Abundant flagellates persisted in the abdominal midgut only as long as a portion of the undigested blood meal was present. By contrast, the persistence of *L. donovani* was unaffected by the loss of the blood meal in the presumed natural vector, *P. chinensis*. Similar findings were described by (Killick-Kendrick, 1985), who examined the development of *L. infantum* in wild-caught *P. sergenti* and in a natural vector, *P. perniciosus*. Both species were initially infected in similar proportions, however, the parasites in *P. sergenti* were lost with the faeces passed on days 3–4. Both sets of studies lacked a parallel demonstration that the refractory *P. sergenti* flies were permissive to *L. tropica*, as had been reported early on by (Adler *et al.*, 1938) and more recently by (Killick-Kendrick *et al.*, 1995), using a limited number of flies. In a comprehensive analysis that included reciprocal controls for vector competence, *P. sergenti* supported the full development both of a sympatric strain of *L. tropica* from Jordan, as well as a strain from Afghanistan (Kamhawi *et al.*, 2000). In contrast, while *L. major* and *L. donovani* established late infections in *P. papatasi* and *P. argentipes*, respectively, they failed to do so in *P. sergenti*, and there was a strong correlation between the loss of blood and the loss of promastigotes.

Sandfly molecules controlling vector competence (Table 10.2)

Blood-induced molecules influencing early parasite survival and growth

Whether or not differences in parasite survival and growth within the first hours or days after an infective feed contribute to species-restricted vector competence, there is little doubt that the freshly blood-fed midgut represents a potentially harmful environment to which developing promastigotes must adapt. Following engorgement, a number of physiological changes take place

Table 10.2 Sandfly tissue/molecules pertinent to the development of *Leishmania* in the midgut

Tissues/molecule	Sandfly species	Expression[¶]	Function in the sandfly[§]	Postulated role in vector competence[‡]	Reference
Peritrophic matrix	*Phlebotomus papatasi*[1]; *P. perniciosus*[2]	Blood induced at 1–4 hours PBM	Permeability barrier	Protects early *Leishmania* stages from digestive enzymes	(Blackburn *et al.*, 1988)[1¶§]; Walters, 1993 #116][2¶§]; (Pimenta *et al.*, 1997)[1¶§‡]
Chymotrypsin	*P. papatasi*[1]; *P. langeroni*[3]	Blood induced; peak activity at 30–48 h PBM	Blood meal digestion	*Leishmania* delay/inhibit enzymatic activity	(Borovsky and Schlein, 1987)[1¶‡]; (Dillon and Lane, 1993b)[13¶§]; (Ramalho-Ortigao *et al.*, 2003)[1¶§]
Trypsin	*P. papatasi*[1]; *P. langeroni*[3]	Blood induced; peak activity at 24–48 h PBM	As above	As above	(Schlein and Romano, 1986)[1§‡]; (Borovsky and Schlein, 1987)[1¶‡]; (Dillon and Lane, 1993b)[13¶§]; (Dillon and Lane, 1993a)[13‡]
Aminopeptidase	*P. papatasi*[1]; *P. langeroni*[3]		As above	As above	(Dillon and Lane, 1993b)[13¶§]; (Dillon and Lane, 1993a)[13‡]
Glycosidases	*P. papatasi*[1]; *P. langeroni*[3]	Levels increase 1–48 h PBM and 1 h PSM	Digestion of complex carbohydrates	Provision of nutrients	(Dillon and el-Kordy, 1997)[3¶§]; (Jacobson and Schlein, 2001)[1§‡]
Lectins	*P. papatasi*[1]; *P. perniciosus*[2]; *P. perfiliewi*[4]; *P. duboscqi*[5]; *P. halepensis*[6]; *P. neglectus*[7]; *P. tobbi*[8]; *P. argentipes*[9] *Lutzomyia longipalpis*[10]; *Lu carmelinoi*[11]; *Lu. youngi*[12]; *P. veruucarum*[13]	Constitutive; elevated PBM; peak at 48 h PBM	Agglutination of RBCs, amino sugars and some glycoconjugates	Agglutination of *Leishmania*	(Wallbanks *et al.*, 1986)[1§‡]; (Ingram and Molyneux, 1991)[1,2,4§‡]; (Volf *et al.*, 1994)[1,2,5,6,7,8,,10,11,12,13§]; (Volf and Killick-Kendrick, 1996)[1,2,5,6,10,11¶§]; (Svobodova *et al.*, 1996)[1,2,5,6,9,10,11§‡]; (Palanova and Volf, 1997)[1,2,5,6 ,10¶§]

PpChit1	*P. papatasi*	Blood induced at 1–4 h PBM	Degradation of peritrophic membrane	Assists in escape of *Leishmania* into the ectoperitrophic membrane	(Ramalho-Ortigao *et al.*, 2005)
LlChit 1	*Lu. longipalpis*	As above	As above	As above	(Ramalho-Ortigao and Traub-Cseko, 2003)[¶]; (Ramalho-Ortigao *et al.*, 2005)[¶§‡]
PpGalec	*P. papatasi*	Constitutive	May be related to absorption of nutrients or innate immunity ?	*Leishmania major* LPG midgut receptor	(Kamhawi *et al.*, 2004)
Defensin	*P. duboscqi*	Infection induced	Innate immunity	Anti-*Leishmania* activity	(Boulanger *et al.*, 2004)
MAP kinase homologue	*Lu. longipalpis*	As above	Signalling molecule; innate immunity (Putative)		(Ramalho-Ortigao *et al.*, 2001)
(V)-ATPase	*Lu. longipalpis*	As above	hydrogen pump; regulates intracellular pH levels (Putative)	May promote metacyclogenesis	(Ramalho-Ortigao *et al.*, 2001)

PBM, post blood meal; PSM, post sugar meal.

within the gut, including as already noted, the release of proteases for digestion of blood meal proteins (Dillon and Lane, 1993a). Adler (1938) was the first to suggest that serum-induced proteolytic factors might prevent the development of an unnatural strain of *Leishmania* in *P. papatasi* based on his observation that lowering the rabbit serum concentration enhanced parasite survival. In the studies by (Borovsky and Schlein, 1987), the addition of soybean trypsin inhibitor to the blood meal enhanced the survival of *L. donovani* in the midgut of *P. papatasi*. Similar experiments were conducted by (Pimenta *et al.*, 1997), which in this case promoted the early survival of the natural strain of *L. major*. These studies also detailed a role for the peritrophic membrane (PM) in reducing the rate of diffusion of digestive enzymes into the blood meal and delaying the exposure of blood cells and parasites to the digestive process. The rapid diffusion of enzymes in chitinase-treated flies with poorly formed PMs was thought to account for the early mortality of parasites during their transition to extracellular promastigotes. Volf *et al.* (2001) also observed an apparent indirect effect on proteolytic enzymes in *P. duboscqi* flies fed on blood meals containing galactosamine. The enhancement of *L. major* infections was associated with depressed proteolytic activity in the gut, suggested to be the result of specific inhibition by the aminosugar of a midgut lectin activity involved in blood meal digestion. Together, these studies support the hypothesis that the concentration and timing of proteolytic enzymes present during parasite development will influence the vector competence of sandflies, and that the modulation of these enzymes by the parasite (discussed below) might confer a selective advantage for their survival.

It should be noted that sandflies also produce digestive glycosidases, including α-amylase and α-glucosidase, which are up-regulated following feeding on plant sugars and in some cases following blood meals (Dillon and el-Kordy, 1997; Gontijo *et al.*, 1998; Jacobson and Schlein, 2001). There is as yet no indication that *Leishmania* infections modulate the levels of digestive glycosidases in the gut or that these enzymes are in any way harmful to parasite growth and development. It is rather more likely that carbohydrate digestion by sandfly enzymes provides a critical source of monosaccharides for parasite metabolism. Finally, there is a single report regarding the up-regulation in whole fly preparations of mRNA encoding an antimicrobial peptide in *P. papatasi* midguts infected with *L. major* promastigtoes (Boulanger *et al.*, 2004). While the recombinant peptide exhibited anti-parasite activity, there is no evidence as yet that the defensin is released into the gut or that it confers any immunity to *Leishmania* infection *in vivo*.

Midgut molecules controlling parasite persistence

The fact that in many studies of species-restricted vectors an abundance of apparently healthy promastigotes was observed at relatively late time points (72–96 hours) in the refractory flies argues against a role for differential killing by digestive enzymes in the gut, particularly since the peak concentration of these enzymes is thought to occur at 24–48 hours post-feeding. The strong correlation between the loss of blood meal and the loss of promastigotes suggests that the inappropriate strains are not killed, but are passed out of the midgut along with the blood meal remnants. Two general mechanisms have been proposed to explain the failure of promastigtoes to persist in the gut during this time: (1) inability to escape the confines of the peritrophic membrane prior to blood meal excretion, and (2) inability to anchor to the gut lining via specific attachment sites.

As already indicated, the PM is a semi-permeable barrier for digestive enzymes, and protects against certain pathogenic microbes by providing a physical barrier to their development. The loss of *Leishmania* infection in some unnatural vectors was found to be due, at least in part, to the failure of the parasite to escape from the PM prior to blood meal excretion (Feng, 1951; Walters *et al.*, 1992). In comparisons of different *L. major* sub-strains in *P. papatasi*, the strain that appeared the slowest to escape from the PM was effectively lost during the time of blood meal excretion (Cihakova and Volf, 1997). The requirement for a sandfly and/or parasite-derived chitinase, both of which have been described (see below), to break down the PM and permit parasite escape is most clearly indicated by the experiments involving the streptomycete antibiotic, allosamidin, a potent competitive inhibitor of insect, plant, and parasite chitinases. Allosamidin prevented the normal escape of *L. major* promastigotes from the blood meal, and in every case when the blood meals in the allosamidin-treated flies were finally passed, the parasites were also lost (Pimenta *et al.*, 1997). A functional, blood-induced chitinolytic system was described in two sandfly species, *L. longipalpis* and *P. papatasi*, and a midgut-specific chitinase cDNA clone was identified that encodes the protein responsible for chitinase activity in *P. papatasi* (Ramalho-Ortigao *et al.*, 2005). There is no information as yet regarding the up- or down-regulation of sandfly chitinase activity by *Leishmania*, or whether species differences in this regulation might help to account for species-restricted vector competence.

By contrast, there is a large body of evidence to indicate that the molecules involved in promastigote attachment to the midgut vary extensively between parasite species and between different vectors, and that these polymorphisms can in many cases fully account for the species restrictions observed. The attachment of promastigotes to the gut wall was first observed under light microscopy (by Adler and Theodor, 1927), and subsequent ultrastructural images have consistently revealed an intimate association of parasite flagella with epithelial cell microvilli (Killick-Kendrick *et al.*, 1974b; Warburg *et al.*, 1986; Walters *et al.*, 1989b). The variable oligosaccharide domains expressed by the promastigote surface lipophoshoglycan (LPG) shown to mediate species-specific binding (discussed below) have implicated gut lectins or lectin-like molecules that should themselves be variably expressed by different vector species. In sandflies, lectin activity has been demonstrated in lysates of various tissues, including the gut, and in some studies the midgut lysates were able to agglutinate *Leishmania* promastigotes that were inhibited by various aminosugars (Wallbanks *et al.*, 1986; Volf *et al.*, 1994; Svobodova *et al.*, 1996). Importantly, marked differences were found in the agglutination of promastigotes of various species and strains, and high agglutination titres were observed in some natural vector-parasite combinations. The relationship between the midgut lectins mediating these agglutinating activities and the molecules that serve as parasite attachment sites has yet to be established. The fact that galactosamine, shown to inhibit midgut lectin activity, actually promoted *L. major* development in *P. duboscqi*, does not support a role for this lectin in midgut binding (Volf *et al.*, 1998). The first molecular characterization of a putative midgut receptor for parasite attachment in sandflies was reported by Dillon and Lane (1999), who described a microvillar 65kDa protein from *P. papatasi* that bound *L. major* LPG. The involvement of this protein in parasite attachment *in vivo* was not explored. The most well-characterized midgut molecule for which a role in parasite development has been clearly ascribed is a galectin homologue, termed PpGalec, that is expressed only by *P. papatasi* and *P. duboscqi* (Kamhawi *et al.*, 2004). The binding of recombinant PpGalec was restricted to *L. major* promastigotes bearing poly-Gal (β1–3) side chains on their LPG. Moreover, anti-PpGalec antibodies inhibited *in vitro* midgut binding of *L. major* LPG as well as live parasites. Finally, PpGalec antibodies fed to *P. papatasi* severely impaired parasite development and survival in the insect midgut, indicating that interaction of *L. major* promastigotes with PpGalec is an essential condition for the development of transmissible infections. The absence of PpGalec from all other phlebotomine vector species examined to date begs the question as to the nature of the parasite attachment sites in these flies. Preliminary data from mass sequencing analysis of a midgut cDNA library from *P. sergenti* has revealed unique expression of a C-type lectin in this fly that may serve as a species-restricted receptor for *L. tropica* (Kamhawi *et al.*, unpublished observations).

Parasite molecules implicated in sandfly interactions (Table 10.3)

Molecules promoting survival and growth within the freshly blood-fed midgut

The first evidence that *Leishmania* might themselves produce molecules that alter the levels of digestive enzymes in the gut was offered by Schlein and Romano (1986), who reported that *L. major* infection reduced proteolysis by *P. papatasi* midgut homogenates by 28%, while *L. donovani*

Table 10.3 Parasite molecules with postulated roles in sandfly interaction. References in bold indicate those studies for which the postulated role was substantiated by parasite mutants with observable phenotypes in the vector

Parasite molecule	Postulated role in vector interaction	Evidence provided	Reference
Excreted factor (EF)	Inhibition of midgut proteases	Addition of EF to blood meal promoted survival of EF deficient *L. major* sub-strain	(Schlein and Romano, 1986; Schlein *et al.*, 1990)
Phosphoglycan containing molecules (LPG,PPG,SAP)	Resistance to midgut proteases	Early killing of phosphoglycan deficient mutants	(Sacks *et al.*, 2000; Spath *et al.*, 2003; Boulanger *et al.*, 2004)
Leishmanolysin (GP63)	Nutrient utilization in early bloodfed midgut	Decrease in early parasite growth in anti-sense mutants	(Hajmova *et al.*, 2004)
Chitinase	Degradation of peritrophic membrane (PM) and stomodeal valve	No direct evidence for role in degradation of PM or stomodeal valve, or in parasite development	(Schlein *et al.*, 1991; Schlein *et al.*, 1992; Shakarian and Dwyer, 2000)
Flagellar membrane protein	Ligand for midgut attachment to maintain infection during excretion of blood meal remnants	Loss of parasites following blood meal excretion in flies infected with mutants lacking a flagellum Partial inhibition of midgut binding of flagellar preparations using monoclonal Ab	(Cuvillier *et al.*, 2003) (Warburg *et al.*, 1989)
Lipophosphoglycan (LPG)	Ligand for midgut attachment to maintain infection during excretion of blood meal remnants	Loss of infection following blood meal excretion in flies infected with LPG deficient mutants Direct binding of LPG to midguts *in vitro*. Inhibition by purified LPG of promastigote binding to dissected midguts *in vitro*	(Sacks *et al.*, 2000; Spath *et al.*, 2003; Boulanger *et al.*, 2004) (Pimenta *et al.*, 1992; Pimenta *et al.*, 1994; Sacks *et al.*, 1995b; Mahoney *et al.*, 1999; Kamhawi *et al.*, 2000; Soares *et al.*, 2002; Soares *et al.*, 2004)

Developmental modification of LPG terminal sugars	Detachment and/or maintenance of metacyclic promastigotes as unattached forms adapted for anterior migration and transmission by bite	Consistent findings regarding loss of sugars mediating midgut attachment, but no direct evidence for role in detachment, anterior migration, or transmission by bite	(Pimenta *et al.*, 1992; Pimenta *et al.*, 1994; Sacks *et al.*, 1995b; Mahoney *et al.*, 1999; Kamhawi *et al.*, 2000; Soares *et al.*, 2002; Kamhawi *et al.*, 2004; Soares *et al.*, 2004)
Myoinhibitory peptide	Inhibition of gut peristalsis promoting post-blood meal parasite persistence	Peptide produced midgut distension and inhibited hindgut contraction. No direct evidence for role in parasite persistence	(Vaidyanathan, 2004, 2005)
Glycosidases (α-glucosidase; α-amylase; sucrase)	Digestion and metabolism of sandfly sugar meals	No direct evidence for their role in parasite growth or development	(Blum and Opperdoes, 1994; Gontijo *et al.*, 1996; Jacobson and Schlein, 2001)
Glucose transporter	Sugar metabolism and sensing function that promotes anterior migration	Glucose transporter deficient mutants showed impaired growth and development following blood meal excretion	(Burchmore *et al.*, 2003)
Promastigote secretory gel (PSG) containing filamentous PPG	Formation of a biological plug that impairs normal blood feeding and promotes regurgitation of metacyclics	No direct evidence for role in plug formation or facilitating parasite regurgitation and egestion	(Stierhof *et al.*, 1999; Rogers *et al.*, 2002)

actually increased the homogenate activity. In follow-up studies, the overall trypsin activity was actually more reduced in *L. donovani*-infected flies compared to *L. major*, although at early time points (20–30 hours) the levels were relatively higher in the former. Dillon and Lane (1993a) also reported suppression of blood meal-induced protease activities, including alkaline protease, trypsin, and aminopeptidase, throughout the first 30 hours of *L. major* amastigote-initiated infections in *P. papatasi*. Inappropriate *Leishmania* species were not compared. The only other sandfly for which the effect of *Leishmania* on protease activities has been studied is *P. langeroni*, an implicated vector for *L. infantum* transmission in Egypt and an unnatural vector for *L. major*. During the first 48 hours., the proteolytic activity and blood protein digestion were reduced in *P. langeroni* fed on human and canine blood containing *L. infantum* but not *L. major* (Daba, 1997b, a). There was no correlation, however, between the level of the proteolytic activity and the survival of promastigotes in the gut.

Schlein *et al.*, (Schlein and Romano, 1986; Schlein *et al.*, 1990) also provided the first clues as to the nature of the parasite molecules that might act to reduce proteolytic activities in the gut, reporting that released glycoconjugates from cultured *L. major* promastigotes, termed excreted factor (EF), promoted the early survival in *P. papatasi* of a strain of *L. major* that did not normally develop. The glycoconjugates were not characterized at the time, but subsequent biochemical analyses have determined that EF contains a class of related glycolipids and glycoproteins incorporating a common phosphoglycan repeat unit structure. The phosphoglycan repeats are the principle constituents of the cell surface and released lipophosphoglycan (LPG), the dominant surface glycolipid on all *Leishmania* promastigotes studied to date (reviewed in McConville and Ferguson, 1993; Mengeling and Turco, 1998). LPG is organized as a densely packed glycocalyx, which covers the entire surface of the cell, including the flagellum. LPGs from different *Leishmania* species share conserved glycan core region of Galα1, 6Galα1, 3Gal$_f\beta$1, 3Manα1, 3Manα1, 4GlcNα linked to a 1-*O*-alkyl-2-*lyso*phosphatidylinositol anchor. In addition, they express a common backbone consisting of repeating, phosphodiester-linked disaccharide units of PO_4-6Gal(β1–4)Manα1. The remarkable structural polymorphisms among the species lie in the repeating units, where the C-3 or C-2 position of the Gal residue can either be unsubstituted, as in *L. donovani*; partially substituted, as in *L. mexicana*; or almost completely substituted with a variety of sugars, as in *L. major* and *L. tropica*. In addition to LPG, the phosphoglycan repeat units are incorporated into other glycoconjugates, including a cell surface and released proteophosphoglycan (PPG), and a secreted acid phosphatase (sAP) (see Chapter 9).

The most conclusive evidence that phosphoglycan-containing molecules promote early survival of parasites in the midgut comes from studies employing mutants specifically deficient in biosynthesis of phosphoglycans as a consequence of targeted deletion of the *LPG2* gene, resulting in defective transport of GDP-Mannose into the Golgi lumen (Descoteaux *et al.*, 1995). For both *L. donovani* and *L. major*, the targeted null mutants were killed in the blood meals of their respective normally permissive vectors within the first 4 days, prior to blood meal excretion (Sacks *et al.*, 2000; Spath *et al.*, 2003; Boulanger *et al.*, 2004). Interestingly, mutants deficient only in LPG biosynthesis due to targeted deletion of the *LPG1* gene involved in biosynthesis of the glycan core domain, survived relatively well during this time, indicating that it is not essential for early survival, and that other phosphoglycan-containing molecules are involved. The survival of the respective LPG biosynthetic mutants of *L. major* is depicted in Fig. 10.5. In these studies, there was no correlation between the level of protease activity and early parasite survival in *P. papatasi* midguts, and the protection afforded the parasites by surface and released phosphoglycans was suggested to result from an abundance of negatively charged molecules in the vicinity of the parasite that protects the cell surface from enzymatic attack, similar to the function of gastric mucins. Released molecules bearing the phosphoglycan epitope have been detected in abundance in *L. major*-infected *P. papatasi* midguts as early as day 2 (Davies *et al.*, 1990).

Parasite proteinases have been suggested to play a role in nutrient utilization in the blood meal during the early stages of development

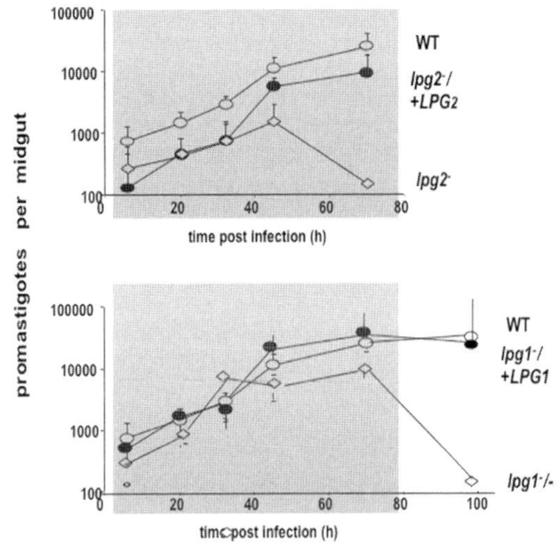

Figure 10.5 The role of phosphoglycan-containing molecules in *L. major* survival in *P. papatasi*. *LPG2*–/– mutants are deficient in all phosphoglycan containing molecules, including LPG, PPG, and SAG. LPG1–/– mutants are deficient only in LPG biosynthesis. Red overlay indicates timing of blood meal retention or loss.

and in resistance to sandfly digestive enzymes (Schlein *et al.*, 1990). The *Leishmania* surface proteinase of 63 kDa, leishmanolysin or GP63, has been reported to be a virulence factor involved in the direct interaction of promastigotes and host macrophage receptors and interaction with the complement system (reviewed in (Yao *et al.*, 2003). *L. major* mutants stripped of the entire 20 kb region containing all seven leishmanolysin genes developed normally within the midgut of a natural vector, *P. duboscqi* (Joshi *et al.*, 2002). Complete development from procyclic to infectious metacyclics occurred by days 8–10. Somewhat different findings were reported for *L. amazonensis* mutants whose GP63 was up- and down-regulated by directional cloning into P6.5 for sense and antisense transcription (Hajmova *et al.*, 2004). There was a significant decrease in the early establishment (day 2) of the parasite in *Lu. longipalpis* midguts when its GP63 was specifically down-regulated. By day 9, however, no difference in the percent of infected flies or the intensity of infection in the midgut was observed between sense and anti-sense transfectants. It was suggested that GP63 assists the proteolysis of haemoglobin to yield haem and other peptides as nutrients essential for the early growth of promastigotes in this New World vector. The discrepancy with the results using the *L. major* GP63 KOs may be due to the vector-parasite combinations examined, or the experimental protocols where amastigotes versus promastigotes were used to initiate infections in *P. duboscqi* and *Lu. longipalpis* respectively. In each case, however, the data are in agreement that the metalloproteinase activity of leishmanolysin is not essential for long term survival in the vector.

Molecules promoting escape from the PM

Initially *Leishmania* parasites ingested with host blood multiply confined within the PM. Schlein *et al.* (1991) found that chitinase and N-acetylglucosaminidase are secreted by cultured *L. major* promastigotes and suggested that these enzymes might catalyse the degradation of the PM permitting parasite escape. Shakarian and Dwyer (2000) identified a single-copy chitinase gene from *L. donovani*, and demonstrated that the gene locus and enzymatic activities are conserved across species lines. Interestingly, *L. mexicana* released/secreted 2- to 4-fold greater chitinase activity when cultured as axenic amastigotes compared to promastigotes, and overexpression of the chitinase gene in *L. mexicana* resulted in enhanced lesion development in mice (Joshi *et al.*, 2005). No phenotype for the episomal constructs, however, was observed in sandflies (M. Joshi, personal communication). Deletion of the gene will be required to establish whether its expression is essential for parasite escape from the PM, or whether the endogenous sandfly chitinase suffices to produce a timely and localized break down of the PM in the vicinity of the mass of promastigotes to permit their release prior to blood meal excretion.

Molecules mediating attachment to midgut epithelial cells

As the PM breaks down during the later stage of blood meal digestion, even the promastigotes that escape the confines of the PM risk being voided from the midgut along with the PM and blood meal remnants. At this stage, the promastigotes of suprapylarian species attach to the midgut lining by inserting their flagella between microvilli. There is general agreement that this attachment serves to anchor the parasite against the peristaltic action of the gut. The requirement for the flagellum in this interaction is supported by the complete loss of *L. amazonensis* from *Lu. longipalpis* midguts just after blood meal excretion in flies infected with promastigotes lacking a flagellum due to overexpression of a mutated G protein (Cuvillier *et al.*, 2003). This striking phenotype might be explained by the inability of the immobilized parasites to encounter the midgut epithelium, and/or the absence of a critical ligand localized to the flagellum. In the first attempt to characterize the parasite molecules that might mediate this attachment, an *in vitro* binding assay was developed in which a monoclonal antibody specific for a 13 kDa flagellar membrane protein was able to inhibit the binding of flagellar preparations to frozen sections of *P. papatasi* midguts (Warburg *et al.*, 1989). The inhibition was only partial, however, and the assay failed to distinguish between the binding of flagellar preparations from *L. major* and *L. b. panamensis*, the latter being a species that clearly fails to persist following blood meal excretion in *P. papatasi*. Thus, while the flagellar protein may contribute to the stability and orientation of parasite binding, other parasite ligands must account for species-specific differences in midgut attachment.

The role for the predominant promastigote surface glycoconjugate, LPG, in mediating attachment to the midgut epithelium has been supported by a number of findings: 1) Procyclic LPG was shown by immunolabelling to bind directly to dissected midguts, with localization to the microvillar lining (Pimenta *et al.*, 1992; Pimenta *et al.*, 1994; Kamhawi *et al.*, 2000), 2) procyclic LPG completely inhibits, in a dose-dependent manner, the binding of promastigotes to the gut in *in vitro* attachment assays (Pimenta *et al.*, 1992; Sacks *et al.*, 1995b); and 3) promastigotes deficient solely in LPG expression as a consequence of targeted deletion of *LPG1*, fail to attach to the midgut *in vitro* or to persist in the sandfly following blood meal excretion *in vivo* (Sacks *et al.*, 2000). The experiments involving the mutant parasites provide especially strong evidence for the role of LPG in mediating midgut binding, as they were replicated using both *L. major* and *L. donovani* mutants and their respective vectors (i.e. *P. papatasi* and *P. argentipes*). Furthermore, for both sets of mutants, when LPG expression was restored by reintroduction of *LPG1*, midgut attachment and persistence *in vivo* were also restored. The timing of the loss of the *L. major LPG1* null mutants in *P. papatasi* is shown in Fig. 10.5.

Perhaps the most compelling aspect of LPG structure as it relates to its function as a parasite ligand for midgut receptors is that the remarkable inter- and intra-specific polymorphisms in the phoshoglycan domains of LPG can in many cases fully account for species-specific vector competence. The ability of *P. papatasi* to transmit only *L. major* sp. can be attributed to the unique, highly substituted nature of *L. major* LPG that provides for multiple, terminally exposed β–linked galactose residues for binding. Thus *Leishmania* species that bear unsubstituted LPGs (i.e. *L. donovani* strains from Sudan), partially substituted or even fully substituted LPGs bearing side-chain oligosaccharides that do not contain β–linked galactose (e.g. *L. mexicana, L. tropica*), fail to bind to *P. papatasi* midguts or to persist in the gut following blood meal excretion (Pimenta *et al.*, 1994). Furthermore, of the LPGs purified from these respective species, only *L. major* LPG stained *P. papatasi* midguts *in vitro*. More critically, an *L. major* mutant deficient in β–linked galactose side-chain biosynthesis (Butcher *et al.*, 1996), and an *L. major* sub-strain naturally deficient in expression of β-linked galactose side-chains (Joshi *et al.*, 2002; Turco *et al.*, unpublished), in each case failed to persist in *P. papatasi* or attach to *P. papatasi* midguts *in vitro*. Dobson *et al.* (2003) used cross-species transfections to recover genes encoding the LPG side chain β1,3-galactosyltransferases (SCGs) in *L. major*. A dispersed family of six SCG genes was recovered, of which at least four encoded LPG side-chain β1,3-galactosyltransferase activity. Importantly, transfection of the *L. major* sub-strain naturally lacking the poly-galactose epitopes with a single active SCG gene rescued their ability to produce persisting and fully developed infections in *P. papatasi* (Dobson *et al.*, unpublished). The vector competence of *P. papatasi* and *P. duboscqi* for *L. major* sub-strains and mutants varying in LPG side-chain galactosylation is depicted in Fig. 10.6.

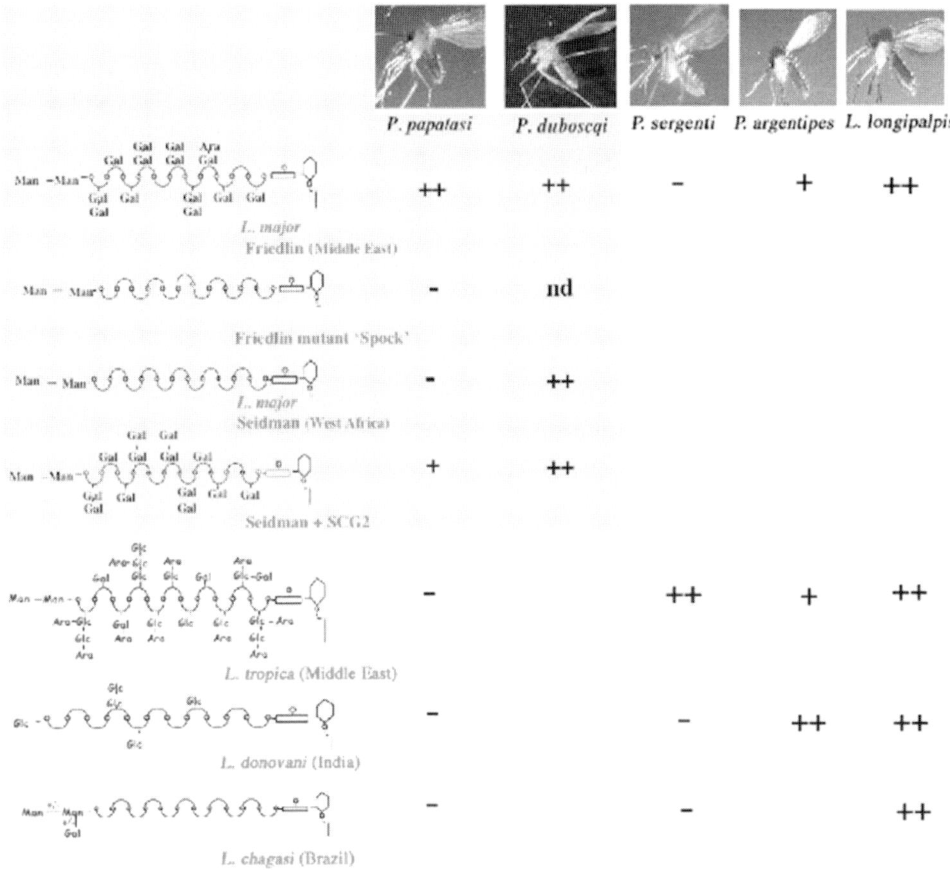

Figure 10.6 The shaded area summarizes the vector competence of *P. papatasi* and *P. duboscqi* for *L. major* substrains and mutants with LPGs expressing conserved lipid anchor, glycan core, and backbone phosphoglycan repeat domains, but varying in expression of β–linked galactose side-chains. Other *Leishmania* species and their natural vectors *(P. sergenti/L. tropica; P. argentipes/L. donovani; L. longopalpis/ L. chagasi)*, including their vector competency for selected inappropriate *Leishmania* species, are shown outside the shaded area.

The prediction that the species-specificity of midgut attachment, and by extension, species-specific vector competence, is controlled by polymorphic structures on the parasite LPG is further supported by the finding that *P. sergenti* midguts were intensely stained following incubation with purified LPG from *L. tropica*, but not LPG from *L. major* or *L. donovani* (Kamhawi *et al.*, 2000). As with *P. papatasi–L. major* interactions, the LPG that controls the specificity of *P. sergenti–L. tropica* interactions is characterized by a highly branched and species-specific structure. Although the backbone repeat units of both *L. tropica* and *L. major* LPGs are fully substituted, the side-chain sugars are distinct, with *L. tropica* synthesizing an abundance of side-chain sugars terminating in arabinose and glucose, distinct from the β–linked galactose side-chains characteristic of most *L. major* strains. The data strongly suggested that at least some phlebotomine vectors differ with respect to the parasite recognition sites that they express, and that midgut adhesion is a sufficiently critical aspect of vector competence as to provide the evolutionary drive for LPG structural polymorphisms. This prediction has essentially been borne out by the identification of PpGalec as the key receptor for *L. major* LPG, and the demonstration that its binding is restricted to LPGs bearing β-linked galactose side-chains, and its expression restricted to the midguts of natural vectors of *L. major*, i.e. *P. papatasi* and *P. duboscqi* (Kamhawi *et al.*, 2004).

It is becoming increasingly clear that *Leishmania* also display considerable intraspecies polymorphisms in their LPG sugars. In particular, significant variation exists in the degree of side-chain galactosylation in *L. major* (Turco *et al.*, unpublished). As already indicated, some *L. major* strains from Africa bear relatively unsubstituted backbone repeats, and as such cannot be transmitted by *P. papatasi*, but rely on *P. duboscqi* or possibly other sandflies species. More interesting is the recent observation that even for *L. major* strains that express a high degree of galactosylation, remarkable heterogeneity exists in the nature of the substitutions, ranging from strains that predominate in the Middle East that are heavily substituted with side chains containing one or two βgals, to strains found in Iran or Southern Russia that bear long, polysaccharide side-chains (5–7 βgals). The strain-specific patterns in *L. major* LPG galactosylation are thought to depend on the pattern of expression of the different SCG genes that encode both initiating and elongating galactosyl transferase activities (Dobson *et al.*, 2003). Of note is the finding that, compared to the prototypical *L. major* strain (NIH Friedlin) isolated from a patient in Israel, which survives well in a *P. papatasi* colony originating from a contiguous biotope in Jordan, the strains bearing elongated galactose side chains survive relatively poorly in this colony, and bind poorly to the PpGalec that was identified and cloned from this fly. These intraspecific LPG modifications predict that *P. papatasi* is a species-complex, and that PpGalec is itself polymorphic with a range of affinities for β galactose-containing sugars that drive the expression of variable conformational poly-galactose epitopes. *L. tropica* LPGs also express a high degree of intraspecies polymorhism, which in one instance was shown to be directly correlated with the fact that the sub-strains were isolated from two distinct vector species, *P. sergenti* and *P. arabicus* (Soares *et al.*, 2004).

It is important to consider the nature of the midgut receptors expressed by the natural vectors of parasites bearing unbranched or relatively unsubstituted phosphoglycan repeats. Interestingly, these flies appear to behave as unrestricted vectors in laboratory-based studies. Thus, *P. halepensis*, *P. argentipes* and *Lu. Longipalpis* can support the growth of several *Leishmania* species, including *L. major*, *L. donovani*, *L. amazonensis* and *L. tropica* (Kamhawi *et al.*, 2000; Sacks, 2001; Pinto-da-Silva *et al.*, 2002; Sadlova *et al.*, 2003; Soares and Turco, 2003; Pinto-da-Silva *et al.*, 2005). Fig. 10.6 summarizes, based on these experimental findings, the species-restricted or -unrestricted vector competence of various sandfly vectors for *Leishmania* species varying in the extent and makeup of their LPG sidechain substitutions. Permissive vectors may express receptors, absent or inaccessible in *P. papatasi* and *P. sergenti*, that recognize conserved structures on LPG, or that more broadly recognize neutral hexoses, such as those that cap the backbone repeats. It should be borne in mind that the role of LPG in midgut attachment involving *Leishmania* species bearing simple backbone repeats and transmitted by broadly permissive vectors has not been conclusively demonstrated using reverse genetics. And while purified LPG from *L. donovani* and *L. chagasi* has been shown to bind to midguts of *P. argentipes* and *L. longipalpis*, respectively, (Mahoney *et al.*, 1999; Sacks *et al.*, 2000; Soares *et al.*, 2002), this does not establish their role in mediating midgut attachment, and it is possible that other ligands are involved. In fact, LPG1−/− mutants of *L. mexicana* were able to establish mature infections in *L. longipalpis* (Rogers *et al.*, 2004), although it should be noted that a high concentration of promastigotes was used to initiate infection in these flies.

Molecules inhibiting gut peristalsis

Some investigators have noted that a large number of unattached promastigotes remain in the gut lumen following blood meal excretion, and Vaidyanathan (2004, 2005) has proposed a novel mechanism to account for this persistence. A myo-inhibitory peptide present in cell lysates of *L. major* promastigotes and released into culture medium was shown to inhibit spontaneous hindgut contractions and produce distension of guts isolated from *P. papatasi* and other diptera. The fact that the myo-inhibitory activity was also detected in species, i.e. *L. braziliensis* and *L. donovani*, that do not persist in *P. papatasi* midguts would indicate that while inhibition of muscle contraction may contribute to and even be necessary for full development within the vector, it is not a sufficient condition to account for species specificity of vector competence. Parasite mutants deficient in synthesis of the peptide will be needed to help clarify if and how its effects co-operate with LPG-mediated midgut attachment.

Molecules associated with metacyclogenesis

There is little evidence that *Leishmania* infections fail during the terminal stages of their development in non-permissive vectors. It is known, for example, that if the deficient growth and/or persistence of *L. donovani* in *P. papatasi* is overcome by inhibiting the early killing (Schlein and Jacobson, 1998) or increasing the initial inoculum of parasites in the blood meal (Adler, 1938) these infections will mature normally. Thus for the final stages of development in the vector involving anterior migration and differentiation to metacyclic promastigotes, it has not been possible to compare permissive and non-permissive vectors in order to help identify the role of specific molecules in these processes. The close coordination of these two events would indicate that either the conditions in the anterior gut provide cues for metacyclogenesis to proceed, or that metacyclic promastigotes are especially well-adapted for anterior migration and for maintaining their forward positioning in the gut. The main argument for the former is that metacyclic promastigotes are found mainly in the thoracic midgut and cardia, and that at no time are the numbers of metacyclics present in the abdominal midgut sufficient to account for the numbers that accumulate in the anterior gut as the infection matures (Lawyer *et al.*, 1990; Saraiva *et al.*, 1995; Rogers *et al.*, 2002). By careful microdissection of different regions of the *Lu. longipalpis* gut at multiple timepoints following infection with *L. mexicana*, Rogers *et al.*, concluded that the majority of metacyclics are generated from leptomonad promastigotes imbedded within a gel-like plug secreted by these forms in the thoracic midgut. The only culture condition that has been clearly shown to enhance metacyclogenesis *in vitro* is acidic pH (Zakai *et al.*, 1998). Culturing promastigotes at pH 5.5 was shown to promote the number of metacyclics arising during growth of diverse *Leishmania* species, and may better reflect acidic conditions in the thoracic midgut that arises following secretion of the negatively charged filamentous proteophosphoglycan (fPPG), which is the main component of the plug (Stierhof *et al.*, 1999). Other extrinsic factors that may operate synchronously to promote metacyclogenesis in the fly may be anaerobic conditions (Mendez *et al.*, 1999), a decline in the levels of tetrahydrobiopterin (Cunningham *et al.*, 2001), absence of haemoglobin (Schlein and Jacobson, 1994), and exposure to sandfly saliva (Charlab *et al.*, 1995). Absence of haemoglobin in culture, reflecting conditions in the midgut following blood meal excretion, increased the infectivity of *L. major* stationary cells. Multiplication of *L. amazonensis* was arrested by the addition of salivary gland homogenate of *Lu. Longipalpis*. This may provide an additional cell-cycle-associated differentiation signal for metacyclogenesis in the thoracic midgut, where exposure to re-ingested salivary components might occur when flies obtain sugar meals.

A strong argument for the pre-adaptation of metacyclic promastigotes for anterior migration and positioning is based on the remarkably consistent findings regarding the loss of intrinsic midgut binding potential of metacyclic promastigotes purified from stationary cultures of different *Leishmania* species, including *L. major*, *L. tropica*, *L. donovani*, *L. chagasi*, *L. amazonensis*, and *L. braziliensis*. The structural modifications of LPG that are responsible for the loss of midgut binding on metacyclics have been studied in most detail for *L. major*, and are of two sorts: (a) a 2–3 fold increase in size due to an increase in the number of phosphorylated saccharide units expressed, and (b) a change in the composition of side-chain terminal sugars involving a pronounced decrease in repeat units with side chains of βGal or Galβ1 3Galβ1-, and a corresponding increase in repeat units with side chains of Ara*p*α1–2 Galβ1- (Sacks *et al.*, 1990; McConville *et al.*, 1992). The elongation of LPG on metacyclics results in an approximate 10nm thickening of the glycocalyx and appears to promote the survival of the parasite in the mammalian host by conferring resistance to complement-mediated damage (Sacks, 1989). The down-regulation of terminal galactose residues, which is reflected by the loss of parasite agglutination with peanut agglutinin and other galactose-binding plant lectins, has a similar consequence in terms of the loss of binding to PpGalec (Kamhawi *et al.*, 2004).

Stage-specific mAb have been used to document identical modification of the LPG during metacyclogenesis of *L. major* promastigotes *in vivo*, beginning on day 5 when most parasites are still in the abdominal midgut (Davies *et al.*, 1990; Lang *et al.*, 1991; Saraiva *et al.*, 1995). Metacyclic LPG was detected on the majority of promastigotes by day 7, with most of these corresponding to anterior metacyclic forms by morphological criteria. Finally, in forced-feeding experiments using 10–14 day infected sandflies, the parasites emerging from the proboscis were all identifiable both morphologically and antigenically as metacyclics. The *in vivo* data thus support a molecular mechanism of parasite attachment and release controlled by developmental modifications of the LPG, which insures that metacyclic promastigotes are well adapted for both arrival and survival in the vertebrate host.

The interspecies polymorphisms in LPG structure will necessarily require that fundamentally different modifications of LPG accompany metacyclogenesis of promastigotes bearing, for instance, unsubstituted or glucosylated backbone repeats. During metacyclogenesis of a Sudanese strain of *L. donovani*, the salient structural feature of LPG is conserved, involving expression of a phosphoglycan chain made up of unsubstituted disaccharide-phosphate repeats (Saraiva *et al.*, 1995). As with *L. major*, the size of the molecule is substantially increased due to a 2-fold increase in the number of phosphorylated disaccharide repeat units expressed. In addition, the display of terminally exposed neutral capping sugars is affected, resulting in the loss of lectin-binding domains, including those involved in mediating binding to midgut epithelial cells of a competent vector, *P. argentipes*. Since the neutral capping sugars expressed by metacyclic LPG still contain both β-Gal- and α-Man- terminating structures shown to mediate binding of the procyclic promastigotes, it was suggested that these epitopes become cryptic as a possible consequence of the elongation and clustering of the phosphoglycan chains. For the LPG of Indian strains of *L. donovani*, which are partially substituted with glucose side-chains and a cap structure that also terminates with β1,3-linked glucose, metacyclogenesis is associated with both chain elongation and the down-regulation of glucose side chain biosynthesis, again resulting in loss of LPG-mediated binding to *P. argentipes* midguts (Mahoney *et al.*, 1999). Identical down-regulation of β1,3-linked glucose residues was shown to accompany metacyclogenesis of *L. chagasi* promastigotes, in this case resulting in loss of binding to *Lu. longipalpis* midguts (Soares *et al.*, 2002). Interestingly, analysis of LPG from a strain of *L. braziliensis* revealed an essentially opposite regulation of glucosyl transferase activity during growth *in vitro*, with procyclic LPG-lacking side chain substitutions, and metacyclic LPG-adding 1–2 (β1–3) glucose residues (Soares *et al.*, 2005). Whether or not these modifications altered the intrinsic binding potential of *L. braziliensis* developmental stages to an appropriate vector midgut, e.g. *Lu. migonei*, was not investigated. By contrast, for *L. tropica* LPG the sugar modifications that accompany metacyclogenesis are not yet known in detail, yet the consequences of these changes in terms of loss of binding to *P. sergenti* midguts has been confirmed (Sacks *et al.*, unpublished). Collectively, these developmental modifications in LPG are the basis for identifying and purifying metacyclic promastigotes from culture using stage-specific lectins and mAb (summarized in Table 10.4). It should be pointed out that there is as yet no conclusive evidence that the developmentally regulated alterations in LPG sugars that accompany metacyclogenesis are directly responsible for detachment of promastigotes from the gut wall. It is possible that the rapid turnover of LPG that is observed during promastigote growth *in vitro* might result in saturation of binding sites by released LPG. Furthermore, it is possible that the developmental modifications of LPG on promastigotes precede their terminal differentiation to metacyclics, and may begin with transitional stage parasites, such as nectomonads or leptomonads, and account for their detachment and anterior migration (Rogers *et al.*, 2002). Certainly metacyclic promastigotes themselves have never been seen in attachment, so at the very least their loss of intrinsic binding potential would maintain them as unattached promastigotes during the critical final stages of maturation, involving their accumulation in the anterior midgut, cardia, and foregut, and their egestion during bloodfeeding.

Other potential virulence factors that show modified or increased expression in metacyclic

Table 10.4 Stage-specific lectins or anti-LPG monoclonal antibodies used for purification of metacyclic promastigotes by negative selection

Species	Lectin	Anti-LPG mAb	Reference
L. major	PNA, RCA		(Sacks *et al.*, 1985)
L. major (West Africa)	ConA		(Joshi *et al.*, 1998)
L. donovani (Sudan)	PNA, ConA		(Sacks *et al.*, 1995b)
L. donovani (India)		MnG	(Mahoney *et al.*, 1999)
L. tropica		1H2	(Lira *et al.*, 1998)
L. chagasi	PNA	MnG	(Soares *et al.*, 2002)
L. amazonensis		3A1	(Courret *et al.*, 1999)
L. braziliensis	*B. purpurea*		(Pinto-da-Silva *et al.*, 2002)

Note that metacyclic promastigotes, regardless of species, can be enriched by buoyant-density centrifugation (Spath and Beverley, 2001).

promastigotes include gp63 (Ramamoorthy *et al.*, 1992), specific cysteine proteinases (Mottram *et al.*, 1998), and a family of novel, hydrophilic proteins encoded by the *LmcDNA16* gene locus (McKean *et al.*, 1997). Of the five genes in this array, *SHERP1* and *SHERP2* are especially interesting as they code for a small peptide that is expressed only in metacyclics (Knuepfer *et al.*, 2001). In no case, however, has the deletion or overexpression of these genes been shown to modify the development or behaviour of metacyclic promastigotes in the vector or to affect the ability of metacyclics to be transmitted by bite. The parasite molecules that have been implicated in these processes are described below, though they too will need to be targeted for deletion or overexpression so that their roles can be better defined.

Molecules involved in sugar metabolism

Since phlebotomine sandflies ingest plant sap or aphid and coccid honeydew rich in sucrose between blood meals, the role of sugar meals in the differentiation and migration of promastigotes needs to be considered. In the earliest studies of laboratory transmission of *Leishmania* by sandfly bite (Shortt and Swaminath, 1928; Shortt, 1931), a source of sucrose was deemed to be highly beneficial. It is likely that sucrose serves as an energy source following blood meal digestion rather than as a cue for differentiation *per se*. In addition, since sugar meals are initially taken into the crop and are gradually delivered into the thoracic midgut, it is likely that they form a concentration gradient against which free-swimming promastigotes will migrate anteriorly. The capacity of promastigotes to move chemotactically in the presence of sugars has been demonstrated *in vitro* (Bray, 1983; Oliveira *et al.*, 2000; Barros *et al.*, 2006). Since sugar meals rich in sucrose provide the main nutritional support for the parasite once the blood meal is voided, and since promastigotes cannot take up sucrose, it is not surprising that they produce an extracellular sucrase that hydrolyses sucrose into its component monosaccharides (Blum and Opperdoes, 1994; Gontijo *et al.*, 1996). Additional glycosidase activities have been reported for *Leishmania*, including α-glucosidase and α-amylase (Jacobson and Schlein, 2001), consistent with the observation that sandfly sugar meals are also rich in starch (Schlein and Jacobson, 1999). Attempts to define the relative contribution of these glycolytic enzymes to parasite growth and development in the fly would again benefit from the generation of mutants specifically deficient in one or more of these enzymes. Thus far the only mutant lines relating to sugar metabolism that have been generated in *Leishmania* are the *L. mexicana* mutants for which the gene family of glucose transporters (LmGT) were knocked out by targeted gene replacement (Burchmore *et al.*, 2003). Each of the three LmGT isoforms is expressed in amastigotes and promastigotes, and as has been reported for some mammalian glucose transporters, may have a sensing function in addition to acquiring substrates for metabolism. When passed through a competent vector, *Lu. longipalpis*, the LmGT-knockout parasites were significantly delayed in their ability to re-populate the gut following blood meal excretion, and there was a log-fold reduction in the peak concentration of parasites, including metacyclics, that accumulated in the thoracic midgut.

Molecules promoting transmission by bite

The presence of a large number of metacyclic promastigotes in the anterior aspects of the gut may not in itself be a sufficient condition for transmission by bite. The prevailing view is that efficient transmission also involves the formation of a biological plug that impairs the intake of blood (Killick-Kendrick *et al.*, 1977; Killick-Kendrick and Molyneux, 1981). This is thought to promote regurgitation of infective promastigotes present in the foregut or contained within the plug itself. One element of the plug is undoubtedly the mass of parasites themselves; however, another is the gel-like substance referred to earlier, secreted by the parasites in the thoracic midgut, and termed promastigote secretory gel (PSG). This gel-like matrix was shown by immuno-EM in *Lu. longipalpis* infected with *L. mexicana* or in *P. papatasi* infected with *L. major*, to be immnologically identical to the fPPG that is produced by these promastigotes *in vitro* (Stierhof *et al.*, 1999). In *Lu. Longipalpis*, the size of the plug was found to impose a strong negative effect on feeding success, thought to be due to mechanical dysfunction of the stomodeal valve forced open by the pressure of the plug (Rogers *et al.*, 2002). Damage to the valve itself has been ascribed to the action of *Leishmania* chitinases (Schlein *et al.*, 1992). Either scenario would result in the valve remaining open, facilitating the regurgitation and egestion of metacyclic promastigotes from the fly. In addition to its role in the formation of the plug, small amounts of the PSG containing fPPG were shown to be egested along with the metacyclic promastigotes and to enhance the development of cutaneous disease (Rogers *et al.*, 2004). Recently, immunization with *L. mexicana* PSG or with a chemically-defined synthetic vaccine containing the glycans found in *L. mexicana* PSG, provided partial protection against challenge by the bite of infected sandflies, but not against needle challenge (Rogers *et al.*, 2006).

Concluding comments

This chapter has reviewed the potential barriers to parasite growth, differentiation and migration that exist within the midguts of phlebotomine sandflies, and the molecular adaptations that the parasite has evolved to permit the development of transmissible infections. As these molecules will in some cases (e.g. LPG) underlie the basis of species-restricted vector competence, their characterization should permit predictions as to the identity of transmitting flies within endemic or emerging disease foci containing a diversity of vector species. For example, the low abundance of galactosylated side-chain repeats on LPG from West African *L. major* isolates (Turco *et al.*, unpublished) implicates vector species other than *P. papatasi* in their transmission, and strengthens the argument that *P. duboscqi*, which is permissive to *L. major* parasites bearing poorly substituted LPGs, is the natural vector of these strains. Similarly, the presence of *L. tropica* within the *L. donovani* endemic region of Bihar (Sacks *et al.*, 1995a), would be predicted to be transmitted by the more broadly permissive *P. argentipes* and not *P. papatasi*, though both species are intensively distributed in the biotope.

The consideration of parasite and vector molecules essential to transmission can of course also lead to the identification of target antigens for transmission-blocking vaccines. A number of preliminary experimental findings have been encouraging with regard to the ability of antibodies raised against sandfly or *Leishmania* molecules to inhibit normal parasite development in the vector midgut. When *P. duboscqi* were fed on blood from animals immunized with both flagella and sandfly gut antigens, development of *L. major* was significantly inhibited (Mbati *et al.*, 2000). When *P. argentipes* were fed *L. donovani* amastigotes in a suspension containing immune sera from hamsters exposed to sandfly bites, the anti-saliva antibodies appeared to promote fly mortality and reduced the migration of parasites in the surviving females (Ghosh and Mukhopadhyay, 1998). In transmission-blocking studies designed to specifically target the molecules involved in midgut attachment, *P. duboscqi* fed on monoclonal anti-LPG antibodies (Tonui *et al.*, 2004), or *P. papatasi* fed on polyclonal antibodies against PpGalec (Kamhawi *et al.*, 2004), were in each case compromised in their ability to support the early growth and/or persistence of *L. major* following blood meal excretion. In no case have these experimental studies been validated by demonstrating that the infective blood meal and transmission-blocking antibodies can be simultaneously obtained following engorgement on an immunized, infected host. Furthermore, as a practical approach to leishmaniasis control, transmission-blocking vaccines can only be applied to transmission cycles involving human or canine reservoirs of infection, e.g. *L. donovani*, *L. infantum*, and *L. tropica*. One great advantage of the approach is that, in contrast to conventional vaccination strategies that require cellular immune responses for protection in the mammalian host and for which no safe and durable non-living vaccine exists, antibodies that confer transmission-blocking effects maybe elicited by defined antigens and maintained over a long period of time. The completion of the *Leishmania* genome, in conjunction with the anticipated completion of the genomes of important vector species (Dillon *et al.*, 2006), should greatly facilitate the search for other molecules involved in parasite-sandfly interactions and their cloning as potential target antigens for transmission-blocking vaccines.

References

Adler, S. (1938). Factors determining the behaviour of *Leishmania* sp. in sandflies. Harefuah *14*, 1–6.

Adler, S., Theodor, O., Witenberg, G. (1938). Investigations on Mediterranean kala-azar. 1. A study of leishmaniasis in Canea. Proc. Roy Soc. Lond., Series B *125*, 491–516.

Adler, S. (1964). *Leishmania*. Advances in Parasitology 2, 35–96.

Adler, S.a.T., O, 1927. The behaviour of cultures of *Leishmania* sp. in *Phlebotomus papatasi*. Annals of Tropical Medicine and Parasitology *21*, 111–134.

Adler, S.a.T., O., 1957. Transmission of disease agents by phlebotomine sandflies. Annu. Rev. Entomol. 2, 203–226.

Arias, J., Freitas, R.A.d. (1978). Sôbre os vetores de leishmaniose cutánea na Amazónia central do Brasil. 2. Incidéncia de flagelados em flebótomos selváticos. Acta Amazónica 8, 387–396.

Ashford, R.W., Bray, M.A., Hutchinson, M.P., Bray, R.S., 1973. The epidemiology of cutaneous leishmaniasis in Ethiopia. Trans R Soc. Trop Med. Hyg 67, 568–601.

Ashford, R.W., 1977. The comparative ecology of *Leishmania aethiopica*. Colloques Internationaux du Centre National de la Recherche Scientifique *239*, 235–240.

Azizi, K., Rassi, Y., Javadian, E., Motazedian, M.H., Rafizadeh, S., Yaghoobi Ershadi, M.R., Mohebali, M. (2006). Phlebotomus (Paraphlebotomus) alexandri: a probable vector of *Leishmania infantum* in Iran. Ann. Trop Med. Parasitol. *100*, 63–68.

Barros, V.C., Oliveira, J.S., Melo, M.N., Gontijo, N.F. (2006). *Leishmania amazonensis*: chemotaxic and osmotaxic responses in promastigotes and their probable role in development in the phlebotomine gut. Exp. Parasitol. *112*, 152–157.

Beach, R., Kiilu, G., Hendricks, L., Oster, C., Leeuwenburg, J. (1984). Cutaneous leishmaniasis in Kenya: transmission of *Leishmania major* to man by the bite of a naturally infected Phlebotomus duboscqi. Trans R Soc. Trop Med. Hyg *78*, 747–751.

Beach, R., Kiilu, G., Leeuwenburg, J. (1985). Modification of sandfly biting behavior by *Leishmania* leads to increased parasite transmission. Am. J. Trop Med. Hyg *34*, 278–282.

Blackburn, K., Wallbanks, K.R., Molyneux, D.H., Lavin, D.R., Winstanley, S.L. (1988). The peritrophic membrane of the female sandfly *Phlebotomus papatasi*. Ann. Trop Med. Parasitol. *82*, 613–619.

Blum, J.J., Opperdoes, F.R. (1994). Secretion of sucrase by *Leishmania donovani*. J. Eukaryot Microbiol. *41*, 228–231.

Borovsky, D., Schlein, Y. (1987). Trypsin and chymotrypsin-like enzymes of the sandfly *Phlebotomus papatasi* infected with *Leishmania* and their possible role in vector competence. Med. Vet. Entomol *1*, 235–242.

Boulanger, N., Lowenberger, C., Volf, P., Ursic, R., Sigutova, L., Sabatier, L., Svobodova, M., Beverley, S.M., Spath, G., Brun, R., Pesson, B., Bulet, P. (2004). Characterization of a defensin from the sandfly *Phlebotomus duboscqi* induced by challenge with bacteria or the protozoan parasite *Leishmania major*. Infect. Immun. *72*, 7140–7146.

Bray, R.S. (1983). Leishmania: chemotaxic responses of promastigotes and macrophages in vitro. J. Protozool *30*, 322–329.

Burchmore, R.J., Rodriguez-Contreras, D., McBride, K., Merkel, P., Barrett, M.P., Modi, G., Sacks, D., Landfear, S.M. (2003). Genetic characterization of glucose transporter function in *Leishmania mexicana*. Proc. Natl. Acad. Sci. USA *100*, 3901–3906.

Butcher, B.A., Turco, S.J., Hilty, B.A., Pimenta, P.F., Panunzio, M., Sacks, D.L. (1996). Deficiency in beta*1*, 3-galactosyltransferase of a *Leishmania major* lipophosphoglycan mutant adversely influences the Leishmania-sandfly interaction. J. Biol. Chem. *271*, 20573–20579.

Chaniotis, B.N., Correa, M.A., Tesh, R.B., Johnson, K.M., 1974. Horizontal and vertical movements of phlebotomine sandflies in a Panamanian rain forest. J. Med. Entomol *11*, 369–375.

Charlab, R., Tesh, R.B., Rowton, E.D., Ribeiro, J.M. (1995). *Leishmania amazonensis*: sensitivity of different promastigote morphotypes to salivary gland homogenates of the sandfly *Lutzomyia longipalpis*. Exp. Parasitol. *80*, 167–175.

Chippaix, J.P., Pajot, F.X., Barbier, D. (1984). Leishmaniasis in French Guinea. 5. Further data on the ecology of the vector in the cacao forest village. ORSTOM Ser Entomol Med. Parasitol. *22*, 313–318.

istensen, H.A., Herrer, A., 1973. Attractiveness of sentinel animals to vectors of leishmaniasis in Panama. Am. J. Trop Med. Hyg *22*, 578–584.

istensen, H.A., Fairchild, G.B., Herrer, A., Johnson, C.M., Young, D.G., de Vasquez, A.M. (1983). The ecology of cutaneous leishmaniasis in the Republic of Panama. J. Med. Entomol *20*, 463–484.

Cihakova, J., Volf, P. (1997). Development of different *Leishmania major* strains in the vector sandflies *Phlebotomus papatasi* and *P. duboscqi*. Ann. Trop Med. Parasitol. *91*, 267–279.

Courret, N., Prina, E., Mougneau, E., Saraiva, E.M., Sacks, D.L., Glaichenhaus, N., Antoine, J.C. (1999). Presentation of the *Leishmania* antigen LACK by infected macrophages is dependent upon the virulence of the phagocytosed parasites. Eur. J. Immunol. *29*, 762–773.

Cunningham, M.L., Titus, R.G., Turco, S.J., Beverley, S.M. (2001). Regulation of differentiation to the infective stage of the protozoan parasite *Leishmania major* by tetrahydrobiopterin. Science *292*, 285–287.

Cuvillier, A., Miranda, J.C., Ambit, A., Barral, A., Merlin, G. (2003). Abortive infection of *Lutzomyia longipalpis* insect vectors by aflagellated LdARL-3A-Q70L overexpressing *Leishmania amazonensis* parasites. Cell Microbiol. 5, 717–728.

Daba, S., Mansour, N.S., Youssef, F., Shanbaky, N.M., Shehata, M., and el Sawaf, B.M., 1997a. Vector-host–parasite inter-relationships in leishmaniasis. II Influence ofblood meal from natural vertebrate hosts with or without *Leishmania infantum* and *L. major* on thee proteolytic activity in the gut of *Phlebotomus langeroni*. J. Egypt Soc. Parasitol. *27*, 639–649.

Daba, S., Mansour, N.S., Youssef, F.G., Shanbaky, N.M., Shehata, M.G., and El Sawaf, B.M., 1997b. Vector-host–parasite inter-relationships in Leishmaniasis. III. Impact of blood meal from natural vertebrate hosts on the survival and the development of *Leishmania infantum* and *L. major* in *Phlebotomus langeroni*. J. Egypt. Soc. Parasitol. *27*, 781–794.

Davidson, I.H. (1987). New species of the subgenus Sergentomyia (Diptera: Psychodidae: Phlebotominae) from southern Africa. J. Ent. Soc. S. Africa *50*, 343–349.

Davies, C.R., Cooper, A.M., Peacock, C., Lane, R.P., Blackwell, J.M. (1990). Expression of LPG and GP63 by different developmental stages of *Leishmania major* in the sandfly *Phlebotomus papatasi*. Parasitology 101 Pt 3, 337–343.

Davies, C.R., Fernandez, M., Paz, L., Roncal, N., Llanos-Cuentas, A. (1993). *Lutzomyia verrucarum* can transmit *Leishmania peruviana*, the aetiological agent of Andean cutaneous leishmaniasis. Trans R Soc. Trop Med. Hyg *87*, 603–606.

de Queiroz, R.G., Vasconcelos Ide, A., Vasconcelos, A.W., Pessoa, F.A., de Sousa, R.N., David, J.R. (1994). Cutaneous leishmaniasis in Ceara state in northeastern Brazil: incrimination of Lutzomyia whitmani (Diptera: Psychodidae) as a vector of *Leishmania braziliensis* in baturite municipality. Am. J. Trop Med. Hyg *50*, 693–698.

de Souza, A., Ishikawa, E., Braga, R., Silveira, F., Lainson, R., Shaw, J. (1996). Psychodopygus complexus, a new vector of *Leishmania braziliensis* to humans in Para State, Brazil. Trans R Soc. Trop Med. Hyg *90*, 112–113.

Deane, L.M., Deane, M.P., 1964. [Visceral leishmaniasis in Central and South America]. Arq Hig Saude Publica *29*, 89–94.

Deane, L.M., Grimaldi, G. (1985). Leishmaniasis in Brazil. Elsevier, Amsterdam.

Dedet, J.P. (1979). Les leishmanioses en Afrique du nord. Bulletin de l'Institut Pasteur *77*, 413–434.

Descoteaux, A., Luo, Y., Turco, S.J., Beverley, S.M. (1995). A specialized pathway affecting virulence glycoconjugates of *Leishmania*. Science *269*, 1869–1872.

Dhanda, V., Shetty, P.S., Dhiman, R.C. (1983). Studies on phlebotomine sandflies as vectors of kala-azar in Bihar. In Proceedings of the Indo-UK Workshop on Leishmaniasis, ICMR, pp. 128–137.

Dillon, R.J., Lane, R.P., 1993a. Influence of *Leishmania* infection on blood-meal digestion in the sandflies *Phlebotomus papatasi* and *P. langeroni*. Parasitol. Res. *79*, 492–496.

Dillon, R.J., Lane, R.P., 1993b. Blood meal digestion in the midgut of *Phlebotomus papatasi* and *Phlebotomus langeroni*. Med. Vet. Entomol. *7*, 225–232.

Dillon, R.J., el-Kordy, E. (1997). Carbohydrate digestion in sandflies: alpha-glucosidase activity in the midgut of *Phlebotomus langeroni*. Comp. Biochem. Physiol. B Biochem. Mol. Biol. *116*, 35–40.

Dillon, R.J., Lane, R.P. (1999). Detection of *Leishmania* lipophosphoglycan binding proteins in the gut of the sandfly vector. Parasitology *118*, 27–32.

Dillon, R.J., Ivens, A.C., Churcher, C., Holroyd, N., Quail, M.A., Rogers, M.E., Soares, M.B., Bonaldo, M.F., Casavant, T.L., Lehane, M.J., Bates, P.A. (2006). Analysis of ESTs from *Lutzomyia longipalpis* sandflies and their contribution toward understanding the insect-parasite relationship. Genomics *88*, 831–840.

Dobson, D.E., Scholtes, L.D., Valdez, K.E., Sullivan, D.R., Mengeling, B.J., Cilmi, S., Turco, S.J., Beverley, S.M. (2003). Functional identification of galactosyltransferases (SCGs) required for species-specific modifications of the lipophosphoglycan adhesin controlling *Leishmania major*–sandfly interactions. J. Biol. Chem. *278*, 15523–15531.

el Sawaf, B., Kassem, H.A., el Said, S. (1985). Description of the hitherto unknown female of *Phlebotomus langeroni* (Diptera: Psychodidae). J. Med. Entomol *22*, 312–314.

Fairchild, G.B., Theodor, O., 1971. On Lutzomyia flaviscutellata (Mangabeira) and l. Olmeca (Vargas and Diaz-Najera) (Diptera: Psychodidae). J. Med. Entomol *8*, 153–159.

Feliciangeli, M.D., Reyes, R.M., Limongi, J.E. (1988). Natural infection of *Lutzomyia ovallesi* (Diptera:

Psychodidae) with parasites of the *Leishmania braziliensis* complex in a restricted focus of cutaneous leishmaniasis in northern Venezuela. Mem Inst Oswaldo Cruz *83*, 393–394.

Feliciangeli, M.D., Rabinovich, J. (1998). Abundance of *Lutzomyia ovallesi* but not Lu. gomezi (Diptera: Psychodidae) correlated with cutaneous leishmaniasis incidence in north-central Venezuela. Med. Vet. Entomol *12*, 121–131.

Feng, L.C., 1951. The role of the peritrophicmembrane in *Leishmania* and trypanosome infection of sandflies. Pek. Nat. His. Bull. *19*, 327–334.

Forattini, O.P., Pattoli, D.B., Rabello, E.X., Ferreira, O.A., 1972. [Natural infection of phlebotominae in an enzootic focus of cutaneous leishmaniasis in Sao Paulo State, Brazil]. Rev. Saude Publica 6, 431–433.

Foster, W.A., 1972. Studies on leishmaniasis in Ethiopia. V. Distribution, infections and assessment of vector potential of *Phlebotomus longipes* (Diptera: Psychodidae). Ann. Trop Med. Parasitol. 66, 445–455.

Gebre-Michael, T., Lane, R.P. (1996). A new sandfly species, *Phlebotomus* (*Larroussius*) *ashfordi* (Diptera, Psychodidae) from Ethiopia, previously confused with *P.* (*l.*). *aculeatus*. Ann. Trop Med. Parasitol. *90*, 523–531.

Ghosh, K.N., Mukhopadhyay, J. (1998). The effect of anti-sandfly saliva antibodies on *Phlebotomus argentipes* and *Leishmania donovani*. Int. J. Parasitol. *28*, 275–281.

Gontijo, N.F., Melo, M.N., Riani, E.B., Almeida-Silva, S., Mares-Guia, M.L. (1996). Glycosidases in *Leishmania* and their importance for *Leishmania* in phlebotomine sandflies with special reference to purification and characterization of a sucrase. Exp. Parasitol. *83*, 117–124.

Gontijo, N.F., Almeida-Silva, S., Costa, F.F., Mares-Guia, M.L., Williams, P., Melo, M.N. (1998). *Lutzomyia longipalpis*: pH in the gut, digestive glycosidases, and some speculations upon *Leishmania* development. Exp. Parasitol.*90*, 212–219.

Gossage, S.M., Rogers, M.E., Bates, P.A. (2003). Two separate growth phases during the development of *Leishmania* in sandflies: implications for understanding the life cycle. Int. J. Parasitol. 33, 1027–1034.

Guan, L.R., Xu, Y.X., Li, B.S., Dong, J. (1986). The role of Phlebotomus alexandri Sinton, 1928 in the transmission of kala-azar. Bull World Health Organ *64*, 107–112.

Hailu, A., Balkew, M., Berhe, N., Meredith, S.E., Gemetchu, T. (1995). Is Phlebotomus (Larroussius) orientalis a vector of visceral leishmaniasis in south-west Ethiopia? Acta Trop. *60*, 15–20.

Hajmova, M., Chang, K.P., Kolli, B., Volf, P. (2004). Downregulation of gp63 in *Leishmania amazonensis* reduces its early development in *Lutzomyia longipalpis*. Microbes Infect. 6, 646–649.

Herrer, A. (1982). [Lutzomyia peruensis Shannon, *1929*, possible natural vector of uta (cutaneous leishmaniasis)]. Rev. Inst Med. Trop Sao Paulo *24*, 168–172.

Heyneman, D., 1963. Comparison of experimental *Leishmania donovani* infections in *Phlebotomus papatasi* with natural infections found in man-baited *P. orientalis* captured in a kala-azar endemic region of the Sudan. Am. J. Trop. Med. Hyg. *12*, 725–740.

Hindle, E., 1931. The development of various strains of *Leishmania* in Chinese sandflies. Proc. Roy. Soc. Ser. B *108*, 366–383.

Ingram, G.A., Molyneux, D.H. (1991). Insect lectins: Role in parasite–vector interactions. Sigma Chemical Company, St. Louis, MO, USA.

Jacobson, R.L., Schlein, Y. (2001). *Phlebotomus papatasi* and *Leishmania major* parasites express alpha-amylase and alpha-glucosidase. Acta Trop *78*, 41–49.

Jarvis, E.K., Rutledge, L.C. (1992). Laboratory observations on mating and leklike aggregations in *Lutzomyia longipalpis* (Diptera: Psychodidae). J. Med. Entomol *29*, 171–177.

Johnson, P.T., 1961. Autogeny in Panamanian *Phlebotomus* sandflies (Diptera: Psychodidae). Ann. Ent Soc. Amer *54*, 116–118.

Johnson, P.T., McConnell, E., Hertig, M., 1963. Natural infections of leptomonad flagellates in Panamanian *Phlebotomus* sandflies. Exp. Parasitol. *14*, 107–122.

Johnson, R.N., Ngumbi, P.M., Robert, L.L., Anjili, C.O., Killick-Kendrick, R., Meredith, S.E. (1993). Phlebotomine sandflies of Kenya (Diptera: Psychodidae). II. Phlebotomus aculeatus as a probable vector of *Leishmania tropica* s.l. Ann. Trop Med. Parasitol. *87*, 541–543.

Joshi, M.B., Rogers, M.E., Shakarian, A.M., Yamage, M., Al-Harthi, S.A., Bates, P.A., Dwyer, D.M. (2005). Molecular characterization, expression, and in vivo analysis of LmexCht*1*, the chitinase of the human pathogen, *Leishmania mexicana*. J. Biol. Chem. *280*, 3847–3861.

Joshi, P.B., Sacks, D.L., Modi, G., McMaster, W.R. (1998). Targeted gene deletion of *Leishmania major* genes encoding developmental stage-specific leishmanolysin (GP63). Mol. Microbiol. *27*, 519–530.

Joshi, P.B., Kelly, B.L., Kamhawi, S., Sacks, D.L., McMaster, W.R. (2002). Targeted gene deletion in *Leishmania major* identifies leishmanolysin (GP63) as a virulence factor. Mol. Biochem. Parasitol. *120*, 33–40.

Kamhawi, S., Modi, G.B., Pimenta, P.F., Rowton, E., Sacks, D.L. (2000). The vectorial competence of phlebotomus sergenti is specific for leishmania tropica and is controlled by species-specific, lipophosphoglycan-mediated midgut attachment [In Process Citation]. Parasitology *121*, 25–33.

Kamhawi, S., Ramalho-Ortigao, M., Pham, V.M., Kumar, S., Lawyer, P.G., Turco, S.J., Barillas-Mury, C., Sacks, D.L., Valenzuela, J.G. (2004). A role for insect galectins in parasite survival. Cell *119*, 329–341.

Killick-Kendrick, R., Molyneux, D.H., Ashford, R.W., 1974a. *Leishmania* in phlebotomid sandflies. I. Modifications of the flagellum associated with attachment to the mid-gut and oesophageal valve of the sandfly. Proc. R Soc. Lond B Biol. Sci. *187*, 409–419.

Killick-Kendrick, R., Molyneux, D.H., Ashford, R.W., 1974b. Ultrastructural observations on the attachment of *Leishmania* in the sandfly. Trans R Soc. Trop Med. Hyg *68*, 269.

Killick-Kendrick, R., Leaney, A.J., Ready, P.D., Molyneux, D.H., 1977. *Leishmania* in phlebotomid sandflies. IV.

The transmission of *Leishmania mexicana* amazonensis to hamsters by the bite of experimentally infected *Lutzomyia longipalpis*. Proc. R Soc. Lond B Biol. Sci. *196*, 105–115.

Killick-Kendrick, R., Molyneux, D.H., Leaney, A.J., Rioux, J.A. (1979). Aspects of the life-cycle of *Leishmania* in the sandfly. Prosveta, Belgrade, Yugoslavia.

Killick-Kendrick, R., Molyneux, D.H. (1981). Transmission of leishmaniasis by the bite of phlebotomine sandflies: possible mechanisms. Trans. R. Soc. Trop. Med. Hyg *75*, 152–154.

Killick-Kendrick, R., Rioux, J.A. (1981). The Cévennes focus of leishmaniasis in southern France and the biology of the vector Phlebtomomus ariasi. Society of Protozoologists Special Publication

Killick-Kendrick, R., Rioux, J.A., Bailly, M., Guy, M.W., Wilkes, T.J., Guy, F.M., Davidson, I., Knechtli, R., Ward, R.D., Guilvard, E., *et al.* (1984). Ecology of leishmaniasis in the south of France. 20. Dispersal of *Phlebotomus ariasi* Tonnoir, 1921 as a factor in the spread of visceral leishmaniasis in the Cevennes. Ann. Parasitol. Hum. Comp *59*, 555–572.

Killick-Kendrick, R. (1985). Some epidemiological consequences of the evolutionary fit between Leishmaniae and their phlebotomine vectors. Bull Soc. Pathol. Exot. Filiales *78*, 747–755.

Killick-Kendrick, R., Killick-Kendrick, M. (1987). Honeydew of aphids as a source of sugar for *Phlebotomus ariasi*. Med. Vet. Entomol *1*, 297–302.

Killick-Kendrick, R., 1990a. Phlebotomine vectors of the leishmaniases: a review. Med. Vet. Entomol *4*, 1–24.

Killick-Kendrick, R., 1990b. The life-cycle of *Leishmania* in the sandfly with special reference to the form infective to the vertebrate host. Ann. Parasitol. Hum. Comp *65*, 37–42.

Killick-Kendrick, R., Killick-Kendrick, M., Tang, Y. (1994). Anthroponotic cutaneous leishmaniasis in Kabul, Afghanistan: the low susceptibility of *Phlebotomus papatasi* to *Leishmania tropica*. Trans R Soc. Trop Med. Hyg *88*, 252–253.

Killick-Kendrick, R., Killick-Kendrick, M., Tang, Y. (1995). Anthroponotic cutaneous leishmaniasis in Kabul, Afghanistan: the high susceptibility of Phlebotomus sergenti to *Leishmania tropica*. Trans R Soc. Trop Med. Hyg *89*, 477.

Killick-Kendrick, R. (1999). The biology and control of phlebotomine sandflies [In Process Citation]. Clin. Dermatol *17*, 279–289.

Knuepfer, E., Stierhof, Y.D., McKean, P.G., Smith, D.F. (2001). Characterization of a differentially expressed protein that shows an unusual localization to intracellular membranes in *Leishmania major*. Biochem. J. *356*, 335–344.

Kreutzer, R.D., Corredor, A., Grimaldi, G., Jr., Grogl, M., Rowton, E.D., Young, D.G., Morales, A., McMahon-Pratt, D., Guzman, H., Tesh, R.B. (1991). Characterization of *Leishmania colombiensis* sp. N. (Kinetoplastida: Trypanosomatidae), a new parasite infecting humans, animals, and phlebotomine sandflies in Colombia and Panama. Am. J. Trop Med. Hyg *44*, 662–675.

Lainson, R., Shaw, J.J., 1968. Leishmaniasis in Brazil: I. Observations on enzootic rodent leishmaniasis – incrimination of *Lutzomyia flaviscutellata* (Mangabeira) as the vector in the Lower Amazonian Basin. Trans R Soc. Trop Med. Hyg *62*, 385–395.

Lainson, R., Shaw, J.J., Ward, R.D., Fraiha, H., 1973. Leishmaniasis in Brazil. IX. Considerations on the *Leishmania braziliensis* complex. Importance of sandflies of the genus Psychodopygus (Mangabeira) in the transmission of *L. braziliensis braziliensis* in north Brazil. Trans R Soc. Trop Med. Hyg *67*, 184–196.

Lainson, R., Ward, R.D., Shaw, J.J., 1976. Letter: Cutaneous leishmaniasis in North Brazil: Lutzomyia anduzei as a major vector. Trans R Soc. Trop Med. Hyg *70*, 171–172.

Lainson, R., Ward, R.D., Shaw, J.J., 1977. Experimental transmission of *Leishmania chagasi*, causative agent of neotropical visceral leishmaniasis, by the sandfly *Lutzomyia longipalpis*. Nature *266*, 628–630.

Lainson, R., Shaw, J.J., Ward, R.D., Ready, P.D., Naiff, R.D. (1979). Leishmaniasis in Brazil: XIII. Isolation of *Leishmania* from armadillos (*Dasypus novemcinctus*), and observations on the epidemiology of cutaneous leishmaniasis in north Para State. Trans. R. Soc. Trop Med. Hyg. *73*, 239–242.

Lainson, R., Shaw, J.J., Ready, P.D., Miles, M.A., Povoa, M. (1981). Leishmaniasis in Brazil: XVI. Isolation and identification of *Leishmania* species from sandflies, wild mammals and man in north Para State, with particular reference *to L. braziliensis guyanensis* causative agent of 'pian-bois'. Trans. R. Soc. Trop. Med. Hyg. *75*, 530–536.

Lainson, R. (1982). Leishmaniasis. CRC Press, Boca Raton, Florida.

Lainson, R., Shaw, J.J. (1987). Evolution, classification and geographical distribution. The Leishmaniasis in Biology and Medicine. Academic Press, London.

Lainson, R., Braga, R.R., De Souza, A.A., Povoa, M.M., Ishikawa, E.A., Silveira, F.T. (1989). *Leishmania (Viannia) shawi* sp. n., a parasite of monkeys, sloths and procyonids in Amazonian Brazil. Ann. Parasitol. Hum. Comp *64*, 200–207.

Lainson, R., Shaw, J.J., Souza, A.A., Silveira, F.T., Falqueto, A. (1992). Further observations on *Lutzomyia ubiquitalis* (Psychodidae: Phlebotominae), the sandfly vector of *Leishmania (Viannia) lainsoni*. Mem Inst Oswaldo Cruz *87*, 437–439.

Lane, R., Fritz, G.N. (1986). The differentiation of the leishmaniasis vector *Phlebotomus papatasi* from the suspected vector *Phlebotomus bergeroti* (Diptera: Psychodidae). Systematic Entomology *11*, 439–445.

Lane, R.P. (1993). Sandflies (Phlebotominae). Chapman & Hall, London.

Lang, T., Warburg, A., Sacks, D.L., Croft, S.L., Lane, R.P., Blackwell, J.M. (1991). Transmission and scanning EM-immunogold labeling of *Leishmania major* lipophosphoglycan in the sandfly *Phlebotomus papatasi*. Eur. J. Cell Biol. 55, 362–372.

Lawyer, P.G., Young, D.G., Butler, J.F., Akin, D.E. (1987). Development of *Leishmania mexicana* in *Lutzomyia diabolica* and *Lutzomyia shannoni* (Diptera: Psychodidae). J. Med. Entomol *24*, 347–355.

Lawyer, P.G., Ngumbi, P.M., Anjili, C.O., Odongo, S.O., Mebrahtu, Y.B., Githure, J.I., Koech, D.K., Roberts, C.R. (1990). Development of *Leishmania major* in *Phlebotomus duboscqi* and *Sergentomyia schwetzi* (Diptera: Psychodidae). Am. J. Trop Med. Hyg *43*, 31–43.

Lawyer, P.G., Mebrahtu, Y.B., Ngumbi, P.M., Mwanyumba, P., Mbugua, J., Kiilu, G., Kipkoech, D., Nzovu, J., Anjili, C.O. (1991). *Phlebotomus guggisbergi* (Diptera: Psychodidae), a vector of *Leishmania tropica* in Kenya. Am. J. Trop Med. Hyg *44*, 290–298.

Lawyer, P.G., Perkins, P.V. (2000). Leishmaniasis and Trypanosomiasis. Kluwer Academic Publishers, Boston.

Le Pont, F., Desjeux, P. (1986). Leishmaniasis in Bolivia. II. The involvement of *Psychodopygus yucumensis* and *Psychodopygus llanosmartinsi* in the selvatic transmission cycle of *Leishmania braziliensis* braziliensis in a lowland subandean region. Mem Inst Oswaldo Cruz *81*, 311–318.

Leger, N., Saratsiotis, A., Pesson, B., Leger, P. (1979). [Leishmaniasis in Greece. Results of an entomological survey during June 1977 (author's transl)]. Ann. Parasitol. Hum. Comp *54*, 11–29.

Lewis, D.J. (1982). A taxonomic review of the genus *Phlebtomus* (Diptera: Psychodidae). Bull. Br. Mus. Nat. Hist (Entomol) *45*, 121–209.

Lewis, D.J., Ward, R.D. (1987). Transmission and vectors. Academic Press, London.

Lira, R., Mendez, S., Carrera, L., Jaffe, C., Neva, F., Sacks, D. (1998). *Leishmania tropica*: the identification and purification of metacyclic promastigotes and use in establishing mouse and hamster models of cutaneous and visceral disease. Exp. Parasitol. *89*, 331–342.

McConville, M.J., Turco, S.J., Ferguson, M.A., Sacks, D.L. (1992). Developmental modification of lipophosphoglycan during the differentiation of *Leishmania major* promastigotes to an infectious stage. Embo J. *11*, 3593–3600.

McConville, M.J., Ferguson, M.A. (1993). The structure, biosynthesis and function of glycosylated phosphatidylinositols in the parasitic protozoa and higher eukaryotes. Biochem. J. *294*, 305–324.

McHugh, C.P., Grogl, M., Kreutzer, R.D. (1993). Isolation of *Leishmania mexicana* (Kinetoplastida: Trypanosomatidae) from Lutzomyia anthophora (Diptera: Psychodidae) collected in Texas. J. Med. Entomol *30*, 631–633.

McKean, P.G., Delahay, R., Pimenta, P.F., Smith, D.F. (1997). Characterisation of a second protein encoded by the differentially regulated LmcDNA16 gene family of *Leishmania major*. Mol. Biochem. Parasitol. *85*, 221–231.

Magill, A.J. (1995). Epidemiology of the leishmaniases. Dermatol. Clin. *13*, 505–523.

Mahoney, A.B., Sacks, D.L., Saraiva, E., Modi, G., Turco, S.J. (1999). Intra-species and stage-specific polymorphisms in lipophosphoglycan structure control *Leishmania donovani*-sandfly interactions. Biochemistry *38*, 9813–9823.

Maroli, M., Gramiccia, M., Gradoni, L. (1987). Natural infection of Phlebotomus perfiliewi with *Leishmania infantum* in a cutaneous leishmaniasis focus of the Abruzzi region, Italy. Trans R Soc. Trop Med. Hyg *81*, 596–598.

Mbati, P.A., Anjili, C.O., Odongo, S., Ogaja, P., Tonui, W. (2000). *Leishmania major* infections in *Phlebotomus duboscqi* fed on murine models immunized with *L. major* subcellular antigens and sandfly gut antigens. Onderstepoort J. Vet. Res. *67*, 57–63.

Mendez, S., Fernandez-Perez, F.J., de la Fuente, C., Cuquerella, M., Gomez-Munoz, M.T., Alunda, J.M. (1999). Partial anaerobiosis induces infectivity of *Leishmania infantum* promastigotes. Parasitol. Res. *85*, 507–509.

Mengeling, B.J., Turco, S.J. (1998). Microbial glycoconjugates. Curr. Opin. Struct Biol. *8*, 572–577.

Molyneux, D.H., 1977. Vector relationships in the Trypanosomatidae. Adv. Parasitol. *15*, 1–82.

Morales, A., Corredor, A., Cáceres, E., Ibagos, A.L., de Rodriques, C. (1981). Aislamiento de tres cepas de *Leishmania* a partir de Lutzomyia trapidoi en Colombia. Biomédica *1*, 198–207.

Mottram, J.C., Brooks, D.R., Coombs, G.H. (1998). Roles of cysteine proteinases of trypanosomes and *Leishmania* in host–parasite interactions. Curr. Opin. Microbiol. *1*, 455–460.

Mukhopadhyay, J., Ghosh, K. (1999). Vector potential of *Phlebotomus duboscqi* and *P. papatasi*: a comparison of feeding behaviour, reproductive capacity and experimental infection with *Leishmania major*. Ann. Trop Med. Parasitol. *93*, 309–318.

Nieves, E., Pimenta, P.F. (2000). Development of *Leishmania* (*Vianna*) *braziliensis* and *Leishmania* (*Leishmania*) *amazonensis* in the sandfly *Lutzomyia migonei* (Diptera: Psychodidae). J. Med. Entomol *37*, 134–140.

Oliveira, J.S., Melo, M.N., Gontijo, N.F. (2000). A sensitive method for assaying chemotaxic responses of *Leishmania* promastigotes. Exp. Parasitol. *96*, 187–189.

Palanova, L., Volf, P. (1997). Carbohydrate-binding specificities and physico-chemical properties of lectins in various tissue of phlebotominae sandflies. Folia Parasitol. *44*, 71–76.

Perfil'ev, P.P., 1966. Diptera: Family Phlebotomidae. [In Russian]. Israel Program of Scientific Translations, Jerusalem.

Perkins, P.V., Githure, J.I., Mebrahtu, Y., Kiilu, G., Anjili, C., Ngumbi, P.S., Nzovu, J., Oster, C.N., Whitmire, R.E., Leeuwenburg, J., *et al.* (1988). Isolation of *Leishmania donovani* from Phlebotomus martini in Baringo district, Kenya. Trans R Soc. Trop Med. Hyg *82*, 695–700.

Pifano, F., Romero, J., Alvarez, A., 1973. Comprobación de une cepa de *Leishmania dermotropa* en *Phlebotomus flaviscutellata* Mangabeira, 1942 de la Sierra Palima, Territoria Federal Amazonás, Venezuela. Archivos Venezolanos de Medicina Tropical y Parasitologia Médica 5, 145–167.

Pimenta, P.F., Turco, S.J., McConville, M.J., Lawyer, P.G., Perkins, P.V., Sacks, D.L. (1992). Stage-specific adhesion of *Leishmania* promastigotes to the sandfly midgut. Science *256*, 1812–1815.

Pimenta, P.F., Saraiva, E.M., Rowton, E., Modi, G.B., Garraway, L.A., Beverley, S.M., Turco, S.J., Sacks, D.L. (1994). Evidence that the vectorial competence of phlebotomine sandflies for different species of

Leishmania is controlled by structural polymorphisms in the surface lipophosphoglycan. Proc. Natl. Acad. Sci. USA *91*, 9155–9156.

Pimenta, P.F., Modi, G.B., Pereira, S.T., Shahabuddin, M., Sacks, D.L. (1997). A novel role for the peritrophic matrix in protecting *Leishmania* from the hydrolytic activities of the sandfly midgut. Parasitology *115*, 359–369.

Pinto-da-Silva, L.H., Camurate, M., Costa, K.A., Oliveira, S.M., da Cunha-e-Silva, N.L., Saraiva, E.M. (2002). *Leishmania* (*Viannia*) *braziliensis* metacyclic promastigotes purified using *Bauhinia purpurea* lectin are complement resistant and highly infective for macrophages in vitro and hamsters in vivo. Int. J. Parasitol. *32*, 1371–1377.

Pinto-da-Silva, L.H., Fampa, P., Soares, D.C., Oliveira, S.M., Souto-Padron, T., Saraiva, E.M. (2005). The 3A1-La monoclonal antibody reveals key features of *Leishmania* (*L*) *amazonensis* metacyclic promastigotes and inhibits procyclics attachment to the sandfly midgut. Int. J. Parasitol. 35, 757–764.

Quate, L.W., 1964. Phlebotomus sandflies of the Paloich area in the Sudan (Diptera, Psychodidae). J. Med. Entomol 39, 213–268.

Ramalho-Ortigao, J.M., Temporal, P., de Oliveira, S.M., Barbosa, A.F., Vilela, M.L., Rangel, E.F., Brazil, R.P., Traub-Cseko, Y.M. (2001). Characterization of constitutive and putative differentially expressed mRNAs by means of expressed sequence tags, differential display reverse transcriptase-PCR and randomly amplified polymorphic DNA-PCR from the sandfly vector *Lutzomyia longipalpis*. Mem Inst Oswaldo Cruz *96*, 105–111.

Ramalho-Ortigao, J.M., Kamhawi, S., Rowton, E.D., Ribeiro, J.M., Valenzuela, J.G. (2003). Cloning and characterization of trypsin- and chymotrypsin-like proteases from the midgut of the sandfly vector *Phlebotomus papatasi*. Insect Biochem. Mol. Biol. 33, 163–171.

Ramalho-Ortigao, J.M., Traub-Cseko, Y.M. (2003). Molecular characterization of Llchit*1*, a midgut chitinase cDNA from the leishmaniasis vector *Lutzomyia longipalpis*. Insect Biochem. Mol. Biol. 33, 279–287.

Ramalho-Ortigao, J.M., Kamhawi, S., Joshi, M.B., Reynoso, D., Lawyer, P.G., Dwyer, D.M., Sacks, D.L., Valenzuela, J.G. (2005). Characterization of a blood activated chitinolytic system in the midgut of the sandfly vectors *Lutzomyia longipalpis* and *Phlebotomus papatasi*. Insect Mol. Biol. *14*, 703–712.

Ramamoorthy, R., Donelson, J.E., Paetz, K.E., Maybodi, M., Roberts, S.C., Wilson, M.E. (1992). Three distinct RNAs for the surface protease gp63 are differentially expressed during development of *Leishmania donovani* chagasi promastigotes to an infectious form. J. Biol. Chem. *267*, 1888–1895.

Rangel, E.F., de Souza, N.A., Wermelinger, E.D., Barbosa, A.F. (1984). [Natural infection of *Lutzomyia intermedia* Lutz and Neiva, *1912*, in an endemic area of visceral leishmaniasis of Rio de Janeiro]. Mem Inst Oswaldo Cruz *79*, 395–396.

Ready, P. (2000). Sandfly evolution and its relationship to *Leishmania* transmission. Memorias do Instituto Oswaldo Cruz 95, 589–590.

Ready, P.D., Lainson, R., J.J., S., Souza, A.A. (1986). The ecology of Lutzomyia umbtatilis Ward and Fraiha (Diptera: Psychodidae), the major vector to man of *Leishmania braziliensis* guyanensis in north-eastern Amazonian, Brazil. Bull Entomol Soc. Am. *76*, 21–40.

Ribeiro, J.M. (1987). Vector salivation and parasite transmission. Mem Inst Oswaldo Cruz *82*, 1–3.

Ribeiro, J.M., Vachereau, A., Modi, G.B., Tesh, R.B. (1989). A novel vasodilatory peptide from the salivary glands of the sandfly *Lutzomyia longipalpis*. Science *243*, 212–214.

Ribeiro, J.M. (1995). Blood-feeding arthropods: live syringes or invertebrate pharmacologists? Infect. Agents Dis. *4*, 143–152.

Rogers, M.E., Chance, M.L., Bates, P.A. (2002). The role of promastigote secretory gel in the origin and transmission of the infective stage of *Leishmania mexicana* by the sandfly *Lutzomyia longipalpis*. Parasitology *124*, 495–507.

Rogers, M.E., Ilg, T., Nikolaev, A.V., Ferguson, M.A., Bates, P.A. (2004). Transmission of cutaneous leishmaniasis by sandflies is enhanced by regurgitation of fPPG. Nature *430*, 463–467.

Rogers, M.E., Sizova, O.V., Ferguson, M.A., Nikolaev, A.V., Bates, P.A. (2006). Synthetic glycovaccine protects against the bite of leishmania-infected sandflies. J. Infect. Dis. *194*, 512–518.

Rowton, E.D., de Mata, M., Rizzo, N., Porter, C.H., Navin, T.R. (1992). Isolation of *Leishmania braziliensis* from *Lutzomyia ovallesi* (Diptera:Psychodidae) in Guatemala. Am. J. Trop Med. Hyg *46*, 465–468.

Ryan, L., Lainson, R., Shaw, J.J. (1987). Leishmaniasis in Brazil. XXIV. Natural flagellate infections of sandflies (Diptera: Psychodidae) in Para State, with particular reference to the role of Psychodopygus wellcomei as the vector of *Leishmania braziliensis* braziliensis in the Serra dos Carajas. Trans R Soc. Trop Med. Hyg *81*, 353–359.

Sacks, D.L. (1989). Metacyclogenesis in *Leishmania* promastigotes. Exp. Parasitol. *69*, 100–103.

Sacks, D.L., Perkins, P.V. (1984). Identification of an infective stage of *Leishmania* promastigotes. Science *223*, 1417–1419.

Sacks, D.L., Perkins, P.V. (1985). Development of infective stage *Leishmania* promastigotes within phlebotomine sandflies. Am. J. Trop Med. Hyg *34*, 456–459.

Sacks, D.L., Hieny, S., Sher, A. (1985). Identification of cell surface carbohydrate and antigenic changes between noninfective and infective developmental stages of *Leishmania major* promastigotes. J. Immunol. *135*, 564–569.

Sacks, D.L., Brodin, T.N., Turco, S.J. (1990). Developmental modification of the lipophosphoglycan from *Leishmania major* promastigotes during metacyclogenesis. Mol. Biochem. Parasitol. *42*, 225–233.

Sacks, D.L., Kenney, R.T., Kreutzer, R.D., Jaffe, C.L., Gupta, A.K., Sharma, M.C., Sinha, S.P., Neva, F.A., Saran, R. (1995a). Indian kala-azar caused by *Leishmania tropica*. Lancet *345*, 959–961.

Sacks, D.L., Pimenta, P.F., McConville, M.J., Schneider, P., Turco, S.J. (1995b). Stage-specific binding of *Leishmania donovani* to the sandfly vector midgut is regulated by conformational changes in the abundant surface lipophosphoglycan. J. Exp. Med. *181*, 685–697.

Sacks, D.L., Modi, G., Rowton, E., Spath, G., Epstein, L., Turco, S.J., Beverley, S.M. (2000). The role of phosphoglycans in Leishmania-sandfly interactions. Proc. Natl. Acad. Sci. USA. *97*, 406–411.

Sacks, D.L. (2001). Leishmania-sandfly interactions controlling species-specific vector competence. Cell Microbiol. *3*, 189–196.

Sadlova, J., Hajmova, M., Volf, P. (2003). *Phlebotomus (Adlerius) halepensis* vector competence for *Leishmania major* and Le. tropica. Med. Vet. Entomol *17*, 244–250.

Saf'janova, V.M. (1979). Petricheva's work on the study of sandflies (Phlebotominae) and further work in that direction. [In Russian] Askhabad.

Saraiva, E.M., Pimenta, P.F., Brodin, T.N., Rowton, E., Modi, G.B., Sacks, D.L. (1995). Changes in lipophosphoglycan and gene expression associated with the development of *Leishmania major* in *Phlebotomus papatasi*. Parasitology *111*, 275–287.

Schlein, Y., Romano, H. (1986). *Leishmania major* and L. donovani: effects on proteolytic enzymes of *Phlebotomus papatasi* (Diptera, Psychodidae). Exp. Parasitol. *62*, 376–380.

Schlein, Y., Warburg, A. (1986). Phytophagy and the feeding cycle of *Phlebotomus papatasi* (Diptera: Psychodidae) under experimental conditions. J. Med. Entomol *23*, 11–15.

Schlein, Y., Schnur, L.F., Jacobson, R.L. (1990). Released glycoconjugate of indigenous *Leishmania major* enhances survival of a foreign *L. major* in *Phlebotomus papatasi*. Trans R Soc. Trop Med. Hyg *84*, 353–355.

Schlein, Y., Jacobson, R.L., Shlomai, J. (1991). Chitinase secreted by *Leishmania* functions in the sandfly vector. Proc. R Soc. Lond. B Biol. Sci. *245*, 121–126.

Schlein, Y., Jacobson, R.L., Messer, G. (1992). *Leishmania* infections damage the feeding mechanism of the sandfly vector and implement parasite transmission by bite. Proc. Natl. Acad. Sci. USA. *89*, 9944–9948.

Schlein, Y., Jacobson, R.L. (1994). Haemoglobin inhibits the development of infective promastigotes and chitinase secretion in *Leishmania major* cultures. Parasitology *109*, 23–28.

Schlein, Y., Jacobson, R.L. (1998). Resistance of *Phlebotomus papatasi* to infection with *Leishmania donovani* is modulated by components of the infective blood meal. Parasitology *117*, 467–473.

Schlein, Y., Jacobson, R.L. (1999). Sugar meals and longevity of the sandfly *Phlebotomus papatasi* in an arid focus of *Leishmania major* in the Jordan Valley. Med. Vet. Entomol *13*, 65–71.

Seccombe, A.K., Ready, P.D., Huddleston, L.M. (1992). A Catalogue of Old World Phlebotomine Sandflies (Diptera: Psychodidae, Phlebotominae). Cromwell Road, London SW7 5BD, U.K.

Shakarian, A.M., Dwyer, D.M. (2000). Pathogenic leishmania secrete antigenically related chitinases which are encoded by a highly conserved gene locus. Exp. Parasitol. *94*, 238–242.

Shatova, S.M., Shulga, M.A., Safjanova, V.M., and Avakjan, A.A. (1984). Comparative electron microscope study of *Leishmania major* and *L. tropica* and the experimental infection of the sandfly *Phlebotomus papatasi*. Parazitologia *18*, 154–158.

Shortt, H., Swaminath, C., 1928. The method of feeding of *Phlebotomus arpentipes* with relation to its bearing on the transmission of kala-azar. Indian J. Med. Res. 15.

Shortt, H.E., Smith, R., Swaminath, C., and Krishnan, K., 1931. Transmission of Indian kala-azar by the bite of *Phlebotomus argentipes*. Indian J. Med. Res. *18*, 1373–1375.

Soares, R.P., Macedo, M.E., Ropert, C., Gontijo, N.F., Almeida, I.C., Gazzinelli, R.T., Pimenta, P.F., Turco, S.J. (2002). *Leishmania chagasi*: lipophosphoglycan characterization and binding to the midgut of the sandfly vector *Lutzomyia longipalpis*. Mol. Biochem. Parasitol. *121*, 213–224.

Soares, R.P., Turco, S.J. (2003). *Lutzomyia longipalpis* (Diptera: Psychodidae: Phlebotominae): a review. An Acad. Bras Cienc *75*, 301–330.

Soares, R.P., Barron, T., McCoy-Simandle, K., Svobodova, M., Warburg, A., Turco, S.J. (2004). *Leishmania tropica*: intraspecific polymorphisms in lipophosphoglycan correlate with transmission by different Phlebotomus species. Exp. Parasitol. *107*, 105–114.

Soares, R.P., Cardoso, T.L., Barron, T., Araujo, M.S., Pimenta, P.F., Turco, S.J. (2005). *Leishmania braziliensis*: a novel mechanism in the lipophosphoglycan regulation during metacyclogenesis. Int. J. Parasitol. *35*, 245–253.

Spath, G.F., Beverley, S.M. (2001). A lipophosphoglycan-independent method for isolation of infective *Leishmania* metacyclic promastigotes by density gradient centrifugation. Exp. Parasitol. *99*, 97–103.

Spath, G.F., Lye, L.F., Segawa, H., Sacks, D.L., Turco, S.J., Beverley, S.M. (2003). Persistence without pathology in phosphoglycan-deficient *Leishmania major*. Science *301*, 1241–1243.

Stierhof, Y.D., Bates, P.A., Jacobson, R.L., Rogers, M.E., Schlein, Y., Handman, E., Ilg, T. (1999). Filamentous proteophosphoglycan secreted by *Leishmania* promastigotes forms gel-like three-dimensional networks that obstruct the digestive tract of infected sandfly vectors. Eur. J. Cell Biol. *78*, 675–689.

Svobodova, M., Volf, P., Killick-Kendrick, R. (1996). Agglutination of *Leishmania* promastigotes by midgut lectins from various species of phlebotomine sandflies. Ann. Trop Med. Parasitol. *90*, 329–336.

Tang, Y., Ward, R.D. (1998). Sugar feeding and fluid destination control in the phlebotomine sandfly *Lutzomyia longipalpis* (Diptera: Psychodidae). Med. Vet. Entomol *12*, 13–19.

Tonui, W.K., Ngumbi, P.M., Mpoke, S.S., Orago, A.S., Mbati, P.A., Turco, S.J., Mkoji, G.M. (2004). *Leishmania major-Phlebotomus duboscqi* interactions: inhibition of anti-LPG antibodies and characterisation of two proteins with shared epitopes. East African Medical Journal *81*, 97–103.

Vaidyanathan, R. (2004). *Leishmania* parasites (Kinetoplastida: Trypanosomatidae) reversibly inhibit visceral muscle contractions in hemimetabolous and holometabolous insects. J. Invertebr Pathol. *87*, 123–128.

Vaidyanathan, R. (2005). Isolation of a myoinhibitory peptide from *Leishmania major* (Kinetoplastida: Trypanosomatidae) and its function in the vector sandfly *Phlebotomus papatasi* (Diptera: Psychodidae). J. Med. Entomol 42, 142–152.

Villaseca, P., Llanos-Cuentas, A., Perez, E., Davies, C.R. (1993). A comparative field study of the relative importance of Lutzomyia peruensis and *Lutzomyia verrucarum* as vectors of cutaneous leishmaniasis in the Peruvian Andes. Am. J. Trop Med. Hyg *49*, 260–269.

Volf, P., Killick-Kendrick, R., Bates, P.A., Molyneux, D.H. (1994). Comparison of the haemagglutination activities in gut and head extracts of various species and geographical populations of phlebotomine sandflies. Ann. Trop Med. Parasitol. *88*, 337–340.

Volf, P., Killick-Kendrick, R. (1996). Post-engorgement dynamics of haemagglutination activity in the midgut of phlebotomine sandflies. Med. Vet. Entomol *10*, 247–250.

Volf, P., Kiewegova, A., Svobodova, M. (1998). Sandfly midgut lectin: effect of galactosamine on *Leishmania major* infections. Med. Vet. Entomol *12*, 151–154.

Volf, P., Svobodova, M., Dvorakova, E. (2001). Blood meal digestion and *Leishmania major* infections in Phlebotomus duboscqi: effect of carbohydrates inhibiting midgut lectin activity. Med. Vet. Entomol *15*, 281–286.

Volf, P., Hajmova, M., Sadlova, J., Votypka, J. (2004). Blocked stomodeal valve of the insect vector: similar mechanism of transmission in two trypanosomatid models. Int. J. Parasitol. *34*, 1221–1227.

Wallbanks, K.R., Ingram, G.A., Molyneux, D.H. (1986). The agglutination of erythrocytes and *Leishmania* parasites by sandfly gut extracts: evidence for lectin activity. Trop Med. Parasitol. *37*, 409–413.

Walters, L.L., Modi, G.B., Tesh, R.B., Burrage, T. (1987). Host-parasite relationship of *Leishmania mexicana* mexicana and Lutzomyia abonnenci (Diptera: Psychodidae). Am. J. Trop. Med. Hyg *36*, 294–314.

Walters, L.L., Chaplin, G.L., Modi, G.B., Tesh, R.B., 1989a. Ultrastructural biology of *Leishmania* (*Viannia*) *panamensis* (=*Leishmania braziliensis panamensis*) in *Lutzomyia gomezi* (Diptera: Psychodidae): a natural host–parasite association. Am. J. Trop Med. Hyg *40*, 19–39.

Walters, L.L., Modi, G.B., Chaplin, G.L., Tesh, R.B., 1989b. Ultrastructural development of *Leishmania chagasi* in its vector, *Lutzomyia longipalpis* (Diptera: Psychodidae). Am. J. Trop Med. Hyg *41*, 295–317.

Walters, L.L., Irons, K.P., Modi, G.B., Tesh, R.B. (1992). Refractory barriers in the sandfly (Diptera: Psychodidae) to infection with *Leishmania panamensis*. Am. J. Trop Med. Hyg *46*, 211–228.

Warburg, A., Hamada, G.S., Schlein, Y., Shire, D. (1986). Scanning electron microscopy of *Leishmania major* in *Phlebotomus papatasi*. Z Parasitenkd *72*, 423–431.

Warburg, A., Tesh, R.B., McMahon-Pratt, D. (1989). Studies on the attachment of *Leishmania* flagella to sandfly midgut epithelium. J. Protozool. *36*, 613–617.

Ward, R.D., Hamilton, J.G., Dougherty, M., Falcao, A.L. (1991). Pheromones in Old and New World sandflies (Diptera: Psychodidae). Ann. Trop Med. Parasitol. *85*, 667–668.

Yao, C., Donelson, J.E., Wilson, M.E. (2003). The major surface protease (MSP or GP63) of *Leishmania* sp. Biosynthesis, regulation of expression, and function. Mol. Biochem. Parasitol. *132*, 1–16.

Young, D.G., Lawyer, P.G. (1987). New World vectors of the leishmaniases. Springer-Verlag, New York.

Young, D.G., Duncan, M.A. (1994). Guide to the Identification and Geographic Distribution of Lutzomyia Sandflies in Mexico, the West Indies, Central and South America (Diptera: Psychodidae). Associated Publishers, American Entomological Institute, Gainesville, Florida.

Zakai, H.A., Chance, M.L., Bates, P.A. (1998). In vitro stimulation of metacyclogenesis in *Leishmania braziliensis*, *L. donovani*, *L. major* and *L. mexicana*. Parasitology *116*, 305–309.

Interactions Between *Leishmania* and the Host Macrophage

11

Martin Olivier and David J. Gregory

Abstract

In this chapter, we address the relationship between *Leishmania* and its host macrophage at the molecular level. From the initial interactions between receptor proteins on the macrophage membrane and *Leishmania* surface molecules, the parasite modulates the activities of the macrophage's intracellular signalling pathways for its own ends. Protein tyrosine phosphatases, particularly SHP-1, are rapidly activated, and intracellular Ca^{2+} and ceramide concentrations are elevated. These, and other factors, lead to inhibition of JAK2, MAP Kinase and Protein Kinase C pathways. Moreover, recent studies have shown that the transcription factor STAT1α is degraded by the proteasome, and proteolytic mechanisms also regulate the NF-κB family. These pathways allow *Leishmania* to radically alter the behaviour of its host cell without ever leaving the phagolysosome. In particular, production of microbicidal molecules (especially nitric oxide and reactive oxygen intermediates) and activating cytokines (e.g. IL-1, IL-12 and TNFα) is inhibited, whereas secretion of immunosuppressive molecules and certain chemokines (e.g. TGFβ, PGE_2 and MIP-2) is increased. Infected macrophages become non-responsive to activating stimuli such as IFNγ, and a number of mechanisms interfere with antigen presentation. We discuss the molecular details of all these processes as revealed by recent studies, and their important contribution to the survival of *Leishmania* in the host.

Introduction

Within its host organism, *Leishmania* operates as an obligate intracellular parasite, living almost exclusively inside macrophages. Successful infection therefore depends on surviving the antimicrobial activities of the macrophage, and preventing activation of an effective immune response. In order to do this, *Leishmania* has developed numerous highly successful strategies for manipulating various macrophage functions, including antigen presentation, nitric oxide and oxygen radical generation, and cytokine production. This is generally the result of interactions between molecules on the *Leishmania* surface, particularly gp63 and LPG, with less-well identified macrophage surface receptors, causing the distortion of specific intracellular signalling cascades. In this chapter, we describe some of the macrophage signalling pathways that are activated or repressed by *Leishmania*, as well as some emerging data on how macrophage gene expression is distorted.

Initial interactions and phagocytosis

Having entered their mammalian host during the blood meal of a female sandfly, *Leishmania* promastigotes must first evade complement-mediated lysis until engulfed by a macrophage. *L. major* procyclic promastigotes cannot resist complement action whereas the metacyclic form, specialized for transmission to the host, can fully avoid complement-driven lysis (Puentes *et al.*, 1989). This difference in complement resistance has been shown to depend upon branched lipophosphoglycans (LPG) on the parasite surface. LPG is longer on the surface of metacyclic promastigotes (see Chapter 9), and seems to prevent the attachment of C5b-C9 subunits of the complement complex, which are the effectors of cellular lysis, to the promastigote. The surface glycoprotein gp63, a protease, has also been reported to protect *L. amazonensis* and *L. major* against cellular lysis by converting the C3b complement subunit into C3bi (Brittingham *et al.*, 1995), and *L. donovani* promastigotes have been shown to fix C3bi on their surface (Puentes *et al.*, 1989). C3bi does not promote C5 convertase formation and subsequent cellular lysis, but instead opsonizes the parasite for phagocytosis by macrophages. It has also been proposed that a *L. major* surface protein kinase may inactivate complement by phosphorylating members of the cascade (Hermoso *et al.*, 1991).

Parasite surface molecules also play an important role during attachment to the macrophage. *In vivo* opsonization of *Leishmania* metacyclic promastigotes by C3b and C3bi permits the interaction with the macrophage complement receptor (CR) 1 and 3, respectively. However, since C3b is rapidly converted to C3bi by gp63, it appears that CR3 is the more important receptor, and interaction with CR1 is only transient (Kane and Mosser, 2000). Attachment via CR3 rather than CR1 is advantageous to the parasite since it will not trigger the oxidative burst during phagocytosis (Mosser and Edelson, 1987). Promastigotes can also attach to the macrophage via the mannose-fucose receptor, which binds to mannan residues of LPG (Blackwell *et al.*, 1985). LPG can also interact with C Reactive Protein (CRP, an early inflammatory product) and thus triggers phagocytosis via the CRP Receptor (Culley *et al.*, 1996), but without leading to the macrophage activation usually seen following CRP Receptor-mediated phagocytosis (Bodman-Smith *et al.*, 2002). Furthermore, gp63 and LPG interact with fibronectin receptor and CR4, respectively (Talamas-Rohana *et al.*, 1990; Brittingham *et al.*, 1999), although LPG appears to play only a minor role during attachment and internalization (reviewed in Descoteaux and Turco, 1999). More recently, a number of *Leishmania* surface molecules that play a role during the initial *Leishmania*-macrophage interaction have been identified, although the macrophage receptors for these molecules are not yet clear. For instance, an ICAM-related molecule, ICAM-L, may be involved in the interaction between the *L. amazonensis* and the murine macrophage cell line J774 (Chiang *et al.*, 2001). A recent study has reported that the greater infectiousness of metacyclic promastigotes is due in part to elevated surface phosphatidylserine (Tripathi and Gupta, 2003), while blocking antibodies against another parasite surface molecule, GIPL, have been shown to inhibit attachment of *L. major* (Suzuki *et al.*, 2002). Finally, amastigotes can also be internalized in an Fc Receptor-dependent fashion, following opsonization with specific antibodies (Guy and Belosevic, 1993). The large number of receptors that have been implicated suggests a degree of redundancy among parasite-macrophage interactions, although it does appear that several interactions are necessary for internalization (Brittingham *et al.*, 1999).

Following their attachment to the macrophage, *Leishmania* promastigotes are internalized to the relatively benign environment of the phagosome, where they begin to differentiate into amastigotes. Unlike amastigotes, promastigotes are vulnerable to degradation by the acidic and hydrolytic environment of the phagolysosome. They must therefore retard phagosome maturation and phagosome/endosome fusion, a process dependent on LPG (Desjardins and Descoteaux, 1997). This retardation has been observed by the absence or delayed arrival of late endosomal markers such as rab7 and LAMP-1 (Scharton-Kersten *et al.*, 1995; Scianimanico *et al.*, 1999), and may be related to the LPG-dependent accumulation of F-actin (Holm *et al.*, 2001). The mechanism is not completely understood, but has been shown to depend upon calcium presence (Tejle *et al.*, 2002) and PKC inhibition (Holm *et al.*, 2003). Furthermore, LPG appears to change the shape of membranes, leading to steric repulsion between the phagosome and the endosome (Miao *et al.*, 1995). The delay in phagolysosome maturation provides a window during which promastigotes can differentiate into the more resistant amastigotes. This is consistent with our earlier observation that parasite survival and infectiousness in different strains of mice correlates with macrophage phagocytic activity (Olivier and Tanner, 1987). Interestingly, while LPG is important for the survival of promastigotes in the early stages of uptake, it is not expressed by amastigotes, indicating that its role is transient, and confined only to the early stages of infection.

Another survival strategy used by *Leishmania* parasites is the inhibition of the hydrolytic enzymes and other destructive molecules that are secreted into the phagolysosome. Two recently discovered *Leishmania* molecules, named peroxidoxin LcPxn1 and 2 (Barr and Gedamu, 2003), and a superoxide dismutase (Ghosh *et al.*, 2003) are believed to deplete nitrite derivatives and reactive oxygen intermediates (ROI), which are the most important microbicidal small molecules. Furthermore, there is evidence that LPG itself can enhance promastigote survival by neutralizing vacuolar ROI (Chan *et al.*, 1989). LPG may also protect against lysosomal enzymes, perhaps by its strong negative charge and galactose-mannose repeating units (Eilam *et al.*, 1985). The proteolytic activity of the surface molecule gp63 is optimal at the acidic pH found in phagolysosomes, supporting the suggestion that it targets lysosomal enzymes (Seay *et al.*, 1996). However, its importance is debatable as parasites with mutations in the six gp63 genes are still capable of survival, differentiation and replication within macrophages (Joshi *et al.*, 1998).

Toll-Like receptors

Research in recent years has shown that the recognition of pathogens by phagocytes depends to a great extent on cell-surface receptors of the Toll-Like Receptor (TLR) family. The first indication that TLRs are important for the development of a protective immune response against *Leishmania* came when it was found that mice lacking TLR4 have an impaired ability to control cutaneous infections of *L. major* (Muller, 1997; Kropf *et al.*, 2004b). TLR2 has also been shown to interact with *L. major* LPG, both in a transfected cell system and on the surface of NK cells (Becker *et al.*, 2003; de Veer *et al.*, 2003). However, few studies have directly addressed the significance of interactions between *Leishmania* and TLRs on the macrophage surface. Bone marrow-derived macrophages from TLR4 –/– eradicate *L. major* equally well as wild type macrophages in the presence of IFNγ, but are less effective than wild type macrophages in the presence of IL-4 (Kropf *et al.*, 2004a). This is associated with increased induction of arginase and reduced iNOS induction, suggesting that TLR4 protects against *Leishmania* infection by promoting NO production by infected macrophages. However, no difference in arginase induction was observed in the absence of cytokine stimulation, and the signalling intermediates responsible for this co-operative induction are still unidentified (Kropf *et al.*, 2004a).

The cytosolic adapter protein, MyD88, binds TLRs and is important for propagation and refinement of TLR-dependent signals. Recent data suggests that MyD88 plays an important role in the response of macrophages to *Leishmania*. Aderem and colleagues reported that the increased IL-1α transcription in macrophages infected with *L. major* is dependent on MyD88 (Hawn *et al.*, 2002), and three groups have shown that MyD88 contributes to resistance to infection *in vivo* (de Veer *et al.*, 2003; Debus *et al.*, 2003; Muraille *et al.*, 2003). However, our recent work shows that both *L. major* and *L. donovani* are able to induce various chemokines (MIP-1α, MIP1β, MCP-1, MIP-2) in a TLR4-dependent but MyD88-independent fashion (Matte and Olivier, 2002; Godbout and Olivier, submitted). This suggests that TLRs play a mixed role during infection – MyD88-dependent signalling contributes to resistance and parasite eradication, whereas MyD88-independent signalling may favour a milder, selective inflammatory response, which involves recruitment of a mixed population of inflammatory cells. These recruited cells include macrophages, which are then available as hosts for the subsequent infection (Matte and Olivier, 2002). Exposing the molecular tricks used by *Leishmania* to trigger MyD88-independent TLR signalling while avoiding the MyD88-dependent pathway therefore represents a most interesting area of current research. Work in our own laboratory has shown that *L. donovani* rapidly and completely inactivates the pivotal MyD88-downstream kinase, IRAK-1 (I. Abu-Dayyeh and M. Olivier, unpublished). Inactivation of IRAK-1 will thus suppress MyD88-dependent signalling, so avoiding macrophage activation and production of microbicidal molecules such as NO, while leaving the MyD88-independent TLR signalling, and resultant chemokine induction, intact (Fig. 11.1). Identification of the signalling intermediates employed by *Leishmania* to achieve this is an important topic of ongoing investigation.

Repression of macrophage signalling pathways

The activity of a given intracellular signalling cascade normally depends on the balance between both positive and negative regulatory influences. Thus, repression of a signalling pathway can be achieved either by direct interference, for example blocking an interaction between intermediates, or by activating a negative regulator, such as a phosphatase. Importantly, an increase in the activity of a given signalling intermediate will often result in the downstream activation of a negative regulator, which will counter-balance the original activating stimulus to prevent uncontrolled activation of the pathway. This negative feedback is of obvious importance in cases such as IFNγ signalling in macrophages, which causes the production of toxic molecules such as nitric oxide, and would thus be extremely harmful both to the macrophage itself and to neighbouring cells if allowed to proceed unchecked. The emerging molecular data indicates that *Leishmania* are particularly adept at activating the macrophage's own negative regulators to repress hostile signalling pathways (Fig. 11.2).

SHP-1 protein tyrosine phosphatase

Protein tyrosine phosphatases (PTPs), such as SHP-1 (Src homology 2 domain-containing tyrosine phosphatase, *a.k.a.* SHPTP-1, HCP, PTP1C), play an important role in limiting the activity of signalling pathways that involve tyrosine-phosphorylation, such as JAK2/STAT1α. SHP-1 is principally expressed in haematopoietic cells, but also in smooth muscle (Marrero *et al.*, 1998) and epithelial cells (Banville *et al.*, 1995). This phosphatase contains two SH2 domains in its N-terminal portion, a central, conserved phosphatase domain, and a C-terminal tail (Yi *et al.*, 1992). The SH2 domains of SHP-1 play a dual role that involves both substrate recognition and PTP auto-regulation. These domains specifically bind proteins that are phosphorylated on a specific tyrosyl residue within a conserved motif of three to six amino acids (Songyang *et*

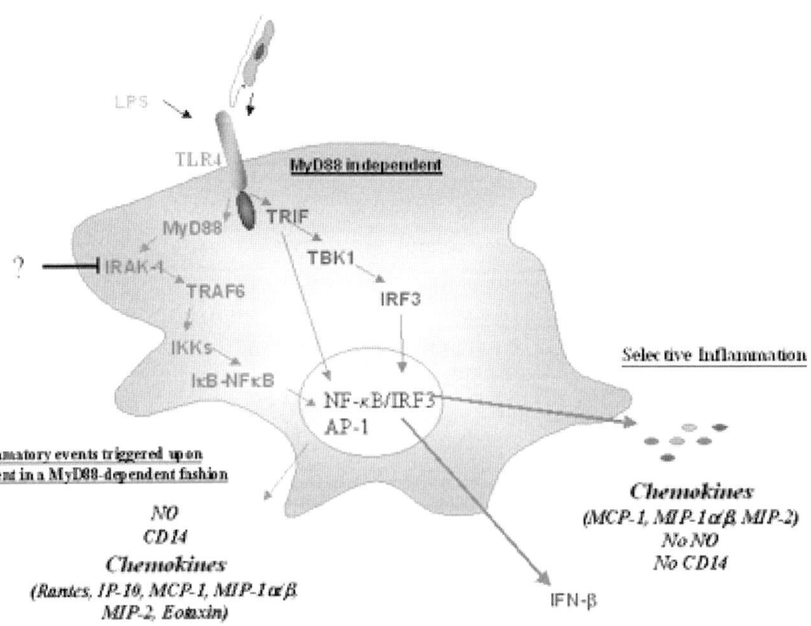

Figure 11.1 Selective inflammation triggered by *Leishmania*. Stimulation of TLR4 by *Leishmania* activates MyD88-independent signalling via intermediates such as TRIF, leading to induction of some chemokines. Inactivation of IRAK-1 blocks MyD88-dependent signalling, and so prevents a full inflammatory response.

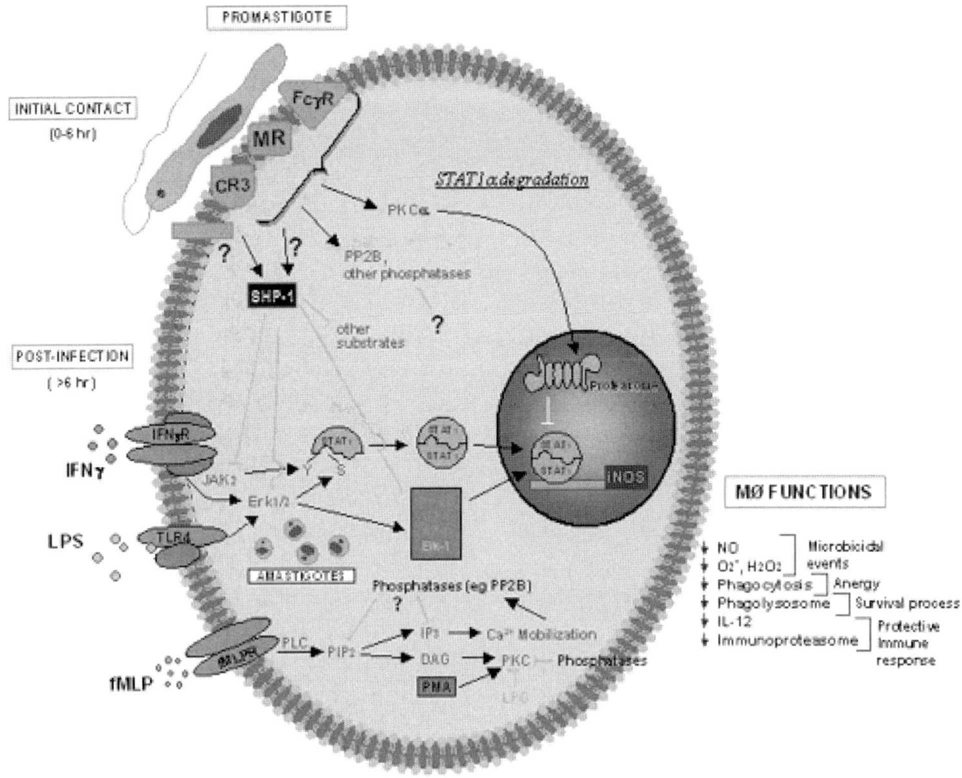

Figure 11.2 Mechanisms by which *Leishmania* represses macrophage functions. See text for details.

al., 1993; Pawson, 1995). The target motif for the two SH2 domains of SHP-1 is an ITIM motif (Immunoreceptor Tyrosine-based Inhibitory Motif), with the consensus sequence I/V/LxYxxL/V (Burshtyn *et al.*, 1997).

SHP-1 is responsible for the negative regulation of many signalling pathways in all haematopoietic cell types, by acting in a variety of fashions. For instance, SHP-1 can bind to receptors and dephosphorylate them directly, or bind the receptor and dephosphorylate other members of the receptor binding complex; or it can interact with cytosolic proteins and tyrosine dephosphorylate them or their associated proteins (Frearson and Alexander, 1997). The majority of documented SHP-1 effects are the result of the inhibition by dephosphorylation of different kinases and their signalling pathways. SHP-1 plays a vital role in limiting the activation of the JAK/STAT pathways following cytokine receptor stimulation. Dephosphorylation of JAK1/2, TYK2, STAT1α, 2, 3, 5α/β and 6 have all been documented (David *et al.*, 1995; Klingmuller *et al.*, 1995; Ram and Waxman, 1997; Haque *et al.*, 1998; Yeung *et al.*, 1998). In viable motheaten mice, whose SHP-1 phosphatase activity is deficient, increased nuclear translocation of the transcription factor NF-κB has been reported (Khaled *et al.*, 1998; Massa and Wu, 1998; Forget *et al.*, 2001), which seems to provoke an exacerbated inflammatory response (Forget *et al.*, 2005a). SHP-1 has been also reported to activate (Krautwald *et al.*, 1996; Xie *et al.*, 2000) or repress MAPKs (Nandan *et al.*, 1999; Forget *et al.*, 2006), as well as to induce Src kinase activity (Somani *et al.*, 1997).

Some studies have demonstrated that the inhibition of IFNγ-dependent phosphorylation cascades following infection is due to activation of host cell tyrosine phosphatases (Olivier *et al.*, 1998; Blanchette *et al.*, 1999). Injection of PTP inhibitors, the bis-peroxovanadium (bpV) compounds, into mice infected with *L. major* or *L. donovani* results in control of the infection. Both cutaneous lesions and parasitic load were significantly reduced in *L. major*-infected mice, whereas *L. donovani*-infected mice were completely protected (Olivier *et al.*, 1998). That *in vitro* and *in vivo* study was the first demonstration that modulation of host PTPs by the parasite plays a pivotal role in the progression of the infection. The bpV-mediated protection was shown thereafter to depend upon iNOS induction and NO generation, as well as to involve augmented innate inflammatory response that must have played an important role in the containment of the infection (Matte *et al.*, 2000).

In a parallel study, we have demonstrated that macrophage PTP activity is activated very rapidly, within 5 minutes of exposure to *L. donovani* promastigotes, and this correlates with a rapid, general tyrosine dephosphorylation of high molecular weight proteins (Blanchette *et al.*, 1999; Forget *et al.*, 2006). SHP-1 is the most important component of this PTP activation (Fig. 11.2, and Forget *et al.*, 2006). Furthermore, SHP-1 directly associates with JAK2 following *L. donovani* infection, and this interaction could explain in part the inability of IFNγ to trigger JAK2 signalling (Blanchette *et al.*, 1999; Forget *et al.*, 2006). More recently, using immortalized macrophage cell lines deficient in SHP-1, we have firmly established that this PTP is responsible for inactivation of JAK2 and ERK1/2 MAP kinases. Protection of these kinases in SHP-1 deficient macrophages allows these cells to respond to IFNγ stimulation by generating NO, even when infected (Forget *et al.*, 2006). Corroborating our findings, another group has reported that SHP-1 interacts strongly with MAPKs upon *L. donovani* infection (Nandan *et al.*, 1999). They showed that infected macrophages did not respond to PMA, an artificial PKC activator, as reflected in inhibition of ERK1/2 phosphorylation, Elk-1 inactivation and c-fos mRNA expression. A striking difference between these two studies is that the inhibition of ERK was observed 17 hours post-infection, whereas interaction between JAK2 and SHP-1 and consequent downregulation of JAK2 was observable within 1 hour. Differences between cell lines and their status of differentiation may partially explain this discrepancy. However, the capacity of the parasite to rapidly inhibit the host cell functions is essential for its survival. In addition to our studies using promastigotes, amastigotes of *L. amazonensis* have also been shown to rapidly inactivate ERK1/2 (Martiny *et al.*, 1999). Comparing the capacity of *L. donovani infantum* promastigotes and axenic amastigotes to trigger host PTP activity, we found that both can rapidly and similarly induce this negative regulatory activity in macrophages (I. Abu-Dayyeh and M. Olivier, unpublished). Recently, it has been proposed that the elongation factor EF-1α of *L. donovani* could be responsible for the later induction of SHP-1, observed 16 hours post-infection (Nandan *et al.*, 2002). However, SHP-1 is activated so rapidly that a receptor-mediated process must also be necessary in the early stages (Blanchette *et al.*, 1999; Martiny *et al.*, 1999). The particular receptors involved may be revealed by a strategy using siRNAs against different negative regulatory receptors of the immunoglobulin family.

As well as these *in vitro* studies, we have firmly demonstrated that SHP-1 plays a crucial role in the progression of *Leishmania* infection *in vivo*. Using the SHP-1-deficient, viable, motheaten mice, we showed that *L. major*-infected animals did not develop footpad swelling and that parasitic load was significantly reduced (Forget *et al.*, 2001). Histochemical and *in situ* hybridization analysis of the infected footpads showed that, in contrast to wild type animals, iNOS gene expression and various signalling molecules known to be involved in its expression (STAT1, NF-κB), were not repressed in SHP-1 deficient mice. In fact, infection resulted in increased NO generation, which must have contributed to killing of the parasite. This was supported by the finding that aminoguanidine, an inhibitor of iNOS, completely reversed the protection observed toward *L. major* infection in SHP-1 deficient animals (Forget *et al.*, 2001). In addition, we recently obtained similar *in vivo* observations concerning the role of SHP-1 in the control of visceral leishmaniasis conferred by *L. donovani* infection. Of interest, using a pathogenomic approach, we found that a majority of the host genes involved in signalling and the immune response that were repressed following infection *in vivo* were almost all negatively regulated by PTP SHP-1 (Olivier *et al.*, unpublished).

Inhibition of IFNγ-dependent signalling pathways

While dephosphorylation of JAK1 and JAK2 by SHP-1 undoubtedly plays a central role in blocking signal transduction in response to IFNγ in infected macrophages, other mechanisms may also contribute. Two studies have suggested that less of the IFNγ receptor itself is available on the surface of macrophages infected with *L. donovani* promastigotes (Ray *et al.*, 2000; Bertholet *et al.*, 2003), perhaps because of reduced synthesis, degradation, or compartmentalization of the receptor. However, an earlier study showed no such reduction in cells infected with *L. donovani* amastigotes (Reiner *et al.*, 1988). Furthermore, IFNγ-dependent regulation of MHC class I is not affected by infection with *L. mexicana* amastigotes, indicating no reduction of IFNγR expression (Kima *et al.*, 1997). The reasons for this conflict are not immediately clear. It may simply be a result of the very different methodologies used by the different groups, particularly the different macrophage cell lines used. However, it may indicate that promastigotes and amastigotes employ different mechanisms to inhibit IFNγ signalling. Further studies to directly address this issue would be extremely useful.

A new family of negative regulatory proteins has recently been the subject of much interest: the SOCS (Suppressors Of Cytokine Signalling) family plays an important role in the inhibition of signalling induced by several cytokines, principally through its action on the JAK/STAT pathway. The first member to be discovered was CIS (cytokine-inducible SH2-containing protein). CIS is an early gene induced in response to stimulation of cytosine receptors, and its over-expression has been shown to inhibit IL-3-mediated signalling (Yoshimura *et al.*, 1995; Endo *et al.*, 1997; Naka *et al.*, 1997; Starr *et al.*, 1997). It now seems that 8 members constitute this family of negative regulator: CIS itself and SOCS-1–7 (Masuhara *et al.*, 1997; Hilton *et al.*, 1998; Starr *et al.*, 1998). Proteins from this family contain two important conserved sites: a central SH2 domain and a 40 amino acid SOCS box motif towards the C-terminus (Kile *et al.*, 2001).

SOCS are not expressed in naïve macrophages. Their transcription is induced in response to various cytokines, hormones and growth factors such as IFNγ, that activate the JAK/STAT pathways (reviewed in Greenhalgh and Hilton, 2001). Once transcribed, the SOCS proteins will inhibit the JAK/STAT pathways in a negative feedback loop. Their modes of action vary, but as an example on how this type of regulator works, it has been reported that SOCS-3 seems to bind to phosphorylated JAK via its SH2 motif. This may then block binding, and subsequent phosphorylation, of STAT (Matsumoto *et al.*, 1997). Interestingly, a recent report has suggested a complex interplay between the different SOCS family members, with SOCS-2 causing ubiquitination, and thus degradation, of both SOCS-1 and SOCS-3 (Tannahill *et al.*, 2005).

A number of recent studies have suggested that members of the SOCS family play a role during *Leishmania* infection. Studies using knockout mice have shown increased susceptibility to *L. major* infection *in vivo* in SOCS-1 $^{+}/^{-}$, but not SOCS-2 –/– or SOCS-5 –/– mice (Bullen *et al.*, 2003; Brender *et al.*, 2004). However, as is always the case with such *in vivo* studies, it is not clear whether the increased susceptibility is due to a specific disruption in the interaction between the parasite and macrophage, or merely supports the importance of SOCS-1 in the development of an immune response. Furthermore, if induction of SOCS proteins were a mechanism by which *Leishmania* suppresses the immune response, mice deficient in SOCS genes would be expected to show *decreased* susceptibility to infection.

Two published reports have addressed the induction of SOCS genes specifically in infected macrophages. In the first, a rapid, but very transient, induction of SOCS-3 mRNA was observed in *L. donovani*-infected human phagocytic cells (Bertholet *et al.*, 2003). The induction was observed following exposure to live or heat-killed parasites, and was not dependent on phagocytosis, suggesting that contact with a surface receptor is a sufficient trigger. Interestingly, the induction was also observed following exposure to *L. major* or *L. amazonensis*, but to a lesser extent (Bertholet *et al.*, 2003). Another report showed that SOCS-1 and SOCS-3 mRNA can be induced by *L. major* LPG and peptidoglycan in a Raw 264.7-derived cell line, although this was in the context of an inflammatory response, which is not found in other systems, and so may have been a secondary effect (de Veer *et al.*, 2003). In contrast to these reports, work in our laboratory has shown no induction of SOCS-1 or SOCS-3 mRNA following infection with *L. donovani* promastigotes,

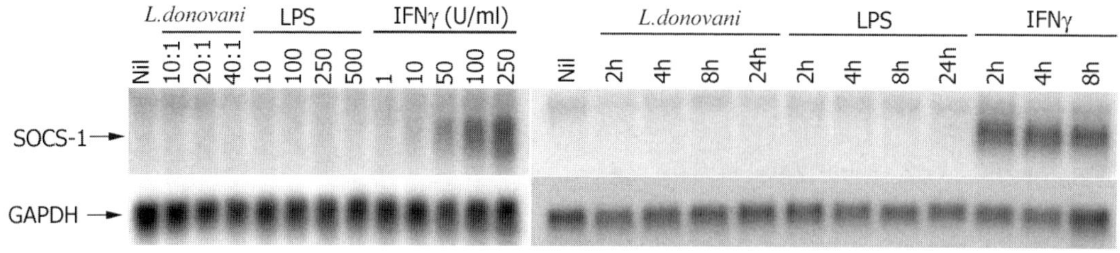

Figure 11.3 SOCS-1 is not induced by *Leishmania*. B10R macrophages were exposed to *Leishmania donovani*, IFNγ or LPS, and SOCS-1 induction studied by Northern blot. Similar results were obtained for SOCS-3 (J.-F. Marquis and M. Olivier).

despite effective induction by IFNγ (Fig. 11.3). Furthermore, given that induction of SOCS is mediated by STAT1α, and that this transcription factor is rapidly degraded following infection (see below), no mechanism for induction of SOCS by *Leishmania* is immediately apparent. Thus it is premature to say that the SOCS family plays a role in the negative regulation of macrophage functions by *Leishmania*, and further studies, using macrophages from SOCS-deficient mice, are necessary to clarify the issue.

Inhibition of MAP kinase cascades

Inhibition of MAP kinases following *Leishmania* infection has been reported by different studies, following stimulation with LPS or PMA (Martiny *et al.*, 1999; Nandan *et al.*, 1999). In the first study, the use of *L. amazonensis* amastigotes was shown to rapidly alter ERK1 MAP kinase phosphorylation in response to LPS (Martiny *et al.*, 1999). In the second, the authors used *L. donovani* amastigotes to demonstrate that ERK1/2 MAP kinase inactivation was accompanied by the inhibition of the transcription factor Elk-1 and c-fos expression (Nandan *et al.*, 1999). In both studies, it was suggested that PTPs are responsible for the dephosphorylation of ERK1/2 MAP kinases. However, Martiny *et al.* found that the PTP in question was derived from the parasite (Martiny *et al.*, 1999), whereas Nandan *et al.*, suggested an endogenous macrophage PTP (Nandan *et al.*, 1999). Our recent data supports the latter study, by showing that the macrophage PTP SHP-1 dephosphorylates and inactivates ERK1/2, as macrophages deficient for SHP-1 showed normal JAK2 and ERK1/2 MAP kinase activity, and IFNγ-dependent NO generation (Forget *et al.*, 2006). Ghosh *et al.*, reported that infection of macrophages with *L. donovani* results in increased synthesis of ceramide (Ghosh *et al.*, 2001). This appears to prevent phosphorylation of ERK1/2 in response to LPS, apparently via activation of an endogenous phosphatase (Ghosh *et al.*, 2001; Ghosh *et al.*, 2002).

These studies are supported by the observation that all three MAP kinases subfamilies (ERK1/2, p38 and JNK) are not induced when *L. donovani* is phagocytosed by naive macrophages (Privé and Descoteaux, 2000). This is due in part to LPG, at least in the case of ERK1/2, as LPG-deficient promastigotes do trigger ERK1/2 phosphorylation (Privé and Descoteaux, 2000). However, this does not necessarily mean that LPG is directly responsible for the inactivation of these kinases, merely that, in the absence of LPG, the interaction of *Leishmania* with its host cell is modified. Using LPG-deficient *L. donovani* promastigotes, we have observed that SHP-1 was still induced by these parasites, resulting in inactivation of signalling and transcription factors related to JAK2 and MAP kinase signalling pathways (G. Forget and M. Olivier, unpublished). The observation that ERK1/2 is activated in the absence of LPG (Privé and Descoteaux, 2000) may be a secondary effect of other, LPG-independent processes, such as chemokine and PGE_2 secretion.

Leishmania can also interfere with other MAP kinase family members. For example, stimulation of *L. major*-infected macrophages with a specific α-CD40 antibody, which mimics the phagocyte–T lymphocyte interaction fails to activate p38 MAP kinase (Awasthi *et al.*, 2003). Interestingly, reversal of this inhibition using anisomycin, an activator of p38, restores NO production in infected cells (Awasthi *et al.*, 2003).

A recent study has proposed another potential mechanism where *L. mexicana* amastigotes cause the rapid, cysteine peptidase-dependent degradation of ERK and JNK, but not p38 (Cameron *et al.*, 2004). At first sight this new report contrasts remarkably with the earlier studies, in which ERK1/2 phosphorylation, but not abundance, was affected by infection. However, these studies all used promastigotes of *L. major* or *L. donovani*, or amastigotes of *L. donovani*, and the only other report of ERK phosphorylation following infection with New World *Leishmania* (*L. amazonensis*) amastigotes did not address total ERK abundance (Martiny *et al.*, 1999). It is therefore possible that a marked difference in mechanism of ERK inactivation between *Leishmania* species has been revealed. The choice of mechanism may depend on divergent evolution between New and Old World strains, or another factor. Direct comparative studies involving a range of species, differing in geographical origin and clinical outcome, are necessary to resolve this question.

Intracellular calcium

The intracellular Ca^{2+} concentration of phagocytes is increased following infection with *L. major* or *L. donovani* (Eilam *et al.*, 1985; Olivier *et al.*, 1992a). This appears to result from depletion of intracellular Ca^{2+} stores, triggering increased Ca^{2+}uptake by infected cells (Olivier *et al.*, 1992a; Mansfield and Olivier, 2002). The surface molecule LPG both chelates Ca^{2+} (Eilam *et al.*, 1985) and contributes to the rapid elevation of intracellular Ca^{2+} concentration (Mansfield and Olivier, 2002), but a causal link between these two properties has not been established. The increased Ca^{2+}concentration coincides with a defect in calcium mobilization in response to agonists such as fMLP (Eilam *et al.*, 1985; Olivier *et al.*, 1992a). This appears to result from defective generation of inositol phosphate signalling intermediates in infected cells (Olivier *et al.*, 1992a). While the mechanism of this inhibition has not been clearly demonstrated, an early study suggested that a *Leishmania* surface acid phosphatase contributes (Das *et al.*, 1986), while we have hypothesized that an unidentified host inositol phosphatase is activated by the elevated intracellular Ca^{2+} concentration (Olivier, 1996).

A number of downstream regulatory enzymes require Ca^{2+} in order to be fully active, for example the serine/threonine phosphatase calcineurin (PP-2B). Elevated intracellular Ca^{2+} concentrations are therefore consistent with increased calcineurin activity (Mansfield and Olivier, 2002). Some isoforms of PKC (Protein Kinase C) are also Ca^{2+} dependent, so it is perhaps surprising that there is reduced PKC activity in *L donovani*-infected macrophages, and this contributes to parasite survival (McNeely and Turco, 1987; Descoteaux *et al.*, 1992; Olivier *et al.*, 1992b). Promastigote LPG has been shown to cause this inhibition (Descoteaux *et al.*, 1992); however amastigotes, which lack LPG, can also inhibit PKC in infected human monocytes, indicating the availability of alternative mechanisms (Olivier *et al.*, 1992b). LPG-mediated repression may be explained by the observations that it interferes with binding of regulators including Ca^{2+} and diacyl glycerol (DAG), and can also block PKC membrane insertion (reviewed in Descoteaux and Turco, 1999). Interestingly, reduced sensitivity of PKC to DAG was first demonstrated biochemically in cells that have been infected with *L. donovani* amastigotes, and correlated with reduced generation of oxygen radicals (Olivier *et al.*, 1992b). Various species of *Leishmania* and purified gp63 have been reported to inhibit MARCKs-Related Protein, a substrate of PKC associated with the cytoskeleton and involved in vacuole dispersal (Corradin *et al.*, 1999a, 1999b). GIPL, another *Leishmania* surface glycolipid, may also inactivate PKC, but this process needs further investigation (McNeely *et al.*, 1989).

Interestingly, inhibition of Ca^{2+}-dependent isoforms of PKC, in particular PKCβ, is accompanied by induction of calcium independent PKCξ (Bhattacharyya *et al.*, 2001b). Both inhibition and activation are apparently a result of increased ceramide synthesis following infection (Ghosh *et al.*, 2001). It has been speculated that this ceramide also contributes to activation of PTP and inhibition of MAP kinases, as discussed below.

One study has also demonstrated that addition of recombinant IL-10 favours macrophage PKC inhibition during *L. donovani* infection, and that addition of anti-IL-10 antibodies prior to infection prevented inhibition of PKC (Bhattacharyya *et al.*, 2001a). This is surprising, since PKC modulation was previously shown to be a direct result of interaction between the macrophage and the parasite (Matte *et al.*, 2001; Freitas Balanco *et al.*, 2001; Bhattacharyya *et al.*, 2001b). However, it is very possible that IL-10 secreted from infected macrophages inhibits the PKC activity of neighbouring, naïve macrophages, so weakening their ability to take part in an immune activation, and perhaps making them more prone to subsequent infection.

Manipulation of macrophage transcription factors

Two families of transcription factors are particularly important in regulating macrophage-specific gene expression: NF-κB and STAT. A number of recent studies have therefore addressed the affect of *Leishmania* infection on their stability and activity.

Inactivation of STAT1α

The STAT family, in particular STAT1α, play a central role in IFNγ signalling, macrophage activation and the induction of nitric oxide synthase (iNOS), which are detrimental to *Leishmania*. Its significance for host resistance is underlined by the marked susceptibility of STAT1 knockout mice to *L. major* infection, associated with reduced IFNγ, nitric oxide and IL-12 production (Rosas, 2003). STAT1α phosphorylation and DNA binding activity is markedly reduced in macrophages following infection with *L. donovani* promastigotes *in vitro* (Nandan and Reiner, 1995; Ray *et al.*, 2000; Forget *et al.*, 2005b). Interestingly, this inhibition is not a result of the elevated SHP-1 and other PTP activity, but results from rapid degradation of STAT1α by the proteasome (Forget *et al.*, 2005b). The degradation is also found following infection with *L. major*, *L. mexicana* and *L. braziliensis* promastigotes, suggesting a general *Leishmania* virulence mechanism, but is specific to STAT1, since STAT3 levels are unaffected by infection (Forget *et al.*, 2005b). The rapidity of the process, occurring within half an hour of infection with virulent promastigotes *in vitro*, underlines the importance of STAT1α inactivation for *Leishmania* survival. A similar, proteasome-mediated degradation of STAT protein has been suggested as a negative feedback mechanism to curtail activation following cytokine stimulation (Kim and Maniatis, 1996; Malek and Halvorsen, 1999), so it is tempting to suppose that *Leishmania* hijacks this mechanism during infection. It will be most interesting to determine what macrophage signalling pathways are employed to achieve this.

An alternate mechanism of STAT1α inactivation has been proposed, that may operate during the later stages of macrophage infection. Bhardwaj and colleagues found that tyrosine phosphorylation of a dominant negative splice variant, STAT1β, was increased following IFNγ treatment (Bhardwaj *et al.*, 2005). This presumably results in increased nuclear localization and DNA binding of STAT1β, and thus inhibition of STAT1α, similar to a mechanism employed by the bacterial pathogen *Mycobacterium avis* (Alvarez *et al.*, 2003). So it is possible that STAT1α-dependent transcription is repressed by proteasome-mediated degradation immediately following interaction of the macrophage with *Leishmania*, but once the infection is established, by the more subtle mechanism of STAT1β.

Modulation of the NF-κB family

While STAT1α is critical for interferon-induced transcription, the NF-κB family is important for expression of genes involved in a wide range of macrophage functions, ranging from suppression of apoptosis to cytokine production. So while *Leishmania*'s survival is well served by inactivation of STAT1α, a more complex modulation of the NF-κB family is necessary in order to inhibit unfavourable genes such as iNOS and IL-12, while simultaneously inducing genes such as MIP2 and IL-10. Interestingly, some recent data suggests that promastigotes and amastigotes may employ different mechanisms to regulate NF-κB. Infection of macrophages with amastigotes of *L. mexicana* causes a rapid degradation of the NF-κB subunit p65 (RelA), in a process dependent on the amastigote cysteine peptidase family, CPB (Cameron *et al.*, 2004). Remarkably, the authors of this report also observed degradation of the inhibitory proteins IκBα and IκBβ, as well as the upstream kinases ERK and JNK. IκB degradation is normally a step in the activation of NF-κB, but in this case appears to be part of a cysteine protease-dependent destruction of the entire NF-κB pathway. This mechanism of inactivation seems to be restricted to amastigotes, since promastigotes, which do not express CPB did not degrade IκBα, at least (Cameron *et al.*, 2004).

Another study also showed inhibition of p65 following infection of macrophages with amastigotes of *L. major* (Guizani-Tabbane *et al.*, 2004). In this case, however, the authors observed induction of another NF-κB species, a heterodimer of c-Rel and p50 (NF-κB1). This species may account for the increased NF-κB reporter gene activity observed by authors, and for the fact that the induction of IL-10 and TNFα is reversed by inhibitors of NF-κB (Guizani-Tabbane *et al.*, 2004). While p65/p50 and c-Rel/p50 heterodimers bind essentially the same promoter sequences, the switch from p65 to c-Rel will have a marked effect on the expression of NF-κB responsive genes, presumably resulting in a profile more favourable to the parasite.

In contrast to amastigotes, promastigotes of *L. donovani* or *L. major* do not cause degradation of IκBα (Privé and Descoteaux, 2000; Cameron *et al.*, 2004). This implies not only that promastigotes employ a different regulatory mechanism, but also that promastigotes do not activate NF-κB. However, a recent report has shown that both c-Rel and p65 are activated by infection of J774A.1 cells with *L. donovani* promastigotes, and that this results from IκBβ degradation (Singh *et al.*, 2004), which previous studies had not addressed. However, this activation was dependent on the MAP Kinases ERK and p38, which are known to be inhibited by both promastigotes and amastigotes, and contrasts with our current data, which shows downregulation of p65 and no activation of c-Rel by promastigotes of various species in B10R macrophages (D.J. Gregory and M. Olivier, unpublished). Ghosh and colleagues observed an ERK-dependent activation of NF-κB by *L. donovani* promastigotes in primary (bone marrow-derived) murine macrophages only following inhibition of macrophage ceramide synthase. Thus it seems that the activation of NF-κB by promastigotes acting through MAP Kinases in J774A.1 cells represents a host-defence pathway that is inactivated via ceramide in other systems (Ghosh *et al.*, 2002).

While promastigotes do not activate conventional NF-κB forms, our recent data show that a novel form of NF-κB does accumulate in the nuclei of infected macrophages. Exposure of B10R murine macrophages to promastigotes of a range of species causes proteolysis of p65 (D. J. Gregory and M. Olivier, unpublished). In contrast to the proteasome-mediated degradation that inactivates STAT1α, this appears to be a specific cleavage, which removes the C-terminal portion of p65 but leaves the N-terminal DNA binding domain intact. The truncated p65 accumulates in the nucleus, and binds DNA as a heterodimer with p50. The exact role of this novel NF-κB form is not yet clear. It lacks a recognized transcriptional activation domain, so may act as a dominant-negative repressor. However, despite the lack of activation of other NF-κB family members, induction of chemokines such as MIP2 and MIP1β by *L. donovani* and *L. major* promastigotes is at least partially dependent on NF-κB (Godbout and Olivier, submitted). So it is also possible that, instead of being a repressor, the cleaved NF-κB interacts with specific co-activators to induce certain genes. This situation is similar to that of p50 homodimers, which also lack a transcriptional activation domain and are therefore normally seen as inhibitory, but can nonetheless induce transcription in certain circumstances via interaction with other transcription factors and co-activators (Fujita *et al.*, 1993; Na *et al.*, 1998; Dechend *et al.*, 1999; Agrawal *et al.*, 2001; Kurland *et al.*, 2001).

Inactivation of macrophage functions

Mechanisms such as those described above allow *Leishmania* to radically alter the macrophage's normal pattern of gene expression and behaviour. One study of 245 macrophage genes showed that 37% were repressed at least 2-fold following *in vitro* infection with amastigotes (Buates and Matlashewski, 2001), although larger scale microarray studies, have suggested that promastigotes induce and repress to similar extents (Chaussabel *et al.*, 2003; Rodriguez *et al.*, 2004). However, these studies, and our unpublished observations following infection *in vivo*, all show reduced expression of numerous genes with important roles in various immune, cell physiological and signalling functions.

In fact, prior to these genomic studies, it had been already reported that *Leishmania* can efficiently manipulate macrophage functions detrimental to its survival. For instance, two types of microbicidal molecules are recognized for their efficacy against *Leishmania*: nitric oxide (NO) (Liew *et al.*, 1990) and ROI (Murray, 1982). NO is critical for parasite clearance, since mice lacking inducible nitric oxide synthase (iNOS, NOS2) are unable to control infection, and macrophages derived from these mice are incapable of eliminating promastigotes in culture (Wei *et al.*, 1995). Infected macrophages, or macrophages incubated with purified *Leishmania* surface molecules, LPG or GIPL, lose their ability to induce iNOS or to generate NO in response to IFNγ and/or LPS (Proudfoot *et al.*, 1995; Proudfoot *et al.*, 1996). However, it seems that IFNγ and LPG can synergize to generate NO when administered simultaneously to naïve macrophages (Proudfoot *et al.*, 1995; Proudfoot *et al.*, 1996). This suggests that contact between the parasite and the macrophage prevents the macrophage from responding to subsequent exposure to IFNγ produced by lymphoid cells. Inhibition of NO production may result from the production of IL-10 and/or TGFβ, inactivation of the JAK/STAT pathway, activation of phosphotyrosine phosphatases, and/or ceramide production, as discussed later.

In contrast to NO, mice deficient for the generation of ROI can ultimately control the infection, after an initial period of increased susceptibility (Murray and Nathan, 1999), indicating that ROI play a less important role in parasite clearance. However, ROI generation is also inhibited by *L. donovani* infection (Buchmuller-Rouiller and Mauel, 1987; Olivier *et al.*, 1992a, 1992b). Inhibition appears to be dependent on the surface molecules LPG and gp63 (Sorensen *et al.*, 1994; Descoteaux and Turco, 1999), and has been shown to involve abnormal PKC activity (Olivier *et al.*, 1992b).

In addition to repressing the microbicidal activities of the host macrophage, *Leishmania* inhibits the ability of the host cell to display parasite antigens to other components of the immune system (Reiner *et al.*, 1987). This appears to be related to the infectiousness of the parasite: macrophages infected with insect adapted, procyclic promastigote cultures are initially able to present the parasite LACK antigen, but lose this ability as the parasite begins to differentiate (Courret *et al.*, 1999). Macrophages infected with the infectious, metacyclic form of promastigotes present very little LACK antigen, and amastigote infected cells present none at all (Courret *et al.*, 1999).

Some studies have shown that *L. donovani* inhibits antigen presentation by repressing MHC class II gene expression, both basal and particularly following stimulation with IFNγ (Reiner *et al.*, 1987; Reiner *et al.*, 1988; De Almeida *et al.*, 2003). In contrast, macrophages infected with *L. amazonensis* have been shown to express normal levels of MHC class II (Prina *et al.*, 1993; Lang *et al.*, 1994a). Antigen presentation may be inhibited in this case by interfering with the loading of antigens onto MHC class II molecules (Fruth *et al.*, 1993; Prina *et al.*, 1993), or by sequestration of the MHC class II molecules and/or antigens within the phagolysosome (Lang *et al.*, 1994b; Kima *et al.*, 1996). De Souza Leao and colleagues demonstrated a third level of inhibition, at least for the case of *L. amazonensis*, when they observed direct endocytosis of MHC class II molecules by amastigotes themselves, followed by cysteine peptidase-dependent degradation (De Souza Leao S. *et al.*, 1995). Consistent with the phagosomal location of the *Leishmania*, MHC class II appears to be more important than class I for resistance, although class I does have a role to play, at least in some situations (Uzonna *et al.*, 2004). Mice defective in MHC class I presentation are resistant to infection with *L. major*, but mice with MHC class II−/− mice are susceptible (Locksley *et al.*, 1993; Huber *et al.*, 1998).

Antigen presentation depends upon cellular communication through co-stimulatory molecules such as B7/CD28 and CD40/CD40l. It has been demonstrated that B7−1 of *L. donovani*-infected macrophages could not be further expressed in response to LPS stimulation (Kaye *et al.*, 1994) and that this inactivation process was prostaglandin-dependent (Saha *et al.*, 1995). It seems that inhibition of CD40/CD40L ligation is responsible for the absence of iNOS and macrophage microbicidal activities (Kamanaka *et al.*, 1996; Soong *et al.*, 1996), and that cure of *L. major* infection depends upon an active CD40/CD40L ligation (Campbell *et al.*, 1996;

Heinzel *et al.*, 1998). Recent findings suggest that p38-dependent signalling triggered by CD40 interaction is altered in infected macrophages, and this may lead to diminished iNOS expression (Awasthi *et al.*, 2003). In contrast, repression of MHC class II gene expression appears to involve a cyclic AMP-independent mechanism (Kwan *et al.*, 1992).

Another survival strategy used by *Leishmania* is to prevent the activation of an effective immune response by inhibiting production of a number of cytokines, particularly those involved in the inflammatory response (IL-1 and TNFα) or in T lymphocyte activation (IL-12) (Bodman-Smith *et al.*, 2002). LPS-induced IL-1β secretion has been reported to be inhibited in *L. donovani*-infected (Reiner *et al.*, 1987; Reiner *et al.*, 1990) and LPG-exposed (Frankenburg *et al.*, 1990) macrophages. LPG seems to repress IL-1β transcription by acting through a promoter repression sequence (Hatzigeorgiou *et al.*, 1996). In contrast, IL-1α transcription is induced by *L. major*, but this is not reflected in increased secretion, indicating that a downstream repression mechanism counteracts the induction (Hawn *et al.*, 2002). Interestingly, the induction of IL-1α appears to be MyD88-dependent, suggesting a role for Toll-Like receptors (TLRs). TNFα production is also repressed in infected macrophages treated with LPS (Descoteaux and Matlashewski, 1989). More recently, this has been shown to involve IL-10 and PKC inhibition (Bhattacharyya *et al.*, 2001a).

The capacity of *Leishmania* to infect macrophages without inducing pro-inflammatory cytokines, and then inhibiting their induction in response to various agonists probably represents a survival mechanism whereby the parasites can abrogate a harmful inflammatory reaction. Nevertheless, these studies have been mainly performed in an *in vitro* context. Recent studies performed *in vivo* have clearly demonstrated that pro-inflammatory cytokines (IL-1, IL-6 and TNFα), as well as various chemokines, a family of cytokines responsible for recruitment of inflammatory cells, were induced in the early stages of *L. donovani* and *L. major* infection (Matte *et al.*, 2001). Of interest, *L. major* promastigotes were shown to be better activators of pro-inflammatory events than *L. donovani*, as shown by a greater transient recruitment of inflammatory cells. This may reflect the different pathologies caused by the two strains. In addition, both species recruit a heterogeneous population of host inflammatory cells, including neutrophils and monocytes/macrophages (Matte *et al.*, 2001; Forget *et al.*, 2005a). This is of particular interest from a host-defence point of view, since neutrophils have been shown to be important for controlling *L. major* infection (Lima *et al.*, 1998; Forget *et al.*, 2001). Interestingly, the tyrosine phosphatase SHP-1 limits the recruitment of inflammatory cells, particularly neutrophils, following infection with *L. major*. *Leishmania* thus strikes a balance, where a harmful inflammatory response is prevented during infection, but some limited inflammation is maintained and contributes to leukocyte recruitment (Forget *et al.*, 2005a).

The cytokine IL-12 plays a critical role in the regulation of the cellular immune response. It is essential for T lymphocyte activation and subsequent IFNγ secretion leading to macrophage activation and production of microbicidal molecules. It is therefore not surprising that *Leishmania* has developed the ability to inhibit IL-12 production. This has been shown for promastigotes of *L. donovani* and *L. major* (Carrera *et al.*, 1996), *L. mexicana* amastigotes (Weinheber *et al.*, 1998) and the phosphoglycan portion of LPG *in vitro* (Piedrafita *et al.*, 1999). IL-12 inhibition has been also reported in *L. major*-infected mice (Belkaid *et al.*, 1998). The intracellular mechanism is still unclear. Macrophage complement receptors and FcγR, which are known to interact with *Leishmania* during phagocytosis, have been shown to repress IL-12 (Marth and Kelsall, 1997; Sutterwala *et al.*, 1997). Furthermore, Piedraftia and colleagues showed that *L. major* LPG-mediated IL-12 repression was independent of the NF-κB transcription factor family, despite IL-12 being NF-κB responsive (Piedrafita *et al.*, 1999). Alternatively, repression may be a result of increased ERK1/2 phosphorylation (Feng *et al.*, 1999). However, this latter report is problematical, because it contrasts with the well-documented dephosphorylation of ERK during infection. A recent report has shown that the ability of *L. mexicana* amastigotes to degrade NF-κB and to repress IL-12 are both dependent on cysteine peptidase B activity (Cameron *et al.*,

2004). While this latter is purely correlative, there is no reason to assume that promastigotes of Old World species and New World amastigotes employ the same mechanisms, especially given that both species and developmental stages differ markedly in both cysteine protease and LPG type and expression.

Induction of immunosuppressive molecules by *Leishmania* infection

In addition to inhibiting the functions of their host macrophages, *Leishmania* parasites can induce the production and/or secretion of various immunosuppressive signalling molecules, such as arachidonic acid metabolites and the cytokines TGFβ and IL-10. These affect numerous different cell types, directly and indirectly, so distorting the normal immune response and favouring parasite survival.

TGFβ production is induced by several *Leishmania* species *in vitro* and *in vivo* (reviewed in Bogdan and Rollinghoff, 1998). Augmentation of TGFβ secretion correlates with retarded iNOS expression and reduced NK cell activity in lymph nodes (Stenger *et al.*, 1994; Scharton-Kersten *et al.*, 1995). This is consistent with the idea that TGFβ inhibits macrophage microbicidal action and the production of IFNγ by NK cells, although the exact role of NK cells during leishmaniasis is somewhat controversial (Kirkpatrick *et al.*, 1985; Kaye and Bancroft, 1992; Scharton and Scott, 1993; Maasho *et al.*, 1998; Wakil *et al.*, 1998; Satoskar *et al.*, 1999). A recent study demonstrated that *L. chagasi* induces TGFβ production in the immediate environment of the infected human macrophage, and this may permit the local inhibition of immune responses (Gantt *et al.*, 2003). Interestingly, at least for the case of *L. chagasi*, the increased production does not appear to be a result of increased gene expression, but of cleavage of pro-TGFβ by amastigote cysteine proteases to produce active TGFβ (Somanna *et al.*, 2002; Gantt *et al.*, 2003). Interaction between the macrophage and phosphatidylserine motifs on the amastigote surface has also been proposed to trigger this induction (Freitas Balanco *et al.*, 2001).

IL-10 is another anti-inflammatory cytokine produced by *Leishmania*-infected macrophages *in vitro*, apparently via interaction with the FcγR (Sutterwala *et al.*, 1998). Its production may be responsible for the suppression of macrophage microbicidal activity involving NO production of several cytokines (IL-1, IL-12 and TNFα), and the expression of co-stimulatory molecules such as B7–1/2. Its importance *in vivo* is illustrated by the observation that transgenic mice constitutively expressing IL-10 are unable to control *Leishmania* infection (Kane and Mosser, 2000). As in the case of TGFβ, IL-10 is apparently induced following recognition of amastigote surface phosphatidyl serine residues by the macrophage (Freitas Balanco *et al.*, 2001).

Prostaglandin E2 (PGE_2) is produced by *Leishmania*-infected macrophages, and appears to favour parasite survival and progression (Reiner *et al.*, 1985; Reiner and Malemud, 1985; Farrell and Kirkpatrick, 1987; Matte *et al.*, 2001). This arachidonic acid metabolite has been reported to cause inhibition of macrophage proliferation, and to suppress production of TNFα, IL-1 and reactive oxygen intermediates (Belley and Chadee, 1995). The increased PGE_2 production results from a PKC-dependent induction of cyclooxygenase-2 (COX-2) gene expression (Matte *et al.*, 2001). Interestingly, one study has correlated increased visceralization of *L. donovani* in malnourished mice with increased in PGE_2 production in the lymph nodes (Anstead *et al.*, 2001).

Comparison with other trypanosomatids

The three main genera of pathogenic trypanosomatids have markedly different life-styles in the mammalian host. While *Leishmania* almost exclusively inhabit macrophages, South American trypanosomes have a much wider range of host cells, including fibroblasts and myocytes, which they enter by at least two pathways (Tardieux *et al.*, 1992; Woolsey *et al.*, 2003). Importantly, instead of staying in the phagolysosome, they escape into the cytosol shortly after phagosome-lysosome fusion, and remain there during the remainder of the infection. African trypanosomes are even more dissimilar: they have no intracellular stage, but remain free in the bloodstream and extravascular fluid throughout the mammalian phase. Thus, major differences are to be expected in the macrophage-parasite interactions between *Leishmania* and other trypanosomes. Furthermore, while much progress has been made during recent years in our understanding of *Leishmania*-macrophage interactions, understanding the molecular events involved in interactions of macrophages with South American, and particularly African, trypanosomes lags behind. However, some common themes are emerging, and the differences so far discovered may highlight the importance of the contrasting outcomes in immune system modulation.

South American trypanosomes: *Trypanosoma cruzi*

The role of calcium signalling during *T. cruzi* infection has been studied well in non-immune cells. Interaction between parasite and macrophage surface molecules results in transient increases in intracellular calcium concentration, which appear to be involved in parasite uptake (reviewed in Burleigh and Woolsey, 2002). Three parasite molecules appear to play a central role in this, but by different mechanisms: oligopeptidase B (opb) indirectly triggers thapsigargin-sensitive calcium mobilization (Burleigh and Woolsey, 2002); cruzipain acts through bradykinin receptors and localized kinin release, leading phospholipase C and inositol-1,4,5 triphosphate production (Rodriguez *et al.*, 1995; Scharfstein *et al.*, 2000); and parasite surface glycoproteins gp83, gp90 and gp35/50 trigger calcium mobilization in both parasite and host cell (Burleigh and Woolsey, 2002). The situation in macrophages is less well studied than in other cell types, but calcium bursts are evident following macrophage-parasite contact, and this is linked to parasite uptake (Wilkowsky *et al.*, 1996). However, it is not clear whether calcium mobilization in macrophages employs any or all of the mechanisms found in non-immune cells. The only mechanistic study available instead implicates binding of the parasite surface molecule tGPI-mucin to L-selectin (de Diego *et al.*, 1997). However, uptake of *T. cruzi* by macrophages has been shown to involve gp83 and PKC (Villalta *et al.*, 1999), which implies that gp83 may trigger calcium busts in macrophages. In contrast, opb-null parasites enter J774 macrophages at a similar rate to wild-type, although the effect on calcium signalling was not studied (Caler *et al.*, 2000). Interestingly, a homology search reveals orthologues of opb in both *L. major* and *L. infantum*, allowing us to speculate that this mechanism may contribute calcium mobilization triggered by *Leishmania*.

In contrast to *Leishmania*, it appears that *T. cruzi* triggers tyrosine phosphorylation in macrophages. A brief report showed that exposure to trypomastigotes causes dephosphorylation in myoblasts, and that phosphatase inhibition reduced parasite uptake (Zhong *et al.*, 1998). However, equivalent studies with macrophages show increased tyrosine phosphorylation (Ruta *et al.*, 1996), and reduced parasite uptake following kinase inhibition (Vieira *et al.*, 1994). Treatment of macrophages with the *T. cruzi* surface molecule gp83 enhances both parasite uptake and protein phosphorylation, including ERK1 phosphorylation (Villalta *et al.*, 1999). Importantly, the effect can be reversed pharmacologically by global kinase inhibition or by specific inhibition of MAPKK (Villalta *et al.*, 1999). Another type of surface molecule, tGPI and tGPI-mucin, also cause MAP kinase phosphorylation in macrophages. Treatment of peritoneal macrophages with either molecule results in increased phosphorylation of ERK1/2, $p38^{SAPK2}$ and MKK4, leading to increased expression of inflammatory mediators leading to increased expression of TNFα and, in conjunction with IFNγ, IL-12 and NO (Ropert *et al.*, 2001). Consistent with activation, rather than repression, of intracellular phosphorylation cascades, uptake of *T. cruzi* does not cause degradation of STAT1α (Forget *et al.*, 2005b). While this study did not address the phosphorylation state or IFNγ responsiveness of STAT1α, it is important in the current context as one of very few to have directly compared the effect of *T. cruzi* and *Leishmania* on macrophages.

A number of reports in the past few years have addressed the role of TLRs. Purified *T. cruzi* surface molecules tGPI and tGPI mucin have been shown to activate macrophages via TLR2, not TLR4, leading to NF-κB translocation via MAP kinase cascades (Campos *et al.*, 2001; Ropert *et al.*, 2001). This NF-κB induces both TNF-α and IL-12 (Ropert *et al.*, 2001). Interestingly, in addition to macrophage activation, stimulation of TLRs with tGPI-mucin causes an NF-κB and (serine/threonine) Protein Phosphatase 2A-mediated hypo-responsiveness to subsequent TLR stimulation (Ropert *et al.*, 2001; Ropert *et al.*, 2003). We can speculate that, in the context of live *T. cruzi*, the interaction between GIPL and TLR2 is actually part of a suppressive mechanism, which renders the macrophage non-responsive to stimulation by other *T. cruzi*-derived TLR ligands. This type of de-sensitization has recently been proposed for the case of the filarial nematode *Acanthocheilonema viteae* (Goodridge *et al.*, 2005): it will be interesting to see if it is employed by *Leishmania* as well.

African trypanosomes

The markedly different lifestyle of African trypanosomes, such as *T. brucei*, compared with *Leishmania*, correlates with a very different impact on macrophage biology. Instead of invading the macrophage and preventing its activation, African trypanosomes remain free in the bloodstream, where they must evade both the innate and acquired immune systems. The prevalent model is that macrophages in infected individuals adopt a 'suppressive' phenotype that inhibits T-cell proliferation by secreting TNFα, NO, and PGE_2 (Schleifer and Mansfield, 1993; Darji *et al.*, 1996; Mansfield and Paulnock, 2005). The principal trigger for this behaviour is apparently the trypanosome secreted polysaccharide sVSG, but reports identifying the macrophage signalling pathways involved are lacking. However, sVSG appears to modulate the response of macrophages to IFNγ. Pre-treatment of Raw 264.7 macrophages with sVSG reduces induction of iNOS and ICSBP by subsequent IFNγ stimulation (Coller *et al.*, 2003). This is associated with reduced STAT1 phosphorylation, but not degradation, suggesting that sVSG may inhibit the response to IFNγ by phosphatase-mediated inhibition of dephosphorylating JAK2 and/or STAT1 (Coller *et al.*, 2003). However, when bone-marrow derived macrophages are primed with IFNγ, treatment with sVSG, or its membrane-bound equivalent, induces NO and cytokine (i.e. IL-6, IL-12 and TNFα) production (Drennan *et al.*, 2005). Furthermore, at higher concentrations, sVSG alone induces a subset of IFNγ-responsive genes, including TNFα, iNOS, IL-6 and IL-12, but not ICSBP or MIG (Paulnock and Coller, 2001). These contrasting results indicate both the complexity of the intracellular signalling response to sVSG, and the resultant need for care in the design and interpretation of studies. Interestingly, the cooperative induction of NO and cytokines by VSG and IFNγ is greatly reduced in MyD88 or TLR9 null macrophages, indicating that VSG is a ligand for TLR9, resulting in activation via MyD88 (Drennan *et al.*, 2005).

Concluding remarks

Intracellular parasites that live in the harsh environment of phagocytes must undergo a rapid physiological adaptation, and have strategies to escape from the host microbicidal activities and to inhibit various functions of their host cells. This chapter has addressed some of the mechanisms by which *Leishmania* achieve these goals by manipulating the signalling pathways of the host macrophage. Such strategies to manipulate the host immune response are by no means unique to *Leishmania*. There is compelling evidence that a great number of unicellular and pluricellular pathogens have also evolved means to inactivate or exacerbate immune cell functions, and this contributes to parasite survival and/or the development of pathogen-specific pathologies. It is also likely that mechanistic details will differ – sometimes markedly – between promastigotes and amastigotes, and also between different species of *Leishmania*. These differences are likely to be of great importance in explaining the widely different clinical manifestations of leishmaniasis. However, it is clear that the common trait of these modulations is the manipulation of the host cell signalling system. As signalling pathways can be pharmacologically manipulated, a better knowledge of their role and the mechanisms whereby they regulate host immune cell functions and pathogens growth should permit the development of new therapy to control infectious agents.

Acknowledgments

M.O. is supported by grants from the Canadian Institutes in Health Research (CIHR) and is member of a CIHR Group in Host-Pathogen Interaction. M.O. is recipient of a Fonds de la Recherche en Santé du Québec (FRSQ) Senior Scholarship and is a Burroughs Wellcome Fund Awardee in Molecular Parasitology. D.J.G. is recipient of CIHR Training Grant post-doctoral award from the Centre for the Study of Host Resistance at McGill University Health Centre.

References

Agrawal, A., Cha-Molstad, H., Samols, D., and Kushner, I. (2001). Transactivation of C-reactive protein by IL-6 requires synergistic interaction of CCAAT/enhancer binding protein beta (C/EBP beta) and Rel p50. J. Immunol. *166*, 2378–2384.

Alvarez, G.R., Zwilling, B.S., and Lafuse, W.P. (2003). *Mycobacterium avium* inhibition of IFN-gamma signaling in mouse macrophages: Toll-like receptor 2 stimulation increases expression of dominant-negative STAT1 beta by mRNA stabilization. J. Immunol. *171*, 6766–6773.

Anstead, G.M., Chandrasekar, B., Zhao, W., Yang, J., Perez, L.E., and Melby, P.C. (2001). Malnutrition alters the innate immune response and increases early visceralization following *Leishmania donovani* infection. Infect. Immun. *69*, 4709–4718.

Awasthi, A., Mathur, R., Khan, A., Joshi, B.N., Jain, N., Sawant, S., Boppana, R., Mitra, D., and Saha, B. (2003). CD40 signaling is impaired in *L. major*-infected macrophages and is rescued by a p38MAPK activator establishing a host-protective memory T cell response. J. Exp. Med. *197*, 1037–1043.

Banville, D., Stocco, R., and Shen, S.H. (1995). Human protein tyrosine phosphatase 1C (PTPN6) gene structure: alternate promoter usage and exon skipping generate multiple transcripts. Genomics *27*, 165–173.

Barr, S.D. and Gedamu, L. (2003). Role of peroxidoxins in *Leishmania chagasi* survival. Evidence of an enzymatic defense against nitrosative stress. J. Biol. Chem. *278*, 10816–10823.

Becker, I., Salaiza, N., Aguirre, M., Delgado, J., Carrillo-Carrasco, N., Kobeh, L.G., Ruiz, A., Cervantes, R., Torres, A.P., Cabrera, N., Gonzalez, A., Maldonado, C., and Isibasi, A. (2003). *Leishmania* lipophosphoglycan (LPG) activates NK cells through toll-like receptor-2. Mol. Biochem Parasitol. *130*, 65–74.

Belkaid, Y., Butcher, B., and Sacks, D.L. (1998). Analysis of cytokine production by inflammatory mouse macrophages at the single-cell level: selective impairment of IL-12 induction in *Leishmania*-infected cells. Eur. J. Immunol. *28*, 1389–1400.

Belley, A. and Chadee, K. (1995). Eicosanoid production by parasites: From pathogenesis to immunomodulation? Parasitol. Today. *11*, 327–334.

Bertholet, S., Dickensheets, H.L., Sheikh, F., Gam, A.A., Donnelly, R.P., and Kenney, R.T. (2003). *Leishmania donovani*-induced expression of suppressor of cytokine signaling 3 in human macrophages: a novel mechanism for intracellular parasite suppression of activation. Infect. Immun. *71*, 2095–2101.

Bhardwaj, N., Rosas, L.E., Lafuse, W.P., and Satoskar, A.R. (2005). *Leishmania* inhibits STAT1-mediated IFN-[gamma] signaling in macrophages: increased tyrosine phosphorylation of dominant negative STAT1[beta] by *Leishmania mexicana*. Int. J. Parasitol. *35*, 75–82.

Bhattacharyya, S., Ghosh, S., Jhonson, P.L., Bhattacharya, S.K., and Majumdar, S. (2001a). Immunomodulatory role of interleukin-10 in visceral leishmaniasis: defective activation of protein kinase C-mediated signal transduction events. Infect. Immun. *69*, 1499–1507.

Bhattacharyya, S., Ghosh, S., Sen, P., Roy, S., and Majumdar, S. (2001b). Selective impairrment of protein kinase C isotypes in murine macrophage by *Leishmania donovani*. Mol. Cell. Biochem. *216*, 47–57.

Blackwell, J.M., Ezekowitz, R.A., Roberts, M.B., Channon, J.Y., Sim, R.B., and Gordon, S. (1985). Macrophage complement and lectin-like receptors bind *Leishmania* in the absence of serum. J. Exp. Med. *162*, 324–331.

Blanchette, J., Racette, N., Faure, R., Siminovitch, K.A., and Olivier, M. (1999). *Leishmania*-induced increases in activation of macrophage SHP-1 tyrosine phosphatase are associated with impaired IFN-gamma-triggered JAK2 activation. Eur. J. Immunol. *29*, 3737–3744.

Bodman-Smith, K.B., Mbuchi, M., Culley, F.J., Bates, P.A., and Raynes, J.G. (2002). C-reactive protein-mediated phagocytosis of *Leishmania donovani* promastigotes does not alter parasite survival or macrophage responses. Parasite Immunol. *24*, 447–454.

Bogdan, C. and Rollinghoff, M. (1998). The immune response to *Leishmania*: mechanisms of parasite control and evasion. Int. J. Parasitol. *28*, 21–134.

Brender, C., Columbus, R., Metcalf, D., Handman, E., Starr, R., Huntington, N., Tarlinton, D., Odum, N., Nicholson, S.E., Nicola, N.A., Hilton, D.J., and Alexander, W.S. (2004). SOCS5 is expressed in primary B and T lymphoid cells but is dispensable for lymphocyte production and function. Mol. Cell. Biol. *24*, 6094–6103.

Brittingham, A., Chen, G., McGwire, B.S., Chang, K.P., and Mosser, D.M. (1999). Interaction of *Leishmania* gp63 with cellular receptors for fibronectin. Infect. Immun. *67*, 4477–4484.

Brittingham, A., Morrison, C.J., Mcmaster, W.R., McGwire, B.S., Chang, K.P., and Mosser, D.M. (1995). Role of the *Leishmania* surface protease gp63 in complement fixation, cell adhesion, and resistance to complement-mediated lysis. J. Immunol. *155*, 3102–3111.

Buates, S. and Matlashewski, G. (2001). General suppression of macrophage gene expression during *Leishmania donovani* infection. J. Immunol. *166*, 3416–3422.

Buchmuller-Rouiller, Y. and Mauel, J. (1987). Impairment of macrophage oxidative-metabolism by *Leishmania*. Experientia *43*, 665–665.

Bullen, D.V.R., Baldwin, T.M., Curtis, J.M., Alexander, W.S., and Handman, E. (2003). Persistence of lesions in suppressor of cytokine signaling-1-deficient mice infected with *Leishmania major*. J. Immunol. *170*, 4267–4272.

Burleigh, B.A. and Woolsey, A.M. (2002). Cell signalling and *Trypanosoma cruzi* invasion. Cell. Microbiol. *4*, 701–711.

Burshtyn, D.N., Yang, W., Yi, T., and Long, E.O. (1997). A novel phosphotyrosine motif with a critical amino acid at position −2 for the SH2 domain-mediated activation of the tyrosine phosphatase SHP-1. J. Biol. Chem. *272*, 13066–13072.

Caler, E.V., Morty, R.E., Burleigh, B.A., and Andrews, N.W. (2000). Dual role of signaling pathways leading to Ca2 + and cyclic AMP elevation in host cell invasion by *Trypanosoma cruzi*. Infect. Immun. *68*, 6602–6610.

Cameron, P., McGachy, A., Anderson, M., Paul, A., Coombs, G.H., Mottram, J.C., Alexander, J., and Plevin, R. (2004). Inhibition of lipopolysaccharide-induced macrophage IL-12 production by *Leishmania mexicana* amastigotes: The role of cysteine peptidases and the NF-κB signaling pathway. J. Immunol. *173*, 3297–3304.

Campbell, K.A., Ovendale, P.J., Kennedy, M.K., Fanslow, W.C., Reed, S.G., and Maliszewski, C.R. (1996). CD40 ligand is required for protective cell-mediated immunity to *Leishmania major*. Immunity. *4*, 283–289.

Campos, M.A.S., Almeida, I.C., Takeuchi, O., Akira, S., Valente, E.P., Procopio, D.O., Travassos, L.R., Smith, J.A., Golenbock, D.T., and Gazzinelli, R.T. (2001). Activation of Toll-like receptor-2 by glycosylphosphatidylinositol anchors from a protozoan parasite. J. Immunol. *167*, 416–423.

Carrera, L., Gazzinelli, R.T., Badolato, R., Hieny, S., Muller, W., Kuhn, R., and Sacks, D.L. (1996). *Leishmania* promastigotes selectively inhibit interleukin 12 induction in bone marrow-derived macrophages from susceptible and resistant mice. J. Exp. Med. *183*, 515–526.

Chan, J., Fujiwara, T., Brennan, P., McNeil, M., Turco, S.J., Sibille, J.C., Snapper, M., Aisen, P., and Bloom, B.R. (1989). Microbial glycolipids: possible virulence factors that scavenge oxygen radicals. Proc. Natl. Acad. Sci. USA *86*, 2453–2457.

Chaussabel, D., Semnani, R.T., McDowell, M.A., Sacks, D., Sher, A., and Nutman, T.B. (2003). Unique gene expression profiles of human macrophages and dendritic cells to phylogenetically distinct parasites. Blood. *102*, 672–681.

Chiang, S.C., Ali, V., Huang, A.L., Chu, K.Y., and Tone, L.S. (2001). Molecular, cellular and functional characterizations of a novel ICAM-like molecule of the immunoglobulin superfamily from *Leishmania mexicana amazonensis*. Mol. Biochem. Parasitol. *112*, 263–275.

Coller, S.P., Mansfield, J.M., and Paulnock, D.M. (2003). Glycosylinositolphosphate soluble variant surface glycoprotein inhibits IFN-{gamma}-induced nitric oxide production via reduction in STAT1 phosphorylation in African Trypanosomiasis. J. Immunol. *171*, 1466–1472.

Corradin, S., Mauel, J., Ransijn, A., Sturzinger, C., and Vergeres, G. (1999a). Downregulation of MARCKS-related protein (MRP) in macrophages infected with *Leishmania*. J. Biol. Chem. *274*, 16782–16787.

Corradin, S., Ransijn, A., Corradin, G., Roggero, M.A., Schmitz, A.A., Schneider, P., Mauel, J., and Vergeres, G. (1999b). MARCKS-related protein (MRP) is a substrate for the *Leishmania major* surface protease leishmanolysin (gp63). J. Biol. Chem. *274*, 25411–25418.

Courret, N., Prina, E., Mougneau, E., Saraiva, E.M., Sacks, D.L., Glaichenhaus, N., and Antoine, J.C. (1999). Presentation of the *Leishmania* antigen LACK by infected macrophages is dependent upon the virulence of the phagocytosed parasites. Eur. J. Immunol. *29*, 762–773.

Culley, F.J., Harris, R.A., Kaye, P.M., McAdam, K.P., and Raynes, J.G. (1996). C-reactive protein binds to a novel ligand on *Leishmania donovani* and increases uptake into human macrophages. J. Immunol. *156*, 4691–4696.

Darji, A., Beschin, A., Sileghem, M., Heremans, H., Brys, L., and De Baetselier, P. (1996). In vitro simulation of immunosuppression caused by *Trypanosoma brucei*: active involvement of gamma interferon and tumor necrosis factor in the pathway of suppression. Infect. Immun. *64*, 1937–1943.

Das, S., Saha, A.K., Remaley, A.T., Glew, R.H., Dowling, J.N., Kajiyoshi, M., and Gottlieb, M. (1986). Hydrolysis of phosphoproteins and inositol phosphates by cell surface phosphatase of *Leishmania donovani*. Mol. Biochem. Parasitol. *20*, 143–153.

David, M., Chen, H.E., Goelz, S., Larner, A.C., and Neel, B.G. (1995). Differential regulation of the alpha/beta interferon-stimulated Jak/Stat pathway by the SH2 domain-containing tyrosine phosphatase SHPTP1. Mol. Cell Biol. *15*, 7050–7058.

De Almeida, M.C., Cardoso, S.A., and Barral-Netto, M. (2003). *Leishmania* (Leishmania) *chagasi* infection alters the expression of cell adhesion and costimulatory molecules on human monocyte and macrophage. Int. J. Parasitol. *33*, 153–162.

de Diego, J., Punzon, C., Duarte, M., and Fresno, M. (1997). Alteration of macrophage function by a *Trypanosoma cruzi* membrane mucin. J. Immunol. *159*, 4983–4989.

De Souza Leao S., Lang, T., Prina, E., Hellio, R., and Antoine, J.C. (1995). Intracellular *Leishmania amazonensis* amastigotes internalize and degrade MHC class II molecules of their host cells. J. Cell Sci. *108*, 3219–3231.

de Veer, M.J., Curtis, J.M., Baldwin, T.M., DiDonato, J.A., Sexton, A., McConville, M.J., Handman, E., and Schofield, L. (2003). MyD88 is essential for clearance of *Leishmania major*: possible role for lipophosphoglycan and Toll-like receptor 2 signaling. Eur. J. Immunol. *33*, 2822–2831.

Debus, A., Glasner, J., Rollinghoff, M., and Gessner, A. (2003). High levels of susceptibility and T helper 2 response in MyD88-deficient mice infected with *Leishmania major* are interleukin-4 dependent. Infect. Immun. *71*, 7215–7218.

Dechend, R., Hirano, F., Lehmann, K., Heissmeyer, V., Ansieau, S., Wulczyn, F.G., Scheidereit, C., and Leutz, A. (1999). The Bcl-3 oncoprotein acts as a bridging factor between NF-kappaB/Rel and nuclear co-regulators. Oncogene. *18*, 3316–3323.

Descoteaux, A. and Matlashewski, G. (1989). c-fos and tumor necrosis factor gene expression in *Leishmania donovani*-infected macrophages. Mol. Cell Biol. *9*, 5223–5227.

Descoteaux, A., Matlashewski, G., and Turco, S.J. (1992). Inhibition of macrophage protein kinase C-mediated protein phosphorylation by *Leishmania donovani* lipophosphoglycan. J. Immunol. *149*, 3008–3015.

Descoteaux, A. and Turco, S.J. (1999). Glycoconjugates in *Leishmania* infectivity. Biochim. Biophys. Acta. (*1455*, 341–352.

Desjardins, M. and Descoteaux, A. (1997). Inhibition of phagolysosomal biogenesis by the *Leishmania* lipophosphoglycan. J. Exp. Med. *185*, 2061–2068.

Drennan, M.B., Stijlemans, B., Van Den Abbeele, J., Quesniaux, V.J., Barkhuizen, M., Brombacher, F., De Baetselier, P., Ryffel, B., and Magez, S. (2005). The Induction of a type 1 immune response following a *Trypanosoma brucei* infection is MyD88 dependent. J. Immunol. *175*, 2501–2509.

Eilam, Y., El On, J., and Spira, D.T. (1985). *Leishmania major*: excreted factor, calcium ions, and the survival of amastigotes. Exp. Parasitol. *59*, 161–168.

Endo, T.A., Masuhara, M., Yokouchi, M., Suzuki, R., Sakamoto, H., Mitsui, K., Matsumoto, A., Tanimura, S., Ohtsubo, M., Misawa, H., Miyazaki, T., *et al.* (1997). A new protein containing an SH2 domain that inhibits JAK kinases. Nature. *387*, 921–924.

Farrell, J.P. and Kirkpatrick, C.E. (1987). Experimental cutaneous leishmaniasis. II. A possible role for prostaglandins in exacerbation of disease in *Leishmania major*-infected BALB/c mice. J. Immunol. *138*, 902–907.

Feng, G.J., Goodridge, H.S., Harnett, M.M., Wei, X.Q., Nikolaev, A.V., Higson, A.P., and Liew, F.Y. (1999). Extracellular signal-related kinase (ERK) and p38 mitogen-activated protein (MAP) kinases differentially regulate the lipopolysaccharide-mediated induction of inducible nitric oxide synthase and IL-12 in macrophages: *Leishmania* phosphoglycans subvert macrophage IL-12 production by targeting ERK MAP kinase. J. Immunol. *163*, 6403–6412.

Forget, G., Matte, C., Siminovitch, K.A., Rivest, S., Pouliot, P., and Olivier, M. (2005a). Regulation of the *Leishmania*-induced innate inflammatory response by the protein tyrosine phosphatase SHP-1. Eur. J. Immunol. *35*, 1906–1917.

Forget, G., Siminovitch, K.A., Brochu, S., Rivest, S., Radzioch, D., and Olivier, M. (2001). Role of host phosphotyrosine phosphatase SHP-1 in the development of murine leishmaniasis. Eur. J. Immunol. *31*, 3185–3196.

Forget, G., Gregory, D.J., and Olivier, M. (2005b). Proteasome-mediated degradation of STAT1alpha following infection of macrophages with *Leishmania donovani*. J. Biol. Chem. *280*, 30542–30549.

Forget, G., Gregory, D.J., Whitcombe, L.A., and Olivier, M. (2006). Role of host protein tyrosine phosphatase SHP-1 in *Leishmania donovani*-induced inhibition of nitric oxide production. Infect. Immun. *74*, 6272–6279

Frankenburg, S., Leibovici, V., Mansbach, N., Turco, S.J., and Rosen, G. (1990). Effect of glycolipids of *Leishmania* parasites on human monocyte activity. Inhibition by lipophosphoglycan. J. Immunol. *145*, 4284–4289.

Frearson, J.A. and Alexander, D.R. (1997). The role of phosphotyrosine phosphatases in haematopoietic cell signal transduction. Bioessays. *19*, 417–427.

Freitas Balanco, J.M., Moreira, M.E., Bonomo, A., Bozza, P.T., Amarante-Mendes, G., Pirmez, C., and Barcinski, M.A. (2001). Apoptotic mimicry by an obligate intracellular parasite downregulates macrophage microbicidal activity. Curr. Biol. *11*, 1870–1873.

Fruth, U., Solioz, N., and Louis, J.A. (1993). *Leishmania major* interferes with antigen presentation by infected macrophages. J. Immunol. *150*, 1857–1864.

Fujita, T., Nolan, G.P., Liou, H.C., Scott, M.L., and Baltimore, D. (1993). The candidate proto-oncogene bcl-3 encodes a transcriptional coactivator that activates through NF-kappa B p50 homodimers. Genes Dev. *7*, 1354–1363.

Gantt, K.R., Schultz-Cherry, S., Rodriguez, N., Jeronimo, S.M., Nascimento, E.T., Goldman, T.L., Recker, T.J., Miller, M.A., and Wilson, M.E. (2003). Activation of TGF-beta by *Leishmania chagasi*: importance for parasite survival in macrophages. J. Immunol. *170*, 2613–2620.

Ghosh, S., Bhattacharyya, S., Das, S., Raha, S., Maulik, N., Das, D.K., Roy, S., and Majumdar, S. (2001). Generation of ceramide in murine macrophages infected with *Leishmania donovani* alters macrophage signaling events and aids intracellular parasitic survival. Mol. Cell. Biochem. *223*, 47–60.

Ghosh, S., Bhattacharyya, S., Sirkar, M., Sa, G.S., Das, T., Majumdar, D., Roy, S., and Majumdar, S. (2002). *Leishmania donovani* suppresses activated protein 1 and NF-kappaB activation in host macrophages via ceramide generation: involvement of extracellular signal-regulated kinase. Infect. Immun. *70*, 6828–6838.

Ghosh, S., Goswami, S., and Adhya, S. (2003). Role of superoxide dismutase in survival of *Leishmania* within the macrophage. Biochem. J. *369*, 447–452.

Goodridge, H.S., Marshall, F.A., Else, K.J., Houston, K.M., Egan, C., Al Riyami, L., Liew, F.Y., Harnett, W., and Harnett, M.M. (2005). Immunomodulation via novel use of TLR4 by the filarial nematode phosphorylcholine-containing secreted product, ES-62. J. Immunol. *174*, 284–293.

Greenhalgh, C.J. and Hilton, D.J. (2001). Negative regulation of cytokine signaling. J. Leukoc. Biol. *70*, 348–356.

Guizani-Tabbane, L., Ben Aissa, K., Belghith, M., Sassi, A., and Dellagi, K. (2004). *Leishmania major* amastigotes induce p50/c-Rel NF-{kappa}B transcription factor in human macrophages: involvement in cytokine synthesis. Infect. Immun. *72*, 2582–2589.

Guy, R.A. and Belosevic, M. (1993). Comparison of receptors required for entry of *Leishmania major* amastigotes into macrophages. Infect. Immun. *61*, 1553–1558.

Haque, S.J., Harbor, P., Tabrizi, M., Yi, T., and Williams, B.R. (1998). Protein-tyrosine phosphatase Shp-1 is a negative regulator of IL-4- and IL-13-dependent signal transduction. J. Biol. Chem. *273*, 33893–33896.

Hatzigeorgiou, D.E., Geng, J., Zhu, B., Zhang, Y., Liu, K., Rom, W.N., Fenton, M.J., Turco, S.J., and Ho,

J.L. (1996). Lipophosphoglycan from *Leishmania* suppresses agonist-induced interleukin 1 beta gene expression in human monocytes via a unique promoter sequence. Proc. Natl. Acad. Sci. USA 93, 14708–14713.

Hawn, T.R., Ozinsky, A., Underhill, D.M., Buckner, F.S., Akira, S., and Aderem, A. (2002). *Leishmania major* activates IL-1 alpha expression in macrophages through a MyD88-dependent pathway. Microbes Infect. *4*, 763–771.

Heinzel, F.P., Rerko, R.M., and Hujer, A.M. (1998). Underproduction of interleukin-12 in susceptible mice during progressive leishmaniasis is due to decreased CD40 activity. Cell. Immunol. *184*, 129–142.

Hermoso, T., Fishelson, Z., Becker, S.I., Hirschberg, K., and Jaffe, C.L. (1991). Leishmanial protein kinases phosphorylate components of the complement system. EMBO J. *10*, 4061–4067.

Hilton, D.J., Richardson, R.T., Alexander, W.S., Viney, E.M., Willson, T.A., Sprigg, N.S., Starr, R., Nicholson, S.E., Metcalf, D., and Nicola, N.A. (1998). Twenty proteins containing a C-terminal SOCS box form five structural classes. Proc. Natl. Acad. Sci. USA. *95*, 114–119.

Holm, A., Tejle, K., Gunnarsson, T., Magnusson, K.E., Descoteaux, A., and Rasmusson, B. (2003). Role of protein kinase C alpha for uptake of unopsonized prey and phagosomal maturation in macrophages. Biochem. Biophys. Res. Commun. *302*, 653–658.

Holm, A., Tejle, K., Magnusson, K.E., Descoteaux, A., and Rasmusson, B. (2001). *Leishmania donovani* lipophosphoglycan causes periphagosomal actin accumulation: correlation with impaired translocation of PKCalpha and defective phagosome maturation. Cell. Microbiol. 3, 439–447.

Huber, M., Timms, E., Mak, T.W., Rollinghoff, M., and Lohoff, M. (1998). Effective and long-lasting immunity against the parasite *Leishmania major* in CD8-deficient mice. Infect. Immun. *66*, 3968–3970.

Joshi, P.B., Sacks, D.L., Modi, G., and Mcmaster, W.R. (1998). Targeted gene deletion of *Leishmania major* genes encoding developmental stage-specific leishmanolysin (GP63). Mol. Microbiol. *27*, 519–530.

Kamanaka, M., Yu, P., Yasui, T., Yoshida, K., Kawabe, T., Horii, T., Kishimoto, T., and Kikutani, H. (1996). Protective role of CD40 in *Leishmania major* infection at two distinct phases of cell-mediated immunity. Immunity. *4*, 275–281.

Kane, M.M. and Mosser, D.M. (2000). *Leishmania* parasites and their ploys to disrupt macrophage activation. Curr. Opin. Hematol. *7*, 26–31.

Kaye, P.M. and Bancroft, G.J. (1992). *Leishmania donovani* infection in scid mice: lack of tissue response and in vivo macrophage activation correlates with failure to trigger natural killer cell-derived gamma interferon production in vitro. Infect. Immun. *60*, 4335–4342.

Kaye, P.M., Rogers, N.J., Curry, A.J., and Scott, J.C. (1994). Deficient expression of co-stimulatory molecules on *Leishmania*-infected macrophages. Eur. J. Immunol. *24*, 2850–2854.

Khaled, A.R., Butfiloski, E.J., Sobel, E.S., and Schiffenbauer, J. (1998). Functional consequences of the SHP-1 defect in motheaten viable mice: role of NF-kappa B. Cell. Immunol. *185*, 49–58.

Kile, B.T., Nicola, N.A., and Alexander, W.S. (2001). Negative regulators of cytokine signaling. Int. J. Hematol. *73*, 292–298.

Kim, T.K. and Maniatis, T. (1996). Regulation of interferon-gamma-activated STAT1 by the ubiquitin-proteasome pathway. Science. *273*, 1717–1719.

Kima, P.E., Ruddle, N.H., and McMahon-Pratt, D. (1997). Presentation via the class I pathway by *Leishmania amazonensis*-infected macrophages of an endogenous leishmanial antigen to CD8$^+$ T cells. J. Immunol. *159*, 1828–1834.

Kima, P.E., Soong, L., Chicharro, C., Ruddle, N.H., and McMahon-Pratt, D. (1996). *Leishmania*-infected macrophages sequester endogenously synthesized parasite antigens from presentation to CD4$^+$ T cells. Eur. J. Immunol. *26*, 3163–3169.

Kirkpatrick, C.E., Farrell, J.P., Warner, J.F., and Denner G (1985). Participation of natural killer cells in the recovery of mice from visceral leishmaniasis. Cell. Immunol. *92*, 163–171.

Klingmuller, U., Lorenz, U., Cantley, L.C., Neel, B.G., and Lodish, H.F. (1995). Specific recruitment of SH-PTP1 to the erythropoietin receptor causes inactivation of JAK2 and termination of proliferative signals. Cell. *80*, 729–738.

Krautwald, S., Buscher, D., Kummer, V., Buder, S., and Baccarini, M. (1996). Involvement of the protein tyrosine phosphatase SHP-1 in Ras-mediated activation of the mitogen-activated protein kinase pathway. Mol. Cell Biol. *16*, 5955–5963.

Kropf, P., Freudenberg, M.A., Modolell, M., Price, H.P., Herath, S., Antoniazi, S., Galanos, C., Smith, D.F., and Muller, I. (2004a). Toll-like receptor 4 contributes to efficient control of infection with the protozoan parasite *Leishmania major*. Infect. Immun. *72*, 1920–1928.

Kropf, P., Freudenberg, N., Kalis, C., Modolell, M., Herath, S., Galanos, C., Freudenberg, M., and Muller, I. (2004b). Infection of C57BL/10ScCr and C57BL/10ScNCr mice with *Leishmania major* reveals a role for Toll-like receptor 4 in the control of parasite replication. J. Leukoc. Biol. *76*, 48–57.

Kurland, J.F., Kodym, R., Story, M.D., Spurgers, K.B., McDonnell, T.J., and Meyn, R.E. (2001). NF-kappaB1 (p50) homodimers contribute to transcription of the bcl-2 oncogene. J. Biol. Chem. *276*, 45380–45386.

Kwan, W.C., Mcmaster, W.R., Wong, N., and Reiner, N.E. (1992). Inhibition of expression of major histocompatibility complex class II molecules in macrophages infected with *Leishmania donovani* occurs at the level of gene transcription via a cyclic AMP-independent mechanism. Infect. Immun. *60*, 2115–2120.

Lang, T., de Chastellier, C., Frehel, C., Hellio, R., Metezeau, P., Leao, S.S., and Antoine, J.C. (1994a). Distribution of MHC class I and of MHC class II molecules in macrophages infected with *Leishmania amazonensis*. J. Cell Sci. *107* (Pt 1), 69–82.

Lang, T., Hellio, R., Kaye, P.M., and Antoine, J.C. (1994b). *Leishmania donovani*-infected macrophages: characterization of the parasitophorous vacuole and potential role of this organelle in antigen presentation. J. Cell Sci. *107* (Pt 8): 2137–2150.

Liew, F.Y., Millott, S., Parkinson, C., Palmer, R.M., and Moncada, S. (1990). Macrophage killing of *Leishmania* parasite in vivo is mediated by nitric oxide from L-arginine. J. Immunol. *144*, 4794–4797.

Lima, G.M., Vallochi, A.L., Silva, U.R., Bevilacqua, E.M., Kiffer, M.M., and Abrahamsohn, I.A. (1998). The role of polymorphonuclear leukocytes in the resistance to cutaneous leishmaniasis. Immunol Lett. *64*, 145–151.

Locksley, R.M., Reiner, S.L., Hatam, F., Littman, D.R., and Killeen, N. (1993). Helper T cells without CD4, control of leishmaniasis in CD4-deficient mice. Science. *261*, 1448–1451.

Maasho, K., Sanchez, F., Schurr, E., Hailu, A., and Akuffo, H. (1998). Indications of the protective role of natural killer cells in human cutaneous leishmaniasis in an area of endemicity. Infect. Immun. *66*, 2698–2704.

McNeely, T.B., Rosen, G., Londner, M.V., and Turco, S.J. (1989). Inhibitory effects on protein kinase C activity by lipophosphoglycan fragments and glycosylphosphatidylinositol antigens of the protozoan parasite *Leishmania*. Biochem. J. *259*, 601–604.

McNeely, T.B. and Turco, S.J. (1987). Inhibition of protein kinase C activity by the *Leishmania donovani* lipophosphoglycan. Biochem. Biophys. Res. Commun. *148*, 653–657.

Malek, R.L. and Halvorsen, S.W. (1999). Ciliary neurotrophic factor and phorbol ester each decrease selected STAT3 pools in neuroblastoma cells by proteasome-dependent mechanisms. Cytokine. *11*, 192–199.

Mansfield, J.M. and Olivier, M. (2002). Immune Evasion by Parasites. In: Immunology of infectious diseases. S.H.E. Kaufmann, A. Sher, and R. Ahmed, Washington, D.C ed. (ASM Press) pp. 379–392.

Mansfield, J.M. and Paulnock, D.M. (2005). Regulation of innate and acquired immunity in African trypanosomiasis. Parasite Immunol. *27*, 361–371.

Marrero, M.B., Venema, V.J., Ju, H., Eaton, D.C., and Venema, R.C. (1998). Regulation of angiotensin II-induced JAK2 tyrosine phosphorylation: roles of SHP-1 and SHP-2. Am. J. Physiol. *275*, C1216-C1223.

Marth, T. and Kelsall, B.L. (1997). Regulation of interleukin-12 by complement receptor 3 signaling. J. Exp. Med. *185*, 1987–1995.

Martiny, A., Meyer-Fernandes, J.R., de Souza, W., and Vannier-Santos, M.A. (1999). Altered tyrosine phosphorylation of ERK1 MAP kinase and other macrophage molecules caused by *Leishmania* amastigotes. Mol. Biochem. Parasitol. *102*, 1–12.

Massa, P.T. and Wu, C. (1998). Increased inducible activation of NF-kappaB and responsive genes in astrocytes deficient in the protein tyrosine phosphatase SHP-1. J. Interferon Cytokine Res. *18*, 499–507.

Masuhara, M., Sakamoto, H., Matsumoto, A., Suzuki, R., Yasukawa, H., Mitsui, K., Wakioka, T., Tanimura, S., Sasaki, A., Misawa, H., Yokouchi, M., Ohtsubo, M., and Yoshimura, A. (1997). Cloning and characterization of novel CIS family genes. Biochem. Biophys. Res. Commun. *239*, 439–446.

Matsumoto, A., Masuhara, M., Mitsui, K., Yokouchi, M., Ohtsubo, M., Misawa, H., Miyajima, A., and Yoshimura, A. (1997). CIS, a cytokine inducible SH2 protein, is a target of the JAK-STAT5 pathway and modulates STAT5 activation. Blood. *89*, 3148–3154.

Matte, C., Maion, G., Mourad, W., and Olivier, M. (2001). *Leishmania donovani*-induced macrophages cyclooxygenase-2 and prostaglandin E2 synthesis. Parasite Immunol. *23*, 177–184.

Matte, C., Marquis, J.F., Blanchette, J., Gros, P., Faure, R., Posner, B.I., and Olivier, M. (2000). Peroxovanadium-mediated protection against murine leishmaniasis: role of the modulation of nitric oxide. Eur. J. Immunol. *30*, 2555–2564.

Matte, C. and Olivier, M. (2002). *Leishmania*-induced cellular recruitment during the early inflammatory response: modulation of proinflammatory mediators. J. Infect. Dis. 185, 673–681.

Miao, L., Stafford, A., Nir, S., Turco, S.J., Flanagan, T.D., and Epand, R.M. (1995). Potent inhibition of viral fusion by the lipophosphoglycan of *Leishmania donovani*. Biochemistry. *34*, 4676–4683.

Mosser, D.M. and Edelson, P.J. (1987). The third component of complement (C3) is responsible for the intracellular survival of *Leishmania major*. Nature. *327*, 329–331.

Muller, I. (1997). *Leishmania major* infection in C57BL/10 mice differing at the Lps locus: a new non healing phenotype. Med. Microbiol. Immunol. (Berl) *186*, 75–81.

Muraille, E., De Trez, C., Brait, M., De Baetselier, P., Leo, O., and Carlier, Y. (2003). Genetically resistant mice lacking MyD88-adapter protein display a high susceptibility to *Leishmania major* infection associated with a polarized Th2 response. J. Immunol. *170*, 4237–4241.

Murray, H.W. (1982). Cell-mediated immune response in experimental visceral leishmaniasis. II. Oxygen-dependent killing of intracellular *Leishmania donovani* amastigotes. J. Immunol. *129*, 351–357.

Murray, H.W. and Nathan, C.F. (1999). Macrophage microbicidal mechanisms in vivo: reactive nitrogen versus oxygen intermediates in the killing of intracellular visceral *Leishmania donovani*. J. Exp. Med. *189*, 741–746.

Na, S.Y., Lee, S.K., Han, S.J., Choi, H.S., Im, S.Y., and Lee, J.W. (1998). Steroid receptor coactivator-1 interacts with the p50 subunit and coactivates nuclear factor kappaB-mediated transactivations. J. Biol Chem. *273*, 10831–10834.

Naka, T., Narazaki, M., Hirata, M., Matsumoto, T., Minamoto, S., Aono, A., Nishimoto, N., Kajita, T., Taga, T., Yoshizaki, K., Akira, S., and Kishimoto, T. (1997). Structure and function of a new STAT-induced STAT inhibitor. Nature. *387*, 924–929.

Nandan, D., Lo, R., and Reiner, N.E. (1999). Activation of phosphotyrosine phosphatase activity attenuates mitogen-activated protein kinase signaling and inhibits c-FOS and nitric oxide synthase expression in macrophages infected with *Leishmania donovani*. Infect. Immun. *67*, 4055–4063.

Nandan, D. and Reiner, N.E. (1995). Attenuation of gamma-interferon-induced tyrosine phosphorylation in Mmononuclear phagocytes infected with

Leishmania donovani. Selective inhibition of signaling through Janus Kinases and Stat1. Infect. Immun. *63*, 4495–4500.

Nandan, D., Yi, T.L., Lopez, M., Lai, C., and Reiner, N.E. (2002). *Leishmania* EF-1 alpha activates the Src homology 2 domain containing tyrosine phosphatase SHP-1 leading to macrophage deactivation. J. Biol. Chem. *277*, 50190–50197.

Olivier, M. (1996). Modulation of host cell intracellular Ca2 $^+$. Parasitol. Today. *12*, 145–150.

Olivier, M., Baimbridge, K.G., and Reiner, N.E. (1992a). Stimulus-response coupling in monocytes infected with *Leishmania*. Attenuation of calcium transients is related to defective agonist-induced accumulation of Inositol Phosphates. J. Immunol. *148*, 1188–1196.

Olivier, M., Brownsey, R.W., and Reiner, N.E. (1992b). Defective stimulus-response coupling in human monocytes infected with *Leishmania* donovani is associated with altered activation and translocation of protein-kinase-C. Proc. Natl. Acad. Sci. USA. *89*, 7481–7485.

Olivier, M., Romero-Gallo, B.J., Matte, C., Blanchette, J., Posner, B.I., Tremblay, M.J., and Faure, R. (1998). Modulation of interferon-gamma-induced macrophage activation by phosphotyrosine phosphatases inhibition. Effect on murine Leishmaniasis progression. J. Biol. Chem. *273*, 13944–13949.

Olivier, M. and Tanner, C.E. (1987). Susceptibilities of macrophage populations to infection in vitro by *Leishmania donovani*. Infect. Immun. 55, 467–471.

Paulnock, D.M. and Coller, S.P. (2001). Analysis of macrophage activation in African trypanosomiasis. J. Leukoc. Biol. 69, 685–690.

Pawson, T. (1995). Protein modules and signalling networks. Nature. *373*, 573–580.

Piedrafita, D., Proudfoot, L., Nikolaev, A.V., Xu, D.M., Sands, W., Feng, G.J., Thomas, E., Brewer, J., Ferguson, M.A.J., Alexander, J., and Liew, F.Y. (1999). Regulation of macrophage IL-12 synthesis by *Leishmania* phosphoglycans. Eur. J. Immunol. *29*, 235–244.

Prina, E., Jouanne, C., de Souza, L.S., Szabo, A., Guillet, J.G., and Antoine, J.C. (1993). Antigen presentation capacity of murine macrophages infected with *Leishmania amazonensis* amastigotes. J. Immunol. *151*, 2050–2061.

Privé, C. and Descoteaux, A. (2000). *Leishmania donovani* promastigotes evade the activation of mitogen-activated protein kinases p38, c-Jun N-terminal kinase, and extracellular signal-regulated kinase-1/2 during infection of naive macrophages. Eur. J. Immunol. *30*, 2235–2244.

Proudfoot, L., Nikolaev, A.V., Feng, G.J., Wei, W.Q., Ferguson, M.A., Brimacombe, J.S., and Liew, F.Y. (1996). Regulation of the expression of nitric oxide synthase and leishmanicidal activity by glycoconjugates of *Leishmania* lipophosphoglycan in murine macrophages. Proc. Natl. Acad. Sci. USA. 93, 10984–10989.

Proudfoot, L., O'Donnell, C.A., and Liew, F.Y. (1995). Glycoinositolphospholipids of *Leishmania major* inhibit nitric oxide synthesis and reduce leishmanicidal activity in murine macrophages. Eur. J. Immunol. *25*, 745–750.

Puentes, S.M., Dwyer, D.M., Bates, P.A., and Joiner, K.A. (1989). Binding and release of C3 from *Leishmania donovani* promastigotes during incubation in normal human serum. J. Immunol. *143*, 3743–3749.

Ram, P.A. and Waxman, D.J. (1997). Interaction of growth hormone-activated STATs with SH2-containing phosphotyrosine phosphatase SHP-1 and nuclear JAK2 tyrosine kinase. J. Biol. Chem. *272*, 17694–17702.

Ray, M., Gam, A.A., Boykins, R.A., and Kenney, R.T. (2000). Inhibition of interferon-gamma signaling by *Leishmania donovani*. J. Infect. Dis. *181*, 1121–1128.

Reiner, N.E. and Malemud, C.J. (1985). Arachidonic acid metabolism by murine peritoneal macrophages infected with *Leishmania donovani*: in vitro evidence for parasite-induced alterations in cyclooxygenase and lipoxygenase pathways. J. Immunol. *134*, 556–563.

Reiner, N.E., Ng, W., Ma, T., and Mcmaster, W.R. (1988). Kinetics of gamma interferon binding and induction of major histocompatibility complex class II mRNA in *Leishmania*-infected macrophages. Proc. Natl. Acad. Sci. USA. *85*, 4330–4334.

Reiner, N.E., Ng, W., and Mcmaster, W.R. (1987). Parasite-accessory cell interactions in murine leishmaniasis.II. *Leishmania* donovani suppresses macrophage expression of class I and class II major histocompatibility complex gene products. J. Immunol. *138*, 1926–1932.

Reiner, N.E., Ng, W., Wilson, C.B., Mcmaster, W.R., and Burchett, S.K. (1990). Modulation of in vitro monocyte cytokine responses to *Leishmania donovani*. Interferon gamma prevents parasite-induced inhibition of interleukin-1 production and primes monocytes to respond to *Leishmania* by producing both tumor necrosis factor-Alpha and interleukin-1. J. Clin. Invest. 85, 1914–1924.

Reiner, N.E., Schultz, L., Malemud, C.J., and Deal, H.E. (1985). Arachidonic acid metabolism by *Leishmania donovani* infected macrophages. Evidence for mouse strain differences in phospholipase-C and lipoxygenase activities. Clin. Invest. Med. *8*, A135-A135.

Rodriguez, A., Rioult, M.G., Ora, A., and Andrews, N.W. (1995). A trypanosome-soluble factor induces IP3 formation, intracellular Ca2 $^+$ mobilization and microfilament rearrangement in host cells. J. Cell Biol. *129*, 1263–1273.

Rodriguez, N.E., Chang, H.K., and Wilson, M.E.(2004). Novel Program of Macrophage Gene Expression Induced by Phagocytosis of *Leishmania chagasi*. Infect. Immun. 72, 2111–2122.

Ropert, C., Almeida, I.C., Closel, M., Travassos, L.R., Ferguson, M.A.J., Cohen, P., and Gazzinelli, R.T. (2001). Requirement of mitogen-activated protein kinases and I{{kappa}}B phosphorylation for induction of proinflammatory cytokines synthesis by macrophages indicates functional similarity of receptors triggered by glycosylphosphatidylinositol anchors from parasitic protozoa and bacterial lipopolysaccharide. J. Immunol. *166*, 3423–3431.

Ropert, C., Closel, M., Chaves, A.C.L., and Gazzinelli, R.T. (2003). Inhibition of a p38/stress-activated protein kinase-2-dependent phosphatase restores function of IL-1 receptor-associated kinase-1 and reverses Toll-Like Receptor 2- and 4-dependent tolerance of macrophages. J. Immunol. *171*, 1456–1465.

Rosas, L.E. (2003). Development of protective immunity against cutaneous Leishmaniasis is dependent on STAT1-mediated IFN signaling pathway. Eur. J. Immunol. *33*, 1799–1805.

Ruta, S., Plasman, N., Zaffran, Y., Capo, C., M ⁺¿ge, J.L., and Vray, B. (1996). *Trypanosoma cruzi*-induced tyrosine phosphorylation in murine peritoneal macrophages. Parasitol. Res. *82*, 481–484.

Saha, B., Das, G., Vohra, H., Ganguly, N.K., and Mishra, G.C. (1995). Macrophage-T cell interaction in experimental visceral Leishmaniasis: failure to express costimulatory molecules on *Leishmania*-infected macrophages and its implication in the suppression of cell-mediated immunity. Eur. J. Immunol. *25*, 2492–2498.

Satoskar, A.R., Stamm, L.M., Zhang, X., Satoskar, A.A., Okano, M., Terhorst, C., David, J.R., and Wang, B. (1999). Mice lacking NK cells develop an efficient Th1 response and control cutaneous *Leishmania major* infection. J. Immunol. *162*, 6747–6754.

Scharfstein, J., Schmitz, V., Morandi, V., Capella, M.M.A., Lima, A.P., Morrot, A., Juliano, L., and Muller-Esterl, W. (2000). Host cell invasion by *Trypanosoma cruzi* is potentiated by activation of bradykinin B2 receptors. J. Exp. Med. *192*, 1289–1300.

Scharton, T.M. and Scott, P. (1993). Natural killer cells are a source of interferon gamma that drives differentiation of CD4 ⁺ T cell subsets and induces early resistance to *Leishmania major* in mice. J. Exp. Med. *178*, 567–577.

Scharton-Kersten, T., Afonso, L.C., Wysocka, M., Trinchieri, G., and Scott, P. (1995). IL-12 is required for natural killer cell activation and subsequent T helper 1 cell development in experimental Leishmaniasis. J. Immunol. *154*, 5320–5330.

Schleifer, K.W. and Mansfield, J.M. (1993). Suppressor macrophages in African trypanosomiasis inhibit T cell proliferative responses by nitric oxide and prostaglandins. J. Immunol. *151*, 5492–5503.

Scianimanico, S., Desrosiers, M., Dermine, J.F., Meresse, S., Descoteaux, A., and Desjardins, M. (1999). Impaired recruitment of the small GTPase rab7 correlates with the inhibition of phagosome maturation by *Leishmania donovani* promastigotes. Cell Microbiol. *1*, 19–32.

Seay, M.B., Heard, P.L., and Chaudhuri, G. (1996). Surface Zn-proteinase as a molecule for defense of *Leishmania mexicana amazonensis* promastigotes against cytolysis inside macrophage phagolysosomes. Infect. Immun. *64*, 5129–5137.

Singh, V., Balaraman, S., Tewary, P., and Madhubala, R. (2004). *Leishmania donovani* activates nuclear transcription factor-[kappa]B in macrophages through reactive oxygen intermediates. Biochem. Biophys. Res. Commun. *322*, 1086–1095.

Somani, A.K., Bignon, J.S., Mills, G.B., Siminovitch, K.A., and Branch, D.R. (1997). Src kinase activity is regulated by the SHP-1 protein-tyrosine phosphatase. J. Biol. Chem. *272*, 21113–21119.

Somanna, A., Mundodi, V., and Gedamu, L. (2002). Functional analysis of cathepsin B-like cysteine proteases from *Leishmania donovani* complex. Evidence for the activation of latent transforming growth factor beta. J. Biol. Chem. *277*, 25305–25312.

Songyang, Z., Shoelson, S.E., Chaudhuri, M., Gish, G., Pawson, T., Haser, W.G., King, F., Roberts, T., Ratnofsky, S., Lechleider, R.J., *et al.* (1993). SH2 domains recognize specific phosphopeptide sequences. Cell. *72*, 767–778.

Soong, L., Xu, J.C., Grewal, I.S., Kima, P., Sun, J., Longley, B.J., Ruddle, N.H., McMahon-Pratt, D., and Flavell, R.A. (1996). Disruption of CD40-CD40 ligand interactions results in an enhanced susceptibility to *Leishmania amazonensis* infection. Immunity. *4*, 263–273.

Sorensen, A.L., Hey, A.S., and Kharazmi, A. (1994). *Leishmania major* surface protease Gp63 interferes with the function of human monocytes and neutrophils in vitro. APMIS. *102*, 265–271.

Starr, R., Metcalf, D., Elefanty, A.G., Brysha, M., Willson, T.A., Nicola, N.A., Hilton, D.J., and Alexander, W.S. (1998). Liver degeneration and lymphoid deficiencies in mice lacking suppressor of cytokine signaling-1. Proc. Natl. Acad. Sci. USA. *95*, 14395–14399.

Starr, R., Willson, T.A., Viney, E.M., Murray, L.J., Rayner, J.R., Jenkins, B.J., Gonda, T.J., Alexander, W.S., Metcalf, D., Nicola, N.A., and Hilton, D.J. (1997). A family of cytokine-inducible inhibitors of signalling. Nature. *387*, 917–921.

Stenger, S., Thuring, H., Rollinghoff, M., and Bogdan, C. (1994). Tissue expression of inducible nitric oxide synthase is closely associated with resistance to *Leishmania major*. J. Exp. Med. *180*, 783–793.

Sutterwala, F.S., Noel, G.J., Clynes, R., and Mosser, D.M. (1997). Selective suppression of interleukin-12 induction after macrophage receptor ligation. J. Exp. Med. *185*, 1977–1985.

Sutterwala, F.S., Noel, G.J., Salgame, P., and Mosser, D.M. (1998). Reversal of proinflammatory responses by ligating the macrophage Fcgamma receptor type I. J. Exp. Med. *188*, 217–222.

Suzuki, E., Tanaka, A.K., Toledo, M.S., Takahashi, H.K., and Straus, A.H. (2002). Role of beta-D-galactofuranose in *Leishmania major* macrophage invasion. Infect. Immun. *70*, 6592–6596.

Talamas-Rohana, P., Wright, S.D., Lennartz, M.R., and Russell, D.G. (1990). Lipophosphoglycan from *Leishmania mexicana* promastigotes binds to members of the CR3, p150, 95 and LFA-1 family of leukocyte integrins. J. Immunol. *144*, 4817–4824.

Tannahill, G.M., Elliott, J., Barry, A.C., Hibbert, L., Cacalano, N.A., and Johnston, J.A. (2005). SOCS2 can enhance interleukin-2 (IL-2) and IL-3 signaling by accelerating SOCS3 degradation. Mol. Cell. Biol. *25*, 9115–9126.

Tardieux, I., Webster, P., Ravesloot, J., Boron, W., Lunn, J.A., Heuser, J.E., and Andrews, N.W. (1992). Lysosome recruitment and fusion are early events required for trypanosome invasion of mammalian cells. Cell. *71*, 1117–1130.

Tejle, K., Magnusson, K.E., and Rasmusson, B. (2002). Phagocytosis and phagosome maturation are regulated by calcium in J774 macrophages interacting with unopsonized prey. Biosci. Rep. *22*, 529–540.

Tripathi, A. and Gupta, C.M. (2003). Transbilayer translocation of membrane phosphatidylserine and its role in macrophage invasion in *Leishmania* promastigotes. Mol. Biochem. Parasitol. *128*, 1–9.

Uzonna, J.E., Joyce, K.L., and Scott, P. (2004). Low dose *Leishmania major* promotes a transient T helper cell type 2 Response that is downregulated by interferon {gamma}-producing CD8$^+$ T cells. J. Exp. Med. *199*, 1559–1566.

Vieira, M.C.F., Decarvalho, T.U., and Desouza, W. (1994). Effect of protein kinase inhibitors on the invasion process of macrophages by *Trypanosoma cruzi*. Biochem. Biophys. Res. Commun. *203*, 967–971.

Villalta, F., Zhang, Y., Bibb, K.E., Pratap, S., Burns, J., and Lima, M.F. (1999). Signal transduction in human macrophages by gp83 ligand of *Trypanosoma cruzi*: Trypomastigote gp83 ligand upregulates trypanosome entry through protein kinase C activation. Mol. Cell Biol. Res. Commun. *2*, 64–70.

Wakil, A.E., Wang, Z.E., Ryan, J.C., Fowell, D.J., and Locksley, R.M. (1998). Interferon gamma derived from CD4$^+$ T cells is sufficient to mediate T helper cell type 1 development. J. Exp. Med. *188*, 1651–1656.

Wei, X.Q., Charles, I.G., Smith, A., Ure, J., Feng, G.J., Huang, F.P., Xu, D., Muller, W., Moncada, S., and Liew, F.Y. (1995). Altered immune responses in mice lacking inducible nitric oxide synthase. Nature *375*, 408–411.

Weinheber, N., Wolfram, M., Harbecke, D., and Aebischer, T. (1998). Phagocytosis of *Leishmania mexicana* amastigotes by macrophages leads to a sustained suppression of IL-12 production. Eur. J. Immunol. *28*, 2467–2477.

Wilkowsky, S.E., Wainszelbaum, M.J., and Isola, E.L.D. (1996). *Trypanosoma cruzi*: Participation of Intracellular Ca^{2+} during metacyclic trypomastigote–macrophage interaction. Biochem. Biophys. Res. Commun. *222*, 386–389.

Woolsey, A.M., Sunwoo, L., Petersen, C.A., Brachmann, S.M., Cantley, L.C., and Burleigh, B.A. (2003). Novel PI 3-kinase-dependent mechanisms of trypanosome invasion and vacuole maturation. J. Cell Sci. *116*, 3611–3622.

Xie, Z.H., Zhang, J., and Siraganian, R.P. (2000). Positive regulation of c-Jun N-terminal kinase and TNF-alpha production but not histamine release by SHP-1 in RBL-2H3 mast cells. J. Immunol. *164*, 1521–1528.

Yeung, Y.G., Wang, Y., Einstein, D.B., Lee, P.S., and Stanley, E.R. (1998). Colony-stimulating factor-1 stimulates the formation of multimeric cytosolic complexes of signaling proteins and cytoskeletal components in macrophages. J. Biol. Chem. *273*, 17128–17137.

Yi, T.L., Cleveland, J.L., and Ihle, J.N. (1992). Protein tyrosine phosphatase containing SH2 domains: characterization, preferential expression in hematopoietic cells, and localization to human chromosome 12p12-p13. Mol. Cell Biol. *12*, 836–846.

Yoshimura, A., Ohkubo, T., Kiguchi, T., Jenkins, N.A., Gilbert, D.J., Copeland, N.G., Hara, T., and Miyajima, A. (1995). A novel cytokine-inducible gene CIS encodes an SH2-containing protein that binds to tyrosine-phosphorylated interleukin 3 and erythropoietin receptors. EMBO J. *14*, 2816–2826.

Zhong, L., Lu, H.G., Moreno, S.N.J., and Docampo, R. (1998). Tyrosine phosphate hydrolysis of host proteins by *Trypanosoma cruzi* is linked to cell invasion. FEMS Microbiol. Lett. *161*, 15–20.

Host Responses to infections with *Leishmania*

12

Fabienne Tacchini-Cottier and Pascal Launois

Abstract

The role of cytokines in human infectious diseases is becoming more and more recognized. Most of the knowledge gained on their importance in *Leishmania* infections has been obtained with animal models. In the murine model of infection with *Leishmania major*, a clear dichotomy is observed between cytokine production by draining lymph node cells of susceptible versus resistant mouse strains. In most inbred mouse strains such as C57BL/6 mice, subcutaneous infection with *L. major* results in the development of a protective T helper-1 (Th1) immune response with high levels of IFN-γ, and resistance to re-infection. In contrast, infection of susceptible strains such as BALB/c mice leads to the development of a Th2 immune response characterized by the production of IL-4 by draining lymph node cells. If the Th1 cytokine, IFN-γ, has been associated with protection against infection with most strains of *Leishmania* in both human and mice, the role of cytokines such as IL-4, IL-13 and IL-10 differ according to the *Leishmania* species. Understanding differences in the role of cytokines in infections with different strains of *Leishmania* should contribute to the design of better and more efficient therapies.

Introduction

Human dermal leishmaniasis is caused by many species of *Leishmania* with different geographic distribution and spectra of disease. Dermal manifestations of leishmaniasis can result from infection with either viscerotropic or dermatotrophic species. In the Old World, *L. major, L. tropica and L. aethiopica* cause dermal leishmaniasis, which is associated with diffuse and mucosal presentation, as well as localized cutaneous leishmaniasis. In the New World, *L. mexicana, L. amazonensis, L. braziliensis, L. panamensis, L. guyanensis* and *L. peruviana* are the most frequent causes of dermal leishmaniasis.

Visceral leishmaniasis (VL) is caused by *L. donovani* and *L. infantum* in the Old World and by *L. chagasi* in the New World. Infection results in a variety of manifestations ranging from subclinical infection to overwhelming infection with hepatosplenomegaly, cachexia, pancytopenia, hyperglobulinaemia and death. Visceralizing *Leishmania* on the other hand can also cause cutaneous manifestations. Dermal lesions at the inoculation site of *L. infantum* or *L chagasi* resemble lesions caused by dermatotrophic species, whereas *L. donovani* is unique in producing disseminated non-ulcerating dermal lesions in some individuals following visceral leishmaniasis, called post-kala-azar dermal leishmaniasis (PKDL). The potential of PKDL and other dermal presentations of leishmaniasis in the contribution to anthroponotic cycles of transmission is unknown and remains clearly an important issue in targeting control measures.

Given the high variety of epidemiological situations and manifestations of the disease described above, the different factors influencing disease transmission and control are difficult to achieve. Furthermore, since effective prevention is not available and actual curative therapy is costly and only moderately tolerated, alternative therapies including vaccination against leishmaniasis is the priority to control *Leishmania* infection. To this end, the host immune responses allowing survival or growth of the parasites have to be characterized, since they are the crucial step in the rationale of immunointerventions.

In this context, experimental murine models of leishmaniasis, particularly infection of genetically resistant and susceptible mice has provided important insights into the immunological mechanisms and pathways leading to pathogenesis.

Experimental models of leishmaniasis

One of the most widely used experimental hosts of *Leishmania* has been the mouse. The outcome of *Leishmania* infection in mice may differ according to, among other factors, the parasite strain, the genetic background of the mice, and the site of parasite inoculation.

Mouse models of cutaneous leishmaniasis following infection with *L. major* (Old World)

The murine model of infection with *L. major* has been extensively studied. Subcutaneous infections of *L. major* stationary phase promastigotes in mice of most strains (e.g. C57BL/6, 129/Sv, C3H/He) leads to self-healing cutaneous lesions. There is slow growth of parasites at the site of inoculation and within the draining lymph nodes, with no further parasite spreading. A small lesion develops, but parasite replication is progressively controlled and the lesion heals slowly; the inoculation site being back to normal size by 5–6 weeks after *L. major* inoculation. In susceptible mouse strains such as BALB/c, sustained inflammation at the site of infection persists, and non-healing lesions develop. Parasites replicate within lesions and draining lymph nodes, spreading via the lymphatics to other lymph nodes, spleen and liver. This spreading is associated with progressive disease. Resistance and susceptibility to infection were shown to correlate with the expansion of $CD4^+$ Th1 cells producing high levels of IFN-γ, and Th2 cells producing high levels of IL-4, respectively. Thus, the development of $CD4^+$ Th1 response in mice of most inbred strains is associated with healing, while the development of Th2 responses in susceptible strains such as BALB/c is associated with non-healing lesions and inability to control parasite growth [reviewed in (Sacks *et al.*, 2002)]. In this model of infection, a relatively high dose (10^6–10^7) of stationary phase *L. major* promastigotes is injected subcutaneously in the footpad or at the base of the tail. Recently, alternative routes of inoculation, delivery of a low dose of *L. major* promastigotes, as well as the use of purified metacyclic *L. major* promastigotes have been performed leading to different outcomes (Belkaid *et al.*, 2000). Although relevant and deserving more interest, the majority of experimental data have been obtained with the high-dose model.

Mouse models of cutaneous leishmaniasis following infection with New World *Leishmania* parasites

Experimental murine infections with *Leishmania* of the *L. mexicana* complex, *L. mexicana and L. amazonensis*, result in a different pattern of diseases following infection in mice when compared with *L. major* infections. Unlike *L. major*, subcutaneous injection of parasites of the *L. mexicana* complex in most mouse strains induces chronic infections with non-healing lesions similar to that in the *L. major*-susceptible BALB/c strain of mice. The susceptibility in BALB/c mice has been correlated with the presence of IL-4 and an important role of IL-13 has also been revealed in this model of infection, while virulence of *L. mexicana* in C57BL/6 mice has been associated with the inhibition of $CD4^+$ Th1 response rather than the promotion of a Th2 response (McMahon-Pratt *et al.*, 2004).

Members of the *Leishmania* (*Vianna*) subgenus do not infect most mouse strains; however, mice deficient in IL-12p40 develop, following subcutaneous injection of *L. braziliensis*, chronic lesions with partial parasite control, providing a putative mouse experimental model (de Souza-Neto *et al.*, 2004).

Mouse models of visceral leishmaniasis

There are several murine models of visceral leishmaniasis that share some similarities with human visceral leishmaniasis. There is a genetic basis for susceptibility to infection with *L. donovani*, based on the presence of the *Sc11 1a1* locus: mice with a wild type locus (CBA) have an earlier (more vigorous) parasite growth than those with the mutated *Scl1 1a1* locus (BALB/c and C57BL/6) (Blackwell, 1989).

The majority of experimental work has been performed following intravenous (i.v.) inoculation of *L. donovani* in mice. Resident macrophages of the liver, spleen and bone marrow become rapidly parasitized. In most mouse strains, including BALB/c and C57BL/6 mice, acute lesions develop in the liver that ultimately resolve, but in contrast, spenomegaly and parasite persistence in the spleen is observed (Kaye *et al.*, 2004).

No dichotomy of a protective CD4⁺ Th1 versus non-curative CD4⁺ Th2 is clearly defined in the visceral model of infection. Protective liver immunity depends on granuloma formation involving cytokines such as IFN-γ, IL-12 and TNF.

The *L. infantum* and *L. chagasi* parasite species of the *L. donovani* complex have also been injected i.v. in mice of different genetic backgrounds. Results very similar to those obtained following injection i.v. of *L. donovani* were obtained, with an efficient granulomatous tissue response in the liver, and the development of chronic infection in the spleen, leading to destruction of the splenic architecture (Kaye *et al.*, 1992; Leclercq *et al.*, 1996).

Recently a new experimental model of infection has been developed involving a low-dose skin dermal model of experimental VL using intradermal or subcutaneous injection of *L. infantum*, which more closely approximates the natural infection route of the parasite. Following infection, parasite clearance was observed in the liver and at the site of parasite infection, but parasites persisted in the draining lymph nodes and the spleen. Hence this model should provide a new mean of investigation to study how the parasites visceralize (Ahmed *et al.*, 2003).

Role of Th1 cytokines in murine models of cutaneous leishmaniasis

CD4⁺ Th1 responses contribute in an essential way to the development of a protective immune response following infection of mice with *Leishmania* species inducing cutaneous or visceral leishmaniasis.

Role of IL-12 in the development of CD4⁺ Th1 cellular response following infection with *Leishmania*

L. Major

IL-12 is a heterodimer formed by two covalently linked chains of 40 and 35 kDa forming the biologically active IL-12p70 (70 kDa). IL-12 is produced mainly by macrophages, polymorphonuclear leukocytes, dendritic and B cells. Two of the main biological functions of IL-12 are the induction of IFN-γ production and the polarization of CD4⁺ T cells towards Th1 differentiation (Trinchieri, 2003). The role of IL-12 in the polarization of CD4⁺ Th1 cells *in vivo* has been extensively studied in the murine model of infection with *L. major*. Injection of recombinant IL-12 during the first week of infection in otherwise susceptible BALB/c mice promoted resistance to infection, characterized by the development of a CD4⁺ Th1 response with the increased IFN-γ production and a decreased IL-4 production by lymph node T cells, and resolution of their lesions (Heinzel *et al.*, 1993; Sypek *et al.*, 1993). The reverse approach consisting in the neutralization of IL-12 (by the injection of polyclonal anti-sera against IL-12) in otherwise resistant mice led, following infection with *L. major*, to progressive disease associated with an increase in IL-4 production (Heinzel *et al.*, 1993; Sypek *et al.*, 1993). The importance of IL-12 in the development of CD4⁺ Th1 cells *in vivo* has been further confirmed using mice genetically deleted in either the p35 or p40 subunits of IL-12. Upon infection with *L. major*, these mice on a resistant genetic background, developed progressive lesions associated with a significant increase in IL-4 mRNA transcription in their draining lymph node cells (Mattner *et al.*, 1996).

Naïve T cells do not express a functional IL-12 receptor (IL-12R). They express the IL-12Rβ1, but not the signalling subunit IL-12Rβ2. IL-12Rβ2 expression is regulated by cytokines and is expressed only after activation in presence of IFN-γ, while IL-4 prevents its expression (Szabo *et al.*, 1997). An early burst of IL-4 in BALB/c mice was shown to down-regulate IL-12Rβ2 expression on CD4⁺ T cells the first days after infection with *L. major*, while IL-12Rβ2 expression was up-regulated in CD4⁺ T cells from resistant C57BL/6 mice (Himmelrich *et al.*, 1998). However, BALB/c mice bearing a transgenic IL-12 receptor β2 gene remained susceptible to infection with *L. major*, and developed non-healing lesions associated with high levels of IL-4 secreted by $CD4^+$ T cells. Thus, although expression of the IL-12Rβ2 chain is an important component of the $CD4^+$ Th1 response, maintenance of IL-12Rβ2 chain is not sufficient to redirect a Th2 response to a $CD4^+$ Th1 response (Nishikomori *et al.*, 2001).

The transcription factor T-bet has been shown to be essential in $CD4^+$ Th1 differentiation and was reported as a master switch in Th1 development (Szabo *et al.*, 2000). T-bet is induced by IFN-γ signalling during T cell activation and is involved in the induction of the IL-12Rβ2 in $CD4^+$ T cells (Afkarian *et al.*, 2002). Mice on a resistant genetic background made genetically deficient for T-bet (Szabo *et al.*, 2002) were susceptible to infection with *L. major* and mounted a typical Th2 cell response, with high levels of IL-4 and the downregulation of the IL-12 production.

The precise kinetics of IL-12 production *in vivo* following infection with *L. major* remains to be determined. Results from one study reported that no IL-12 transcripts was detected during the first week of infection in both susceptible and resistant mice (Reiner *et al.*, 1994), while in another study, low levels of IL-12 were detected in lymph nodes from either susceptible BALB/c or resistant C3H mice, but not in the resistant C57BL/6 strain of mice (Vieira *et al.*, 1994; Scharton-Kersten *et al.*, 1995).

Leishmania *other than L. major*

IL-12 induction of a $CD4^+$ Th1 response has been reported to promote protective immunity against most species of *Leishmania*. IL-12 induces the secretion of IFN-γ mainly by $CD4^+$ T cells but also by $CD8^+$ T cells and NK cells in *L. amazonensis* and *L. donovani* infections, even if secretion of IFN-γ by $CD8^+$ T cells can be IL-12 independent. Exogenous injection of IL-12 was unable to induce healing of lesions in *L. amazonensis*-infected C3H mice, a process associated with defective IL-12R expression (Jones *et al.*, 2002). It was recently reported that during the early phase of infection (first six weeks) IL-12p40 $^{-/-}$ mice infected with *L. mexicana* developed cutaneous lesions, with immune response and parasite burden similar to that measured in wild type mice. In contrast, IFN-γ $^{-/-}$, STAT4 $^{-/-}$ and iNOs $^{-/-}$ mice developed more severe unhealing lesions with uncontrolled parasite growth, suggesting that control of parasite and cutaneous pathology occurs via an IL-12 independent activation of iNOS, IFN-γ and STAT4; this later factor being required for IL-12 signalling (Aguilar Torrentera *et al.*, 2002; Buxbaum *et al.*, 2002). However, from 7 weeks after infection, non-healing lesions develop in IL-12p40-deficient mice or in mice treated with IL-12p40 neutralizing mAb, suggesting that IL-12 is not essential in the early control of *L. mexicana* infection while, in contrast, it plays a key role in the control of the late stage of *L. mexicana* infection (Aguilar Torrentera *et al.*, 2002).

Role of new IL-12-family members in the development of $CD4^+$ Th1 cellular response following infection with Leishmania

IL-12p40 has been recently shown to dimerize with IL-23p19 forming the IL-23 cytokine. IL-23 uses the IL-12Rβ1, but not the IL-12Rβ2, signalling receptor (Hunter, 2005). IL-12 p35 $^{-/-}$ mice and IL-12p40 $^{-/-}$ mice, both on a resistant genetic background, are susceptible to infection with *L. major* and do not develop a $CD4^+$ Th1 protective response (Mattner *et al.*, 1996). As IL-12 p35 $^{-/-}$ mice are able to produce IL-23, these data suggest that IL-23 alone is not sufficient to drive $CD4^+$ Th1 differentiation. Recently IL-23 has been reported to induce naïve precursors cell to differentiate into IL-17-producing T cells (Th17), a T cell lineage that is distinct from the classical $CD4^+$ Th1 and Th2 (Harrington *et al.*, 2005; Park *et al.*, 2005). Further work with IL-23p19 deficient mice may clarify its importance during *Leishmania* infection.

IL-27 is formed by the p40-related EBI3 protein and p28, a molecule related to p35. Mice deficient for the IL-27 receptor (WSX-1 $^{-/-}$) on a resistant genetic background are susceptible to infection with *L. major*, produce elevated levels of IL-4 and are unable to develop a protective $CD4^+$ Th1 response (Yoshida *et al.*, 2001). Further studies revealed that IL-27 was essential for $CD4^+$ Th1 development in the early phase of the *L. major* immune response; however infected mice were able to generate a robust CD4$^+$ Th1 response with high levels of IFN-γ and eventually cured their lesions. These results suggest that IL-27 has only a transient role in the CD4$^+$ Th1 development (Hunter *et al.*, 2004). The exact role of IL-27 during the early phase following *Leishmania* infection remains to be determined.

Role of IL-18 in the development of CD4$^+$ Th1 cellular response following infection with *Leishmania*

IL-18, a member of the IL-1 family, was originally described as an IFN-γ inducible factor (IGIF) in T and NK cells (Okamura *et al.*, 1995). IL-12 upregulates IL-18R suggesting that IL-18 could be a cofactor of Th1 cell development (Ahn *et al.*, 1997; Tomura *et al.*, 1998). However, IL-18 was shown to synergize with IL-12 in the induction of CD4$^+$ Th1 cells but on its own, it was not able to polarize a CD4$^+$ Th1 cellular response or to instruct protective immunity to *L. major*. Infection of mice deficient in IL-18 with *L. major* provided conflicting results as IL-18$^{-/-}$ mice on a 129xCD1 genetic background were demonstrated to be susceptible to infection with *L. major* (Wei *et al.*, 1999), while IL-18$^{-/-}$ mice on a C57BL/6 genetic background were shown to be resistant to infection with *L. major* (Monteforte *et al.*, 2000). It is now established that *in vivo*, following infection with *L. major*, IL-18 can regulate either CD4$^+$ Th1 or Th2 cell differentiation, depending on the cytokine environment found in mice of different genetic backgrounds (Xu *et al.*, 2000).

Role of IFN-γ in the development of CD4$^+$ Th1 cellular response following infection with *Leishmania*

IFN-γ, the major effector cytokine produced by Th1 cells, plays a key function in activating macrophages, enhancing their leishmanicidal and leishmaniastatic properties. IFN-γ could influence Th1 differentiation by enhancing either IL-12 expression or expression of the IL-12Rβ2.

IFN-γ was shown to be critical for the cure of leishmaniasis in mice. However the exact role of IFN-γ in favouring Th1 cell response during infection with *L. major* is not defined yet. Administration of anti-IFN-γ mAb within the first 2 days of infection in mice of resistant genetic background such as C57BL/6 or C3H/He rendered these mice unable to resolve their lesions and led to the appearance of a Th2 response (Belosevic *et al.*, 1989; Scott, 1991; Launois *et al.*, 2002). Similarly, IFN-γ-deficient C57BL/6 mice developed a Th2 response after *L. major* infection (Wang *et al.*, 1994). In contrast, 129/Sv/Ev (a resistant genetic background) deficient in the IFN-γ receptor developed non-healing lesions, but nevertheless developed a CD4$^+$ Th1 response (Swihart *et al.*, 1995). These differences may originate from distinct IL-12Rβ2 chain regulation (Afkarian *et al.*, 2002).

Role of Th2 cytokines in murine models of cutaneous leishmaniasis

The role of cytokines in the development of polarized CD4$^+$ Th2 responses following infection with *L. major*

Studies using the murine model of infection with *L. major* have established that IL-4, during the early stages of infection, has an important role on the subsequent development of specific CD4$^+$ Th2 cells. In this context, the correlation between resistance to infection and the absence of a Th2 response was confirmed by results showing that susceptible mice treated with an anti-IL-4 mAb during the first days of infection with *L. major* were unable to mount a Th2 response and were resistant to infection (Sadick *et al.*, 1990).

We have reported IL-4 mRNA expression in CD4$^+$ T cells draining lymph nodes of susceptible mice within the first day of infection with *L. major*. This early IL-4 was not detectable in resistant C57BL/6 mice (Launois *et al.*, 1995). A particular subpopulation of CD4$^+$ T cells, bearing the Vβ4-Vα8 TCR, has been shown to provide the early IL-4 necessary for Th2 maturation. Thus it has been suggested that in BALB/c mice, a single antigen, *Leishmania* Activated C Kinase (LACK), from this highly complex microorganism drives the early IL-4 response that underlies subsequent Th2 cell maturation resulting in progressive disease. The contribution of these cells in susceptibility to infection with *L. major* was demonstrated by experiments showing that susceptible BALB/c rendered selectively deficient in Vβ4 $CD4^+$ T cells by infection with a mouse mammary tumour virus (MMTV SIM) became resistant to *L. major* infection (Launois *et al.*, 1997). Furthermore, BALB/c mice rendered tolerant to LACK, as a result of the transgenic expression of this molecule in the thymus under CMH class II promoters are resistant to infection with *L. major* and develop a Th1 response (Julia *et al.*, 1996). These results were recently confirmed using depletion of specific LACK cells. BALB/c mice reconstituted with $CD4^+$ T cells depleted of LACK-specific cells have a reduced footpad swelling and parasite burden as compared to control mice receiving total $CD4^+$ T cells (Stetson *et al.*, 2002).

However, results using multivalent MHC/peptides and IL-4 reporter mice have shown a similar precursor frequency and increase of IL-4 mRNA transcription in B10.D2 (resistant phenotype) and BALB/c mice (Stetson *et al.*, 2002), confirming previous results showing comparable early IL-4 response to *L. major* in these mice (Julia *et al.*, 1999). Interestingly, *L. major* parasites expressing a mutated LACK in the epitope recognized by susceptible BALB/c mice ($I\text{-}A^d$ peptide containing the aa 156–173) elicited a reduced expansion of LACK-specific T cells. However, this parasite was still able to induce the development of a Th2 response in BALB/c mice. These results suggest that this LACK epitope is not essential for susceptibility to infection with *L. major* (Kelly *et al.*, 2004).

Although these studies confirmed the pathogenic role of LACK-reactive cells, they suggest that IL-4 might not be the only signal necessary for Th2 differentiation in susceptible mice. This conclusion is also supported by the conflicting results obtained using IL-4-deficient susceptible mice. One group reported that IL-4 $^{-/-}$ mice still develop a susceptible phenotype following infection with *L. major* (Noben-Trauth *et al.*, 2003), whereas another group reported that IL-4-deficient BALB/c mice were resistant to infection with *L. major* (Kopf *et al.*, 1996). Understanding the basis of these conflicting results requires a better understanding of the role of other Th2 cytokines such as IL-13 in the development of Th cell responses following infection with *L. major*.

Several observations indicate that IL-13, which shares biological functions with IL-4 (Chomarat *et al.*, 1998), is involved in susceptibility to infection with *L. major*. The relative importance of IL-13 in the induction of susceptibility to *L. major* infection varies depending on the parasite strain used. IL-13 transgenic mice on an $IL\text{-}4^{-/-}$ deficient C57BL/6 background are susceptible to *L. major* (LV39) infection and $IL\text{-}13^{-/-}$ BALB/c mice control infection with *L. major* (Matthews *et al.*, 2000). However, $IL\text{-}4^{-/-}$ BALB/c mice infected with the *L. major* IR173 strain only partially controlled the infection, but $IL\text{-}4R\alpha^{-/-}$ mice were fully resistant to infection with this strain of parasite (Noben-Trauth *et al.*, 1999). In contrast, $IL\text{-}4R\alpha^{-/-}$ mice developed progressive disease following infection with a LV39 *L. major* strain, suggesting that in the absence of IL-4 and IL-13 biological activity, these mice could be, under some circumstances, fully susceptible (Noben-Trauth *et al.*, 1999). Since in $IL\text{-}4R\alpha^{-/-}$ mice treated with anti-IL-10 mAb or in double $IL\text{-}4R\alpha^{-/-}IL\text{-}10^{-/-}$ mice infected with *L. major* LV39, the lesions size and number of parasites are reduced, IL-10 should play a role in susceptibility to infection with *L. major* (Noben-Trauth *et al.*, 2003).

Role of Th2 cytokines in infection other than *L. major*

The role of cytokines in the development of disease after *L. mexicana* infection has been analysed using IL-4 and IL-4Rα mice. Whereas wild type mice develop large cutaneous lesions, IL-4-deficient mice do not develop lesions after infection with *L. mexicana* and acquire a Th1 response (Satoskar *et al.*, 1997). In addition, lesions due to infection with *L. mexicana* are identical in IL-13 $^{-/-}$ and wild type mice and a Th2 cell response develops (Sosa *et al.*, 2001), suggesting that IL-13 is not essential for susceptibility to infection with *L. mexicana*.

Association between the development of a Th2 response and non-healing disease is quite difficult to demonstrate in experimental visceral leishmaniasis. Indeed, a stable mixture of Th1 and Th2 cells occurs during infection (Kaye *et al.*, 1991).

Role of IL-10 in murine models of infection with *Leishmania*

The role of IL-10 during infection with *L. major*

IL-10 was first described as a cytokine synthesis inhibitory factor (CSIF), with suppressive and deactivating properties shared by most Th2 cells, as well as some of the recently described regulatory T cells. IL-10 inhibits activation and cytokine production of $CD4^+$ Th1 cells, playing a key role in suppressing the host immune system (Moore *et al.*, 2001; Pestka *et al.*, 2004).

The role of IL-10 in promoting susceptibility to infection with *L. major* is not firmly established. Treatment of susceptible mice with anti-IL-10 mAb at the onset of infection had little effect on T helper differentiation and development of lesion (Chatelain *et al.*, 1999), but a transient up-regulation of IFN–γ production was observed. However, mice on a *L. major*-resistant genetic background, expressing an IL-10 transgene under a MHC II Eα promotor were more susceptible to infection than C57BL/6 mice (Groux *et al.*, 1999). In contrast, transgenic mice expressing IL-10 under the control of the IL-2 promoter on the C57BL/6 healing genetic background retained resistance to *L. major* infection and Th1 response comparable to that of wild type C57BL/6 mice (Hagenbaugh *et al.*, 1997). These results suggest that the cellular source of IL-10 is important to develop susceptibility to infection. In this context, it has been shown that macrophage-derived IL-10 increased susceptibility to infection with *L. major* (Lang *et al.*, 2002) and that IgG induced activation of macrophages to produce IL-10 rather than IL-12 (Miles *et al.*, 2005). Conversely, IL-$10^{-/-}$ BALB/c mice were more resistant to *L. major* than wild type mice, harbouring fewer parasites (Padigel *et al.*, 2005). In addition, lesion size and the number of parasites were reduced in IL-4R$\alpha^{-/-}$ mice treated with anti-IL-10 mAb or in IL-4R$\alpha^{-/-}$IL-$10^{-/-}$ mice infected with *L. major* LV39, indicating that IL-10 plays a role in susceptibility to infection with *L. major* (Noben-Trauth *et al.*, 2003). Interestingly, it has been recently shown that IL-10 is also implicated in the persistence of *Leishmania* in healing lesions of resistant C57BL/6 mice since complete elimination of parasites occurred in these mice only when the IL-10 signalling pathway was abrogated with anti-IL-10R mAb (Belkaid *et al.*, 2001).

$CD4^+$ $CD25^+$ regulatory T cells (Treg) have been shown to regulate immune responses in *L. major*-infected BALB/c and C57BL/6 mice, acting on both $CD4^+$ Th1 and Th2 cellular responses respectively (Aseffa *et al.*, 2002; Belkaid *et al.*, 2002; Xu *et al.*, 2003). Furthermore, $CD4^+$ $CD25^+$ Treg from *L. major* lesions were shown to produce high levels of IL-10 (Belkaid *et al.*, 2002). The Treg-produced IL-10 was shown to be essential in the suppressive activity of these Treg. However, using a different strain of *L. major*, $CD4^+$ $CD25^-$ IFN-γ-producing Th1 cells have recently been shown to be the source of suppressive IL-10 in chronic leishmaniasis (Anderson *et al.*, 2007). Thus, the exact origin of secreted IL-10 may differ according to the infecting strain of *L. major*.

Role of IL-10 in infection other than *L. major*

The role of IL-10 following infection with New World species of *Leishmania* is not clear. BALB/c IL-$10^{-/-}$ mice infected with *L. amazonensis* and *L. mexicana* failed to heal their lesion, but lower levels of infection were observed. These results suggest that IL-10 does not play a major role in susceptibility of BALB/c mice to infection with these New World parasites, but contributes to it by regulating the development of Th1 protective response (Padigel *et al.*, 2005). It was further shown that neutralization of IL-10 was not sufficient to achieve resistance to infection with *L. mexicana*, and neutralization of both IL-4 and IL-10 was required to achieve resistance to *L. mexicana* infection in BALB/c mice (Padigel *et al.*, 2005). C57BL/6 IL-10-deficient mice failed to exhibit increased resistance to *L. amazonensis*, despite an increased $CD4^+$ Th1 response (Jones *et al.*, 2002), but recent studies showed that IL-10 is required to suppress healing response in C57BL/6 mice infected with *L. mexicana*, as *L. mexicana*-infected B6 IL-$10^{-/-}$ mice resolved their lesions and exhibited increased IFN-γ and nitric oxide production (Buxbaum *et al.*, 2005).

A role for IL-10 in infections with *L. donovani* was investigated using BALB/c and C57BL/6

IL-10$^{-/-}$ mice. Both IL-10 $^{-/-}$ mice were highly resistant to *L. donovani* infection with lower liver parasite burdens, increased IFN-γ and nitric oxide production. In addition, treatment with anti-IL-10 mAb has also been shown to promote increased resistance to infection with this parasite (Murphy *et al.*, 2001; Murray *et al.*, 2003).

Role of transforming growth factor-β (TGF-β) in murine models of infection with *Leishmania*

TGF-β is considered as a potent immunosuppressive cytokine, involved in the control of inflammatory responses as well as differentiation and proliferation of cells from both innate and adaptive immunity. TGF-β also participates in CD4$^+$ Th cell differentiation, either promoting CD4$^+$ Th1 development (Laouar *et al.*, 2005), or suppressing it by inhibition of T-bet expression (Letterio *et al.*, 1998).

The role of TGF-β during infection with *L. major* has been explored using CB6F1 mice that normally display intermediate susceptibility to *L. major*. Treatment of these infected mice with a mAb against TGF-β promoted rapid healing by enhancing *in vivo* nitric oxide production (Li *et al.*, 1999; Gorelik *et al.*, 2002). BALB/c mice transgenic for the dominant negative TGF-receptor II expressed on T cells (thereby blocking TGF-β signalling in T cells) were resistant to infection with *L. major* with increased production of Th1 cytokines, but these mice also displayed an increase in Th2 cytokine production (Gorelik *et al.*, 2002).

The role of TGF-β in the progression of infection was most convincingly demonstrated in mice infected with *L. amazonensis*, or *L. braziliensis* in which local injection of TGF-β at the site of parasite inoculation in mice on a resistant genetic background resulted in non-healing of the lesions. In addition, susceptible BALB/c mice treated with a neutralizing anti-TGF-β antibody during the course of infection with *L. amazonensis* decreased markedly their lesion size within 5 weeks post-infection. IFN-γ mRNA was observed in the draining lymph node of these mice while IL-4 (but no IFN-γ) was detected in the draining lymph nodes of infected mice treated with a control antibody (Barral-Netto, 1996). TGF-β production may act by preventing *Leishmania* destruction by host macrophages. Indeed it was reported to inhibit NO production by macrophages. Notably, intramuscular vaccination of BALB/c mice with *L. amazonensis* antigens promotes enhanced susceptibility to infection by increasing the secretion of TGF-β, an effect abrogated by treatment with a specific anti-TGF-β mAb (Pinheiro *et al.*, 2005).

TGF-β also plays a role in the induction of susceptibility in *L. chagasi* infection. Administration of an adenovirus expressing TGF-β caused genetically resistant C3H mice to become more susceptible to *L. chagasi* infection. It is thought that in susceptible BALB/c mice, locally secreted TGF-β in liver granuloma inhibits associated cure of murine visceral leishmaniasis (Wilson *et al.*, 1998). Thus, TGF-β contributes to the down regulation of Th1 cellular responses favouring susceptibility to disease.

CD8$^+$ T cells in murine models of leishmaniasis

Although the CD4$^+$ T cell population plays a central role in immunity to infection with *L. major*, specific CD8$^+$ T cells are triggered during infection of mice with *L. major* and are involved in the healing of lesions and in immunity to re-infection.

Infections with *L. major*

The role of specific CD8$^+$ T cells in the development of primary lesions

Expansion of CD8$^+$ T cells contributing to *L. major* DTH reaction and producing IFN-γ was noticed in lymphoid tissues of mice immune to re-infection with *L. major*. Draining lymph nodes of resistant mice contained more CD8$^+$ T cells able to transfer specific DTH reaction to normal syngeneic recipients (Titus *et al.* 1987; Muller *et al.*, 1993). Furthermore, the lesions of resistant mice treated with an anti-CD8 mAb were significantly larger than controls, and healing was delayed (Farrell *et al.*, 1989). A contribution of CD8$^+$ T cells to immunity can be demonstrated under well-defined conditions such as immunological interventions that turn susceptible mice into mice resistant to *L. major* infection. Indeed, susceptible BALB/c can be rendered resistant to infection with *L. major* by treatment with anti-CD4 mAb at the beginning of infection. First, mice treated with both anti-CD4 and anti-CD8 mAbs at the beginning of infection developed larger lesions than mice treated with only anti-CD4 mAb. Moreover, compared to mice treated only with anti-CD4, the number of viable parasites in lesions of mice treated with both anti-CD4 and anti-CD8 mAbs was significantly higher (Muller *et al.*, 1991).

$CD8^+$ T cells contributed also significantly to the IFN-γ production measured after specific *in vitro* stimulation of draining lymph node cells of BALB/c mice treated with anti-CD4 mAb at the time of parasite inoculation (Muller *et al.*, 1991). Lastly, increased numbers of specific $CD8^+$ T cells, able to transfer *Leishmania*-specific DTH responses, were found in organs of BALB/c mice treated with anti-CD4mAbs (Muller *et al.*, 1991). In the same context, elimination of $CD4^+$ T cells in thymectomized susceptible mice also induced an inhibition of lesion development associated with the activity of $CD8^+$ T cells (Hill *et al.*, 1989). Thus it seems that $CD8^+$ T cells play a role in the control of cutaneous leishmaniasis due to *L. major* in C57BL/6 mice and in BALB/c mice rendered resistant by administration of an anti-CD4 mAb at the beginning of infection. However, resistant mice deficient in β_2-microglobulin or in CD8 maintained their ability to heal their primary lesions (Wang *et al.*, 1993; Huber *et al.*, 1998). These results indicate that although involved in the evolution of the primary lesions, $CD8^+$ T cells are not required for the control of primary lesions. However, all these results were obtained with infection involving high dose of *L. major* ($3x10^6$) by subcutaneous infection. Recent data using a low dose model of infection with dermal injection clearly show that $CD8^+$ T cells, in addition to $CD4^+$ T cells, are required to control primary lesions (Belkaid *et al.*, 2002).

The role of specific $CD8^+$ T cells in the development of secondary lesions

There is also evidence for the contribution of the $CD8^+$ T cells in control of secondary infection. This was first demonstrated in a 'vaccination' system. The resistance to infection by prior administration of killed *L. major* intravenously was abrogated by depletion of $CD8^+$ T cells at the time of re-infection (Farrell *et al.*, 1989). Furthermore, resistant C57BL/6 mice immunized with autoclaved *L. major* (ALM) or specific LACK antigen plus CpGODNs were protected against *L. major* challenge and the protection correlated with both $CD4^+$ Th1 and $CD8^+$ T cell responses. Indeed, protection was abrogated in mice depleted in $CD8^+$ T cells at time of infection (Gurunathan *et al.*, 1997; Rhee *et al.*, 2002). Lesions developed during secondary infection with *L. major* in resistant mice are enhanced by treatment with anti-CD8 mAbs (Muller *et al.*, 1991). Conversely, the resistance to re-infection in BALB/c mice treated with anti-CD4 mAb at the beginning of infection was prevented by treatment with anti-CD8 mAb (Muller *et al.*, 1991). A secondary IFN-γ-producing $CD8^+$ T cell response was also elicited by challenge of both resistant C57BL/6 mice and in susceptible BALB/c mice rendered resistant by treatment with anti-CD4 mAb. Furthermore, since these cells are able to produce IFN-γ, and thus active macrophage microbicidal activity, these data provide a rationale role for $CD8^+$ T cells in control of secondary infection (Muller *et al.*, 1994).

Interestingly, recent data showed the role of CD8-derived IFN-γ on the modulation of $CD4^+$ T cell responses (Uzonna *et al.*, 2004). In this line, depletion of $CD8^+$ T cells in mice treated with anti-CD4 mAb at the beginning of infection did not alter proliferation of $CD4^+$ T cells but decreased the production of IFN-γ and the number of IFN-γ producing $CD4^+$ T cells, allowing the reversal of healing in these mice (Herath *et al.*, 2003).

Nevertheless, a direct cytotoxic effect of $CD8^+$ T cells on parasitized macrophages cannot be excluded. Indeed, lymph nodes from resistant mice that have healed from primary lesions contain high level of *L. major*-specific $CD8^+$ T cells that are able to expand after challenge (da Conceicao-Silva *et al.*, 1994). In this context, the relative contribution of IFN-γ production versus cytotoxicity in the effector role of CD8 T cells has been analysed in perforin-deficient mice in which $CD8^+$ T cells are poorly cytotoxic but release normal amount of IFN-γ. Interestingly, perforin-deficient mice on a resistant genetic background healed their lesions, suggesting that cytotoxicity is not involved in the immunity of *Leishmania* (Conceicao-Silva *et al.*, 1998). However, syngeneic *gld* and *lpr* mice that lack a functional Fas system fail to heal their lesions (Conceicao-Silva *et al.*, 1998; Huang *et al.*, 1998). In this context, it has been reported that the IFN-γ produced by CD4 Th cells can up-regulate Fas expression on macrophages, inducing their apoptotic death (Conceicao-Silva *et al.*, 1998). A similar up-regulation of macrophage Fas expression could also result from the IFN-γ produced by $CD8^+$ T cells.

Infection with other species of *Leishmania*

$CD8^+$ T cells are also required in the protective immune response against *L. amazonensis* infection in mice vaccinated with an antigen from *L. amazonensis* amastigotes (P8), suggesting that despite diverse immunological responses elicited by different *Leishmania* species causing cutaneous leishmaniasis, the protective effect of $CD8^+$ T cells is common to these species (Colmenares *et al.*, 2003).

$CD8^+$ T cells have been shown to control parasite replication and to contribute to the hepatic granuloma induced by *L. donovani* infection in BALB/c mice (Stern *et al.*, 1988). C57BL/6 mice infected with *L. infantum* also mount a specific $CD8^+$ T cells response through perforin-mediated cytotoxicity and IFN-γ production (Tsagozis *et al.*, 2005). Altogether these results showed that $CD8^+$ T cells play a role in resistance to visceralizing *Leishmania* strain.

Human leishmaniasis

Th1/Th2 responses during human infection

Human leishmaniasis is a spectrum of clinical forms with different immunopathogenesis. Localized cutaneous leishmaniasis (LCL) presents limited and ulcerated skin lesions and the growth of the parasite is restricted. Diffuse cutaneous leishmaniasis (DCL) is characterized by the presence of progressive non-ulcerated nodules in which parasites are abundant. In mucocutanoeus leishmaniasis (MCL), chronic destructive and progressive lesions occur in the oral and pharyngeal tracts. With regards to the immunology of cutaneous leishmaniasis, investigations have been mainly done on the profile of $CD4^+$ and $CD8^+$ T cell subsets and the cytokines produced by these cells in the periphery (PBMC) or in the lesions.

In humans, the role of Th1 and Th2 $CD4^+$ T cells and the cytokines they produce is not yet well understood. However, since the role of Th1 and Th2 cell responses has been clearly associated with resistance and susceptibility, respectively in the murine model of infection with *L. major*, one way to demonstrate such differences in human is comparison of LCL and MCL patients with patients with DCL.

It is admitted that LCL patients present an adequate cell immune response with a predominance of Th1 immune response but the type of T cells (CD4 or CD8) within the lesions is still not clear. Indeed, a high number of $CD8^+$ T cells has been demonstrated in lesions and in peripheral blood during the acute stage of lesion formation and during the healing process of New and Old World cutaneous leishmaniasis such as *L. braziliensis* (Da-Cruz *et al.*, 1994), *L. major* (Gaafar *et al.*, 1999) and *L. aethiopica* (Maasho *et al.*, 1998). However, in other studies, no difference was found in the percentage of $CD8^+$ T cells within *Leishmania*-stimulated PBMC from LCL, MCL and DCL patients (Carvalho *et al.*, 1985; Castes *et al.*, 1993). Furthermore, an expansion of both $CD4^+$ and $CD8^+$ T cells has been detected in PBMC from healthy subjects stimulated with *L. major* (Rogers *et al.*, 2004), *L. aethiopica* (Maasho *et al.*, 2000), *L. guyanensis* (Bourreau *et al.*, 2002) and *L. amazonensis* (Russo *et al.*, 1999).

Whereas the Th1 cytokine IFN-γ has been shown to be associated with protective immunity, the exact role of regulatory Th2 cytokines including IL-4, IL-10 and IL-13 in disease susceptibility is still under debate in LCL patients. Although, in mice, IL-4 is found as a potent inducer of Th2 cell development, a predominance of IL-13 over IL-4 in T cell functions in humans has been recently shown. Indeed IL-13 was demonstrated to be the predominant Th2 cytokine in localized cutaneous leishmaniasis (LCL) lesions. The IL-13 prevented IL-12 responsiveness on $CD4^+$ T cells, thereby not allowing Th1 development and favour Th2 cell development and disease susceptibility (Bourreau *et al.*, 2001b).

High IL-10 level was shown to be associated with active visceral and mucosal leishmaniasis (Bacellar *et al.*, 2000; Bacellar *et al.*, 2002). Some IL-10 production has been also demonstrated in human LCL due to *L. braziliensis* and *L. mexicana* (Melby *et al.*, 1996) and its down-regulation of IFN-γ production has been proposed to correlate with disease (Rocha *et al.*, 1999). Furthermore, IL-10 expression was significantly higher in the more slowly healing lesions in patients with LCL caused by *L. major* (Louzir *et al.*, 1998) and was a promoter of disease prolongation in patients infected with *L. mexicana* (Melby *et al.*, 1996). In human LCL due to *L. guyanensis*, IL-10 was poorly produced by PBMC (Bourreau *et al.*, 2002), in spite of the observed association of intra-lesional IL-10 mRNA expression with unresponsiveness to chemotherapy (Bourreau *et al.*, 2001a).

In MCL, lesions are associated with strong T cell immune responses, as demonstrated by exacerbated DTH (Silveira *et al.*, 1998), and it was demonstrated that $CD4^+$ T cells are in elevated amount in MCL lesions as compared to other forms of cutaneous leishmaniasis (Barral *et al.*, 1987; Pirmez *et al.*, 1990; Martinez-Arends *et al.*, 1991). A mixture of $CD4^+$ Th1 and Th2 cell responses has been demonstrated in MCL patients, since PBMC from these patients produce IFN-γ and IL-5, a typical Th2 cytokine in response to *Leishmania* antigens (Caceres-Dittmar *et al.*, 1993; Castes *et al.*, 1993); but a high IFN-γ expression was demonstrated (Silveira *et al.*, 1998); suggesting a uncontrolled Th1 response in MCL patients.

In terms of immune responses, DCL patients demonstrate a lack of cellular immune responsiveness either *in vivo* or *in vitro*. Indeed DTH to *Leishmania* antigens is always negative in DCL patients and PBMC from these patients are unable to produce IFN-γ when they are stimulated with *Leishmania* antigens (Castes *et al.*, 1988). This is in accordance with results showing low levels of $CD4^+$ and $CD8^+$ T cells within DCL lesions as compared to other forms of cutaneous leishmaniasis (Silveira *et al.*, 2004). In contrast, the PBMC from DCL patients are able to produce high levels of IL-5 upon restimulation with *L. braziliensis* (Castes *et al.*, 1993). The predominance of a Th2 cell response in these patients was further confirmed by the fact that lesions expressed more IL-4 and IL-10 mRNA than IFN-γ mRNA (Barral-Netto, 1996; Silveira *et al.*, 2004).

TGF-β and Treg cells in human leishmaniasis

TGF-β is another regulatory cytokine with anti-proliferative and APC-deactivating properties, which has also been implicated in human leishmaniasis. In human infections, TGF-β has been detected in lesions of either active or chronic human cutaneous leishmaniasis (Melby *et al.*, 1994) and macrophages from LCL patients produced TGF-β after infection with different *Leishmania* species (Barral *et al.*, 1995). Since the amount of TGF-β1 produced by macrophages upon infection with *Leishmania* correlates with both the strain virulence and the multiplication of parasites within the macrophage, TGF-β production was associated with virulence mechanism during infection with different *Leishmania* species (Barral *et al.*, 1995). However, recent data have demonstrated that *in vitro* TGF-β can poorly inhibit IFN-γ production by PBMC from either cutaneous or mucosal leishmaniasis patients stimulated with soluble antigens from *L. braziliensis*, suggesting that exogenous TGF-β was not able to reverse IFN-γ anti-parasite effector mechanisms (Bacellar *et al.*, 2002). Therefore, its exact role in disease susceptibility is still unclear.

Both IL-10 and TGF-β are secreted by mouse and human regulatory T cells (Treg). Treg represent 1–3% of the peripheral $CD4^+$ T cells. The suppressive activity of $CD4^+$ $CD25^+$ Treg cells has been reported both *in vitro* and *in vivo*. Suppression was shown to depend on cell contact, at least *in vitro*, while the cytokines IL-10 and TGF-β appeared critical in their suppressive activity *in vivo*. $CD4^+$ $CD25^+$ Treg cells are restricted by self-antigens, but their role in the control of immune responses against several pathogens has been recently described (Mittrucker *et al.*, 2004). An important role for $CD4^+$ $CD25^+$ Treg has been reported in the control of *L. major* specific immune responses in both susceptible BALB/c and resistant

C57BL/6 mice (Aseffa *et al.*, 2002; Belkaid *et al.*, 2002; Xu *et al.*, 2003). In these studies, $CD4^+$ $CD25^+$ Treg cells were shown to control both $CD4^+$ Th1 and Th2 cellular immune responses. Recently, we demonstrated TGF-β1 production in PBMC from healthy subjects never exposed to *Leishmania* in response to live *L. guyanensis* and the TGF-β1-producing cells were described as a distinct subpopulation of $CD4^+$ $CD25^+$ regulatory T cells. The suppressive properties of $CD4^+$ $CD25^+$ T cells were demonstrated *in vitro*. Although neutralization of TGF-β1 did not reverse the suppressive activity of $CD4^+$ $CD25^+$ T cells activated with anti-CD3, it reversed the suppressive activity of $CD4^+$ $CD25^+$ T cells activated by *L. guyanensis* (Kariminia *et al.*, 2005). Recently, $CD4^+$ $CD25^+$ T cells with suppressive functions were found in active skin lesions caused by *L. braziliensis* (Campanelli *et al.*, 2006) suggesting a role for $CD4^+$ $CD25^+$ Treg cells in human cutaneous leishmaniasis. However, the role of these cells in human leishmaniasis deserves further investigations.

References

Afkarian, M., Sedy, J.R., Yang, J., Jacobson, N.G., Cereb, N., Yang, S.Y., Murphy, T.L., and Murphy, K.M. (2002). T-bet is a STAT1-induced regulator of IL-12R expression in naive $CD4^+$ T cells. Nat. Immunol. *3*, 549–557.

Aguilar Torrentera, F., Laman, J.D., Van Meurs, M., Adorini, L., Muraille, E., and Carlier, Y. (2002). Endogenous interleukin-12 is critical for controlling the late but not the early stage of *Leishmania mexicana* infection in C57BL/6 mice. Infect. Immun. *70*, 5075–5080.

Ahmed, S., Colmenares, M., Soong, L., Goldsmith-Pestana, K., Munstermann, L., Molina, R., and McMahon-Pratt, D. (2003). Intradermal infection model for pathogenesis and vaccine studies of murine visceral leishmaniasis. Infect. Immun. *71*, 401–410.

Ahn, H.J., Maruo, S., Tomura, M., Mu, J., Hamaoka, T., Nakanishi, K., Clark, S., Kurimoto, M., Okamura, H., and Fujiwara, H. (1997). A mechanism underlying synergy between IL-12 and IFN-gamma-inducing factor in enhanced production of IFN-gamma. J. Immunol. *159*, 2125–2131.

Anderson, C.F., Oukka, M., Kuchroo, V.J., and Sacks, D. (2007). $CD4^+$ $CD25^-$ $Foxp3^-$ Th1 cells are the source of IL-10-mediated immune suppression in chronic cutaneous leishmaniasis. J. Exp. Med. *204*, 285–297.

Aseffa, A., Gumy, A., Launois, P., MacDonald, H.R., Louis, J.A., and Tacchini-Cottier, F. (2002). The early IL-4 response to *Leishmania major* and the resulting Th2 cell maturation steering progressive disease in BALB/c mice are subject to the control of regulatory $CD4^+$ $CD25^+$ T cells. J. Immunol. *169*, 3232–3241.

Bacellar, O., D'Oliveira, A., Jr., Jeronimo, S., and Carvalho, E.M. (2000). IL-10 and IL-12 are the main regulatory cytokines in visceral leishmaniasis. Cytokine *12*, 1228–1231.

Bacellar, O., Lessa, H., Schriefer, A., Machado, P., Ribeiro de Jesus, A., Dutra, W.O., Gollob, K.J., and Carvalho, E.M. (2002). Up-regulation of Th1-type responses in mucosal leishmaniasis patients. Infect. Immun. *70*, 6734–6740.

Barral, A., Jesus, A.R., Almeida, R.P., Carvalho, E.M., Barral-Netto, M., Costa, J.M., Badaro, R., Rocha, H., and Johnson, W.D. (1987). Evaluation of T-cell subsets in the lesion infiltrates of human cutaneous and mucocutaneous leishmaniasis. Parasite Immunol. *9*, 487–497.

Barral, A., Teixeira, M., Reis, P., Vinhas, V., Costa, J., Lessa, H., Bittencourt, A.L., Reed, S., Carvalho, E.M., and Barral-Netto, M. (1995). Transforming growth factor-beta in human cutaneous leishmaniasis. Am. J. Pathol. *147*, 947–954.

Barral-Netto, M., Brodskyn, C.I., Bonfim, G., and Barral, A. (1996). Cytokine regulation in American cutaneous leishmaniasis. Molecular and Immune Mechanisms in the Pathogenesis of Cutaneous Leishmaniasis F.J. Tapia, G. Cáceres-Dittmar, and M.A. Sanchez. (Org), RG Landes Co., Biomedical Publs, Austin, Texas, USA. *vol 1*, pp. 153–172.

Belkaid, Y., Hoffmann, K.F., Mendez, S., Kamhawi, S., Udey, M.C., Wynn, T.A., and Sacks, D.L. (2001). The role of interleukin IL-10 in the persistence of *Leishmania major* in the skin after healing and the therapeutic potential of anti-IL-10 receptor antibody for sterile cure. J. Exp. Med. *194*, 1497–1506.

Belkaid, Y., Mendez, S., Lira, R., Kadambi, N., Milon, G., and Sacks, D. (2000). A natural model of *Leishmania major* infection reveals a prolonged 'silent' phase of parasite amplification in the skin before the onset of lesion formation and immunity. J. Immunol. *165*, 969–977.

Belkaid, Y., Piccirillo, C.A., Mendez, S., Shevach, E.M., and Sacks, D.L. (2002). $CD4^+$ $CD25^+$ regulatory T cells control *Leishmania major* persistence and immunity. Nature. *420*, 502–507.

Belosevic, M., Finbloom, D.S., Van Der Meide, P.H., Slayter, M.V., and Nacy, C.A. (1989). Administration of monoclonal anti-IFN-gamma antibodies in vivo abrogates natural resistance of C3H/HeN mice to infection with *Leishmania major*. J. Immunol. *143*, 266–274.

Blackwell, J.M. (1989). The macrophage resistance gene Lsh/Ity/Bcg. Res. Immunol. *140*, 767–769.

Bourreau, E., Collet, M., Prevot, G., Milon, G., Ashimoff, D., Hasagewa, H., Parra-Lopez, C., and Launois, P. (2002). IFN-gamma-producing $CD45RA^+$ $CD8^+$ and IL-10-producing CD45RA-$CD4^+$ T cells generated in response to LACK in naive subjects never exposed to Leishmania. Eur. J. Immunol. *32*, 510–520.

Bourreau, E., Prevot, G., Gardon, J., Pradinaud, R., and Launois, P. (2001a). High intralesional interleukin-10 messenger RNA expression in localized cutaneous leishmaniasis is associated with unresponsiveness to treatment. J. Infect. Dis. *184*, 1628–1630.

Bourreau, E., Prevot, G., Pradinaud, R., and Launois, P. (2001b). Interleukin IL-13 is the predominant Th2 cytokine in localized cutaneous leishmaniasis lesions and renders specific CD4$^+$ T cells unresponsive to IL-12. J. Infect. Dis. *183*, 953–959.

Buxbaum, L.U., and Scott, P. (2005). Interleukin 10- and Fcgamma receptor-deficient mice resolve *Leishmania mexicana* lesions. Infect. Immun. *73*, 2101–2108.

Buxbaum, L.U., Uzonna, J.E., Goldschmidt, M.H., and Scott, P. (2002). Control of New World cutaneous leishmaniasis is IL-12 independent but STAT4 dependent. Eur. J. Immunol. *32*, 3206–3215.

Caceres-Dittmar, G., Tapia, F.J., Sanchez, M.A., Yamamura, M., Uyemura, K., Modlin, R.L., Bloom, B.R., and Convit, J. (1993). Determination of the cytokine profile in American cutaneous leishmaniasis using the polymerase chain reaction. Clin. Exp. Immunol. *91*, 500–505.

Campanelli, A.P., Roselino, A.M., Cavassani, K.A., Pereira, M.S., Mortara, R.A., Brodskyn, C.I., Goncalves, H.S., Belkaid, Y., Barral-Netto, M., Barral, A., and Silva, J.S. (2006). CD4$^+$ CD25$^+$ T cells in skin lesions of patients with cutaneous leishmaniasis exhibit phenotypic and functional characteristics of natural regulatory T cells. J. Infect. Dis. *193*, 1313–1322.

Carvalho, E.M., Johnson, W.D., Barreto, E., Marsden, P.D., Costa, J.L., Reed, S., and Rocha, H. (1985). Cell mediated immunity in American cutaneous and mucosal leishmaniasis. J. Immunol. *135*, 4144–4148.

Castes, M., Cabrera, M., Trujillo, D., and Convit, J. (1988). T-cell subpopulations, expression of interleukin-2 receptor, and production of interleukin-2 and gamma interferon in human American cutaneous leishmaniasis. J. Clin. Microbiol. *26*, 1207–1213.

Castes, M., Trujillo, D., Rojas, M.E., Fernandez, C.T., Araya, L., Cabrera, M., Blackwell, J., and Convit, J. (1993). Serum levels of tumor necrosis factor in patients with American cutaneous leishmaniasis. Biol. Res. *26*, 233–238.

Chatelain, R., Mauze, S., and Coffman, R.L. (1999). Experimental *Leishmania major* infection in mice: role of IL-10. Parasite Immunol. *21*, 211–218.

Chomarat, P., and Banchereau, J. (1998). Interleukin-4 and interleukin-13, their similarities and discrepancies. Int. Rev. Immunol. *17*, 1–52.

Colmenares, M., Kima, P.E., Samoff, E., Soong, L., and McMahon-Pratt, D. (2003). Perforin and gamma interferon are critical CD8$^+$ T-cell-mediated responses in vaccine-induced immunity against *Leishmania amazonensis* infection. Infect. Immun. *71*, 3172–3182.

Conceicao-Silva, F., Hahne, M., Schroter, M., Louis, J., and Tschopp, J. (1998). The resolution of lesions induced by *Leishmania major* in mice requires a functional Fas (APO-1, CD95) pathway of cytotoxicity. Eur. J. Immunol. *28*, 237–245.

da Conceicao-Silva, F., Perlaza, B.L., Louis, J.A., and Romero, P. (1994). *Leishmania major* infection in mice primes for specific major histocompatibility complex class I-restricted CD8$^+$ cytotoxic T cell responses. Eur. J. Immunol. *24*, 2813–2817.

Da-Cruz, A.M., Conceicao-Silva, F., Bertho, A.L., and Coutinho, S.G. (1994). Leishmania-reactive CD4$^+$ and CD8$^+$ T cells associated with cure of human cutaneous leishmaniasis. Infect. Immun. *62*, 2614–2618.

de Souza-Neto, S.M., Carneiro, C.M., Vieira, L.Q., and Afonso, L.C. (2004). *Leishmania braziliensis*: partial control of experimental infection by interleukin-12 p40 deficient mice. Mem. Inst. Oswaldo Cruz. *99*, 289–294.

Farrell, J.P., Muller, I., and Louis, J.A. (1989). A role for Lyt-2$^+$ T cells in resistance to cutaneous leishmaniasis in immunized mice. J. Immunol. *142*, 2052–2056.

Gaafar, A., Veress, B., Permin, H., Kharazmi, A., Theander, T.G., and el Hassan, A.M. (1999). Characterization of the local and systemic immune responses in patients with cutaneous leishmaniasis due to *Leishmania major*. Clin. Immunol. *91*, 314–320.

Gorelik, L., Constant, S., and Flavell, R.A. (2002). Mechanism of transforming growth factor beta-induced inhibition of T helper type 1 differentiation. J. Exp. Med. *195*, 1499–1505.

Groux, H., Cottrez, F., Rouleau, M., Mauze, S., Antonenko, S., Hurst, S., McNeil, T., Bigler, M., Roncarolo, M.G., and Coffman, R.L. (1999). A transgenic model to analyze the immunoregulatory role of IL-10 secreted by antigen-presenting cells. J. Immunol. *162*, 1723–1729.

Gurunathan, S., Sacks, D.L., Brown, D.R., Reiner, S.L., Charest, H., Glaichenhaus, N., and Seder, R.A. (1997). Vaccination with DNA encoding the immunodominant LACK parasite antigen confers protective immunity to mice infected with *Leishmania major*. J. Exp. Med. *186*, 1137–1147.

Hagenbaugh, A., Sharma, S., Dubinett, S.M., Wei, S.H., Aranda, R., Cheroutre, H., Fowell, D.J., Binder, S., Tsao, B., Locksley, R.M., Moore, K.W., and Kronenberg, M. (1997). Altered immune responses in interleukin 10 transgenic mice. J. Exp. Med. *185*, 2101–2110.

Harrington, L.E., Hatton, R.D., Mangan, P.R., Turner, H., Murphy, T.L., Murphy, K.M., and Weaver, C.T. (2005). Interleukin 17-producing CD4$^+$ effector T cells develop via a lineage distinct from the T helper type 1 and 2 lineages. Nat. Immunol. *6*, 1123–1132.

Heinzel, F.P., Schoenhaut, D.S., Rerko, R.M., Rosser, L.E., and Gately, M.K. (1993). Recombinant interleukin 12 cures mice infected with *Leishmania major*. J. Exp. Med. *177*, 1505–1509.

Herath, S., Kropf, P., and Muller, I. (2003). Cross-talk between CD8$^+$ and CD4$^+$ T cells in experimental cutaneous leishmaniasis: CD8$^+$ T cells are required for optimal IFN-gamma production by CD4$^+$ T cells. Parasite Immunol. *25*, 559–567.

Hill, J.O., Awwad, M., and North, R.J. (1989). Elimination of CD4$^+$ suppressor T cells from susceptible BALB/c mice releases CD8$^+$ T lymphocytes to mediate protective immunity against Leishmania. J. Exp. Med. *169*, 1819–1827.

Himmelrich, H., Parra-Lopez, C., Tacchini-Cottier, F., Louis, J.A., and Launois, P. (1998). The IL-4 rapidly produced in BALB/c mice after infection with *Leishmania major* downregulates IL-12 receptor beta 2-chain expression on CD4$^+$ T cells resulting in a state of unresponsiveness to IL-12. J. Immunol. *161*, 6156–6163.

Huang, F.P., Xu, D., Esfandiari, E.O., Sands, W., Wei, X.Q., and Liew, F.Y. (1998). Mice defective in Fas are highly susceptible to *Leishmania major* infection despite elevated IL-12 synthesis, strong Th1 responses, and enhanced nitric oxide production. J. Immunol. *160*, 4143–4147.

Huber, M., Timms, E., Mak, T.W., Rollinghoff, M., and Lohoff, M. (1998). Effective and long-lasting immunity against the parasite *Leishmania major* in CD8$^-$deficient mice. Infect. Immun. *66*, 3968–3970.

Hunter, C.A. (2005). New IL-12-family members: IL-23 and IL-27, cytokines with divergent functions. Nat. Rev. Immunol. *5*, 521–531.

Hunter, C.A., Villarino, A., Artis, D., and Scott, P. (2004). The role of IL-27 in the development of T-cell responses during parasitic infections. Immunol. Rev. *202*, 106–114.

Jones, D.E., Ackermann, M.R., Wille, U., Hunter, C.A., and Scott, P. (2002). Early enhanced Th1 response after *Leishmania amazonensis* infection of C57BL/6 interleukin-10-deficient mice does not lead to resolution of infection. Infect. Immun. *70*, 2151–2158.

Julia, V., and Glaichenhaus, N. (1999). CD4$^+$ T cells which react to the *Leishmania major* LACK antigen rapidly secrete interleukin-4 and are detrimental to the host in resistant B10.D2 mice. Infect. Immun. *67*, 3641–3644.

Julia, V., Rassoulzadegan, M., and Glaichenhaus, N. (1996). Resistance to *Leishmania major* induced by tolerance to a single antigen. Science. *274*, 421–423.

Kariminia, A., Bourreau, E., Pascalis, H., Couppie, P., Sainte-Marie, D., Tacchini-Cottier, F., and Launois, P. (2005). Transforming growth factor beta 1 production by CD4$^+$ CD25$^+$ regulatory T cells in peripheral blood mononuclear cells from healthy subjects stimulated with *Leishmania guyanensis*. Infect. Immun. *73*, 5908–5914.

Kaye, P.M., and Bancroft, G.J. (1992). *Leishmania donovani* infection in scid mice: lack of tissue response and in vivo macrophage activation correlates with failure to trigger natural killer cell-derived gamma interferon production in vitro. Infect. Immun. *60*, 4335–4342.

Kaye, P.M., Curry, A.J., and Blackwell, J.M. (1991). Differential production of Th1- and Th2-derived cytokines does not determine the genetically controlled or vaccine-induced rate of cure in murine visceral leishmaniasis. J. Immunol. *146*, 2763–2770.

Kaye, P.M., Svensson, M., Ato, M., Maroof, A., Polley, R., Stager, S., Zubairi, S., and Engwerda, C.R. (2004). The immunopathology of experimental visceral leishmaniasis. Immunol. Rev. *201*, 239–253.

Kelly, B.L., and Locksley, R.M. (2004). The *Leishmania major* LACK antigen with an immunodominant epitope at amino acids 156 to 173 is not required for early Th2 development in BALB/c mice. Infect. Immun. *72*, 6924–6931.

Kopf, M., Brombacher, F., Kohler, G., Kienzle, G., Widmann, K.H., Lefrang, K., Humborg, C., Ledermann, B., and Solbach, W. (1996). IL-4-deficient BALB/c mice resist infection with *Leishmania major*. J. Exp. Med. *184*, 1127–1136.

Lang, R., Rutschman, R.L., Greaves, D.R., and Murray, P.J. (2002). Autocrine deactivation of macrophages in transgenic mice constitutively overexpressing IL-10 under control of the human CD68 promoter. J. Immunol. *168*, 3402–3411.

Laouar, Y., Sutterwala, F.S., Gorelik, L., and Flavell, R.A. (2005). Transforming growth factor-beta controls T helper type 1 cell development through regulation of natural killer cell interferon-gamma. Nat. Immunol. *6*, 600–607.

Launois, P., Gumy, A., Himmelrich, H., Locksley, R.M., Rocken, M., and Louis, J.A. (2002). Rapid IL-4 production by *Leishmania* homolog of mammalian RACK1-reactive CD4$^+$ T cells in resistant mice treated once with anti-IL-12 or -IFN-gamma antibodies at the onset of infection with *Leishmania major* instructs Th2 cell development, resulting in nonhealing lesions. J. Immunol. *168*, 4628–4635.

Launois, P., Maillard, I., Pingel, S., Swihart, K.G., Xenarios, I., Acha-Orbea, H., Diggelmann, H., Locksley, R.M., MacDonald, H.R., and Louis, J.A. (1997). IL-4 rapidly produced by Vbeta4 Valpha8 CD4$^+$ T cells instructs Th2 development and susceptibility to *Leishmania major* in BALB/c mice. Immunity. *6*, 541–549.

Launois, P., Ohteki, T., Swihart, K., MacDonald, H.R., and Louis, J.A. (1995). In susceptible mice, *Leishmania major* induce very rapid interleukin-4 production by CD4$^+$ T cells which are NK1.1. Eur. J. Immunol. *25*, 3298–3307.

Leclercq, V., Lebastard, M., Belkaid, Y., Louis, J., and Milon, G. (1996). The outcome of the parasitic process initiated by *Leishmania infantum* in laboratory mice: a tissue-dependent pattern controlled by the Lsh and MHC loci. J. Immunol. *157*, 4537–4545.

Letterio, J.J., and Roberts, A.B. (1998). Regulation of immune responses by TGF-beta. Annu. Rev. Immunol. *16*, 137–161.

Li, J., Hunter, C.A., and Farrell, J.P. (1999). Anti-TGF-beta treatment promotes rapid healing of *Leishmania major* infection in mice by enhancing in vivo nitric oxide production. J. Immunol. *162*, 974–979.

Louzir, H., Melby, P.C., Ben Salah, A., Marrakchi, H., Aoun, K., Ben Ismail, R., and Dellagi, K. (1998). Immunologic determinants of disease evolution in localized cutaneous leishmaniasis due to *Leishmania major*. J. Infect. Dis. *177*, 1687–1695.

Maasho, K., Sanchez, F., Schurr, E., Hailu, A., and Akuffo, H. (1998). Indications of the protective role of natural killer cells in human cutaneous leishmaniasis in an area of endemicity. Infect. Immun. *66*, 2698–2704.

Maasho, K., Satti, I., Nylen, S., Guzman, G., Koning, F., and Akuffo, H. (2000). A *Leishmania* homologue of receptors for activated C-kinase (LACK) induces both interferon-gamma and interleukin-10 in natural killer cells of healthy blood donors. J. Infect. Dis. *182*, 570–578.

Martinez-Arends, A., Tapia, F.J., Caceres-Dittmar, G., Mosca, W., Valecillos, L., and Convit, J. (1991). Immunocytochemical characterization of immune cells in lesions of American cutaneous leishmaniasis using novel T cell markers. Acta Trop. *49*, 271–280.

Matthews, D.J., Emson, C.L., McKenzie, G.J., Jolin, H.E., Blackwell, J.M., and McKenzie, A.N. (2000). IL-13 is a susceptibility factor for *Leishmania major* infection. J. Immunol. *164*, 1458–1462.

Mattner, F., Magram, J., Ferrante, J., Launois, P., Di Padova, K., Behin, R., Gately, M.K., Louis, J.A., and Alber, G. (1996). Genetically resistant mice lacking interleukin-12 are susceptible to infection with *Leishmania major* and mount a polarized Th2 cell response. Eur. J. Immunol. *26*, 1553–1559.

McMahon-Pratt, D., and Alexander, J. (2004). Does the *Leishmania major* paradigm of pathogenesis and protection hold for New World cutaneous leishmaniases or the visceral disease? Immunol. Rev. *201*, 206–224.

Melby, P.C., Andrade-Narvaez, F., Darnell, B.J., and Valencia-Pacheco, G. (1996). In situ expression of interleukin-10 and interleukin-12 in active human cutaneous leishmaniasis. FEMS Immunol. Med. Microbiol. *15*, 101–107.

Melby, P.C., Andrade-Narvaez, F.J., Darnell, B.J., Valencia-Pacheco, G., Tryon, V.V., and Palomo-Cetina, A. (1994). Increased expression of proinflammatory cytokines in chronic lesions of human cutaneous leishmaniasis. Infect. Immun. *62*, 837–842.

Miles, S.A., Conrad, S.M., Alves, R.G., Jeronimo, S.M., and Mosser, D.M. (2005). A role for IgG immune complexes during infection with the intracellular pathogen Leishmania. J. Exp. Med. *201*, 747–754.

Mittrucker, H.W., and Kaufmann, S.H. (2004). Mini-review: regulatory T cells and infection: suppression revisited. Eur. J. Immunol. *34*, 306–312.

Monteforte, G.M., Takeda, K., Rodriguez-Sosa, M., Akira, S., David, J.R., and Satoskar, A.R. (2000). Genetically resistant mice lacking IL-18 gene develop Th1 response and control cutaneous *Leishmania major* infection. J. Immunol. *164*, 5890–5893.

Moore, K.W., de Waal Malefyt, R., Coffman, R.L., and O'Garra, A. (2001). Interleukin-10 and the interleukin-10 receptor. Annu. Rev. Immunol. *19*, 683–765.

Muller, I., Kropf, P., Etges, R.J., and Louis, J.A. (1993). Gamma interferon response in secondary *Leishmania major* infection: role of $CD8^+$ T cells. Infect. Immun. *61*, 3730–3738.

Muller, I., Kropf, P., Louis, J.A., and Milon, G. (1994). Expansion of gamma interferon-producing $CD8^+$ T cells following secondary infection of mice immune to *Leishmania major*. Infect. Immun. *62*, 2575–2581.

Muller, I., Pedrazzini, T., Kropf, P., Louis, J., and Milon, G. (1991). Establishment of resistance to *Leishmania major* infection in susceptible BALB/c mice requires parasite-specific $CD8^+$ T cells. Int. Immunol. *3*, 587 597.

Murphy, M.L., Wille, U., Villegas, E.N., Hunter, C.A., and Farrell, J.P. (2001). IL-10 mediates susceptibility to *Leishmania donovani* infection. Eur. J. Immunol. *31*, 2848–2856.

Murray, H.W., Moreira, A.L., Lu, C.M., DeVecchio, J.L., Matsuhashi, M., Ma, X., and Heinzel, F.P. (2003). Determinants of response to interleukin-10 receptor blockade immunotherapy in experimental visceral leishmaniasis. J. Infect. Dis. *188*, 458–464.

Nishikomori, R., Gurunathan, S., Nishikomori, K., and Strober, W. (2001). BALB/c mice bearing a transgenic IL-12 receptor beta 2 gene exhibit a nonhealing phenotype to *Leishmania major* infection despite intact IL-12 signaling. J. Immunol. *166*, 6776–6783.

Noben-Trauth, N., Lira, R., Nagase, H., Paul, W.E., and Sacks, D.L. (2003). The relative contribution of IL-4 receptor signaling and IL-10 to susceptibility to *Leishmania major*. J. Immunol. *170*, 5152–5158.

Noben-Trauth, N., Paul, W.E., and Sacks, D.L. (1999). IL-4- and IL-4 receptor-deficient BALB/c mice reveal differences in susceptibility to *Leishmania major* parasite substrains. J. Immunol. *162*, 6132–6140.

Okamura, H., Tsutsi, H., Komatsu, T., Yutsudo, M., Hakura, A., Tanimoto, T., Torigoe, K., Okura, T., Nukada, Y., Hattori, K., and *et al.* (1995). Cloning of a new cytokine that induces IFN-gamma production by T cells. Nature *378*, 88–91.

Padigel, U.M., and Farrell, J.P. (2005). Control of infection with *Leishmania major* in susceptible BALB/c mice lacking the common gamma-chain for FcR is associated with reduced production of IL-10 and TGF-beta by parasitized cells. J. Immunol. *174*, 6340–6345.

Park, H., Li, Z., Yang, X.O., Chang, S.H., Nurieva, R., Wang, Y.H., Wang, Y., Hood, L., Zhu, Z., Tian, Q., and Dong, C. (2005). A distinct lineage of CD4 T cells regulates tissue inflammation by producing interleukin 17. Nat. Immunol. *6*, 1133–1141.

Pestka, S., Krause, C.D., Sarkar, D., Walter, M.R., Shi, Y., and Fisher, P.B. (2004). Interleukin-10 and related cytokines and receptors. Annu. Rev. Immunol. *22*, 929–979.

Pinheiro, R.O., Pinto, E.F., Lopes, J.R., Guedes, H.L., Fentanes, R.F., and Rossi-Bergmann, B. (2005). TGF-beta-associated enhanced susceptibility to leishmaniasis following intramuscular vaccination of mice with *Leishmania amazonensis* antigens. Microbes Infect. *7*, 1317–1323.

Pirmez, C., Cooper, C., Paes-Oliveira, M., Schubach, A., Torigian, V.K., and Modlin, R.L. (1990). Immunologic responsiveness in American cutaneous leishmaniasis lesions. J. Immunol. *145*, 3100–3104.

Reiner, S.L., Zheng, S., Wang, Z.E., Stowring, L., and Locksley, R.M. (1994). *Leishmania* promastigotes evade interleukin 12 (IL-12) induction by macrophages and stimulate a broad range of cytokines from $CD4^+$ T cells during initiation of infection. J. Exp. Med. *179*, 447–456.

Rhee, E.G., Mendez, S., Shah, J.A., Wu, C.Y., Kirman, J.R., Turon, T.N., Davey, D.F., Davis, H., Klinman, D.M., Coler, R.N., Sacks, D.L., and Seder, R.A. (2002). Vaccination with heat-killed *Leishmania* antigen or recombinant leishmanial protein and CpG oligodeoxynucleotides induces long-term memory $CD4^+$ and $CD8^+$ T cell responses and protection against *Leishmania major* infection. J. Exp. Med. *195*, 1565–1573.

Rocha, P.N., Almeida, R.P., Bacellar, O., de Jesus, A.R., Filho, D.C., Filho, A.C., Barral, A., Coffman, R.L., and Carvalho, E.M. (1999). Downregulation of Th1 type of response in early human American cutaneous leishmaniasis. J. Infect. Dis. *180*, 1731–1734.

Rogers, K.A., and Titus, R.G. (2004). Characterization of the early cellular immune response to *Leishmania major* using peripheral blood mononuclear cells from Leishmania-naive humans. Am. J. Trop. Med. Hyg. *71*, 568–576.

Russo, D.M., Chakrabarti, P., and Higgins, A.Y. (1999). Leishmania: naive human T cells sensitized with promastigote antigen and IL-12 develop into potent Th1 and CD8$^+$ cytotoxic effectors. Exp. Parasitol. *93*, 161–170.

Sacks, D., and Noben-Trauth, N. (2002). The immunology of susceptibility and resistance to *Leishmania major* in mice. Nat. Rev. Immunol. *2*, 845–858.

Sadick, M.D., Heinzel, F.P., Holaday, B.J., Pu, R.T., Dawkins, R.S., and Locksley, R.M. (1990). Cure of murine leishmaniasis with anti-interleukin 4 monoclonal antibody. Evidence for a T cell-dependent, interferon gamma-independent mechanism. J. Exp. Med. *171*, 115–127.

Satoskar, A., Brombacher, F., Dai, W.J., McInnes, I., Liew, F.Y., Alexander, J., and Walker, W. (1997). SCID mice reconstituted with IL-4-deficient lymphocytes, but not immunocompetent lymphocytes, are resistant to cutaneous leishmaniasis. J. Immunol. *159*, 5005–5013.

Scharton-Kersten, T., Afonso, L.C., Wysocka, M., Trinchieri, G., and Scott, P. (1995). IL-12 is required for natural killer cell activation and subsequent T helper 1 cell development in experimental leishmaniasis. J. Immunol. *154*, 5320–5330.

Scott, P. (1991). IFN-gamma modulates the early development of Th1 and Th2 responses in a murine model of cutaneous leishmaniasis. J. Immunol. *147*, 3149–3155.

Silveira, F.T., Blackwell, J.M., Ishikawa, E.A., Braga, R., Shaw, J.J., Quinnell, R.J., Soong, L., Kima, P., McMahon-Pratt, D., Black, G.F., and Shaw, M.A. (1998). T cell responses to crude and defined leishmanial antigens in patients from the lower Amazon region of Brazil infected with different species of *Leishmania* of the subgenera *Leishmania* and *Viannia*. Parasite Immunol. *20*, 19–26.

Silveira, F.T., Lainson, R., and Corbett, C.E. (2004). Clinical and immunopathological spectrum of American cutaneous leishmaniasis with special reference to the disease in Amazonian Brazil: A review. Mem. Inst. Oswaldo Cruz. *99*, pp. 239–251.

Sosa, M.R., Rosas, L.E., McKenzie, A.N., and Satoskar, A.R. (2001). IL-13 gene-deficient mice are susceptible to cutaneous *L. mexicana* infection. Eur. J. Immunol. *31*, 3255–3260.

Stern, J.J., Oca, M.J., Rubin, B.Y., Anderson, S.L., and Murray, H.W. (1988). Role of L3T4$^+$ and LyT-2$^+$ cells in experimental visceral leishmaniasis. J. Immunol. *140*, 3971–3977.

Stetson, D.B., Mohrs, M., Mallet-Designe, V., Teyton, L., and Locksley, R.M. (2002). Rapid expansion and IL-4 expression by Leishmania-specific naive helper T cells in vivo. Immunity. *17*, 191–200.

Swihart, K., Fruth, U., Messmer, N., Hug, K., Behin, R., Huang, S., Del Giudice, G., Aguet, M., and Louis, J.A. (1995). Mice from a genetically resistant background lacking the interferon gamma receptor are susceptible to infection with *Leishmania major* but mount a polarized T helper cell 1-type CD4$^+$ T cell response. J. Exp. Med. *181*, 961–971.

Sypek, J.P., Chung, C.L., Mayor, S.E., Subramanyam, J.M., Goldman, S.J., Sieburth, D.S., Wolf, S.F., and Schaub, R.G. (1993). Resolution of cutaneous leishmaniasis: interleukin 12 initiates a protective T helper type 1 immune response. J. Exp. Med. *177*, 1797–1802.

Szabo, S.J., Dighe, A.S., Gubler, U., and Murphy, K.M. (1997). Regulation of the interleukin IL-12R beta 2 subunit expression in developing T helper 1 (Th1) and Th2 cells. J. Exp. Med. *185*, 817–824.

Szabo, S.J., Kim, S.T., Costa, G.L., Zhang, X., Fathman, C.G., and Glimcher, L.H. (2000). A novel transcription factor, T-bet, directs Th1 lineage commitment. Cell. *100*, 655–669.

Szabo, S.J., Sullivan, B.M., Stemmann, C., Satoskar, A.R., Sleckman, B.P., and Glimcher, L.H. (2002). Distinct effects of T-bet in Th1 lineage commitment and IFN-gamma production in CD4 and CD8 T cells. Science. *295*, 338–342.

Titus, R.G., Milon, G., Marchal, G., Vassalli, P., Cerottini, J.C., and Louis, J.A. (1987). Involvement of specific Lyt-2$^+$ T cells in the immunological control of experimentally induced murine cutaneous leishmaniasis. Eur J. Immunol. *17*, 1429–1433.

Tomura, M., Maruo, S., Mu, J., Zhou, X.Y., Ahn, H.J., Hamaoka, T., Okamura, H., Nakanishi, K., Clark, S., Kurimoto, M., and Fujiwara, H. (1998). Differential capacities of CD4$^+$, CD8$^+$, and CD4$^-$ CD8$^-$ T cell subsets to express IL-18 receptor and produce IFN-gamma in response to IL-18. J. Immunol. *160*, 3759–3765.

Trinchieri, G. (2003). Interleukin-12 and the regulation of innate resistance and adaptive immunity. Nat. Rev. Immunol. *3*, 133–146.

Tsagozis, P., Karagouni, E., and Dotsika, E. (2005). Function of CD8$^+$ T lymphocytes in a self-curing mouse model of visceral leishmaniasis. Parasitol. Int. *54*, 139–146.

Uzonna, J.E., Joyce, K.L., and Scott, P. (2004). Low dose *Leishmania major* promotes a transient T helper cell type 2 response that is downregulated by interferon gamma-producing CD8$^+$ T cells. J. Exp. Med. *199*, 1559–1566.

Vieira, L.Q., Hondowicz, B.D., Afonso, L.C., Wysocka, M., Trinchieri, G., and Scott, P. (1994). Infection with *Leishmania major* induces interleukin-12 production in vivo. Immunol. Lett. *40*, 157–161.

Wang, Z.E., Reiner, S.L., Hatam, F., Heinzel, F.P., Bouvier, J., Turck, C.W., and Locksley, R.M. (1993). Targeted activation of CD8 cells and infection of beta 2-microglobulin-deficient mice fail to confirm a primary protective role for CD8 cells in experimental leishmaniasis. J. Immunol. *151*, 2077–2086.

Wang, Z.E., Reiner, S.L., Zheng, S., Dalton, D.K., and Locksley, R.M. (1994). CD4$^+$ effector cells default to the Th2 pathway in interferon gamma-deficient mice infected with *Leishmania major*. J. Exp. Med. *179*, 1367–1371.

Wei, X.Q., Leung, B.P., Niedbala, W., Piedrafita, D., Feng, G.J., Sweet, M., Dobbie, L., Smith, A.J., and Liew, F.Y. (1999). Altered immune responses and susceptibility to *Leishmania major* and Staphylococcus aureus infection in IL-18-deficient mice. J. Immunol. *163*, 2821–2828.

Wilson, M.E., Young, B.M., Davidson, B.L., Mente, K.A., and McGowan, S.E. (1998). The importance of TGF-beta in murine visceral leishmaniasis. J. Immunol. *161*, 6148–6155.

Xu, D., Liu, H., Komai-Koma, M., Campbell, C., McSharry, C., Alexander, J., and Liew, F.Y. (2003). $CD4^+$ $CD25^+$ regulatory T cells suppress differentiation and functions of Th1 and Th2 cells, *Leishmania major* infection, and colitis in mice. J. Immunol. *170*, 394–399.

Xu, D., Trajkovic, V., Hunter, D., Leung, B.P., Schulz, K., Gracie, J.A., McInnes, I.B., and Liew, F.Y. (2000). IL-18 induces the differentiation of Th1 or Th2 cells depending upon cytokine milieu and genetic background. Eur. J. Immunol. *30*, 3147–3156.

Yoshida, H., Hamano, S., Senaldi, G., Covey, T., Faggioni, R., Mu, S., Xia, M., Wakeham, A.C., Nishina, H., Potter, J., Saris, C.J., and Mak, T.W. (2001). WSX-1 is required for the initiation of Th1 responses and resistance to *L. major* infection. Immunity. *15*, 569–578.

Impact of the *Leishmania* Genome on Vaccine Development 13

Jenefer M. Blackwell, Diane McMahon-Pratt and Mary E. Wilson

Abstract

Leishmaniasis affects 12 million people but there are no vaccines in routine use. The genomic sequence of *Leishmania* is complete, providing a rich source of vaccine candidates. Two recent studies used genome-based approaches to screen for novel vaccine candidates. The first screened 100 randomly-selected amastigote-expressed genes as DNA vaccines against *L. major* infection in mice. Fourteen reproducibly protective novel vaccine candidates were identified; the best was an amastin-like gene with 54 related copies in the genome. Of concern, seven vaccines reproducibly exacerbated disease. This correlated with interleukin-10 production by antigen-driven $CD4^{+}CD25^{+}$ regulatory T cells. Protection correlated with $CD4^{+}$ effector T cells producing interferon-γ in the presence of low interleukin-10. The second study used a two-step procedure to identify T cell antigens. Step one was to screen a *L. chagasi* cDNA library with a pool of sera from visceral leishmaniasis patients. Positive clones were then screened for ability to elicit proliferation and interferon-γ in T cells from immune mice. Six unique clones were identified: glutamine synthetase, a transitional endoplasmic reticulum ATPase, elongation factor 1γ, kinesin K-39, repetitive protein A2, and a hypothetical conserved protein. The 20 antigens identified in these two studies are being further evaluated for vaccine development.

Leishmaniasis disease and control in humans

Leishmania spp. are obligate intracellular protozoan parasites of macrophages. Leishmaniasis affects 12 million people, with an annual incidence of 2 million cases, of which 500,000 are clinical visceral leishmaniasis (VL) that is fatal unless treated. VL is caused by members of the *L. donovani* complex: *L. donovani*, *L. infantum* and *L. chagasi*. Incidence is rising with urbanization and HIV co-infection. Drug therapy is expensive and resistance is increasing. Excluding epidemics there are 57,000 deaths annually. Cutaneous leishmaniasis (CL) is associated with a clinical spectrum that ranges from self-limiting ulcers to debilitating diffuse or mucocutaneous forms of disease. Species of *Leishmania* associated with CL include *L. major* (Old World) and *L. mexicana* and *L. braziliensis* complexes (New World). Overall disease burden is calculated by WHO as 1.98 million DALYs. There are no vaccines in routine use. Spontaneous or drug-induced recovery from leishmaniasis is accompanied by solid immunity against re-infection, providing a rational basis for vaccine development. Indeed, traditional use of live parasites (leishmanization) to induce lesions and scars in a preferred body site was continued until relatively recently (Greenblatt, 1988), and has been reinstated as a live challenge in vaccine trials of killed parasites (Khamesipour *et al.*, 2005). The combination of autoclaved *Leishmania* sp. plus BCG adjuvant looked promising in animal models (Misra *et al.*, 2001; Srivastava *et al.*, 2003), and autoclaved *L. major* or *L. amazonensis* plus BCG elicits a type 1 immune response in humans (Satti *et al.*, 2001; Velez *et al.*, 2000). However, phase II/III trials in humans have been disappointing, eliciting only a weak Th1 response and no protection from cutaneous leishmaniasis in Latin America (Armijos *et al.*, 2004; Mahmoodi *et al.*, 2003) and no protection against visceral leishmaniasis in the Sudan (Khalil *et al.*, 2000). The most promising trial utilized whole-killed promastigotes plus BCG in Ecuador. The vaccine efficacy was 72.9% for preventing cutaneous leishmaniasis during the 12 months after immunization. However, by 13–18 months the efficacy dropped to 56.5%, and by 24–60 months there was no significant protection (Armijos *et al.*, 2003). Amongst several of these early vaccine trials one common thread has been a measure of protection observed in individuals who skin-test convert during the trial (even in controls), but no overall statistically significant protection has been associated with any trial (Antunes *et al.*, 1986; Khalil *et al.*, 2000; Mayrink *et al.*, 1979; Momeni *et al.*, 1999; Sharifi *et al.*, 1998). Skin-test reactivity following natural exposure is also thought to reflect resistance to infection, with genetic epidemiology studies (Shaw *et al.*, 1995) demonstrating a strong genetic component to ability to mount a skin-test response to leishmanial antigen.

The challenges for novel *Leishmania* vaccines are: (1) to induce long-lasting immunity and to protect genetically susceptible people against fatal VL or debilitating CL; (2) to protect against VL and CL where they occur in the same populations; and (3) to identify the key correlates of protective vaccine-induced immunity. Here we examine how far research has progressed towards the development of recombinant vaccines that might meet these criteria, and examine the potential impact that the *Leishmania* genome project might make in attaining these goals.

Defined vaccines for leishmaniasis

Since the development of molecular cloning, a number of potential *Leishmania* vaccine candidates have been identified (reviewed by Mauel, 2002). Subunit vaccines against leishmaniasis have been tried primarily in experimental animals. Approaches have included recombinant proteins, DNA vaccines, leishmania antigens expressed in BCG, and *Leishmania* antigens expressed in *Salmonella* (Connell *et al.*, 1993; Ghosh *et al.*, 2001a; Yang *et al.*, 1990). Some proteins that have shown promise for inclusion in a subunit vaccine include GP46, P-4, P-2 (cysteine proteinases), P-8, D-13, GP63, A2, P36/LACK, Lcr1, and HASPB1 (Campbell *et al.*, 2003; Champsi and McMahon-Pratt, 1988; Ghosh *et al.*, 2001b; Gonzalo *et al.*, 2002; Gurunathan *et al.*, 1997; Melby *et al.*, 2001; Paraguai de Souza *et al.*, 2001; Rafati *et al.*, 2001; Soong *et al.*, 1995; Stager *et al.*, 2003). A novel approach is the immunization with sandfly salivary components, which induce a protective local inflammatory response (Morris *et al.*, 2001; Valenzuela *et al.*, 2001). Because leishmaniasis is a disease in which there is a clear need for the development of a type 1 (Th1-type) cellular immune response, adjuvants eliciting the type 1 cytokine interferon-γ (IFN-γ) such as monophosphoryl lipid A (MPL Corixa), a chemically modified derivative of lipopolysaccharide, plus squalene (MPL®-SE) or interleukin-12 (IL-12) have been useful in animal models (Reed and Campos-Neto, 2003; Reed *et al.*, 2003; Scott and Trinchieri, 1997). A newer approach is the use of recombinant *Listeria monocytogenes* expressing the *L. major* antigen LACK (Saklani-Jusforgues *et al.*, 2003; Soussi *et al.*, 2000; Soussi *et al.*, 2002). Unfortunately, LACK is effective in preventing *L. major* disease if introduced with IL-12, but it has not been promising for immunization against other species of *Leishmania* (Gurunathan *et al.*, 1997; Melby *et al.*, 2001; Stober *et al.*, 2005).

Interesting among the recombinant antigens tested to date are candidates that have been screened first as vaccines against murine *L. major* infection, and then been shown to be protective in a rhesus monkey nonhuman primate model (Campos-Neto *et al.*, 2001). This precedent in using murine models to screen for antigens that are protective against leishmaniasis in primates is critical to our ability to undertake large-scale screens for novel vaccine candidate. One of the antigens already shown to be effective in both mice and primates is a homologue of a eukaryote thiol-specific antioxidant (TSA), a 22.1 kDa protein identified by screening an expression library with murine immune sera (Webb *et al.*, 1998). This is the same molecule that had earlier been cloned, sequenced and characterized as tryparedoxin peroxidase (TryP) during expressed sequence tag (EST) analysis of *L. major* cDNA libraries undertaken to look for novel vaccine candidates (Levick *et al.*, 1996; Levick *et al.*, 1998). DNA/DNA (Stober *et al.*, 2005) or DNA/MVA (Stober *et al.*, 2005; Stober *et al.*, 2007) TryP vaccines were shown to protect against *L. major* in mice, with the DNA/MVA delivery resulting in long term memory compared to DNA/DNA (Stober *et al.*, 2007). Other antigens effective in mice and nonhuman primates are *L. major* stress-inducible protein 1 (LmSTI1) (Webb *et al.*, 1997; Webb *et al.*, 1996) and *Leishmania* elongation initiation factor (LeIF) (Skeiky *et al.*, 1998), the latter also identified as an EST (Almeida *et al.*, 2004). Di-gene or tri-gene (designated Leish-111f (Skeiky *et al.*, 2002)) poly-protein combinations of these antigens with MPL adjuvant are protective against *L. major* (Coler *et al.*, 2002; Skeiky *et al.*, 2002) and *L. mexicana* (Coler and Reed, 2005) in mice, and *L. infantum* in mice and hamsters (Coler *et al.*, 2007). Plasmid DNA encoding the di-gene TSA/LmSTI1 also protects against high-dose *L. major* infection in BALB/c mice (Campos-Neto *et al.*, 2002), as does a pool of LACK (Gurunathan *et al.*, 1997; Mougneau *et al.*, 1995), TSA and LmSTI1 DNA against low-dose intradermal challenge in C57BL6 mice (Mendez *et al.*, 2001). Recombinant LmSTI1 and TSA induce good protection against *L. major* infection in non-human primate (rhesus monkey) models (Campos-Neto *et al.*, 2001), and the poly-protein Leish-111f administered with IL-12 and alum also provides partial protection in this rhesus monkey/*L. major* model (G. Grimaldi, personal communication). The combination of antigens in the tri-gene poly-protein Leish-111f have been used as immunotherapy against drug-resistant mucosal leishmaniasis (Badaro *et al.*, 2006), and Leish-111f administered with MPL®-SE is currently in phase I/II trials in human volunteers in the USA (Coler and Reed, 2005; Reed *et al.*, 2003).

Another vaccine candidate of specific interest is HASPB1, a recombinant stage-specific hydrophilic acylated surface protein that protects against experimental *L. donovani* LV9 in mice (Stager *et al.*, 2000a; Stager *et al.*, 2000b). Protection induced by rHASPB1 does not require adjuvant and controls parasites in the spleen which is usually refractory to immunological and chemotherapeutic interventions (Stager *et al.*, 2000b). Vaccinated and control-infected mice had equivalent frequencies of rHASPB1-specific $CD4^+$ T cells making IFN-γ. Vaccine-induced protection correlated with rHASPB1-specific, IFN-γ-producing $CD8^+$ T cells, similar to immunity seen with DNA vaccination (Gurunathan *et al.*, 2000b). Adjuvant activity is thought to involve IgM antibody to HASPB1, activating complement and inducing $CD8^+$ memory T cells (Stager *et al.*, 2003).

Adjuvant considerations

MPL®-SE and IL-12 have already been alluded to as potential adjuvants to elicit Th1-type immunity that might be critical for protection against leishmaniasis. MPL®-SE works by ligating toll-like receptor 4 (TLR) (Evans *et al.*, 2003; Stover *et al.*, 2004) and inducing an innate immune response including IL-12. Different TLRs appear to elicit different responses *in vitro* (Jones *et al.*, 2001; Re and Strominger, 2001; Rhee *et al.*, 2003), and the potential for the interaction/synergism between various TLRs is beginning to be explored (Dalpke and Heeg, 2003; Qiao *et al.*, 2003; Yi *et al.*, 2001). TLR4 and TLR9 can act to provide enhanced TNF responses. $CD4^+$ regulatory cells have been shown to down-regulate vaccine responses (Ahlers *et al.*, 2002; Stober *et al.*, 2005), but TLR4 engagement has been reported to overcome $CD4^+CD25^+$ T_{reg} cells (Pasare and Medzhitov, 2003). CD1d-dependent NKT cell activation (by αGalCer) has been shown to result in the bystander activation of CD8 T cells (CTL/IFN-γ), and NKT cells have been demonstrated to activate CD4 and CD8 memory cells (Eberl *et al.*, 2000; Eberl and MacDonald, 2000; Nishimura *et al.*, 2000). Interestingly, tumour necrosis factor-α (TNF-α) induced by various TLRs has been shown to be responsible for the recruitment of CD1d-restricted NKT cells (Mempel, 2002). Hence, it is likely that vaccine delivery systems utilizing TLR4 (e.g. MPL®-SE) or TLR9 (DNA) could be enhanced by activating NKT using α-C-GalCer (Gansert *et al.*, 2003).

Vaccine delivery systems

Recombinant protein-based vaccines fail to elicit long-lived memory T cell responses (Gurunathan *et al.*, 1998), so novel approaches are required. Novel adjuvants that promote Th1 immune responses, including IL-12 (Gurunathan *et al.*, 1998), CpG motifs (Stacey and Blackwell, 1999) and MPL (Reed *et al.*, 2003), provide one approach. Naked DNA, which may also be used with these adjuvants, and viral or bacterial delivery systems are another. DNA vaccines elicit long-lived humoral and cellular ($CD4^+$ and $CD8^+$ T cell) immunity against a variety of pathogens in rodents and non-human primates (Gurunathan *et al.*, 2000a; Gurunathan *et al.*, 2000c), including murine (Gurunathan *et al.*, 1997; Mendez *et al.*, 2001) and canine (Ramiro *et al.*, 2003) leishmaniasis. In human trials (Wang *et al.*, 1998) DNA vaccines evoked specific immunity and were safe in immunocompromised individuals. Heterologous prime/boost strategies, e.g. priming with DNA and boosting with recombinant MVA, elicit stronger cellular and humoral immunity in dogs (Ramiro *et al.*, 2003), non-human primates (Schneider *et al.*, 2001) and humans (McConkey *et al.*, 2003) than DNA alone. In the mouse (Fig. 13.1) homologous DNA/DNA prime/boost works as efficiently as heterologous DNA/MVA against leishmaniasis when mice were challenged 2–4 weeks post-boost, with the same immunological correlates of protection (see below). This is consistent with DNA alone eliciting strong cellular immunity in mice (Gurunathan *et al.*, 2000c). Interestingly, when mice were challenged 4–6 months post-boost, DNA/MVA was superior to DNA/DNA. Further, studies of visceral leishmaniasis indicate that DNA/DNA prime/boost fails to protect, while a heterologous DNA/Vaccinia virus prime/boost vaccination provides protection (Dondji *et al.*, 2005; Melby *et al.*, 2001) (Dondji, *et al.*,

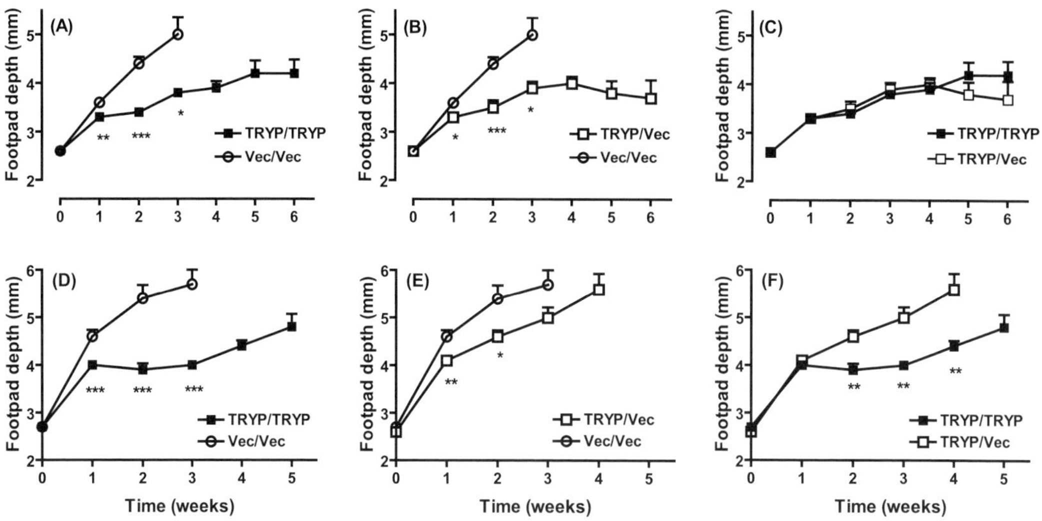

Figure 13.1 Comparison of clinical outcome following *L. major* challenge of BALB/c mice at 2 weeks (A-C) or 16 weeks (D-F) after vaccination with TRYP DNA/TRYP MVA (TRYP/TRYP), TRYP DNA/Vector MVA (TRYP/Vec) or Vector DNA/Vector MVA (Vec/Vec). Mice were vaccinated twice with 100μg of *L. major* TRYP or Vec DNA 3 weeks apart, and boosted with 1×10^6 PFU of TRYP or Vec MVA 2 weeks later. Mice were challenged at 2 (A–C) or 16 (D–F) weeks post-booster with 2×10^6 *L. major* promastigotes into the hind footpad, and footpad depth monitored at weekly intervals with Vernier callipers. Results are mean ± SE for groups of mice within one experiment, which is representative of replicate experiments in which similar results were obtained. Asterisks indicate significant differences to Vec/Vec by Student's *t* test (*$P < 0.05$; **$P < 0.01$; ***$P < 0.001$). These experiments demonstrate that, whilst vaccination with DNA alone is equivalent to prime-boost with DNA/MVA if mice are challenged at 2 weeks in the effector phase of the immune response to vaccination, prime-boost with DNA/MVA provides a better long-term memory response and efficacy against challenge infection 16 weeks after the final booster vaccination. Adapted from (Stober *et al.*, 2007).

unpublished). Therefore, a heterologous boost provides better long term memory for a protective immune response (Stober *et al.*, 2007) (Fig. 13.1) and protection across the species. Studies are in progress comparing homologous DNA/DNA with heterologous DNA/MVA prime/boost using the TryP antigen against *L. major* infection in the rhesus monkey *Macaca mulatta* (R. Porrozzi, G. Grimaldi and J.M. Blackwell, unpublished). DNA vaccines have advantages in terms of ease of preparation, delivery and storage and, if effective, would provide a cost-effective means of vaccination for developing countries.

Immunological correlates of vaccine efficacy

In carrying out trials with novel vaccines it will be essential to monitor immune response and determine the correlates of protection. Vaccine efficacy against *Leishmania* spp., as for *Mycobacterium tuberculosis*, is thought to rest on induction and maintenance of a cellular response that produces IFN-γ to activate macrophages to kill the pathogen. That IFN-γ is crucial to pathogen clearance is not contested, since IFN-γ knockout mice (Swihart *et al.*, 1995; Wang *et al.*, 1994) fail to clear leishmanial infection. It is also clear, however, is that ability to elicit an IFN-γ response is not the sole predictor of vaccine success. For example, Peters and colleagues (Peters *et al.*, 1990) demonstrated that human volunteers vaccinated with avirulent *l. arabica* parasites had strong IFN-γ responses, but were not protected against challenge with virulent *L. major* infection. Similarly, antigen-specific IFN-γ responses have been observed in vaccine trials using BCG and killed leishmanial antigen in Venezuela (Castes *et al.*, 1994), but no significant protection has been observed using this vaccine in Iran (Momeni *et al.*, 1999; Sharifi *et al.*, 1998) or Sudan (Khalil *et al.*, 2000). The murine and human data together tell us that, although IFN-γ is essential, it is either not sufficient or its action may be blocked or down-regulated by another mediator. According to the Th1/Th2 paradigm in *L. major*-infected mice (reviewed in Locksley and Scott, 1991; Reiner *et al.*, 1993), this was thought to be due to production of IL-4. Hence it was assumed that a shift in balance from IL-4 to IFN-γ would be the key to vaccine success. However, demonstration of a role for IL-10 and T_{reg} cells in parasite persistence (Belkaid *et al.*, 2001; Belkaid *et al.*, 2002) has prompted recent re-evaluation of vaccine-induced immunity. When comparing DNA/DNA or DNA/MVA vaccine-induced immune responses (Stober *et al.*, 2005) to the two antigens LACK (Mougneau *et al.*, 1995) and TryP (Levick *et al.*, 1998), high pre-challenge CD4 T cell-derived IL-10 predicted vaccine failure using LACK, whereas low IL-10 predicted protection using TryP. The ratio of IFN-γ:IL-10 was a clear pre-challenge indicator of vaccine success (Table 13.1). Challenge infection caused further polarization to high IL-10/low IFN-γ with LACK and low IL-10/high IFN-γ with TryP. *In vitro* depletion experiments and *ex vivo* quantitative RT/PCR demonstrated that antigen-driven $CD4^{+}CD25^{+}$ T_{reg} cells were the primary source of IL-10 in LACK-vaccinated mice. Anti-IL-10R treatment *in vivo* demonstrated that IL-10 was functional in determining vaccine failure for LACK (Stober *et al.*, 2005) and the novel exacerbatory antigen lmd29 (Roberts *et al.*, 2005), rendering both

Table 13.1 Ratios of IFN-γ:IL-10 in lymph node cells stimulated *in vitro* with recombinant protein (pre- and post-challenge) or crude freeze-thawed parasite antigen (post-challenge) for BALB/c mice vaccinated with either the non-protective LACK antigen or the protective TRYP antigen (Stober *et al.*, 2005)

Vaccine group	Ratio IFN-γ:IL-10		
	Pre-challenge recombinant protein	Post-challenge recombinant protein	Post-challenge FTP
LACK	15.5 ± 1.4	4.0 ± 0.1	3.5 ± 0.2
TryP	41.0 ± 2.8	29.3 ± 1.1	22.3 ± 0.9

These experiments demonstrate that the balance of IFN-γ and IL-10 is a good predictor of vaccine outcome. The deleterious effect of IL-10 in this model is demonstrated in Figure 13.2. Data from experiments presented in (Stober *et al.*, 2005).

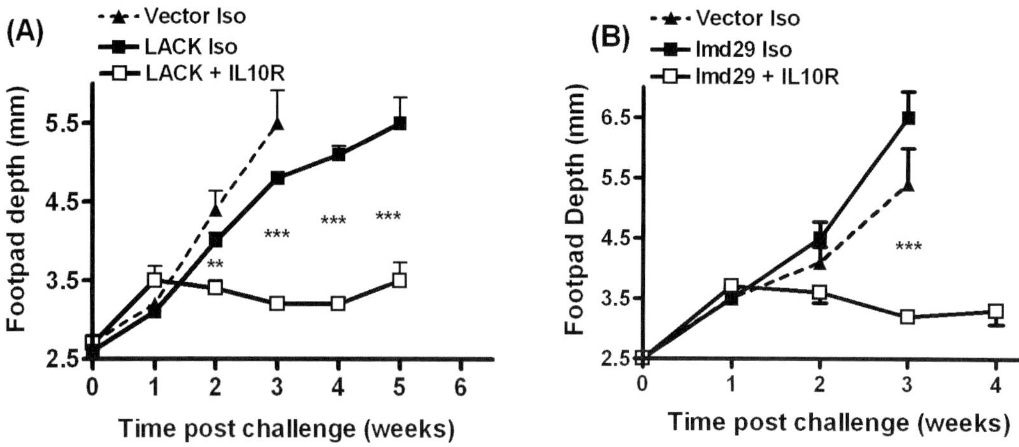

Figure 13.2 Effect of anti-IL-10R treatment *in vivo* prior to challenge infection in (A) LACK-vaccinated or (B) lmd129-vaccinated BALB/c mice. (A) Footpad depths in isotype control mAb treated vector or LACK-vaccinated mice compared to anti-IL-10R treated LACK-vaccinated mice. (B) Footpad depths in isotype control mAb treated vector or lmd129-vaccinated mice compared to anti-IL-10R treated lmd129-vaccinated mice. Asterisks indicate significant differences between LACK/lmd29-vaccinated mice and anti-IL10R-treated LACK/lmd29-vaccinated mice by Student's *t* test (**$P < 0.01$; ***$P < 0.001$). These experiments demonstrate the role of IL-10 in vaccine failure for LACK and exacerbation of disease for lmd29. Adapted from (Roberts *et al.*, 2005; Stober *et al.*, 2005).

protective (Fig. 13.2) in the presence of high IFN-γ/low IL-4/5 responses. Vaccination of BALB background IL4KO mice demonstrates that IL-4 and IL-10 collude in vaccine failure associated with exacerbatory antigens (Roberts *et al.*, 2005). Overall, these data suggest that success of a particular antigen as a vaccine is determined by whether or not it induces an IL-10 response, rather than its ability *per se* to stimulate IFN-γ production. It will be important to monitor IL-10 alongside IFN-γ and IL-4/5/13 in future vaccine studies, as well as to characterize the IFN-γ and IL-10-producing T cell populations.

Do we need more vaccine candidates?

Since there are already potential vaccine candidates going into canine, primate and human trials, why do we need more candidate vaccine antigens? One major concern is that, since the antigens in trial have been mostly selected on the basis of efficacy against *L. major* infection, they are not necessarily proven vaccines against VL or New World species of *Leishmania*. This may be due to differences in the antigens (due to the genetic distances between species complexes in *Leishmania*) and also differences in mechanisms of pathogenesis between species. In addition, the current vaccines in trial may fail to elicit lasting T cell memory, and may not work in genetically heterogeneous humans and dogs in endemic areas. In particular, they may not elicit immunity across all HLA/DLA haplotypes, and they may not be fully protective across *Leishmania* spp. It is also possible that 'resistance' to specific epitopes may develop under vaccine pressure, a problem that is less likely the more antigenically complex novel vaccines can be made. Therefore, the next generation vaccines/vaccine antigens should always be waiting in the wings, and we should continue to improve on methods of delivery that will elicit lasting T cell memory. With this in mind, recent studies have focused on ways in which we might utilize the output of the *Leishmania* genome project to identify potential novel vaccine candidates.

Screening 100 *L. major* genes to identify novel vaccine candidates

The genomic sequence of *Leishmania major* (Ivens *et al.*, 2005), supported by excellent coverage of *L. infantum* and *L. braziliensis* (Peacock *et al.*, 2007) provides a rich source of vaccine candidates. Even at the early stages of EST analysis it became clear that we would quickly become inundated with ready-cloned potential vaccine antigens. To try to assess what proportion of proteins with unknown function might provide novel vaccine candidates, 100 randomly selected amastigote-expressed genes were screened as DNA vaccines, and efficacy determined following high-dose *L. major* footpad challenge in BALB/c mice (Stober *et al.*, 2006b). Fourteen protective novel vaccine candidates were identified, while seven vaccines reproducibly exacerbated disease (Fig. 13.3). Using the computer programme RANKPEP (Reche *et al.*, 2002; Reche *et al.*, 2004) to predict T cell epitopes, no differences were found in the number of predicted MHC H-2^d class I or II epitopes mapping to protective versus exacerbatory antigens. A proportion of both protective (7/14; 50%) and exacerbatory (4/7; 57%) proteins showed short (8- to 18-mer) 100 percent amino acid sequence identities to human, mouse or gut flora proteins. A high proportion of these (4/7 protective; 3/4 exacerbatory) showed full or partial overlap with RANKPEP-predicted H-2^d class I and class II epitopes. The data suggest, therefore, that there may be little difference between antigens/epitopes that drive regulatory versus effector CD4 T cell populations, making it difficult to use epitope prediction as a means of developing a predictive algorithm to search the genome for further protective vaccine candidates. However, this kind of approach really needs to be further evaluated comparing a much larger set of protective versus exacerbatory proteins. The best novel protective antigen found in the 100 antigen vaccine screen was an amastin-like gene that maps to a 17-gene tandem array on *Leishmania* chromosome 8 and is closely related to 37 other amastin-like genes. Two ribosomal proteins, a V-ATPase subunit, and a dynein light chain orthologue were the only other protective genes with putative functions (Table 13.2).

Identifying *L. chagasi* T-cell antigens using a double-library screen

Another genome-based approach to screening for novel vaccine candidates was to generate a cDNA library from the intracellular amastigote form of *L. chagasi*, the cause of South American visceral leishmaniasis, and to use this in a two-step expression screen to systematically identify T cell antigens and T-dependent B cell antigens (Martins *et al.*, 2006). The first step was aimed at identifying the largest possible number of clones producing an epitope-containing polypeptide by screening with a pool of sera from Brazilians with documented visceral leishmaniasis. After removal of clones encoding heat shock proteins, positive clones underwent a second-step screen

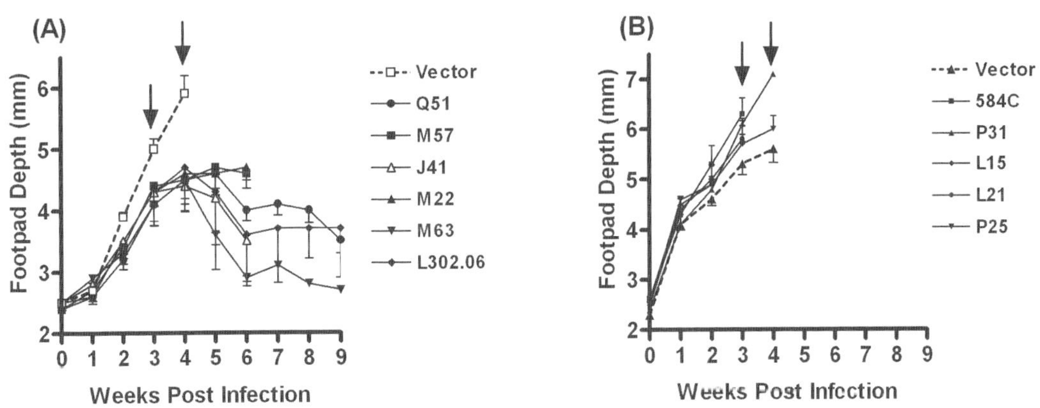

Figure 13.3 Clinical outcome measured as mean ± SEM footpad depth in mm for (A) 6 of 14 reproducibly protective antigens and (B) and 5 of 7 reproducibly exacerbatory antigens identified during a screen of 100 novel vaccines using DNA vaccination and challenge with *L. major* LV39 (Stober *et al.*, 2006a). Footpad depth measurements are shown until >20 % mice euthanized. Arrows indicate time points at which all 6 of the antigens in (A) had significantly lower footpad measurements compared to the control vector-vaccinated mice, while all 5 of the antigens in (B) had significantly higher footpad measurements compared to the control. These experiments demonstrated one concern that arose from screening 100 randomly selected genes as vaccines against *L. major* infection in that reproducibly exacerbatory antigens were identified as well as protective antigens. Adapted from (Stober *et al.*, 2006a).

TTable 13.2 Details of 14 protective and 7 exacerbatory proteins identified in the 101 antigen vaccine screen

Vaccine no.	Clone name	EST accession number	EST pool no.	CDS in vaccine (FL) (bp)	GeneDB accession number	Predicted protein size (kDa)	Predicted TMH	Signal sequence	Putative function
Protective antigens									
v5	N52	AA680824	65	414 (414)	LmjF12.0520	15.4	None	None	V-ATPase subunit F
v9	M18	AA680767	105	189 (600)	LmjF08.0770*	22.2	5–27, 80–102, 109–131, 151–173	None	Amastin-like protein
v12	R32	AA680932	136	240 (339)	LmjF33.2955	11.6	None	None	Hypothetical protein
v15	R71	AA680948	142	390 (390)	LmjF36.4420 LmjF36.3270	15.0	None	None	60S ribosomal L22
v16	J89	AA680690	150	162 (162)	LmjF26.1015	7.2	None	None	Hypothetical protein
v19	L37	AA680742	179	480 (519)	LmjF34.2780	19.3	None	None	40S ribosomal S19
v22	Q24	AA680882	250	276 (276)	LmjF32.0230	10.6	None	None	Dynein light chain
v50	lmk628	W88281	1046	339 (339)	LmjF34.2515	12.7	None	1_38	Hypothetical protein
v51	Q51	AA680894	9	213 (498)	LmjF30.1750	18.6	None	none	Hypothetical protein
v60	M57	AA680781	113	252 (252)	LmjF20.1375	9.4	None	1_28	Hypothetical protein
v79	J41	AA680671	449	468 (567)	LmjF33.1535	21.0	None	None	Hypothetical protein
v82	M22	AA680769	489	237 (555)	LmjF29.1605	8.2	None	None	Hypothetical protein
v83	M63	AA680786	494	144 (144)	LmjF33.2725	5.3	None	None	Hypothetical protein
v95	L302.06	–	4	777 (777)	LmjF04.0230	27.7	7_26; 51_73	None	Hypothetical protein

Exacerbatory antigens									
v14	584C	AI034795	141	549 (567)	LmjF35.3280 LmjF35.3290	21.2	None	None	60S ribosomal L31
v26	P31	AA680854	316	336 (1611)	LmjF34.2490	56.9	None	None	Hypothetical protein
v36	L15	AA680734	477	249 (249)	LmjF35.4500	9.3	None	None	Anaphase promoting complex subunit (Zn finger)
v37	L21	AA680736	478	447 (447)	LmjF34.3980	16.7	None	None	Ribosomal protein L7Ae/ L30e/S12e/Gadd45 family
v41	P25	AA680851	523	354 (621)	LmjF35.4730	–	–	–	Hypothetical protein (pseudogene)
v56	S33	AA680974	55	168 (501)	LmjF25.0730 LmjF25.0720	17.8	None	None	Eukaryotic translation initiation factor 5A-1
v101	lmd29	T67335	260	270 (270)	LmjF18.0595	9.7	None	None	Hypothetical protein

*Also has 100% match to other members of amastin family: LmjF08.0740, LmjF08.0720, LmjF08.0760, LmjF08.0700

Table 13.3 *Leishmania chagasi* protective antigens identified with the double screen. The table provides information on the predicted and observed size and physical characteristics of each *Leishmania chagasi* cDNA clone. Adapted from Martins *et al.* (2006)

Clone	Insert size (bp)	ORF size (bp)	ORF size (aa)	Predicted protein mass	Observed protein mass (immunoblot)	Homologues (% identity*)
314	3076	2355 bp	785 aa	86.8 kDa	98 kDa	*L. major* transitional ER ATPase (99%)
319	1770	1140 bp	380 aa	42.4 kDa	NV **	*L. infantum* glutamine synthetase (100%)
419	2073	1092 bp	442 aa	43.3 kDa	53 kDa	*L. infantum* hypothetical protein (100%)
425	1644	1344 bp	448 aa	49.7 kDa	60 kDa	*L. infantum* K-39 (88%)
503	2009	1473 bp	491 aa	56.2 kDa	50 kDa	*L. infantum* EF-1γ (100%)
648	2886	768 bp	256 aa	23.9 kDa	NV**	*L. infantum* A2 (97%)

*Per cent amino acid identity with sequences from GenBank or the *Leishmania major* and *L. infantum* GeneDB genomic databases at the Sanger Centre. GenBank accession numbers of homologues are: clone 314 (transitional ER ATPase): CAJ09090; clone 319 (glutamine synthetase): CAJ02080; clone 419 (*L. infantum* hypothetical protein): CAJ02805; clone 425 (kinesin-like protein): AAA29254; clone 503: identical to *L. infantum* EF-1γ accession number CAC35543; clone 648 (A2 gene class member): AAB30592.

**NV – bands corresponding to recombinant proteins were not visualized on immunoblots.

for their ability to cause proliferation and IFN-γ responses in T cells from resistant C3H mice that had been infected with *L. chagasi*. Six unique clones were selected from the second screen for further analysis (Table 13.3). The corresponding antigens were derived from glutamine synthetase, a transitional endoplasmic reticulum ATPase, elongation factor 1γ, kinesin K-39, repetitive protein A2, and a hypothetical conserved protein. Importantly, humans naturally infected with *L. chagasi* mounted both cellular and antibody responses to these proteins. Hence, preparations containing one or a cocktail of these antigens may provide novel diagnostic candidates as well as candidates for development of protective vaccines.

Is there a need for live attenuated vaccines?

There are various ways in which we can use our knowledge of the *Leishmania* genome to develop new vaccines. One of these has been the possibility of generating a new genetically modified live attenuated vaccine against leishmaniasis. Historically, injection of live virulent organisms, a process known as leishmanization, has been the only successful way of vaccinating humans (Greenblatt, 1988). As with resistance to re-infection following natural infection, the success of leishmanization is thought to be due to persistent parasites. Adverse effects in susceptible individuals have led to discontinuation of leishmanization as a routine method of immunizing humans. The rise in HIV-induced immuno-compromised individuals is also a concern for continued use of live vaccines. One hope is that we could genetically engineer a live vaccine strain that can induce long-lasting immune memory and protection against natural infection without itself causing disease. Some live attenuated vaccines have been made by genetic engineering, but so far these have worked against *L. major* in mice (Uzonna *et al.*, 2004) but not, where tested, against *L. chagasi* in mice (Streit *et al.*, 2001) or *L. major* in non-human primates (Amaral *et al.*, 2002). Our lack of understanding of the immunological mechanisms that provide long-lived immunity after natural infection has led some workers to return to a mouse model of 'leishmanization' to understand more about the process. Examining *L. major* infection in mice, Scott and coworkers (Zaph *et al.*, 2004) have demonstrated that, although effector CD4(+) T cells are lost in the apparent absence of parasites central memory CD4(+) T cells are maintained. Upon secondary infection, these central memory T cells become tissue-homing effector T cells and mediate protection. These workers conclude that immunity to *L. major* is mediated by at least two distinct populations of CD4(+) T cells: short-lived pathogen-dependent effector cells and long-lived pathogen-independent central memory cells. These observations provide new hope that central memory T cells could be the targets for non-live vaccines against leishmaniasis. Intriguingly, one genetically attenuated *Leishmania* lacking the GDP-mannose transporter (termed δlpg2 parasites) induces protection on its own in susceptible BALB/c mice, but only protects resistant C57BL/6 mice when administered with DNA (Kebaier *et al.*, 2006), indicating that parasite persistence alone is not sufficient to maintain protective immunity to *L. major*.

Concluding remarks

In humans, live leishmanization is the only protocol used to date that elicits long-lasting protection against re-infection with *Leishmania spp*. We know from animal studies that it is possible to elicit long-lasting protection using molecularly defined vaccines, and the *Leishmania* genomes (Ivens *et al.*, 2005; Peacock *et al.*, 2007) have provided (Martins *et al.*, 2006; Stober *et al.*, 2006a), and will continue to provide, a rich source of vaccine candidates. The major road block for development of vaccine that will be effective in humans is to identify a mode of delivery that will elicit long-lived central memory. We remain hopeful that the right combination of antigens and delivery system will be found for the successful development of non-live vaccines against leishmaniasis.

References

Ahlers, J.D., Belyakov, I.M., Terabe, M., Koka, R., Donaldson, D.D., Thomas, E.K. and Berzofsky, J.A. (2002). A push-pull approach to maximize vaccine efficacy: abrogating suppression with an IL-13 inhibitor while augmenting help with granulocyte/macrophage colony-stimulating factor and CD40l. Proc. Natl. Acad. Sci. USA *99*, 13020–130205.

Almeida, A.S., Gilmartin, B.J., McCann, S., Norrish, A., Ivens, A., Lawson, D., Levick, M.P., Smith, D.F., Dyall, S.D., Vetrie, D., Freeman, T.C., Coulson, R.M., Sampaio, I., Schneider, H. and Blackwell, J.M. (2004). Expression profiling of the *Leishmania* life cycle: cDNA arrays identify developmentally regulated genes present but not annotated in the genome. Mol. Biochem. Parasitol. *136*, 87–100.

Amaral, V.F., Teva, A., Oliveira-Neto, M.P., Silva, A.J., Pereira, M.S., Cupolillo, E., Porrozzi, R., Coutinho, S.G., Pirmez, C., Beverley, S.M. and Grimaldi, G., Jr. (2002). Study of the safety, immunogenicity and efficacy of attenuated and killed *Leishmania* (Leishmania) major vaccines in a rhesus monkey (Macaca mulatta) model of the human disease. Mem. Inst. Oswaldo Cruz *97*, 1041–1048.

Antunes, C.M., Mayrink, W., Magalhaes, P.A., Costa, C.A., Melo, M.N., Dias, M., Michalick, M.S., Williams, P., Lima, A.O., Vieira, J.B. and *et al.*, 1986. Controlled field trials of a vaccine against New World cutaneous leishmaniasis. Int. J. Epidemiol. *15*, 572–580.

Armijos, R.X., Weigel, M.M., Calvopina, M., Hidalgo, A., Cevallos, W. and Correa, J. (2004). Safety, immunogenecity, and efficacy of an autoclaved *Leishmania amazonensis* vaccine plus BCG adjuvant against New World cutaneous leishmaniasis. Vaccine *22*, 1320–1326.

Armijos, R.X., Weigel, M.M., Romero, L., Garcia, V. and Salazar, J. (2003). Field trial of a vaccine against new world cutaneous leishmaniasis in an at-risk child population: how long does protection last? J. Infect. Dis. *187*, 1959–1961.

Badaro, R., Lobo, I., Munos, A., Netto, E.M., Modabber, F., Campos-Neto, A., Coler, R.N. and Reed, S.G. (2006). Immunotherapy for drug-refractory mucosal leishmaniasis. J. Infect. Dis. *194*, 1151–1159.

Belkaid, Y., Hoffmann, K.F., Mendez, S., Kamhawi, S., Udey, M.C., Wynn, T.A. and Sacks, D.L. (2001). The role of interleukin (IL)-10 in the persistence of *Leishmania major* in the skin after healing and the therapeutic potential of anti-IL-10 receptor antibody for sterile cure. J. Exp. Med. *194*, 1497–1506.

Belkaid, Y., Piccirillo, C.A., Mendez, S., Shevach, E.M. and Sacks, D.L. (2002). CD4+CD25+ regulatory T cells control *Leishmania major* persistence and immunity. Nature *420*, 502–507.

Campbell, K., Diao, H., Ji, J. and Soong, L. (2003). DNA immunization with the gene encoding P4 nuclease of *Leishmania amazonensis* protects mice against cutaneous Leishmaniasis. Infect. Immun. *71*, 6270–6278.

Campos-Neto, A., Porrozzi, R., Greeson, K., Coler, R.N., Webb, J.R., Seiky, Y.A., Reed, S.G. and Grimaldi, G., Jr. (2001). Protection against cutaneous leishmaniasis induced by recombinant antigens in murine and non-human primate models of the human disease. Infect. Immun. *69*, 4103–4108.

Campos-Neto, A., Webb, J.R., Greeson, K., Coler, R.N., Skeiky, Y.A. and Reed, S.G. (2002). Vaccination with plasmid DNA encoding TSA/LmSTI1 leishmanial fusion proteins confers protection against *Leishmania major* infection in susceptible BALB/c mice. Infect. Immun. *70*, 2828–2836.

Castes, M., Blackwell, J., Trujillo, D., Formica, S., Cabrera, M., Zorrilla, G., Rodas, A., Castellanos, P.L. and Convit, J. (1994). Immune response in healthy volunteers vaccinated with killed leishmanial promastigotes plus BCG. I: Skin-test reactivity, T cell proliferation and interferon-gamma production. Vaccine *12*, 1041–1051.

Champsi, J. and McMahon-Pratt, D. (1988). Membrane glycoprotein M-2 protects against *Leishmania amazonenesis* infection. Infect. Immun. 52, 3272–3279.

Coler, R.N., Goto, Y., Bogatzki, L., Raman, V. and Reed, S.G. (2007). Leish-111f, a recombinant polyprotein vaccine that protects against visceral leishmaniasis by the elicitation of CD4+ T cells. Infect. Immun. *75*, 4648–4654.

Coler, R.N. and Reed, S.G. (2005). Second-generation vaccines against leishmaniasis. Trends Parasitol. *21*, 244–249.

Coler, R.N., Skeiky, Y.A., Bernards, K., Greeson, K., Carter, D., Cornellison, C.D., Modabber, F., Campos-Neto, A. and Reed, S.G. (2002). Immunization with a polyprotein vaccine consisting of the T-Cell antigens thiol-specific antioxidant, *Leishmania major* stress-inducible protein 1, and *Leishmania* elongation initiation factor protects against leishmaniasis. Infect. Immun. *70*, 4215–4225.

Connell, N.D., Medina-Acosta, E., McMaster, W.R., Bloom, B.R. and Russell, D.G. (1993). Effective immunization against cutaneous leishmaniasis with recombinant bacille Calmette-Gu,rin expressing the *Leishmania* surface proteinase gp63. Proc. Natl. Acad. Sci. USA *90*, 11473–11477.

Dalpke, A.H. and Heeg, K. (2003). Synergistic and antagonistic interactions between LPS and superantigens. J. Endotoxin Res. *9*, 51–54.

Dondji, B., Perez-Jimenez, E., Goldsmith-Pestana, K., Esteban, M. and McMahon-Pratt, D. (2005). Prime-boost vaccination using the LACK antigen protects against murine visceral leishmaniasis. Infect Immun *73*, 5286–5289.

Eberl, G., Brawand, P. and MacDonald, H.R. (2000). Selective bystander proliferation of memory CD4+ and CD8+ T cells upon NK T or T cell activation. J. Immunol. *165*, 4305–4311.

Eberl, G. and MacDonald, H.R. (2000). Selective induction of NK cell proliferation and cytotoxicity by activated NKT cells. Eur. J. Immunol. *30*, 985–992.

Evans, J.T., Cluff, C.W., Johnson, D.A., Lacy, M.J., Persing, D.H. and Baldridge, J.R. (2003). Enhancement of antigen-specific immunity via the TLR4 ligands MPL adjuvant and Ribi.529. Expert Rev Vaccines *2*, 219–229.

Gansert, J.L., Kiessler, V., Engele, M., Wittke, F., Rollinghoff, M., Krensky, A.M., Porcelli, S.A., Modlin, R.L. and Stenger, S. (2003). Human NKT cells express granulysin and exhibit antimycobacterial activity. J. Immunol *170*, 3154–3161.

Ghosh, A., Labrecque, S. and Matlashewski, G. (2001a. Protection against *Leishmania donovani* infection by DNA vaccination: increased DNA vaccination efficiency through inhibiting the cellular p53 response. Vaccine *19*, 3169–3178.

Ghosh, A., Zhang, W.W. and Matlashewski, G. (2001b. Immunization with A2 protein results in a mixed Th1/Th2 and a humoral response which protects mice against *Leishmania donovani* infections. Vaccine *20*, 59–66.

Gonzalo, R.M., del Real, G., Rodriguez, J.R., Rodriguez, D., Heljasvaara, R., Lucas, P., Larraga, V. and Esteban, M. (2002). A heterologous prime-boost regime using DNA and recombinant vaccinia virus expressing the *Leishmania infantum* P36/LACK antigen protects BALB/c mice from cutaneous leishmaniasis. Vaccine *20*, 1226–1231.

Greenblatt, C.L. (1988). Cutaneous leishmaniasis: The prospects for a killed vaccine. Parasitol. Today *4*, 53–54.

Gurunathan, S., Klinman, D.M. and Seder, R.A. (2000a. DNA vaccines: immunology, application, and optimization. Annu. Rev. Immunol. *18*, 927–974.

Gurunathan, S., Prussin, C., Sacks, D.L. and Seder, R.A. (1998). Vaccine requirements for sustained cellular immunity to an intracellular parasitic infection. Nat. Med. *4*, 1409–1415.

Gurunathan, S., Sacks, D.L., Brown, D.R., Reiner, S.L., Charest, H., Glaichenhaus, N. and Seder, R.A. (1997). Vaccination with DNA encoding the immunodominant LACK parasite antigen confers protective immunity to mice infected with *Leishmania major*. J. Exp. Med. *186*, 1137–1147.

Gurunathan, S., Stobie, L., Prussin, C., Sacks, D.L., Glaichenhaus, N., Fowell, D.J., Locksley, R.M., Chang, J.T., Wu, C.Y. and Seder, R.A. (2000b. Requirements for the maintenance of Th1 immunity in vivo following DNA vaccination: a potential immunoregulatory role for CD8+ T cells. J. Immunol. *165*, 915–924.

Gurunathan, S., Wu, C.Y., Freidag, B.L. and Seder, R.A. (2000c. DNA vaccines: a key for inducing long-term cellular immunity. Curr. Opin. Immunol. *12*, 442–447.

Ivens, A.C., Peacock, C.S., Worthey, E.A., Murphy, L., Aggarwal, G., Berriman, M., Sisk, E., Hertzfowler, C., Quail, M.A., Harris, D., Rajandream, M.-A., Davies, R.M., Aert, R., Andersson, B., Anupama, A., Attipoe, P., Bason, N., Bauser, C., Beck, A., Beverley, S.M., Bianchettin, G., Borzym, K., Bothe, G., Bruschi, C.V., Collins, M., Cadag, E., Ciarloni, L., Clayton, C., Coulson, R.M.R., Cronin, A., Cruz, A.C., De Gaudenzi, J., Dobson, D.E., Duesterhoeft, A., Fazelina, G., Fosker, N., Frasch, A.C., Fraser, A., Fuchs, M., Gabel, C., Goble, C., Goffeau, A., Hilbert, H., Horn, D., Huang, Y., Klages, S., Knights, A., Kube, M., Larke, N., Litvin, L., Lord, A., Louie, T., Marra, M., Masuy, D., Matthews, K., Michaeli, S., Mottram, J.C., Mueller-Auer, S., Munden, H., Nelson, S., Nilsson, D., Norbertczak, H., Oliver, K., O'Neill, S., Pentony, M., Pohl, T.M., Price, C., Purnelle, B., Rabbinowitsch, E., Reinhardt, R., Rieger, M., Rinta, J., Robben, J., Robertson, L., Ruiz, J., Rutter, S., Saunders, D., Schaefer, M., Schein, J., Schwartz, D.C., Seeger, K., Seyler, A., Sharp, S., Shin, H., Sivam, D., Squares, R., Squares, S., Tosato, V., Tram, A.-N., Vogt, C., Volckaert, G., Wambutt, R., Warren, T., Wedler, H.,

Woodward, J., Zhou, S., Zimmermann, W., Smith, D.F., Blackwell, J.M., Stuart, K.D., Barrell, B. and Myler, P.J. (2005). The genome of the kinetoplastid parasite, *Leishmania major*. Science *309*, 436–442.

Jones, B.W., Means, T.K., Heldwein, K.A., Keen, M.A., Hill, P.J., Belisle, J.T. and Fenton, M.J. (2001). Different Toll-like receptor agonists induce distinct macrophage responses. J. Leuk. Biol. *69*, 1036–1044.

Kebaier, C., Uzonna, J.E., Beverley, S.M. and Scott, P. (2006). Immunization with persistent attenuated Delta lpg2 *Leishmania major* parasites requires adjuvant to provide protective immunity in C57BL/6 mice. Infect. Immun. *74*, 777–780.

Khalil, E.A., El Hassan, A.M., Zijlstra, E.E., Mukhtar, M.M., Ghalib, H.W., Musa, B., Ibrahim, M.E., Kamil, A.A., Elsheikh, M., Babiker, A. and Modabber, F. (2000). Autoclaved *Leishmania major* vaccine for prevention of visceral leishmaniasis: a randomised, double-blind, BCG-controlled trial in Sudan. Lancet *356*, 1565–1569.

Khamesipour, A., Dowlati, Y., Asilian, A., Hashemi-Fesharki, R., Javadi, A., Noazin, S. and Modabber, F. (2005). Leishmanization: use of an old method for evaluation of candidate vaccines against leishmaniasis. Vaccine *23*, 3642–3648.

Levick, M.P., Blackwell, J.M., Connor, V., Coulson, R.M., Miles, A., Smith, H.E., Wan, K.L. and Ajioka, J.W. (1996). An expressed sequence tag analysis of a full-length, spliced-leader cDNA library from *Leishmania major* promastigotes. Mol. Biochem. Parasitol. *76*, 345–348.

Levick, M.P., Tetaud, E., Fairlamb, A.H. and Blackwell, J.M. (1998). Identification and characterisation of a functional peroxidoxin from *Leishmania major*. Mol. Biochem. Parasitol. *96*, 125–137.

Locksley, R.M. and Scott, P. (1991). Helper T-cell subsets in mouse leishmaniasis induction, expansion and effector function. Immunol. Today *12*, A58-A61.

Mahmoodi, M., Khamesipour, A., Dowlati, Y., Rafati, S., Momeni, A.Z., Emamjomeh, M., Hejazi, H. and Modabber, F. (2003). Immune response measured in human volunteers vaccinated with autoclaved *Leishmania major* vaccine mixed with low dose of BCG. Clin. Exp. Immunol. *134*, 303–308.

Martins, D.R., Jeronimo, S.M., Donelson, J.E. and Wilson, M.E. (2006). *Leishmania chagasi* T Cell Antigens Identified Through a Double Library Screen. Infect. Immun. *74*, 6940–6948.

Mauel, J. (2002). Vaccination against *Leishmania* infections. Curr. Drug Targets Immune Endocr. Metabol. Disord. *2*, 201–226.

Mayrink, W., da Costa, C.A., Magalhaes, P.A., Melo, M.N., Dias, M., Lima, A.O., Michalick, M.S. and Williams, P. (1979). A field trial of a vaccine against American dermal leishmaniasis. Trans R. Soc. Trop. Med. Hyg. *73*, 385–387.

McConkey, S.J., Reece, W.H., Moorthy, V.S., Webster, D., Dunachie, S., Butcher, G., Vuola, J.M., Blanchard, T.J., Gothard, P., Watkins, K., Hannan, C.M., Everaere, S., Brown, K., Kester, K.E., Cummings, J., Williams, J., Heppner, D.G., Pathan, A., Flanagan, K., Arulanantham, N., Roberts, M.T., Roy, M., Smith, G.L., Schneider, J., Peto, T., Sinden, R.E., Gilbert, S.C. and Hill, A.V. (2003). Enhanced T-cell immunogenicity of plasmid DNA vaccines boosted by recombinant modified vaccinia virus Ankara in humans. Nat. Med. *9*, 729–735.

Melby, P.C., Yang, J., Zhao, W., Perez, L.E. and Cheng, J. (2001). *Leishmania donovani* p36(LACK) DNA vaccine is highly immunogenic but not protective against experimental visceral leishmaniasis. Infect. Immun. *69*, 4719–4725.

Mempel, M., C. Ronet, F. Suarez, M. Gilleron, G. Puzo, L. Van Kaer, A. Lehuen, P. Kourilsky, and G. Gachelin. (2002). Natural killer T cells restricted by the monomorphic MHC class 1b CD1d1 molecules behave like inflammatory cells. J. Immunol. *168*, 365–371.

Mendez, S., Gurunathan, S., Kamhawi, S., Belkaid, Y., Moga, M.A., Skeiky, Y.A., Campos-Neto, A., Reed, S., Seder, R.A. and Sacks, D. (2001). The potency and durability of DNA- and protein-based vaccines against *Leishmania major* evaluated using low-dose, intradermal challenge. J. Immunol. *166*, 5122–5128.

Misra, A., Dube, A., Srivastava, B., Sharma, P., Srivastava, J.K., Katiyar, J.C. and Naik, S. (2001). Successful vaccination against *Leishmania donovani* infection in Indian langur using alum-precipitated autoclaved *Leishmania major* with BCG. Vaccine *19*, 3485–3492.

Momeni, A.Z., Jalayer, T., Emamjomeh, M., Khamesipour, A., Zicker, F., Ghassemi, R.L., Dowlati, Y., Sharifi, I., Aminjavaheri, M., Shafiei, A., Alimohammadian, M.H., Hashemi-Fesharki, R., Nasseri, K., Godal, T., Smith, P.G. and Modabber, F. (1999). A randomised, double-blind, controlled trial of a killed *L. major* vaccine plus BCG against zoonotic cutaneous leishmaniasis in Iran. Vaccine *17*, 466–472.

Morris, R.V., Shoemaker, C.B., David, J.R., Lanzaro, G.C. and Titus, R.G. (2001). Sandfly maxadilan exacerbates infection with *Leishmania major* and vaccinating against it protects against *L. major* infection. J. Immunol. *167*, 5226–5230.

Mougneau, E., Altare, F., Wakil, A.E., Zheng, S., Coppola, T., Wang, Z.-E., Waldmann, R., Locksley, R.M. and Glaichenhaus, N. (1995). Expression cloning of a protective *Leishmania* antigen. Science *268*, 563–566.

Nishimura, T., Kitamura, H., Iwakabe, K., Yahata, T., Ohta, A., Sato, M., Takeda, K., Okumura, K., Van Kaer, L., Kawano, T., Taniguchi, M., Nakui, M., Sekimoto, M. and Koda, T. (2000). The interface between innate and acquired immunity: glycolipid antigen presentation by CD1d-expressing dendritic cells to NKT cells induces the differentiation of antigen-specific cytotoxic T lymphocytes. Int. Immunol. *12*, 987–994.

Paraguai de Souza, E., Bernardo, R.R., Palatnik, M. and Palatnik de Sousa, C.B. (2001). Vaccination of Balb/c mice against experimental visceral leishmaniasis with the GP36 glycoprotein antigen of *Leishmania donovani*. Vaccine *19*, 3104–3115.

Pasare, C. and Medzhitov, R. (2003). Toll pathway-dependent blockade of CD4+CD25+ T cell-mediated suppression by dendritic cells. Science *299*, 1033–1036.

Peacock, C.S., Seege, K., Harris, D., Murphy, L., Ruiz, J.C., Quail, M.A., Peters, N., Adlem, E., Tivey, A., Aslett, M., Kerhornou, A., Ivens, A., Fraser, A., Rajandream,

M.A., Carver, T., Norbertczak, H., Chillingworth, T., Hance, Z., Jagels, K., Moule, S., Ormond, D., Rutter, S., Squares, R., Whitehead, S., Rabbinowitsch, E., C., A., White, B., Thurston, S., Bringaud, F., Baldauf, S.L., Faulconbridge, A., Jeffares, D., Depledge, D.P., Oyola, S.O., Hilley, J.D., Brito, L.O., Tosi, L.R.O., Barrell, B., Cruz, A.K., Mottram, J.C., Smith, D.F. and Berriman, M. (2007). Analysis of genome architecture and content between three *Leishmania* species causing diverse human disease reveals only a small subset of species-specific genes. Nat. Genet. 39, 839–847.

Peters, W., Bryceson, A., Evans, D.A., Neal, R.A., Kaye, P., Blackwell, J., Killick-Kendrick, R. and Liew, F.Y. (1990). *Leishmania* infecting man and wild animals in Saudi Arabia. 8. The influence of prior infection with *Leishmania arabica* on challenge with *L. major* in man. Trans R. Soc. Trop. Med. Hyg. *84*, 681–689.

Qiao, M., Murata, K., Davis, A.R., Jeong, S.H. and Liang, T.J. (2003). Hepatitis C virus-like particles combined with novel adjuvant systems enhance virus-specific immune responses. Hepatology *37*, 52–59.

Rafati, S., Salmanian, A.H., Taheri, T., Vafa, M. and Fasel, N. (2001). A protective cocktail vaccine against murine cutaneous leishmaniasis with DNA encoding cysteine proteinases of *Leishmania major*. Vaccine *19*, 3369–3375.

Ramiro, M.J., Zarate, J.J., Hanke, T., Rodriguez, D., Rodriguez, J.R., Esteban, M., Lucientes, J., Castillo, J.A. and Larraga, V. (2003). Protection in dogs against visceral leishmaniasis caused by *Leishmania infantum* is achieved by immunization with a heterologous prime-boost regime using DNA and vaccinia recombinant vectors expressing LACK. Vaccine *21*, 2474–2484.

Re, F. and Strominger, J.L. (2001). Toll-like receptor 2 (TLR2) and TLR4 differentially activate human dendritic cells. J. Biol. Chem. *276*, 37692–37699.

Reche, P.A., Glutting, J.P. and Reinherz, E.L. (2002). Prediction of MHC class I binding peptides using profile motifs. Hum Immunol *63*, 701–709.

Reche, P.A., Glutting, J.P., Zhang, H. and Reinherz, E.L. (2004). Enhancement to the RANKPEP resource for the prediction of peptide binding to MHC molecules using profiles. Immunogenetics 56, 405–419.

Reed, S.G. and Campos-Neto, A. (2003). Vaccines for parasitic and bacterial diseases. Curr. Opin. Immunol. *15*, 456–460.

Reed, S.G., Coler, R.N. and Campos-Neto, A. (2003). Development of a leishmaniasis vaccine: the importance of MPl. Expert Rev. Vaccines 2, 239–252.

Reiner, S.L., Wang, Z.-E., Hatam, F., Scott, P. and Locksley, R.M. (1993). TH1 and TH2 cell antigen receptors in experimental leishmaniasis. Science *259*, 1457–1460.

Rhee, S.H., Jones, B.W., Toshchakov, V., Vogel, S.N. and Fenton, M.J. (2003). Toll-like receptors 2 and 4 activate STAT1 serine phosphorylation by distinct mechanisms in macrophages. J. Biol. Chem. *278*, 22506–22512.

Roberts, M.T.M., Stober, C., McKenzie, A.N. and Blackwell, J.M. (2005). IL-4 and IL-10 collude in vaccine failure for novel exacerbatory antigens in murine *Leishmania major* infection. Infect. Immun. *73*, 7620–7628.

Saklani-Jusforgues, H., Fontan, E., Soussi, N., Milon, G. and Goossens, P.L. (2003). Enteral immunization with attenuated recombinant Listeria monocytogenes as a live vaccine vector: organ-dependent dynamics of CD4 T lymphocytes reactive to a *Leishmania major* tracer epitope. Infect. Immun. *71*, 1083–1090.

Satti, I.N., Osman, H.Y., Daifalla, N.S., Younis, S.A., Khalil, E.A., Zijlstra, E.E., El Hassan, A.M. and Ghalib, H.W. (2001). Immunogenicity and safety of autoclaved *Leishmania major* plus BCG vaccine in healthy Sudanese volunteers. Vaccine *19*, 2100–2106.

Schneider, J., Langermans, J.A., Gilbert, S.C., Blanchard, T.J., Twigg, S., Naitza, S., Hannan, C.M., Aidoo, M., Crisanti, A., Robson, K.J., Smith, G.L., Hill, A.V. and Thomas, A.W. (2001). A prime-boost immunisation regimen using DNA followed by recombinant modified vaccinia virus Ankara induces strong cellular immune responses against the Plasmodium falciparum TRAP antigen in chimpanzees. Vaccine *19*, 4595–4602.

Scott, P. and Trinchieri, G. (1997). IL-12 as an adjuvant for cell-mediated immunity. Semin. Immunol. *9*, 285–291.

Sharifi, I., FeKri, A.R., Aflatonian, M.R., Khamesipour, A., Nadim, A., Mousavi, M.R., Momeni, A.Z., Dowlati, Y., Godal, T., Zicker, F., Smith, P.G. and Modabber, F. (1998). Randomised vaccine trial of single dose of killed *Leishmania major* plus BCG against anthroponotic cutaneous leishmaniasis in Bam, Iran. Lancet *351*, 1540–1543.

Shaw, M.A., Davies, C.R., Llanos-Cuentas, E.A. and Collins, A. (1995). Human genetic susceptibility and infection with *Leishmania peruviana*. Am. J. Hum. Genet. *57*, 1159–1168.

Skeiky, Y.A., Coler, R.N., Brannon, M., Stromberg, E., Greeson, K., Crane, R.T., Campos-Neto, A. and Reed, S.G. (2002). Protective efficacy of a tandemly linked, multi-subunit recombinant leishmanial vaccine (Leish-111f) formulated in MPL adjuvant. Vaccine *20*, 3292–3303.

Skeiky, Y.A., Kennedy, M., Kaufman, D., Borges, M.M., Guderian, J.A., Scholler, J.K., Ovendale, P.J., Picha, K.S., Morrissey, P.J., Grabstein, K.H., Campos-Neto, A. and Reed, S.G. (1998). LeIF: a recombinant *Leishmania* protein that induces an IL-12-mediated Th1 cytokine profile. J. Immunol. *161*, 6171–6179.

Soong, L., Duboise, S.M., Kima, P. and McMahon-Pratt, D. (1995). *Leishmania pifanoi* amastigote antigens protect mice against cutaneous leishmaniasis. Infect. Immun. *63*, 3559–3566.

Soussi, N., Milon, G., Colle, J.H., Mougneau, E., Glaichenhaus, N. and Goossens, P.L. (2000). Listeria monocytogenes as a short-lived delivery system for the induction of type 1 cell-mediated immunity against the p36/LACK antigen of *Leishmania major*. Infect. Immun. *68*, 1498–1506.

Soussi, N., Saklani-Jusforgues, H., Colle, J.H., Milon, G., Glaichenhaus, N. and Goossens, P.L. (2002). Effect of intragastric and intraperitoneal immunisation with attenuated and wild-type LACK-expressing Listeria monocytogenes on control of murine *Leishmania major* infection. Vaccine *20*, 2702–2712.

Srivastava, J.K., Misra, A., Sharma, P., Srivastava, B., Naik, S. and Dube, A. (2003). Prophylactic potential of autoclaved *Leishmania donovani* with BCG against experimental visceral leishmaniasis. Parasitology *127*, 107–114.

Stacey, K.J. and Blackwell, J.M. (1999). Immunostimulatory DNA as an adjuvant in vaccination against *Leishmania major*. Infect. Immun. *67*, 3719–3726.

Stager, S., Alexander, J., Kirby, A.C., Botto, M., Rooijen, N.V., Smith, D.F., Brombacher, F. and Kaye, P.M. (2003). Natural antibodies and complement are endogenous adjuvants for vaccine-induced CD8+ T-cell responses. Nat. Med. *9*, 1287–1292.

Stager, S., Smith, D.F. and Kaye, P.M. (2000a. Immunization with a recombinant stage-regulated surface protein from *Leishmania donovani* induces protection against visceral leishmaniasis. J. Immunol. *165*, 7064–7071.

Stager, S., Smith, D.F. and Kaye, P.M. (2000b. Immunization with a recombinant stage-regulated surface protein from *Leishmania donovani* induces protection against visceral leishmaniasis. J. Immunol. *165*, 7064–7071.

Stober, C., Lange, U., Roberts, M.T., Gilbmartin, B., Francis, R., Almeida, R., Peacock, C.S., McCann, S. and Blackwell, J.M. (2006a. From genome to vaccines for leishmaniasis: Screening 100 novel vaccine candidates against murine *Leishmania major* infection. Vaccine *24*, 2602–2614.

Stober, C., Lange, U., Roberts, M.T.M., Alcami, A. and Blackwell, J.M. (2005). IL-10 from regulatory T cells determines vaccine efficacy in murine *Leishmania major* infection. J. Immunol. *175*, 2517–2514.

Stober, C.B., Lange, U.G., Roberts, M.T., Alcami, A. and Blackwell, J.M. (2007). Heterologous priming-boosting with DNA and modified vaccinia virus Ankara expressing tryparedoxin peroxidase promotes long-term memory against *Leishmania major* in susceptible BALB/c mice. Infect. Immun. *75*, 852–860.

Stober, C.B., Lange, U.G., Roberts, M.T., Gilmartin, B., Francis, R., Almeida, R., Peacock, C.S., McCann, S. and Blackwell, J.M. (2006b. From genome to vaccines for leishmaniasis: screening 100 novel vaccine candidates against murine *Leishmania major* infection. Vaccine *24*, 2602–2616.

Stover, A.G., Da Silva Correia, J., Evans, J.T., Cluff, C.W., Elliott, M.W., Jeffery, E.W., Johnson, D.A., Lacy, M.J., Baldridge, J.R., Probst, P., Ulevitch, R.J., Persing, D.H. and Hershberg, R.M. (2004). Structure-activity relationship of synthetic toll-like receptor 4 agonists. J. Biol. Chem. *279*, 4440–4449.

Streit, J.A., Recker, T.J., Filho, F.G., Beverley, S.M. and Wilson, M.E. (2001). Protective immunity against the protozoan *Leishmania chagasi* is induced by subclinical cutaneous infection with virulent but not avirulent organisms. J. Immunol. *166*, 1921–1929.

Swihart, K., Fruth, U., Messmer, N., Hug, K., Behin, R., Huang, S., Del Giudice, G., Aguet, M. and Louis, J.A. (1995). Mice from a genetically resistant background lacking the interferon gamma receptor are susceptible to infection with *Leishmania major* but mount a polarized T helper cell 1-type CD4+ T cell response. J. Exp. Med. *181*, 961–971.

Uzonna, J.E., Spath, G.F., Beverley, S.M. and Scott, P. (2004). Vaccination with phosphoglycan-deficient *Leishmania major* protects highly susceptible mice from virulent challenge without inducing a strong Th1 response. J. Immunol. *172*, 3793–3797.

Valenzuela, J.G., Belkaid, Y., Garfield, M.K., Mendez, S., Kamhawi, S., Rowton, E.D., Sacks, D.L. and Ribeiro, J.M. (2001). Toward a defined anti-*Leishmania* vaccine targeting vector antigens: characterization of a protective salivary protein. J. Exp. Med. *194*, 331–342.

Velez, I.D., del Pilar Agudelo, S., Arbelaez, M.P., Gilchrist, K., Robledo, S.M., Puerta, J.A., Zicker, F., Berman, J. and Modabber, F. (2000). Safety and immunogenicity of a killed *Leishmania* (l.) amazonensis vaccine against cutaneous leishmaniasis in Colombia: a randomized controlled trial. Trans R. Soc. Trop. Med. Hyg. *94*, 698–703.

Wang, R., Doolan, D.L., Le, T.P., Hedstrom, R.C., Coonan, K.M., Charoenvit, Y., Jones, T.R., Hobart, P., Margalith, M., Ng, J., Weiss, W.R., Sedegah, M., de Taisne, C., Norman, J.A. and Hoffman, S.L. (1998). Induction of antigen-specific cytotoxic T lymphocytes in humans by a malaria DNA vaccine. Science *282*, 476–480.

Wang, Z.E., Reiner, S.L., Zheng, S., Dalton, D.K. and Locksley, R.M. (1994). CD4+ effector cells default to the Th2 pathway in interferon gamma-deficient mice infected with *Leishmania major*. J. Exp. Med. *179*, 1367–1371.

Webb, J.R., Campos-Neto, A., Ovendale, P.J., Martin, T.I., Stromberg, E.J., Badaro, R. and Reed, S.G. (1998). Human and murine immune responses to a novel *Leishmania major* recombinant protein encoded by members of a multicopy gene family. Infect. Immun. *66*, 3279–3289.

Webb, J.R., Campos-Neto, A., Skeiky, Y.A. and Reed, S.G. (1997). Molecular characterization of the heat-inducible LmSTI1 protein of *Leishmania major*. Mol. Biochem. Parasitol. *89*, 179–193.

Webb, J.R., Kaufmann, D., Campos-Neto, A. and Reed, S.G. (1996). Molecular cloning of a novel protein antigen of *Leishmania major* that elicits a potent immune response in experimental murine leishmaniasis. J. Immunol. *157*, 5034–5041.

Yang, D.M., Fairweather, N., Button, L.L., McMaster, W.R., Kahl, L.P. and Liew, F.Y. (1990). Oral Salmonella typhimurium (AroA-) vaccine expressing a major leishmanial surface protein (gp63) preferentially induces T helper 1 cells and protective immunity against leishmaniasis. J. Immunol. *145*, 2281–2285.

Yi, A.K., Yoon, J.G., Hong, S.C., Redford, T.W. and Krieg, A.M. (2001). Lipopolysaccharide and CpG DNA synergize for tumor necrosis factor-alpha production through activation of NF-kappaB. Int. Immunol. *13*, 1391–1404.

Zaph, C., Uzonna, J., Beverley, S.M. and Scott, P. (2004). Central memory T cells mediate long-term immunity to *Leishmania major* in the absence of persistent parasites. Nat. Med. *10*, 1104–1110.

Index

A

J

K

L

M

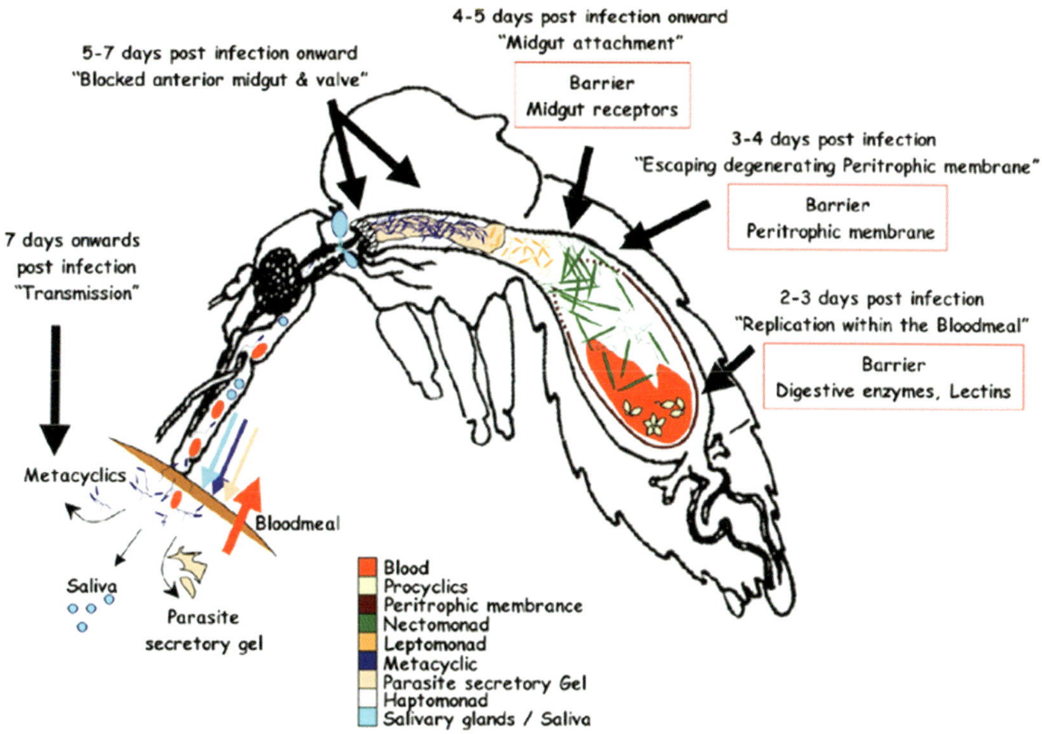

Figure 10.2 The life cycle of *Leishmania* from amastigotes to infective metacyclics in the digestive tract of a competent vector sandfly.

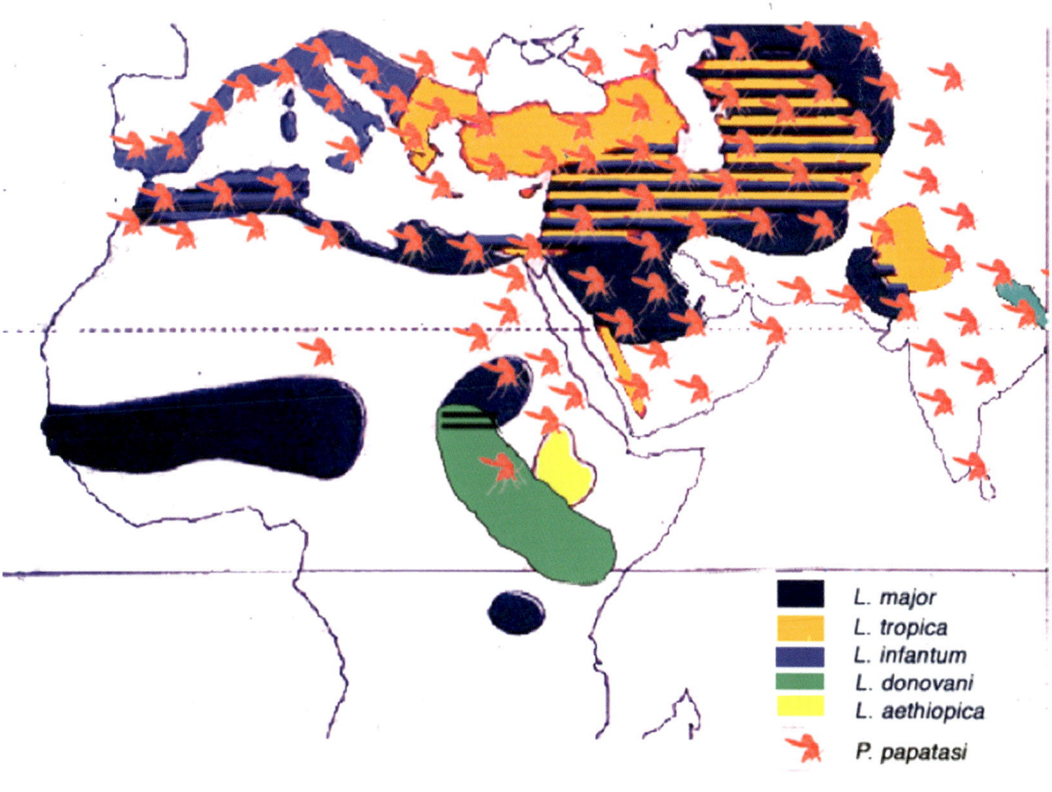

Figure 10.4 Geographic distribution of *Phlebotomus papatasi* compared with the distribution of Old World leishmanias other than *L. major*.